SEPARATION METHODS
in MICROANALYTICAL
SYSTEMS

Edited by
Jörg P. Kutter
Yolanda Fintschenko

CRC Press
Taylor & Francis Group
Boca Raton London New York

CRC Press is an imprint of the
Taylor & Francis Group, an **informa** business
A TAYLOR & FRANCIS BOOK

T0239578

Cover photography courtesy of Detlef Snakenborg and Mihalis Kampanis, MIC.

CRC Press
Taylor & Francis Group
6000 Broken Sound Parkway NW, Suite 300
Boca Raton, FL 33487-2742

First issued in paperback 2020

© 2006 by Taylor & Francis Group, LLC
CRC Press is an imprint of Taylor & Francis Group, an Informa business

No claim to original U.S. Government works

ISBN 13: 978-0-367-57803-9 (pbk)
ISBN 13: 978-0-8247-2530-3 (hbk)

Library of Congress Card Number 2005012833

Library of Congress Cataloging-in-Publication Data

Kutter, Jorg P.
    Separation methods in microanalytical systems / Jorg P. Kutter and Yolanda Fintschenko.
        p. cm.
    Includes bibliographical references and index.
    ISBN-13: 978-0-8247-2530-3 (alk. paper)
    ISBN-10: 0-8247-2530-1 (alk. paper)
    1. Separation (Technology) 2. Microelectromechanical systems. I. Fintschenko, Yolanda. II. Title.

TP156.S45K88 2006
620.1'06--dc22   2005012833

Visit the Taylor & Francis Web site at
http://www.taylorandfrancis.com

and the CRC Press Web site at
http://www.crcpress.com

Dedication

For those who really know me.

—JPK

For Andy, Peter, and Lydia.

—YF

Editors

Jörg P. Kutter received his B.S. in chemistry in 1991, and his Ph.D. in analytical chemistry in 1995, both from the University of Ulm, Germany. Both theses focused on chromatographic and electrophoretic separation techniques. After graduation, from January 1996 to May 1998, he worked as a postdoctoral research fellow in the Laser Spectroscopy and Microinstrumentation Group at Oak Ridge National Laboratory (Oak Ridge, Tennessee) developing microchip-based analytical tools. In June 1998 he joined the Bio/Chemical Microsystems Group at the Institute for Micro and Nanotechnology (MIC) of the Technical University of Denmark in Lyngby. Presently he is an associate professor at MIC, where he is also project leader of the internal micro-TAS project. His research interests include liquid phase separation and sample pretreatment techniques and the development of chemical analysis systems based on microfabricated structures, with emphasis on the integration of optical and fluidic elements.

Yolanda Fintschenko co-manages the Microfluidics Department at Sandia National Labs. She received her B.S. in chemistry from Trinity University in San Antonio, Texas, and her Ph.D. in bioanalytical chemistry from The University of Kansas. She has been active in the field of microfluidics for approximately 7 years, first as a post-doc at The University of Twente, then at Sandia National Laboratories, Livermore, California, as a post-doc and staff member. Her recent technical contributions include on-chip electrochromatographic separations using porous polymer monoliths and demonstrating insulator-based dielectrophoresis for selective pathogen concentration.

Preface

Undoubtedly, the development from vacuum tubes to transistors and integrated circuits is one of the biggest success stories in the history of technology. The ability to miniaturize functional elements and combine thousands and millions of them on a small piece of silicon real estate allows the production of powerful yet cheap microprocessors, which have now become ubiquitous in our daily lives. A similar success is envisioned for technologies that have emerged in the wake of the microelectronic revolution: micro(electro)mechanical systems (MEMS) and microfluidic systems. This book concerns itself with certain aspects of microfluidics — the behavior of fluids in confined spaces and the manipulation of these fluids — namely, the possibility to perform chemical analyses, biochemical assays, and similar processes. The products of this kind of research are often dubbed micro-Total Analysis Systems (μ-TAS) or, more generally, lab-on-a-chip (LOC) devices.

Research and development within the field of LOC is a challenging task because of the multidisciplinary nature of this field. It has not developed linearly from one given discipline, despite what some die-hard chemists or engineers might tell you. The field has more appropriately crystallized, after the different fields became saturated with new ideas about miniaturization, from different locations within the space that is science and technology. The scope of the challenges encountered while striving to realize LOC makes it indispensable to work in cross-disciplinary teams. The result is cross-fertilization and synergy. Nonetheless, a common language needs to be established to improve mutual understanding. One of the apparent indications that this communication still needs improvement is the absence of common monographs accessible to all researchers working in the field of LOC. This book is one step to fill the existing gap. As it is intended for a wide audience, it was also written by contributors from many of the disciplines that constitute the backbone of the LOC community.

Of course, this book cannot attempt to cover the entire field of LOC. Instead it focuses on what has been one of the main driving forces behind the development of LOC for the last 15 years: miniaturized separation systems. Separation units are still at the heart of many micro-TAS and LOC devices, and modern separation techniques are indispensable tools for analytical chemists. This book gives an overview of separation techniques on microfabricated devices: theoretical background information, design and understanding, fabrication and material issues, implementations, and separation systems in relation to other parts of LOC applications (sample preparation, detection, etc.). It is intended as a one-stop shopping guide for your questions concerning separation techniques in microanalytical devices. It is, however, not so much meant only as a quick reference guide, but rather as a place to linger and browse. It is very likely that the information you are seeking is provided in several locations within the book. A multiauthor volume gives you different styles, different approaches, and different opinions — make good use of this! Many topics are so common that they reappear in different chapters, showing you different angles to approach a given problem, reflecting the different backgrounds from which researchers attack the same issues.

We hope that this book will become a reference and help in understanding, designing, and working with miniaturized separation devices. It has been a tremendous experience for us to put this book together and, while it has taken much longer than anticipated, we feel that the result justifies the hard work, the enthusiasm, and above all the patience that the contributing authors have put into this monograph. We wish to thank all the authors, the editors at Taylor & Francis/CRC Press, and our colleagues for supporting this project.

Enjoy!

Jörg P. Kutter
MIC
Department of Micro and Nanotechnology
Technical University of Denmark
Lyngby, Denmark

Yolanda Fintschenko
Sandia National Laboratories
Livermore, California

Contributors

Per Andersson
Gyros AB
Uppsala Science Park
Uppsala, Sweden

Catherine Cabrera
MIT Lincoln Laboratory
Lexington, Massachusetts

Brian Carlson
Department of Pharmaceutical Chemistry
University of Kansas
Lawrence, Kansas

Laura Ceriotti
European Commission
JRC
Ispra, Italy

Gabriela S. Chirica
Sandia National Laboratories
Livermore, California

Emil Chmela
Polymer Analysis Group
Universiteit van Amsterdam
Amsterdam, The Netherlands

Christopher T. Culbertson
Department of Chemistry
Kansas State University
Manhattan, Kansas

Nico F. de Rooij
University of Wenchâtel
Institute of Microtechnology
Neuchâtel, Switzerland

Gert Desmet
Department of Chemical Engineering
Vrije Universiteit Brussel
Brussels, Belgium

Vladislav Dolnik
Alcor BioSeparations
Mountain View, California

Carlo S. Effenhauser
Roche Diagnostics Microtechnology Center
Rotkreuz, Switzerland

Barbara Fogarty
Tyndall National Institute
Cork, Ireland

Han J.G.E. Gardeniers
MESA+ Institute for Nanotechnology
University of Twente
Enschede, The Netherlands

Ernest F. Hasselbrink, Jr.
Department of Mechanical Engineering
University of Michigan
Ann Arbor, Michigan

Bryan Huynh
Department of Pharmaceutical Chemistry
University of Kansas
Lawrence, Kansas

Stephen C. Jacobson
Department of Chemistry
Indiana University
Bloomington, Indiana

Andrew Kamholz
Tecan-Boston
Medford, Massachusetts

Sander Koster
Groningen Research Institute of Pharmacy
Pharmaceutical Analysis Group (FA)
University of Groningen
Groningen, The Netherlands

Richard J. Kottenstette
Sandia National Laboratories
Albuquerque, New Mexico

Nathan Lacher
Pfizer
La Jolla, California

Jan Lichtenberg
Physical Electronics Laboratory
Swiss Federal Institute of Technology
Zurich, Switzerland

Shaorong Liu
Department of Chemistry and Biochemistry
Texas Tech University
Lubbock, Texas

Susan Lunte
Department of Pharmaceutical Chemistry
University of Kansas
Lawrence, Kansas

Sneha Madhavan-Reese
Department of Mechanical Engineering
University of Michigan
Ann Arbor, Michigan

Scott Martin
Department of Chemistry
St. Louis University
St. Louis, Missouri

Curtis D. Mowry
Sandia National Laboratories
Albuquerque, New Mexico

Rikke P.H. Nikolajsen
MIC
Department of Micro and Nanotechnology
Technical University of Denmark
Lyngby, Denmark

Stephanie Pasas
Merck Research Laboratories
West Point, Pennsylvania

Gunnar Thorsén
Gyros AB
Uppsala Science Park
Uppsala, Sweden

Robert Tijssen
Department of Chemical Engineering
Vrije Universiteit Brussel
Brussels, Belgium

Walter Vandaveer IV
Midwest Research Institute
Kansas City, Missouri

Albert van den Berg
MESA+ Institute for Nanotechnology
University of Twente
Enschede, The Netherlands

Elisabeth Verpoorte
Groningen Research Institute of Pharmacy
Pharmaceutical Analysis Group (FA)
University of Groningen
Groningen, The Netherlands

Susanne R. Wallenborg
Gyros AB
Uppsala Science Park
Uppsala, Sweden

Chung-Nin Channy Wong
Sandia National Laboratories
Albuquerque, New Mexico

Paul Yager
Department of Bioengineering
University of Washington
Seattle, Washington

Table of Contents

1 Analytical Microsystems: A Bird's Eye View

Carlo S. Effenhauser

CONTENTS

1.1 INTRODUCTION

The controlled transport, manipulation, and analysis of liquid solutions, suspensions, or individual microscopic objects in a size regime ranging from approximately 1 femtoliter ($1\,fl = 1\,\mu m^3$) to microliters ($1\,\mu l = 1\,mm^3$) are the fundamental tasks with many applications in modern life sciences and beyond (e.g., in chemical process engineering, printers, miniaturized power cells). The size range indicated, commonly denoted as "microfluidics," covers about 10 to 12 orders of magnitude in terms of volume and 3 to 4 orders on the length scale. The lower end of this range is often imposed by light diffraction, either indirectly as in the case of photolithographic generation of structural features, or directly, e.g., in a microanalysis cavity defined by the focal volume of a focused laser beam. Microfluidics represents one of the pillars, liquid-phase microanalytical systems are based upon. The term "nanofluidics" is commonly used if at least one functionally relevant dimension of the fluidic system is further miniaturized to about 100 nm or below. Not at least due to issues of compatibility and interfacing, the size regime of microfluidics stretches at its upper end to typical dimensions of the conventional manual lab world ($1\,\mu l$ to 1 ml).

A variety of terms is currently used to denote the special classes of analytical systems based on miniaturized and integrated device approaches. Terms emphasizing purpose, function, or fabrication methods such as μ-TAS (micro-total-analysis systems), lab-on-a-chip, BioMEMS (bio-micro-electro-mechanical systems), microfluidic devices, biochips, find their use. The difficulty in defining a clear and precise terminology is to some extent indicative of a new, quickly developing, and highly interdisciplinary area of research. However, it also reflects the fact that neither the fabrication technology itself nor the size or physical function can serve as unique indicators that would facilitate classification of the whole range of approaches. A wide range of micromachining techniques coexists with traditional (mechanical and other) fabrication methods, flow-based separation systems with immobilized arrays, and monolithic architectures with hybrid systems. In addition, this variety in nomenclature also reflects the various origins of this field that can be traced back to analytical separation science, semiconductor microfabrication technology, chemical sensor research, and others. An authoritative and comprehensive review of the historical origins, the early developments, and the various roots of the field was recently given by Manz et al. [1, 2].

A simple and straightforward definition of "microfluidics" is further complicated by the fact that, in certain cases, even volumes in the milliliter range have to be handled, e.g., for the analysis of extremely low abundant analytes: the determination of a concentration of $10^{-18} M$ (about 1,000 molecules in 1 ml, well within the scope of target amplification based detection schemes such as polymerase chain reaction) with a relative standard deviation of 1% requires at least 10 ml of sample $((10,000)^{1/2}/10,000 = 1\%$, "molecular shot noise"). Even in cases where detection systems with single-molecule sensitivity are available [3] — and the history of analytical microsystems is very tightly linked to the development of highly sensitive detection systems for obvious reasons — an "ocean of zeros" has to be screened for some rare events. The demands for extremely low limits of detection and miniaturization are clearly contradictory in this regime. Analyte preconcentration as a way of "data compression" will alleviate this situation; however, any technique used will still have to handle the correspondingly large volumes, with consequences for the degree of miniaturization that can be principally achieved for the whole system.

A look at a modern life science laboratory readily reveals that the way we handle liquids for generating biochemical information (e.g., molecular identity, concentration, rate constants, etc.) still depends heavily on manual labor and relatively simple automation concepts. This is in striking contrast to the extreme degree of automation, parallelization, speed, miniaturization, reliability, and very low fabrication costs encountered in electronic data processing. Nevertheless, major trends in the modern life sciences emphasize exactly these qualities. For example, "high-throughput" methods and massively parallel concepts play a central role in the quest for new lead compounds in drug development [4, 5], in whole genome sequencing [6], or for the identification of genetic markers [7]. The latter could find applications in medical diagnostics, e.g., for a genetic risk analysis of a patient in view of a certain disease (predisposition screening), or for an individual gene-based drug therapy adjustment (pharmacogenomics).

For instance, at its very front end, modern drug discovery starts with molecular target identification and validation. This task is currently dominated by protein/peptide separation techniques, such as electrophoresis (gel or capillary) or chromatography, coupled to mass spectrometry (matrix-assisted laser desorption ionization or electrospray ionization mass spectrometry) for protein identification [8–11]. After selection of promising targets, up to 10^6 compounds of a library are screened for desired biomolecular interactions against a single target using ultrahigh throughput technologies (UHTS) with a routine capacity in excess of 100,000 assays per day per system. In order to save precious reagents such as purified target proteins, miniaturization of test volumes is important [12]. From the beginning of HTS in the 1980s, test volumes have been reduced from typically 100 µl in 96-well plates to around 2 to 10 µl in the impending standard of 1536-well plates [13]. Currently, this demand is mostly met with "conventional" process automation technology relying on robotic multiwell plate handling, parallel dispensing units, translation stages, etc. It is certainly dangerous to extrapolate the demands into the future, as high-throughput-lead generation still has to prove its value in terms of the rate of novel drugs entering the marketplace, and library quality and incorporation of rationale design is generally regarded more important than "playing" sheer number games [14]. Nevertheless, if further assay miniaturization proves to be a key issue in the future, it is likely that more radical concepts capitalizing on advantages and features of chip-based integrated systems will prevail in the long run, and the first commercially available chip-based systems have been already introduced (see Section 1.2).

Medical diagnostics is currently not yet a field of massively parallel approaches, although, as already mentioned, they play an important role in the search for diagnostic markers. Identification of marker profiles is expected to lead to an increasingly differentiated understanding of an individual's disease, subsequently requiring multiplex routine diagnostic formats. Apart from a few exceptions like genotyping of viruses with high mutation rates (e.g., HIV), diagnostic target panels in the order of tens to hundreds are expected to cover most of the envisaged applications in the foreseeable future. Examples can be found in the diagnosis of autoimmune diseases, genetic diseases like cystic fibrosis, and molecular profiling of tumors for therapy prediction and selection.

1.2 MICROANALYTICAL SEPARATION SYSTEMS

Separation-based systems have played a key role in the renaissance of mini-aturized analytical devices starting from the early 1990s. According to an elegant and concise definition by Giddings, "separation is the art and science of maximizing separative (differential) transport relative to dispersive trans-port" [15]. Maximizing differential, separative mass transport leads to opti-mized spatial separation of the various constituents of a sample at a given point of time. Minimizing dispersive mass transport helps to conserve this spatial separation pattern in the best possible way until the pattern or a relevant part of it has been recorded.

Based on fundamental physical–chemical considerations of the transport of mass, momentum, and energy on a small length scale, it has been known for a long time that minimizing dispersive transport in separation techniques requires miniaturization of at least one decisive dimension of the system [15]. This could be the diameter of a capillary in an electrophoresis system or the use of small-diameter packing materials in chromatography. Capillary elec-trophoresis has conceptually been proposed as early as 1937 by Tiselius from the conclusion that the transfer of Joule heating in electrophoresis will be improved by reducing the diameter of the tube [16].

With hindsight, the initial use of electrokinetic effects for the demonstra-tion of the benefits of a miniaturized and integrated separation system has been a particularly fortunate choice for a number of reasons [17–19]. Electro-kinetics inherently comprises of two important features required for this purpose:

1. An efficient pumping mechanism (electro-osmosis) very well suited for application at the microscale, that (a) enables collective mass transport in a microchannel virtually free of dispersion caused by the flow profile ("plug profile"), and (b) is efficient at a small length scale and almost independent of the cross section of the microchannel (in contrast to the r^{-4} dependence of the volume flow rate in Hagen–Poiseuille flow).
2. A separation mechanism for differential mass transport (electrophor-esis).

Both phenomena depend linearly on the driving gradient, the external electric field, and require only external voltage sources and electrodes as compara-tively simple instrumental means.

Most importantly, the combination of these two features allowed for integration of all fluid handling steps that critically impact function into a device simply featuring a network of interconnected microchannels. This is in contrast to other attempts, where at critical sites additional fluidic interfaces had to be designed, either between the micro- and macroworlds or between microcomponents that could not be fabricated together in an integrated way. In chip-based electrophoresis, all critical components that determine the analytical function of the system can be completely "hard-wired" into a planar substrate, and the only function left is the transfer of sample, buffer, and reagents into macroscopic reservoirs. Dead volumes at channel intersections typically are in the femtoliter range, while sample size and detection volumes typically measure about 10 to 100 pl or more. Thus, the whole fluidic logistics is performed in a virtually dead volume free environment through switching of externally applied electrical potentials.

Arguably, functional integration of the critical components has played a more important role than miniaturization itself, as the micromachined channel cross sections (typically around $50 \times 20 \mu m$) have not been significantly different from common commercially available capillaries (typical i.d. 75 μm). In the early work of chip-based electrophoresis, the integrated functions comprised an injection means, channels as ducts or reactors for derivatization reactions, and "invisibly" also all required pumps and valves. Analytical separations almost at the theoretical lower limit of band broadening could be performed, proving the effective handling of all avoidable sources of dispersion during the separation process [18]. Up to 160,000 theoretical plates (for a definition of terms, see Chapter 2) could be obtained with a separation length of only 50 mm, resulting in a height equivalent to a theoretical plate of only 0.3 μm. One should not forget, however, that electrokinetically controlled microfluidics also has significant drawbacks, e.g., the strong dependence and delicate nature of the surface properties of the channel walls.

It is interesting to note that flow- and separation-based systems are also starting to enter the high-throughput screening arena. For instance, ACLARA Biosciences has developed a system that relies on the electrophoretic separation of so-called e-Tag reagents that can be coupled to molecular recognition elements [20]. Separation of the complexes formed with the molecular targets is performed in parallel channels, either on a polymer chip or with a multiplex capillary electrophoresis system. The number of channels (e.g., 96) times the number of e-Tag labels resolved (up to something like 100) results in thousands of assays carried out in parallel (basically, the $x–y$ coordinates in a microarray or multi-well plate are replaced by a set of $x–t$ data, with t being the electrophoretic migration time that has been "designed" into the e-Tag). Caliper Technologies has developed a continuous flow based HTS system that features a special world-to-chip interface (the so-called "sipper" technology) [21]. Both pressure and electrokinetic flow controls are applied. Reagent consumption is particularly low in this approach, as it does not depend on conventional dispensing technology, and can reach a few nanoliters per assay.

An impressive recent example of the impact of miniaturized, automated, and parallelized microanalytical separation technology (although not "micromachined") is the completion of the human genome project on a timescale faster than originally proposed. The use of a few hundred 96-lane capillary electrophoresis instruments provided a capacity of a few tens of thousands of sequencing capillaries at simultaneously reduced analysis time (about 500 base pairs in less than 2 h per capillary). A total sequencing capacity in the order of 10^8 base pairs/day resulted, as the instruments can be operated in an automated manner with very little hands-on time [6].

The virtues of miniaturization, however, can also be exploited in microarrays, usually consisting of immobilized spots of ligand binding biomolecules arranged in two dimensions [22], as has been worked out by Ekins [23]. Spot miniaturization leads to a sufficiently small number of capturing sites, and thus, low-concentration analytes can be analyzed under the so-called "ambient analyte conditions," where the measurement does not disturb the analyte concentration. In this regime, the fractional occupancy of capturing sites is directly related to the analyte concentration, independent of the actual sample volume and the actual amount of capturing sites present. In addition, with decreasing spot size, equilibrium occupancy of binding sites will be reached at an increased rate. Spatial analyte separation is predefined in microarrays, and efficient mass transfer, high detection sensitivity, minimizing unspecific and mismatch binding, and reproducible and cost-efficient fabrication of the array "content" are major issues in the development of these systems.

1.3   STRATEGIC BENEFITS OF ANALYTICAL MICROSYSTEMS

The state-of-the-art in separation-based microanalytical systems has been reviewed in many articles over the past 5 years [24–30]. Traditionally, these reviews discuss the current state-of-the-art of the field in terms of achievements in the various analytical subdisciplines. However, in an attempt to point out the progress toward the general promises on a more fundamental level, another approach will be followed here. The strategic goals of analytical microsystems include:

- Functional integration
- Faster-time-to-result
- Parallelization
- Miniaturization
- Novel enabled analytical functions.

In the following sections, exemplary accomplishments will be briefly highlighted. Of course, every assignment to a certain category of strategic benefits is of subjective nature and often not possible without ambiguity. In most cases, analytical microsystems are designed to realize more than one strategic benefit at the same time, however, to different extents.

1.3.1 FUNCTIONAL INTEGRATION

Briefly, the basic idea underlying the term "functional integration" is a kind of analytical "microprocessor" integrating common analytical functions such as sampling, metering of liquids, filtration, reaction chambers, separation processes, detection, etc., in a compact device, which may be either reusable or, in many cases, disposable. Functional integration has been a major driver in the development of microanalytical systems, and is arguably more important than miniaturization itself.

From an user's point of view, the associated advantages include mainly compact instrument size (desktop or even portable, with associated point-of-care capability for environmental monitoring, near patient testing in physician's office or hospital, user self-testing, military applications), and ease of use (reduced workload, operation through untrained personnel). Almost all of the typical individual protocol steps such as sample pretreatment (particle [cell] separation, dilution, preconcentration), metering, chemical reaction, separation, detection, and waste collection have been demonstrated as single-unit operations or in different combinations in a format suitable for integration. Nevertheless, the integration of any specific subset of these tasks under given boundary conditions such as required analytical performance (precision, dynamic range, menu of analytical parameters, fabrication costs, etc.) is still a formidable task. In any actual development, the question of which functions to integrate into the device and what to perform in the instrument is of prime importance and cannot be answered in a general way, but rather will depend on cost target, market drivers, intended user, etc.

The integrated nanoliter DNA analysis device developed by Burns et al. [31] represents a particularly remarkable achievement with an outstanding degree of functional integration (Figure 1.1). It is based on silicon and glass micromachining and features an integrated sample mixing and positioning system, a temperature controlled reaction chamber including heaters and temperature sensors, an electrophoresis unit with integrated electrodes, and an integrated fluorescence detection system composed of a cutoff filter and a photodiode. Metering occurs through a combination of hydrophobic stop patches and external pneumatic actuation. In order to perform an isothermal strand displacement amplification on the device, two 120-nl volumes of sample template and enzyme solutions were metered, mixed, and transported to a separate chamber, and, after heating to a controlled constant temperature of 50°C, merged with a 25-nl volume of an intercalating dye. Finally, the amplified genetic material was detected by means of an integrated gel filled microchannel using a cross-type injector.

1.3.2 FASTER TIME-TO-RESULT

Due to fast mass transport along short ducts and fast diffusive mixing on very small length scales, miniaturization and functional integration tend to generate

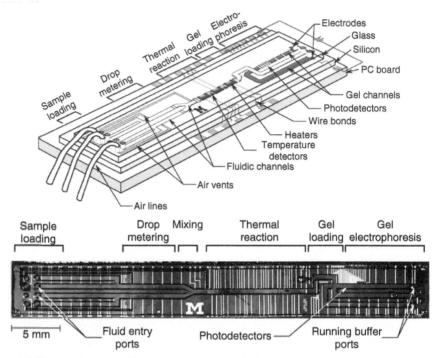

FIGURE 1.1 (See color insert following page 208) Integrated silicon-based device for DNA analysis. (From Burns MA, Johnson BN, Brahmasandra SN, Handique K, Webster JR, Krishnan M, Sammarco TS, Man FP, Jones D, Heldsinger D, Namasivayam V, Mastrangelo CH, Burke DT. *Science* 1998, *282*, 484–487. With permission.)

faster results. Minimizing dispersive mass transport through miniaturization in many cases allows running the separation under conditions of increased differential mass transport. This either leads to improved separation power or to faster separations at a given, sufficient resolution. Chip-based capillary electrophoresis is an instructive example, where the efficient suppression of excessive Joule heating allows for the application of comparably high electric field strengths. In combination with precise sample injection schemes generating well-defined sample plugs and small volume on-column detection, small separation lengths can be used, leading to faster results without compromising resolution. This example also nicely demonstrates the interplay between functional integration (small and precise sample injection) and miniaturization (Joule heat removal) required for optimal performance.

It is interesting to note that mixing can be performed faster both at very small and relatively large dimensions. In the former, this is due to fast diffusion transport ($t \propto \lambda^{1/2}$), in the latter to the existence of turbulence. However, there exists an intermediate range of dimensions approximately between 10^1 and 10^3 μm, where diffusion mixing has become slow; however, Reynolds numbers under typical flow velocities are still not high enough to allow for turbulent mixing. Indeed, it has been the strategy of many micro-mixers to subdivide flow in this region into a larger number of small portions and rearrange them in a way that reduces the required diffusion pathways to the "fast" dimensions.

Generation of fast analytical results is a prerequisite in order to enable high analytical throughput per time unit. It has to be kept in mind, however, that analytical speed always needs to be seen in the context of the other time scales involved in the total process and in many cases does not represent the bottle-neck for a given application (often sample preparation and device loading are the rate limiting steps). In medical diagnostics, fast time-to-result is particu-larly important in cases where diagnostic information triggers immediate ac-tion, e.g., in emergency cases or in situations where a treatment needs to be monitored on a short time scale and adjusted accordingly.

An extreme example of high-speed analysis has been demonstrated by Jacobson et al. [32] (Figure 1.2). They designed a device that is capable of electrophoretic separations at a field strength of 53 kV/cm, more than two orders of magnitude higher than typical field strengths employed in capillary electrophoresis. This resulted in a sub-millisecond separation of a simple binary mixture of two dyes. The separation length used measured 200 μm, with an injected plug length of about 30 μm. The high-field regions were generated and shaped by the channel network through variation of the channel cross section.

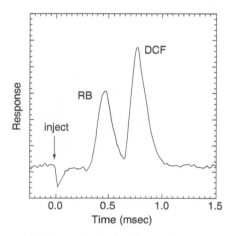

FIGURE 1.2 High-speed chip-based electrophoretic separation of two dyes. (From Jacobson SC, Culbertson CT, Daler JE, Ramsey JM. *Anal. Chem.* 1998, *70*, 3476–3480. With permission.)

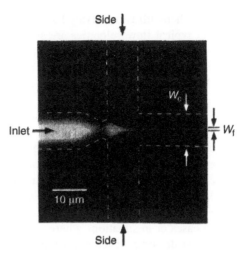

FIGURE 1.3 Hydrodynamic focusing on a microfluidic device. (From Knight JB, Vishwanath A, Brody JP, Austin RH. *Phys. Rev. Lett.* 1998, *80*, 3863–3866. With permission.)

Another impressive example, the principle of hydrodynamic focusing of a sample stream through sheath flow confinement, has been investigated under extreme conditions by Knight et al. [33] (Figure 1.3). By fluorescence detection, these authors could demonstrate the compression of a sample stream to a mere 50-nm width with an associated mixing time constant of 10 μsec.

1.3.3 PARALLELIZATION

As very convincingly demonstrated by the microelectronics industry, a major asset of microtechnology is the straightforward fabrication of arrays and parallel units by batch-type processes. In the case of microanalytical systems, the application of arrays of separation units (identical or different) operated in parallel would increase the outcome in terms of analytical results per device. Additionally, if analysis time can be reduced and the device area is used in a most economic manner, high analytical throughput in terms of results per device area per time unit will result. It goes without saying that the reliable functioning of such a microanalytical array device in routine applications is much more demanding than its mere fabrication.

A beautiful (also from an aesthetic point of view) state-of-the-art demonstration of this capability is the capillary array bioanalyzer with 384 independent separation lanes developed by Mathies et al. for high throughput genetic analysis [34] (Figure 1.4). The 384 capillaries (length 8 cm, cross section 60×30 μm) were radially arranged around a shared electrode in the center of a 200-mm diameter glass plate. For reasons of a very compact design and a

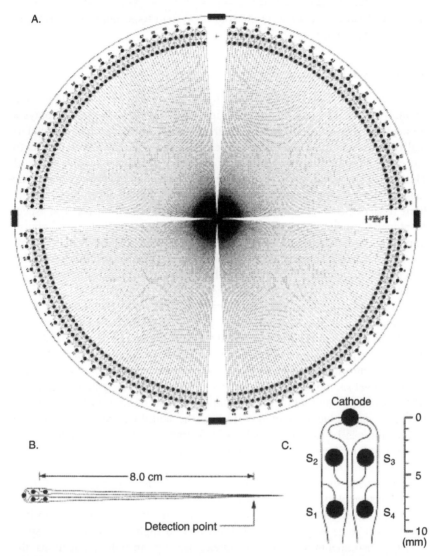

FIGURE 1.4 Microfabricated 384-lane electrophoresis analyzer chip. (From Emrich CA, Tian H, Medintz IL, Mathies RA. *Anal. Chem.* 2002, *74*, 5076–5083. With permission.)

reduction of the number of reservoirs, a novel injection scheme with a so-called "back-biased" protocol was employed in order to compensate partially for the sample biasing in direct electrokinetic injection (see also Chapter 2). Although separation-based genotyping results can be generated with an impressive rate of one result per second, loading of the samples to the individual reservoirs is the bottleneck, as with many similar high-throughput screening systems (multi-well plates). A single-probe robotic system took more than 40 min to load all 480 reservoirs.

1.3.4 MINIATURIZATION

Miniaturization relates to the downscaling of one or more physical dimension of analytical devices and instrumentation. Although miniaturization plays a pronounced role in discussions about the benefits of micromachined analytical systems, it is often somewhat overemphasized, especially with respect to the outer physical dimensions of devices and instruments (in the sense of "honey, I shrunk the lab"). Typical outer dimensions of devices are in the range of several centimeters, enforced not at least by the necessity to handle these devices either manually or by established automated means. The aspect of reduced structural feature size enabling specific functions is of course important, and will be discussed in the subsequent section. Rated against the other strategic benefits, miniaturization in general is certainly second in importance to functional integration, which plays a more important role, e.g., in the development of portable instruments.

An important general aspect of miniaturization is the drastic reduction of consumption of precious samples and reagents in analytical operations, provided that suitable detection systems are available that are capable of detecting the correspondingly reduced amounts of analytes. Similarly, in the field of *in vitro* diagnostics, miniaturization leads to reduced sample requirements and enables novel sampling procedures causing less pain to the patient. Miniaturized instruments can find *in situ* application at point-of-care sites, and in some instances, e.g., glucose spot monitoring, the additional aspect of discrete application can be an important benefit. Ultimately, even implantable diagnostic instrumentation comes within reach, e.g., in the case of continuous monitoring of glucose levels in a patient.

The recent attempts to downsize large analytical instruments, such as mass spectrometers, are maybe a particular illustrative example. A number of groups have started working toward that goal [35, 36] and although it is clearly still some way to go until miniaturized high-resolution mass spectrometry instruments will find widespread use, e.g., in field applications, the progress is remarkable. In addition to the reduction of size and weight, the shorter mean free path that can be tolerated in a miniaturized instrument is a special advantage of downscaling, leading to strongly relieved vacuum requirements. A first concept of a batch surface micromachined instrument has been presented by Siebert et al. [37] featuring a novel mechanism for mass discrimination; however, no actual data were presented. The first experimental proof-of-concept of an instrument based on a conventional Wien filter has been recently published by Sillon and Baptist [38] (Figure 1.5). Although the demonstrated mass resolution is still poor, the approach could find first applications in leak detection.

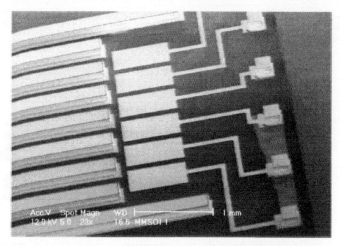

FIGURE 1.5 Chip-based mass analyzer (Wien filter) and Faraday detection plates. (From Sillon N, Baptist R. *Sens. Actuators B* 2002, *83*, 129–137. With permission.)

1.3.5 NOVEL ANALYTICAL FUNCTIONS

Micro- and nanofabrication technology enables the exploitation of completely new functions and mechanisms, e.g., for biomolecule separation. Instead of relying on the pore size range and the statistical pore size distribution of established separation media such as polyacrylamide or cellulose gels, artificial separation gels can be fabricated with well-defined structural features. Micro- and nanotechnology will allow to tailor-design microstructures for the analytical problem under consideration, i.e., for the separation of large DNA molecules. Novel classes of separation devices that have been recently described are based on rectified Brownian motion (so-called "Brownian ratchets"), arrays of entropic traps, and continuously operating DNA fractionation devices with asymmetric pulsed electric fields [39]. An ultimate goal of all these approaches will be tools for the manipulation, analysis, and sorting of biological matter on the molecular level in a way that allows easy integration into microanalytical systems.

Another concept enabling novel types of assays and sample handling has been developed by Yager and Weigl ("H-filter" and "T-sensor") [40]. These devices utilize the particular behavior of flow at small Reynolds numbers that lead to the formation of "virtual membranes" between two or more liquid streams. Mass transport across these well-defined virtual boundaries through diffusion enables novel assay concepts under steady-state flow conditions. In addition, external gradients can be applied in a way resembling field flow fractionation (see Chapter 7).

The so-called "DNA prism" developed by Huang et al. [41] may serve as a state-of-the-art example of a novel separation technology enabled by

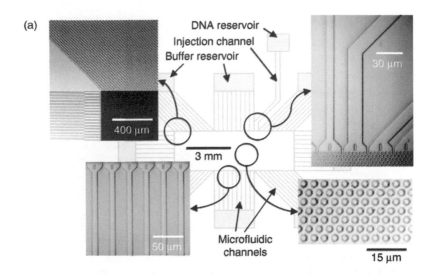

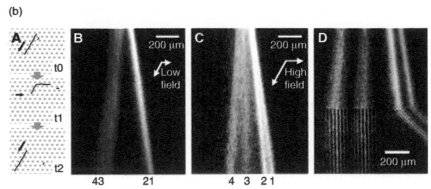

FIGURE 1.6 (a) Microfabricated device for continuous, size-based DNA separation. (b) Schematic and fluorescence micrographs of continuous DNA separation. (From Huang LR, Tegenfeldt JO, Kraeft JJ, Sturm JC, Austin RH, Cox EC. *Nat. Biotechnol.* 2002, *20*, 1048–1051. With permission.)

microfabrication (Figure 1.6). The heart of the device is a hexagonally packed array of microposts (2 μm in diameter in height, 2 μm post-to-post distance). Arrays of microchannels provide access of sample and buffer solutions to the "artificial gel" and help to shape the electric field. Sample DNA is injected and separated continuously through the application of pulsed asymmetric electric fields, which lead to a size-dependent deflection of DNA molecules in a way resembling free flow electrophoresis. After separation on a very short time scale, DNA fractions can be collected for subsequent analysis, preferentially in an integrated manner on the same chip.

1.4 ANALYTICAL MICROSYSTEMS: A SHORT CONCLUSION

The work presented in the previous sections underlines that truly outstanding and remarkable achievements have been accomplished in the past few years and the rate of progress is still impressive. Analytical microsystems represent a very modern field of research and the number of groups contributing to this field grows worldwide. This trend is also reflected by new specialized journals especially devoted to this topic (e.g., "Lab-on-Chip" [42]).

Apart from the strong academic interest in the study of novel phenomena at the micro- and nanoscales, a major driving force is certainly the anticipated extraordinary commercial opportunities. However, from a commercial point of view, progress is experienced by many as slower than expected. A blooming analytical microfluidics industry has not become a reality yet. A strong indication of the difficulties encountered is given by the fact that even pioneering microfluidics companies have recently changed their business model toward more traditional concepts of value creation through "chemical intelligence" rather than "microfluidic intelligence." Even outstanding achievements, such as those presented in this chapter, still have a long way to go before a commercially viable product will enter the marketplace, if at all. For many of the envisaged products, integration of a wide variety of equally reliable and high-performing functional microanalytical modules has to be accomplished using very cost-efficient and robust mass fabrication processes. The perception of "solutions looking for problems" is still widely spread and indicative of a technology foundation still in the process of maturing. To some extent, underestimation of what can be achieved by further development of well-established technologies may also have contributed to the apparent slowness of the paradigm change and the delay in the rise of a new technological S-curve.

For commercial success, a high awareness of these issues and an intimate knowledge of the marketplace and its rules are mandatory. It is encouraging to see an increasing number of promising products that have overcome at least some of the hurdles and will eventually break the ice for more to come. They will be presented in other chapters in this book.

ACKNOWLEDGMENTS

The author would like to thank Drs Wolfgang Fiedler, Peter Finckh, Patrick Griss, and Martin Kopp for critical reading of the manuscript and many helpful suggestions. Dr Michael Hein is gratefully acknowledged for many helpful suggestions and his continued support.

REFERENCES

1. Reyes, D.R.; Iossifidis, D.; Auroux, P.-A.; Manz, A. *Anal. Chem.* 2002, *74*, 2623–2636.

2. Auroux, P.-A.; Iossifidis, D.; Reyes, D.R.; Manz, A. *Anal. Chem.* 2002, *74*, 2637–2652.
3. Zander, C. *Fres. J. Anal. Chem.* 2000, *366*, 745–751.
4. Dunnington, D.; Li, Z.; Binnie, A.; Vollert, H. in *Integrated Microfabricated Biodevices*, Heller, M.J., Guttman, A., eds.; Marcel Dekker, Inc., New York, Basel, 2002, pp. 371–414.
5. Bajorath, J. *Nat. Rev. Drug Discov.* 2002, *1*, 882–894.
6. Venter, J.C. *Science* 2001, *291*, 1304–1351.
7. Christensen, C.B.V. *Talanta* 2002, *56*, 289–299.
8. Kellner, R. *Fresenius J. Anal. Chem.* 2000, *366*, 517–524.
9. Lahm, H.W.; Langen, H. *Electrophoresis* 2000, *21*, 2105–2114.
10. Shen, Y.; Smith, R.D. *Electrophoresis* 2002, *23*, 3106–3124.
11. Figeys, D. *Proteomics* 2002, *2*, 373–382.
12. Cunningham, D.D. *Anal. Chim. Acta* 2001, *429*, 1–18.
13. Dunn, D.A.; Feygin, I. *Drug Discov. Today* 2000, *5*, S92–S103.
14. Fox, S.; Wang, H.; Sopchak, L.; Farr-Jones, S. *J. Biomol. Screening* 2002, *7*, 313–316.
15. Giddings, J.C. *Unified Separation Science*, Wiley, New York, 1991.
16. Tiselius, A. *Trans. Faraday Soc.* 1937, *33*, 524.
17. Harrison, D.J.; Manz, A.; Fan, Z.H.; Lüdi, H.; Widmer, H.M. *Anal. Chem.* 1992, *64*, 1926–1932.
18. Effenhauser, C.S.; Manz, A.; Widmer, H.M. *Anal. Chem.* 1993, *65*, 2637–2642.
19. Jacobson, S.C.; Koutny, L.B.; Hergenröder, R.; Moore, A.W.; Ramsey, J.M. *Anal. Chem.* 1994, *66*, 3472–3476.
20. Gibbons, I. *Drug Discov. Today* 2000, *5*, S33-S37.
21. Sundberg, S.A.; Chow, A.; Nikiforov, T.; Wada, H.G. *Drug Discov. Today* 2000, *5*, S84–S91.
22. Joos, T.O.; Stoll, D.; Templin, M.F. *Curr. Opin. Chem. Biol.* 2001, *6*, 76–80.
23. Ekins, R.P. *Clin. Chem.* 1998, *44*, 2015–2030.
24. Effenhauser, C.S. *Top. Curr. Chem.* 1998, *194*, 51–82.
25. Bruin, G.J.M. *Electrophoresis* 2000, *21*, 3931–3951.
26. Kutter, J.P. *Trends Anal. Chem.* 2000, *19*, 352–363.
27. Bousse, L.; Cohen, C.; Nikiforov, T.; Chow, A.; Kopf-Sill, A.; Dubrow, R.; Parce, J.W. *Annu. Rev. Biophys. Biomol. Struct.* 2000, *29*, 155–181.
28. Jakeway, S.C.; de Mello, A.J.; Russell, E.L. *Fresenius J. Anal. Chem.* 2000, *366*, 525–539.
29. Greenwood, P.A.; Greenway, G.M. *Trends Anal. Chem.* 2002, *21*, 726–740.
30. Weigl, B.H.; Bardell, R.L.; Cabrera, C.R. *Adv. Drug Del. Rev.* 2003, *55*, 349–377.
31. Burns, M.A.; Johnson, B.N.; Brahmasandra, S.N.; Handique, K.; Webster, J.R.; Krishnan, M.; Sammarco, T.S.; Man, F.P.; Jones, D.; Heldsinger, D.; Namasivayam, V.; Mastrangelo, C.H.; Burke, D.T. *Science* 1998, *282*, 484–487.
32. Jacobson, S.C.; Culbertson, C.T.; Daler, J.E.; Ramsey, J.M. *Anal. Chem.* 1998, *70*, 3476–3480.
33. Knight, J.B.; Vishwanath, A.; Brody, J.P.; Austin, R.H. *Phys. Rev. Lett.* 1998, *80*, 3863–3866.
34. Emrich, C.A.; Tian, H.; Medintz, I.L.; Mathies, R.A. *Anal. Chem.* 2002, *74*, 5076–5083.
35. Badman, E.R.; Crooks, R.G. *J. Mass Spectrom.* 2000, *35*, 659–671.

36. de Mello, A. *Lab Chip* 2001, *1*, 7N–12N.
37. Siebert, P.; Petzold, G.; Hellenbart, A.; Müller, J. *Appl. Phys. A* 1998, *67*, 155–160.
38. Sillon, N.; Baptist, R. *Sens. Actuators B* 2002, *83*, 129–137.
39. Chou, C.F.; Austin, R.H.; Bakajin, O.; Tegenfeldt, J.O.; Castelino, J.A.; Chan, S.S.; Cox, E.C.; Craighead, H.; Darnton, N.; Duke, T.; Han J.Y.; Turner, S. *Electrophoresis* 2000, *21*, 81–90.
40. Weigl, B.H.; Yager, P. *Science* 1999, *283*, 346–347.
41. Huang, L.R.; Tegenfeldt, J.O.; Kraeft, J.J.; Sturm, J.C.; Austin, R.H.; Cox, E.C. *Nat. Biotechnol.* 2002, *20*, 1048–1051.
42. http://www.rsc.org/is/journals/current/loc/locpub.htm

2 Microfluidics: Some Basics

*Stephen C. Jacobson and
Christopher T. Culbertson*

CONTENTS

In this chapter, we will discuss general concepts and expressions associated with microfluidic separation systems. Our discussion will begin with a description of a basic microfluidic device and include examples of electrokinetic and pressure-driven methods for transporting fluids through these devices. Our attention will then turn to separation efficiency, resolution, and peak capacity followed by a discussion of the contributions to analyte dispersion in microfluidic separations. The chapter will then discuss microfluidic devices as equivalent electrical circuits, and how this circuit analogy applies to mixing and valving on microfluidic devices. The chapter will conclude with scaling of separations that are limited by the injection plug length. More in-depth treatment of other transport and separation techniques will be addressed in subsequent chapters (e.g., Chapters 5 to 9).

2.1   MICROFLUIDIC DEVICES

To begin our discussion, we first will consider the layout of a basic microfluidic device for separations. Depicted in Figure 2.1 is a microfluidic device with a simple cross intersection design. The device was fabricated

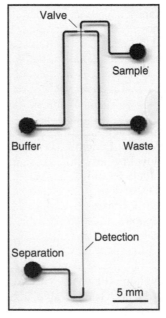

FIGURE 2.1 Scanned image of a microfluidic device with a cross intersection. The channels and fluid reservoirs are labeled buffer, sample, waste, and separation. The narrow channels are 50 μm wide at the top and 15 μm deep, and the wide channels are 230 μm wide at the top and 15 μm deep. The length of the separation channel is 35 mm. The substrate is 25 mm wide, 50 mm long, and 2.5 mm thick. The channels and reservoirs are filled with black ink for contrast. (From Oak Ridge National Laboratory. With permission.)

using standard photolithographic, etching, and bonding techniques described in Chapter 3. In this design, the analysis or separation channel is 35 mm long from the cross intersection, e.g., valve, to where the channel widens at the bottom of the device. The buffer, sample, and waste channels also expand to the wider channels 0.5 mm from the cross intersection. At the terminals of each channel is a 3-mm diameter hole that has been ultrasonically drilled in the substrate to provide fluid access to the channels. The separation length is typically defined as the distance between the injection valve and the detection point, and with on-channel detection techniques, e.g., fluorescence, the detection point can be positioned anywhere along the analysis channel. The velocity of a component, u, traveling down the separation channel is

$$u = \frac{L}{t}$$ (2.1)

where L is the separation length and t is the time from injection to detection. In order to operate the microchip, the appropriate fluids are placed in the buffer, sample, waste, and analysis reservoirs. In most applications, electrical potentials or pressures are applied at the channel terminals to facilitate fluid transport. Microfluidic valving on such devices is discussed below, and in Figure 2.1, the reservoirs are arranged for the gated injection.

For many microfluidic devices, the channels are wet chemically etched into the substrate. If the substrate is an amorphous material, e.g., glass or fused silica, the channel etches isotropically, producing a channel cross section similar to what is depicted in Figure 2.2. The rectangular region in the center of the channel (between the dashed lines in Figure 2.2) is defined by the initial width, w_i, of the channel prior to etching. This width is often defined by the photomask used to transfer the channel pattern into the photoresist during fabrication. For example, the initial widths of the narrow and wide channels for the microfluidic device in Figure 2.1 were 20 and 200 μm, respectively. The channels etch uniformly in all directions producing

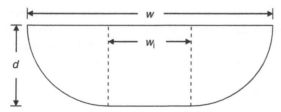

FIGURE 2.2 Schematic of a channel cross section for an isotropically etched channel. The channel depth is d, the channel width is w, and the initial channel width is w_i.

the two quarter circles on each side of the channel. The cross-sectional area, A, of an isotropically etched channel is

$$A = d(w - 2d) + \frac{\pi d^2}{2} \qquad (2.2)$$

where w is the width at the top of the channel and d is the channel depth. The aspect ratio, R_A, of the isotropically etched channel cross section is

$$R_A = \frac{w}{d} \qquad (2.3)$$

In Figure 2.1, the aspect ratios for the narrow and wide channels are 3.3 and 15.3, respectively. As the initial channel width decreases, the aspect ratio approaches 2, thus producing a channel that is semicircular.

For nonetching processes such as molding or embossing, a wide variety of cross sections have been reported from semicircular to rectangular.

2.2 FLUID TRANSPORT

2.2.1 ELECTROOSMOTIC FLOW

A common means to transport fluids on microfluidic devices is electroosmotic flow. Electroosmotic flow originates near the surface of a channel that is partially ionized. The excess charge that accumulates in the diffuse region is mobile and when an electrical potential is applied across a microfluidic channel parallel to the surface, the charges move toward the electrode of opposite sign. Through viscous forces the fluid layers adjacent to the diffuse layer are pulled in same direction. This momentum transfer results in a constant velocity throughout the bulk solution in the channel, resulting in a flow profile that is planar and perpendicular to the direction of transport.

Figure 2.3 shows a schematic of the interface between the channel wall in the substrate and the fluid[1]. The negative charges along the channel wall represent ionized silanol groups and are considered immobile. The adsorbed layer is the layer closest to the substrate, and the ions in this layer are not solvated and are considered specifically adsorbed to the channel wall. This layer is also referred to as the compact layer and the plane cutting through the center of charge is the inner Helmholtz plane. The next layer of ions out from the channel surface is solvated and the plane cutting through this center of charge is referred to as the outer Helmholtz plane (OHP). The diffuse layer is the region of excess charge that lies between the OHP and the bulk liquid. The electrostatic potential across the diffuse layer is referred to as the zeta potential, ζ.

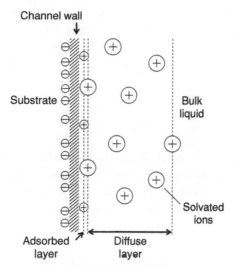

Channel wall

Substrate

Bulk
liquid

Solvated
ions

Adsorbed   Diffuse
layer   layer

FIGURE 2.3 A schematic of the excess charge distribution at and near the channel wall. The excess negative charge on the channel wall is due to partial ionization of substrate material, e.g., silanol groups on glass or silica. A portion of the excess positive charge is adsorbed to the channel surface, and a portion forms the diffuse layer. The diffuse layer extends from the outer Helmholtz plane into the bulk liquid, and the potential across the diffuse layer is the ζ-potential. (Adapted from Bard AJ, Faulkner LR. *Electrochemical Methods*, John Wiley and Sons, New York, 1980.)

For most microfluidic applications, the double layer is thin compared to the smallest dimension of the microchannel, e.g., depth. The inverse of the extent of the double-layer thickness or the Debye–Hückel parameter[1] is

$$\kappa = \left[\frac{2nz^2e^2}{\varepsilon kT}\right]^{0.5} \tag{2.4}$$

where n is the number concentration of each ion, e is the quantity of charge on an electron, z is the charge magnitude of each ion, ε is the permittivity of the dielectric, i.e., the solution in the double-layer region, k is the Boltzmann constant, and T is the temperature. The parameter κ has units of inverse length, typically nm^{-1}. For a 1:1 electrolyte at 25°C in water, Equation (2.4) becomes

$$\kappa = 3.29C^{0.5} \tag{2.5}$$

where C is the concentration of the electrolyte. Consequently, the characteristic thickness of the diffuse layer $(1/\kappa)$ ranges from 1 to 10 nm for buffer concentrations of 0.1 to 0.001 M, respectively. For these buffer concentrations and microfluidic channel depths, d, from 1 to 100 μm, the quantity $\kappa d \gg 1$.

With no external pressure applied, Stokes' equation for the force on the liquid due to an electric field, E, applied parallel to the surface[2] is

$$\eta \nabla^2 u = -\rho_e E \tag{2.6}$$

where η is the viscosity of the fluid, u is the fluid velocity, and ρ_e is the charge density. Because the velocity profile only varies perpendicular to the channel surface, Equation (2.6) can be reduced to a one-dimensional problem. Substituting the Poisson equation relating the charge density and the equilibrium potential, ψ_e, yields

$$\eta \frac{d^2 u_x}{dx^2} = -\varepsilon \frac{d^2 \psi_e}{dx^2} E \tag{2.7}$$

where ε is the permittivity of the solution. Equation (2.7) can then be integrated twice with the following boundary conditions. In the bulk liquid, the potential is zero $(\psi = 0)$, and the velocity is equal to the electroosmotic velocity $(u = u_{eo})$. Also, in the bulk fluid, the potential is uniform $(d\psi/dx = 0)$ and the velocity is constant $(du/dx = 0)$. Near the channel surface, the potential is equal to the zeta potential $(\psi = \zeta)$, and the velocity is zero $(u_x = 0)$. Integration with these conditions yields the Smoluchowski equation for electroosmotic velocity, u_{eo},

$$u_{eo} = -\frac{\varepsilon \zeta}{\eta} E = \mu_{eo} E \tag{2.8}$$

where μ_{eo} is the electroosmotic mobility.

For glass and silica surfaces that are negatively charged ζ is negative, and the positive counterions move toward the negative electrode (cathode). The electrostatic potential is at a maximum absolute value near the wall and goes to zero in the bulk liquid. Conversely, the electroosmotic velocity is zero at the channel wall and reaches a maximum in the bulk liquid. The velocity is constant throughout the bulk liquid in the channel, and the flow profile is planar and perpendicular to the direction of transport as depicted in Figure 2.4(a).

2.2.2 PRESSURE-DRIVEN FLOW

The second mode of transport on microfluidic devices is pressure-driven flow. The linear velocity, u_r, for pressure-driven flow in a circular tube or capillary[3] is

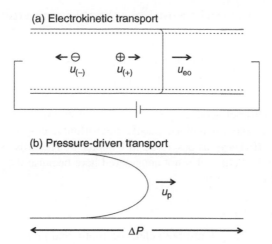

(a) Electrokinetic transport

(b) Pressure-driven transport

FIGURE 2.4 Schematic of the velocity profiles for electrokinetic and pressure-driven transport. (a) Electrokinetic transport is a wall-driven process yielding an axially planar flow profile and velocity u_{eo}. In addition, cations and anions in the electric field move at velocities $u_{(+)}$ and $u_{(-)}$, respectively. The dashed lines depict the diffuse layer from where the electroosmotic flow originates, and the dimensions of the diffuse layer relative to the channel are not to scale. (b) For the pressure-driven transport, the fluid moves at a velocity u_p and has a parabolic flow profile with the maximum velocity in the center of the channel and a velocity of zero at the channel wall.

$$\nabla^2 u_r = \frac{1}{r}\frac{d}{dr}\left(r\frac{du_r}{dr}\right) = \frac{\Delta P}{\eta L_p} \tag{2.9}$$

where r is the distance from the center, ΔP is the pressure drop across the channel, and L_p is the channel length over which the pressure is applied. When the pressure drop along the length of the capillary is constant, the linear velocity at a distance r from the center is

$$u_r = \frac{\Delta P}{4\eta L_p}(r_0^2 - r^2) \tag{2.10}$$

where r_0 is the radius of the capillary. From Equation (2.10), the velocity reaches a maximum at the center of the capillary and is zero at the wall (see Figure 2.4b). The average flow velocity, u_p, is then determined by integrating the flow across the capillary[4] and is

$$u_p = \frac{\int_0^{r_0} 2\pi r u_r\, dr}{\int_0^{r_0} 2\pi r\, dr} = \frac{\Delta P r_0^2}{8\eta L_p} \tag{2.11}$$

Similarly, expressions can be derived for a channel having a rectangular cross section[3] yielding

$$u_p = \frac{\Delta P}{L_p} \frac{d^2}{\eta} \left[\frac{1}{12} - \frac{16d}{\pi^5 w} \tanh\left(\frac{\pi w}{2d}\right) \right]$$

(2.12)

where d is the channel depth and w is the channel width. A rectangular cross section is often seen for molded plastic microfluidic devices, e.g., polydimethylsiloxane (PDMS), as described in Chapter 3. The cross section for an isotropically etched channel is not considered here because the solution must be determined numerically.

2.2.3 COMBINED FLOW

In some microfluidic systems, both electroosmotic and pressure-driven flows are present, and the fluid velocity is the sum of the two sources

$$u = u_{eo} + u_p$$

(2.13)

To illustrate a system of combined flow, electroosmotic flow can be used to generate hydraulic pressures where an electric potential is applied only over a portion of the channel.[5] Figure 2.5 shows a schematic of this configuration. The electrical contacts are placed along the length of the channel to apply an electric potential across one segment of the channel. Electroosmotic flow is used to generate a hydraulic pressure at the junction between the channel segment with an electric field applied, i.e., the pump channel, and the channel segment with no electric field applied, i.e., the field-free channel. The fluid velocity in the pump channel is the sum of the electroosmotic flow and the pressure-driven flow

$$u_E = \frac{\mu_{eo} V}{L_E} + \frac{\Delta P_E k_g}{\eta L_E}$$

(2.14)

and the velocity in the field-free channel is pressure-driven flow

$$u_{FF} = \frac{\Delta P_{FF} k_g}{\eta L_{FF}}$$

(2.15)

The subscript "E" designates the channel with the electric field applied, V is the applied voltage, k_g is the geometric constant, and the subscript "FF" designates the channel segment without an electric field applied. With uniform channel cross sections and incompressible fluids, the fluid velocity in each channel segment is equal

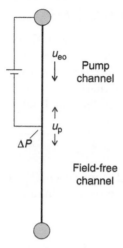

FIGURE 2.5 Schematic of microfluidic device for electroosmotically induced hydraulic pumping. The electrical potential is applied across the channel between the reservoir and a location along the channel. The electroosmotic velocity, u_{eo}, is generated in the pump channel creating a superambient pressure, ΔP, where the electric field is terminated. The resulting pressure-driven velocity, u_p, is equal and opposite in the field-free and pump channels. This design was evaluated in Ref. [5]. (From McKnight TE, Culbertson CT, Jacobson SC, Ramsey JM. *Anal. Chem.* 2001, *73*, in press. With permission.)

$$u_E = u_{FF} \tag{2.16}$$

Also, the pressures generated in the pump and field-free channels are equal and opposite,

$$\Delta P_E = -\Delta P_{FF} \tag{2.17}$$

From Equations (2.13) to (2.15), the fluid velocity in the channel is

$$u_E = u_{FF} = \frac{\mu_{eo} V}{L_E + L_{FF}} \tag{2.18}$$

From Equation (2.18), the fluid velocity depends on the applied potential over the entire channel length $L_E + L_{FF}$. There are several methods for generating hydraulic pressures with electroosmotic flow including varying the surface characteristics of two channels, the buffer composition in two channels, or the cross-sectional areas of two channels. These methods are discussed in Chapter 5.

2.3 SEPARATION MECHANISMS

Two primary separation methods on microfluidic devices are electrophoresis and chromatography. Electrophoretic separations rely on a difference in the mobilities of the ions in an electric field, and chromatographic separations on a difference in the equilibrium of the components between two phases. A more detailed description of these and other separation mechanisms will follow in Chapters 6 to 8.

2.3.1 ELECTROPHORESIS

In an electric field, ions will move depending on the electrical force exerted on the ion and the frictional resistance from the liquid.[6] The electrical force on a particle, f_e, is

$$f_e = QE \qquad (2.19)$$

and the Stokes frictional resistance, f_v, is

$$f_v = 6\pi v_{ep} a \eta \qquad (2.20)$$

where Q is the charge on the particle, v_{ep} is the particle velocity, and a is the particle radius. Balancing the electrical force and frictional resistance in Equations (2.19) and (2.20), respectively, the electrophoretic mobility, μ_{ep}, is

$$\mu_{ep} = \frac{v_{ep}}{E} = \frac{Q}{6\pi a \eta} = \frac{2\varepsilon\zeta}{3\eta}(1 + \kappa a) \qquad (2.21)$$

Equation (2.21) was initially derived by Hückel and is valid for $\kappa a \ll 1$ where the electric field lines are unaffected by the particle. However, Smoluchowski considered $\kappa a \gg 1$, which yields an expression for the electrophoretic mobility similar to the electroosmotic mobility (Equation (2.8))

$$\mu_{ep} = \frac{\varepsilon\zeta}{\eta} \qquad (2.22)$$

To resolve the discrepancy between Equations (2.21) and (2.22), Henry developed the expression

$$\mu_{ep} = \frac{\varepsilon\zeta}{\eta} f_1(\kappa a) \qquad (2.23)$$

where $f_1(\kappa a)$ ranges from 1 to 1.5 to account for the two extreme cases developed by Hückel and Smoluchowski.

When a system has ions that move both electrophoretically and electroosmotically (see Figure 2.4a), the electrokinetic velocity of the ion, u_{ek} (or net velocity), is the sum of the electrophoretic and electroosmotic mobilities multiplied by the electric field

$$u_{ek} = (\mu_{ep} + \mu_{eo})E \tag{2.24}$$

In this case, the electroosmotic flow provides nonselective transport of the ion and does not contribute to the separation.

2.3.2 CHROMATOGRAPHY

The second mode of separation considered in this chapter is chromatography and is based on the distribution of a solute between two phases, e.g., the stationary and mobile phases.[4] The distribution coefficient, K, is defined as

$$K = \frac{c_s}{c_m} \tag{2.25}$$

where c_s is the concentration in the stationary phase and c_m is the concentration in the mobile phase. Consequently, the capacity factor, k, for a solute is defined as the ratio of the amount of solute in the stationary phase to the amount of solute in the mobile phase

$$k = \frac{c_s V_s}{c_m V_m} = K \frac{V_s}{V_m} \tag{2.26}$$

where V_s is the volume of the stationary phase and V_m is the volume of the mobile phase. The retention ratio, R, is the fraction of solute in the mobile phase and is related to the capacity factor as

$$R = \frac{c_m V_m}{c_m V_m + c_s V_s} = \frac{1}{1 + (c_s V_s / c_m V_m)} = \frac{1}{1 + k} \tag{2.27}$$

The velocity of a retained component, u_R, can be calculated by multiplying the retention ratio by the mobile phase velocity, u_0,

$$u_R = R u_0 \tag{2.28}$$

The retention ratio is always less than 1. Combining Equations (2.27) and (2.28) gives

$$\frac{u_R}{u_0} = \frac{1}{1 + k} \tag{2.29}$$

From Equation (2.29) and using $u = L/t$, the capacity factor can be determined experimentally by

$$k = \frac{t_R - t_0}{t_0} \qquad (2.30)$$

where t_R is the elution time of a retained component and t_0 is the elution time of an unretained component.

Another expression that is commonly seen in chromatography is the relative separation between two species and is called the selectivity, α,

$$\alpha = \frac{k_2}{k_1} \qquad (2.31)$$

where 1 and 2 designate the capacity factors for the less and more retained components, respectively. In chromatographic separations, α is defined such that it is always greater than 1.

2.4 SEPARATION EFFICIENCY

2.4.1 PLATE HEIGHT

For a Gaussian profile, the variance, σ^2, can be written as

$$\sigma^2 = 2Dt \qquad (2.32)$$

where D is the effective diffusion coefficient and t is time. Substituting $t = L/u$ into Equation (2.32), the variance is equal to

$$\sigma^2 = \frac{2DL}{u} \qquad (2.33)$$

Rearranging Equation (2.33) yields the plate height[4], H, or the variance per unit separation length,

$$H = \frac{\sigma^2}{L} = \frac{2D}{u} \qquad (2.34)$$

The plate height description in Equation (2.34) does not take into account all of the forms of band broadening that occur over the course of a separation. A more complete description is given by the van Deemter equation

$$H = A + \frac{B}{u} + Cu \qquad (2.35)$$

where A is the contribution of constant sources to the plate height, B/u is the contribution of longitudinal diffusion to the plate height, and Cu is the contribution of mass transfer to the plate height. These contributions to the plate height are plotted as a function of velocity in Figure 2.6. For microfluidic separations, the constant contributions include the injection plug length and the detector path length. The contribution of diffusion decreases with increasing velocity and is the primary means of band broadening in electrophoresis. The mass transfer term includes mass transfer in the mobile and stationary phases and dominates for chromatographic separations. These various contributions are discussed in greater detail below.

2.4.2 PLATE NUMBER

The plate number N,[4] or, the number of plate heights per separation length is a common means of reporting the separation efficiency and is defined as

$$N = \frac{L}{H} = \frac{L^2}{\sigma^2} \tag{2.36}$$

The genesis of the terms plate height and plate number is historical, and the terms are still used although the physical significance related to a distillation column is no longer appropriate.

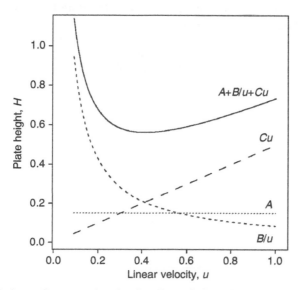

FIGURE 2.6 A van Deemter plot showing the variation of the plate height, H, with the linear velocity, u. Plotted are the total ($A + B/u + Cu$), constant (A), longitudinal diffusion (B/u), and mass transfer (Cu) contributions to the plate height. See Equation (2.35).

2.5 RESOLUTION

The resolution between two bands is a measure of the ability of a separation system to resolve two components and categorizes overlap between adjacent peaks.[4] If there are two component bands migrating down a separation channel, the resolution, R_s, between the two zones is the difference in their position divided by the average width of the peaks,

$$R_s = \frac{2(L_1 - L_2)}{(w_1 + w_2)} = \frac{\Delta L}{4\sigma} \tag{2.37}$$

where L is the separation distance for components 1 and 2, w is the peak width of components 1 and 2, ΔL is the difference in separation distances, and σ is the average standard deviation of the two peaks. Here, 4σ is used as the baseline width of a peak. Figure 2.7 shows two profiles that are separated by 4σ with a resolution of 1.

Assuming uniform separation conditions, the term ΔL in Equation (2.38) grows linearly with distance L,

$$\Delta L = \Delta u \cdot t = \frac{\Delta u L}{u} \tag{2.38}$$

where Δu is the difference in velocity between components 1 and 2 and u is the average velocity for components 1 and 2. Inserting Equations (2.34) and (2.38) into Equation (2.37) yields

$$R_s = \frac{\Delta u L^{0.5}}{4u H^{0.5}} \tag{2.39}$$

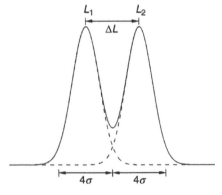

FIGURE 2.7 Schematic of the resolution of two peaks. L_1 and L_2 are the distances the two peaks have traveled, ΔL is the distance between the two peak maxima, σ is the average standard deviation of the two peaks, and 4σ is the baseline width of a peak. See Equation (2.37).

Consequently, the resolution increases as the square root of the separation length and inversely with the square root of the plate height. Therefore, if the separation length is doubled or the plate height is halved, the resolution improves by the square root of 2.

Substituting Equation (2.36) for the plate height gives

$$R_s = \frac{\sqrt{N}}{4} \frac{\Delta u}{u} \tag{2.40}$$

From Equation (2.40), the resolution for a pair of components in electrophoresis is

$$R_s = \frac{\sqrt{N}}{4} \left(\frac{\Delta \mu_{1,2}}{\mu_{1,2} + \mu_{eo}} \right) \tag{2.41}$$

where $\Delta \mu_{1,2}$ is the difference in the mobilities of components 1 and 2 and $\mu_{1,2}$ is the average mobility of the two components. From Equation (2.41), the presence of electroosmotic flow tends to degrade the resolution, but in some cases the electroosmotic flow can be tuned to optimize the resolution for a separation.

Similarly, an equation for resolution between adjacent peaks can be constructed for chromatography,

$$R_s = \frac{\sqrt{N}}{4} \left(\frac{\alpha - 1}{\alpha} \right) \left(\frac{k_2}{1 + k_2} \right) \tag{2.42}$$

Equation (2.42) is optimized when the selectivity for the separation, α, increases, and the capacity factor, k_2, of the second component increases.

2.6   PEAK CAPACITY

The most important criterion for describing the separative performance of a system is the peak capacity, primarily because the peak capacity describes a separation performed under either equilibrium or nonequilibrium conditions. The peak capacity,[4] n_c, is the number of component peaks that can fit into a separation length, L, for a resolution, R_s,

$$n_c = \frac{L}{w} = \frac{L}{4\sigma R_s} \tag{2.43}$$

where w is the average peak width of components and is usually 4σ. If the resolution of the system is 1, then the peak capacity can be estimated from the plate number,

$$n_c \approx \frac{\sqrt{N}}{4} \qquad (2.44)$$

For elution systems, the peak capacity can be estimated using the expression

$$n_c \approx 1 + \frac{\sqrt{N}}{4} \ln\frac{V_{max}}{V_{min}} \qquad (2.45)$$

where V_{max} and V_{min} are the maximum and minimum volumes in which peaks can be eluted and deleted.

The importance of having sufficient peak capacity to resolve components of interest is best illustrated for randomly distributed peaks in the separation space. Davis and Giddings[4] used Poisson statistics to determine the probability of successfully isolating a single component from a mixture. The success ratio (s/m) of separating components having a saturation factor (m/n_c) is

$$\frac{s}{m} = \exp\left(-\frac{2m}{n_c}\right) \qquad (2.46)$$

where s is the number of singlets, m is the number of components in the system, and n_c is the peak capacity. As seen in Figure 2.8, if the saturation factor is 0.5, the probability of isolating a singlet is 0.368. In order to increase the probability to 0.9, the saturation factor needs to be 0.05 or below. This

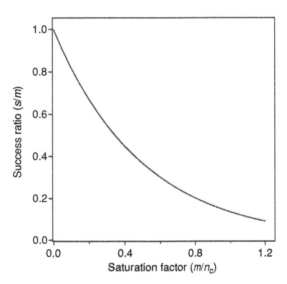

FIGURE 2.8 Variation of the success ratio (s/m) with saturation factor (m/n_c) for a one-dimensional separation. (Adapted from Giddings JC. *Unified Separation Science*, John Wiley and Sons, New York, 1991.)

means that the peak capacity must be at least 20 times the number of components in the sample to be separated.

One approach to enhancing the peak capacity is to perform multidimensional separations. For multidimensional separations, the peak capacity is crucial in determining the total number of components that can be inserted into a separation space. Assuming the separation techniques are orthogonal, the total peak capacity, n_{tot}, is the product of the peak capacities of each dimension. For example, the peak capacity can be used to describe the separation performance of multidimensional separation systems

$$n_{tot} \approx \frac{L_1 L_2}{w_1 w_2} \tag{2.47}$$

where L_1 and L_2 are the separation lengths of the first and second dimensions, respectively, and w_1 and w_2 are the average peak widths in the first and second dimensions, respectively. Assuming the separation mechanisms in each of m dimensions are orthogonal, the total peak capacity, n_{tot}, can be written as the product of the peak capacities in each of the dimensions

$$n_{tot} = n_1 n_2 \ldots n_m \tag{2.48}$$

2.7 DISPERSION

As discussed above, the analyte bandwidth in the efficiency equation is generally expressed in terms of the second moment or variance (σ^2) of the concentration distribution of the band. The variance of the analyte bandwidth incorporates all of the various band dispersion (broadening) sources operative during a separation on a microfluidic device. These sources include molecular (longitudinal) diffusion (σ_{diff}^2), the injection plug length (σ_{inj}^2), the detection window length (σ_{det}^2), Joule heating (σ_{Joule}^2), the channel geometry (σ_{geo}^2), mass transfer effects (σ_{MT}^2), adsorption (σ_{ads}^2), electrodispersion (σ_{edisp}^2), and laminar flow (σ_{flow}^2). With the exception of electrodispersion, these sources of dispersion are generally considered to be independent and so when written as variances are additive. The total variance, σ_{total}^2, can be written as

$$\sigma_{total}^2 = \sigma_{diff}^2 + \sigma_{inj}^2 + \sigma_{det}^2 + \sigma_{Joule}^2 + \sigma_{geo}^2 + \sigma_{MT}^2$$
$$+ \sigma_{ads}^2 + \sigma_{edisp}^2 + \sigma_{flow}^2 \tag{2.49}$$

Many of these sources of dispersion remain poorly characterized on microfluidic devices and become significant only under extreme separation conditions. Below, some of the better characterized and more important sources of dispersion are discussed. All of these sources are also operative in fused silica capillaries; however, as the geometry of the channel is an important consideration, direct parallels to the equivalent dispersion sources occurring in fused silica capillaries frequently cannot be drawn.

2.7.1 DIFFUSION

Longitudinal diffusion represents the irreducible minimum amount of band dispersion expected for any separation process. It arises from the analyte band concentration gradient axially along the separation channel. This concentration, i.e., chemical potential gradient, results in a diffusional mass flux driven by entropy and is described by Fick's second law:

$$\frac{\partial C}{\partial t} = D \frac{\partial^2 C}{\partial x^2} \tag{2.50}$$

where the change in concentration (C) in the analyte band is expressed as a function of both time (t) and space (x). This one-dimensional expression of Fick's second law is, however, not sufficient as it describes diffusion only from a stationary source. Under conditions where electrokinetic migration (flow) is present the change in concentration with time axially along the separation channel is given by the simplified one-dimensional form of the equation of continuity (mass conservation):

$$\frac{\partial C}{\partial t} = -u \frac{\partial C}{\partial x} + D \frac{\partial^2 C}{\partial x^2} = -\mu E \frac{\partial C}{\partial x} + D \frac{\partial^2 C}{\partial x^2} \tag{2.51}$$

where u is the band velocity, D is the diffusion coefficient, μ is the electrokinetic mobility, and E the electric field strength. Assuming that a sample is injected into the separation channel as an infinitely thin band, i.e., a Dirac function, this differential equation can be solved for all x at $t = 0$ with the following result:

$$C(x,t) = \frac{1}{2\sqrt{\pi D t}} e^{-x^2/4Dt} \tag{2.52}$$

This is the equation for a Gaussian distribution with a spatial variance (σ^2) equal to $2Dt$. The analyte band variance due to longitudinal diffusion, therefore, is given by the Einstein–Smoluchowski equation:

$$\sigma^2 = 2Dt \tag{2.53}$$

where D is the analyte diffusion coefficient and t is the separation time. Substituting this equation into the efficiency equation defines the ultimate separation performance. For such a separation, the analyte bandlength will increase with the square root of time and the separation efficiency will increase linearly with time. Many examples of such separations have been demonstrated and reported on microfluidic devices. Some representative diffusion coefficients for both small and large molecules have been reported by Culbertson et al.[7] and can be used to test how well a separation system is functioning.

2.7.2 INJECTION PLUG LENGTH

Generally, samples are injected into separation channels using gated, pinched, or double tee valving techniques on microfluidic devices. This results in a very short spatial extent sample band. The concentration profile of this band can be described using either a Gaussian or impulse (top hat) function depending on the plug length. The variances associated with these distributions as a function of the injection plug length are shown below[8]:

$$\sigma_{inj}^2 = \frac{l_{inj}^2}{12} \quad \text{impulse (top hat)} \tag{2.54}$$

$$\sigma_{inj}^2 = \frac{l_{inj}^2}{16} \quad \text{Gaussian} \tag{2.55}$$

The contribution of the injection plug length to the total band dispersion is constant. Its importance, therefore, decreases as a function of separation time and distance. Generally, we have found the Gaussian function to more realistically model the injection plugs seen for pinched injections.[9,10] The injection plug length is usually an insignificant source of dispersion except in cases of extremely rapid separations.[11]

2.7.3 DETECTION WINDOW LENGTH

For laser-induced fluorescence (LIF) or spectrochemical detection methods the detection window length is generally defined by the interrogation beam (laser beam) spot size or the spatial filter aperture. If the interrogation beam spot size projected on the spatial filter is smaller than the spatial filter then the interrogation beam defines the detection window length. Generally, the interrogation beam will have a Gaussian intensity distribution and so the variance generated would be given by Equation (2.55). If the spatial filter defines the detection window length, i.e., the spatial filter aperture < interrogation beam diameter, then an impulse (top hat) function (Equation (2.54)) should be used to define the detection window length. For electrochemical detection, the axial length of the active electrode surface serves to define the detection window and is best approximated by an impulse response function (Equation (2.54)).

As with the injection plug length, the detection window length is constant. For LIF detection and effective apertures <20 μm (i.e., actual aperture size divided by the collection objective magnification), the dispersion due to the spatial filter is insignificant and can be ignored.[12]

2.7.4 PARABOLIC FLOW PROFILES

Parabolic flow profiles can be generated from a variety of sources including pressure differentials and Joule heating. Pressure differentials, in particular, can be generated by variation in the electroosmotic flow along the length of the channel or by unequal fluid level heights in the reservoirs.[7]

The fluid velocity in a rectangular channel is

$$u_p = \frac{\Delta P}{L_p} \frac{d^2}{\eta} \left[\frac{1}{12} - \frac{16d}{\pi^5 w} \tanh\left(\frac{\pi w}{2d}\right) \right] \qquad (2.56)$$

where η is the viscosity, d is the channel depth, w is the channel width, $\Delta P/L_p$ is the pressure drop axially along the channel.[3,13] If $w > 3d$, then the hyperbolic tan function can be dropped from the second term in the brackets. If $w \gg d$, then the second term in the brackets can be dropped altogether. Equation (2.56) has been used to predict the pressure-induced velocities in microfluidic channels.[5,14] The variance associated with parabolic flows in rectangular channels as derived by Golay[15] is

$$\sigma_{\text{flow}}^2 = \frac{4}{105} \frac{u_p^2 d^2}{D} t \qquad (2.57)$$

Substituting Equation (2.56) in Equation (2.57) gives the variance expected due to pressure-induced flow in rectangular microchip channels

$$\sigma_{\text{flow}}^2 = 4 \frac{(\Delta P)^2 d^6}{105 D \eta^2 L^2} \left[\frac{1}{12} - \frac{16d}{\pi^5 w} \tanh\left(\frac{\pi w}{2d}\right) \right]^2 t \qquad (2.58)$$

To measure the effect of unequal reservoir heights (Δh) in simple straight channels, the pressure difference is calculated using $\Delta P = \rho g \Delta h$ where ρ is the solution density and g is the gravitational constant. Given a reservoir height difference of 1 mm, a channel length of 3 cm, a channel depth of 15 μm, a channel width of 45 μm, a diffusion coefficient of 3×10^{-6} cm²/s, and a separation time of 10 s, this results in $\sigma_{\text{flow}}^2 = 7 \times 10^{-14}$ cm². The variance due to diffusion over the same amount of time is $\sigma_{\text{diff}}^2 = 6 \times 10^{-5}$ cm². Dispersion due to parabolic flow is, therefore, negligible under most conditions.

2.7.5 THERMAL OR JOULE HEATING

When an electric potential is applied across the length of a microchip channel some of the electrical energy is converted to heat energy through the frictional drag of the ionic species that are carrying the current. This is known as Joule heating. The rate at which heat is generated per unit volume (J_Q) is given by the following equation:

$$J_Q = \frac{i^2}{\kappa A^2} = \frac{i^2}{\kappa d^2 w^2} \tag{2.59}$$

where i is the current, κ is the conductivity ($\Omega^{-1} \text{cm}^{-1}$), A is the cross-sectional area of the channel, d is the channel depth, and w is the channel width. The rate at which this heat is dissipated is a function of the thermal conductivity of the buffer, the thermal conductivity of the microchip substrate, the thermal conductivity of the surrounding air, the temperature differentials at the interfaces, the surface areas of the interfaces, and the mass of the substrate. Joule heating can become problematic when the heat generated cannot be dissipated fast enough as it gives rise to a temperature gradient along the depth of the microchannel. The temperature gradient, in turn, gives rise to a viscosity gradient. In fact, a one-degree change in temperature can give rise to a 2 to 3% change in viscosity.[16] Because the electroosmotic velocity (Equation (2.8)) is dependent upon the solution viscosity, the flat flow profile associated with the bulk fluid flow is compromised. In fused silica capillaries, the flow profile of a solution with an axial temperature gradient generally assumes a parabolic shape.[17] A similar situation might be expected in a microchannel although no definitive work in this area has been published. In lieu of a model specific for Joule-heat-generated dispersion in a microchannel, Jacobson et al.[11] estimated the magnitude of this contribution using an expression developed by Knox and Grant[18] and Grushka et al.[17] for the Joule-heat-generated dispersion generated in fused silica capillaries. The expression qualitatively fits the experimental data generated.

While a specific model for heat-generated dispersion in microchannels has not yet been reported, it is apparent that the large mass and large surface area exposed to the surrounding environment allows glass microchips to dissipate heat faster (more effectively) than fused silica capillaries. Fused silica capillaries are generally limited to effectively dissipating the heat generated by a linear power gradient of ~1 W/m. In microchips, however, the effective dissipation of up to 172 W/m has been reported in a small section of a microchannel without adversely affecting the separation performance.[11] Under less extreme conditions, effective dissipation of 5 to 10 W/m can be expected.[19]

Temperature gradients across acrylic microchip channels have been measured using the temperature-dependent fluorescence of Rhodamine B.[20] These
channels showed the same power dissipation characteristics as fused silica
capillaries potted in PDMS. A temperature elevation of ~4°C above the
ambient temperature was seen when the power dissipation was ~1 W/m.
The temperature gradient across the width of the acrylic channels, however,
was greater due to the trapezoidal shape of the channels. No attempt was
made to compare these results with any theoretical predictions.

2.7.6 Geometric Dispersion

Geometric dispersion brought about by adding turns to separation channels
has been intensively studied. Understanding and reducing this type of dispersion is critical for being able to generate long separation channels on devices
with small footprints because analyte bands can be considerably widened as
they migrate around a turn of constant radius and channel width. This band
broadening arises from the difference in migration path lengths for molecules
traversing a turn along the inner versus the outer wall, i.e., the racetrack effect
(Figure 2.9a–c).[12] In addition, the electric field strength along the inner wall is
higher than that along the outer wall. The molecules along the inner wall,
therefore, not only have a shorter distance to migrate but also move faster
than those along the outer wall of the turn. Experimentally, the tighter the turn
and the smaller the analyte diffusion coefficient the greater the problem.

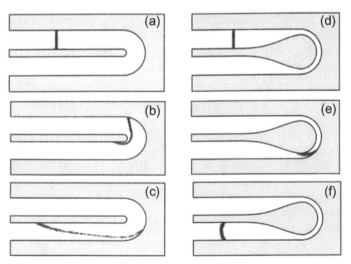

FIGURE 2.9 Profiles of a simulated band traversing (a–c) a standard 180° turn and
(d–f) a 180° low dispersion turn. (Adapted from Griffiths SK, Nilson RH. *Anal. Chem.*
2001, *73*, 272–278.)

Several models to predict the magnitude of the dispersion that occurs during the transit of an analyte band around a turn have been reported.[12,21,22] Griffiths and Nilson, in particular, have developed a closed-form analytical solution to model the dispersion generated by a turn.[22] This model was developed by generating rigorous solutions for turn-induced dispersion at both high and low Peclet numbers, after which a composite solution was generated using a function with the correct limiting behavior. The Peclet number ($Pe = uw/D$) is the ratio of the product of the analyte band velocity (u) and the channel width (w) to the analyte diffusion coefficient (D) and represents the relative magnitudes of convective versus diffusive transport rates. Given the same velocity and channel widths large molecules have much larger Peclet numbers than small molecules. The predicted geometric dispersion introduced by a turn according to the Griffiths and Nilson model is shown in the following equation:

$$\sigma^2_{geo} = \left(\frac{\theta^2 \delta Pe}{15\theta + 3\delta Pe} + \frac{2\theta}{\delta Pe} \right) w^2 \tag{2.60}$$

where θ is the radius of the turn in radians, δ is the ratio of the turn width to mean turn radius (w/r), Pe is the Peclet number, and w is the channel width. This solution was compared with experimental data published by Culbertson et al., and a comparison of the experimental data and the model is plotted in Figure 2.10. Griffiths and Nilson concluded that after correcting for the effective channel widths their model fit the experimental data well. Molho et al.[21] have also developed an analytical expression for predicting dispersion around a turn and tested the model with experimental data. Their model tracks closely that of Griffiths and Nilson. Either model can be used to successfully predict the amount of band broadening that can be expected from a given turn. Using the experimental data generated by Culbertson et al.[12] and Molho et al.[21] both the Griffiths and Nilson[22] and Molho et al.[21] models have been shown to be accurate to within ~10% for all Peclet numbers, all mean channel radii, and all included turn angles up to 180°.[21,22]

A variety of channel designs have now been reported for both minimizing the dispersion introduced by turns and minimizing the overall channel footprint. No particular design is best suited for all applications, so a brief survey of all of the designs is presented.

Prior to the publishing of the Griffiths and Nilson or Molho et al.'s work, Culbertson et al.[12] developed and successfully tested an empirical model for predicting band broadening in turns. This model predicted that the amount of dispersion generated by a turn is governed by the ratio of the analyte diffusion time across the width of the turn to the transit time of the analyte molecule through the turn. Therefore, the smaller the ratio the lower the geometric dispersion contribution. (Griffiths and Nilson arrived at the same result analytically in the limit of high Peclet number for their

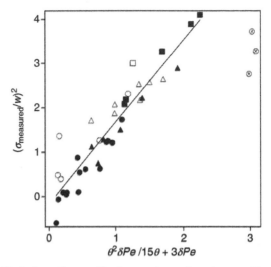

FIGURE 2.10 Variation of normalized experimental variance with estimated variance. Data are taken from Ref. [12] and Equation (2.60) is from Ref. [22]. (From Griffiths SK, Nilson RH. *Anal. Chem.* 2000, 72, 5473–5482. With permission.)

model.) Using this empirically derived model they developed a spiral-shaped separation channel to significantly reduce turn dispersion and to keep the overall chip footprint small.[23] The spiral-shaped separation channel in Figure 2.11 is 25-cm long and fabricated on a piece of glass that was 5 cm × 5 cm. Separation efficiencies of 1.1 million theoretical plates in 46 s for dichlorofluorescein on this device were demonstrated. The geometrical dispersion introduced by the large radii of curvature turns was only 4% of the total dispersion.

Also prior to Griffiths and Nilson's work, Paegal et al. described a symmetrically tapered turn as a method to reduce dispersion.[24] In this design, the channel width in the turn was reduced. As can be seen from Griffiths and Nilson's model, the geometric dispersion is a function of the square of the channel width in the turn. The narrowed turn significantly reduced the band broadening introduced by a turn.

Both Griffiths and Nilson[25] and Molho et al.[21] have used the models that they developed to generate turn geometries that minimize the dispersion introduced. Both groups developed asymmetrically tapered turns (Figure 2.9d–f). These turns are designed to minimize both the race track and electric field effects that can broaden analyte bands. The asymmetrical taper results in the distance traveled by molecules on both the inside and outside walls of the turn to be approximately equal. It also results in the average electric field along both the inside and outside walls of the turn to be equal. Griffiths and Nilson[25] used a nonlinear constrained minimization

FIGURE 2.11 Image of the microchip used for 1D and 2D separations. The first-dimension separation channel (1D) extends from the first valve V1 to the second valve V2. The second-dimension channel (2D) extends from the second valve V2 to the detection point (X) indicated by the arrow. The fluid reservoirs are sample (S), buffer 1 and 2 (B1, B2), and waste 1, 2, and 3 (W1, W2, W3). The 1D separation channel length is 25 cm, and the 2D separation channel length is 2 cm. The channels and reservoirs are filled with black ink for contrast. (Adapted from Culbertson CT, Jacobson SC, Ramsey JM. *Anal. Chem.* 2000, *72*, 5814–5819; Gottschlich N, Jacobson SC, Culbertson CT, Ramsey JM. *Anal. Chem.* 2001, *73*, 2669–2674.)

algorithm to design their turn which minimized dispersion. Practically the optimum geometry can be described as a gradual tapering of the inner channel wall toward the outer channel wall. The smooth taper begins about three channel widths from the turn and terminates when the channel width is about ~28% of the channel width prior to the taper. The narrowed channel segment spans ~85% of the included angle of the turn, and then expands back to the initial channel width using another asymmetrical taper. This design should reduce the dispersion around a turn about 1000-fold, and make it essentially not measurable experimentally. Ramsey et al. have experimentally confirmed the validity of this design. They used seven asymmetric turns to put a 20-cm-long channel on a 2.54 cm × 5.08 cm chip. The number of plates increased linearly with separation distance.[26]

Design parameters for generating folded channels of uniform channel width without significant turn-induced band broadening have also recently been published by Griffiths and Nilson.[27] Using their previously reported model, Griffiths and Nilson set the amount of broadening due to geometry to less than 10% of that due to longitudinal diffusion. The minimum radius of curvature needed to meet this constraint can be calculated. In addition to these design guidelines, they report two novel folded channel designs. The first is a pleated turn that halves the space taken up by a conventional folded (or serpentine) channel pattern. The second is a double-coiled channel that further decreases chip surface area needed for long separation channels while keeping all channel termini on the outside of the coil.

Folded channels potentially offer a method of reducing turn-introduced dispersion without modifying the channel width in the turn or the turn radii of curvature. This method relies on the idea of allowing a second turn, opposite in direction to the first, to remove the skewing of the analyte plug introduced by the first turn. Transverse diffusion, however, must be minimized (i.e., Peclet number maximized) in the straight segment between the two turns. Culbertson et al. attempted to use this idea of "compensating turns" but showed that even for molecules with small diffusion coefficients (600 base pair DNA fragments in a sieving matrix) only about half of the skew introduced in one turn can be removed in a second turn with straight lengths between the turns of 4 mm.[23]

One final chip design recently introduced improves upon spiral channel designs. In this design, the inside wall is "wavy," which allows the transit times for molecules along both the inside and outside walls of the device to be equivalent.[28] Numerical simulations suggest that this design could reduce geometrical dispersion by a factor of up to 64 when compared with a spiral geometry where the channel is of uniform (constant) width.

2.7.7 MASS TRANSFER PROCESSES

Both open-channel and packed bed electrochromatographies have been reported in microchip channels. Using nonequilibrium theory Giddings has developed solutions for analyte band dispersion due to partitioning into and out of a thin, uniform stationary phases attached to microchannel walls under conditions of constant mobile phase velocity.[29] The dispersion introduced by diffusion through the stationary and mobile phases, respectively, are given below:

$$\sigma_{st}^2 = \frac{k}{(1+k)^2} uL \frac{2d_{st}^2}{3D_{st}} \tag{2.61}$$

$$\sigma_{mp}^2 = \frac{1+6k+11k^2}{96(1+k)^2} uL \frac{d_{mp}^2}{D_{mp}} \tag{2.62}$$

where k is the capacity factor, d_{st} is the stationary-phase thickness, d_{mp} is the channel depth, u is the mobile-phase velocity, L is the channel length, D_{st} is the analyte diffusion coefficient in the stationary phase, and D_{mp} is the analyte diffusion coefficient in the mobile phase. Kutter et al.[30] have used these equations to assess open-channel electrochromatography performance in C-18 coated microchip channels. There was an excellent agreement between the expected performance and that predicted.

2.7.8 ELECTRODISPERSION

Electromigration dispersion is generated when there is a significant difference between the conductivity of the sample band and the background electrolyte. It can also arise from a large difference between the mobility of the sample ion and the background electrolyte co-ion. This dispersion generates triangular-shaped peaks. The peak broadening generated by electrodispersion is not additive with the other sources described above as the mobility differences between the analyte and background buffer cause an isotachophoretic effect at one edge of the sample band, which counteracts diffusive band broadening. To minimize electrodispersion effects the sample should be dissolved in the same buffer that comprises the background electrolyte and the sample concentration should be 400- to 1000-fold less than the buffer concentration.[31]

2.8 ANALOGOUS ELECTRICAL CIRCUITS

For electrically driven microfluidic systems, the channel manifold can be treated as an electrical circuit whereby the fluid-filled channels act as resistors. The resistance, R, in a microfluidic channel is

$$R = \frac{L}{A\kappa} = \frac{L}{A\Lambda c} \tag{2.63}$$

where L is the channel length, A is the cross-sectional area of the channel, κ is the conductivity of the fluid in the channel ($\Omega^{-1}\,cm^{-1}$), Λ is the equivalent conductivity of the fluid ($\Omega^{-1}\,cm^{-1}\,M^{-1}$), and c is the concentration of the conducting material in the fluid (M). Analogous to an electrical circuit, Ohm's law is used where the voltage drop, ΔV, axially across a channel j is the product of the current, i, and the channel resistance, R,

$$\Delta V_j = i_j R_j \tag{2.64}$$

For multiple channels intersecting at an intersection, I, Kirchhoff's rules state the sum of the currents at an intersection must equal zero

$$\sum_j i_j = 0 \tag{2.65}$$

In other words, the current flowing into the intersection must equal the current flowing out of the intersection. Using Equations (2.64) and (2.65), the electric potential at the intersection, V_I, can be calculated as

$$V_I = \frac{\left(\sum_j V_j / R_j\right)}{\left(\sum_j 1/R_j\right)} \tag{2.66}$$

where V_j is the potential applied to the channel terminals. Consequently, the absolute value of the electric field strength, E, in channel j is

$$E_j = \frac{|V_j - V_1|}{L_j} \tag{2.67}$$

Similar to the currents in Kirchhoff's rule, the sum of the electric field strengths in the channels intersecting at the intersection is zero.

The resistances within the channels can be measured experimentally. For multiple channels intersecting at a junction, the resistance or current is measured in at least two channels in series. For three channels intersecting at a tee intersection, the resistances are measured across channels 1 and 2, 1 and 3, and 2 and 3. The resistance in the individual channels is determined by solving the equations simultaneously. In this case (for three unknown resistances), three equations are solved simultaneously where

$$R_{i,j} = R_i + R_j \tag{2.68}$$

where the subscripts i and j are any given pair of channels.

2.9 MIXING

To determine whether microfluidic devices operate in a laminar or turbulent flow regime, the Reynolds number, Re, can be calculated using

$$Re = \frac{\rho u d}{\eta} \tag{2.69}$$

where ρ is the fluid density, u is the fluid velocity, d is the characteristic geometric size, and η is the fluid viscosity. For water traveling at a velocity of $1\,cm/s$ in a 10-μm-deep channel, the Reynolds number is approximately 0.1. These are typical conditions in microfluidic experiments, and due to the small channel dimensions of the devices, viscous forces prevail. Thus, the flow is laminar.

Using these circuit models described above, mixing of two fluid streams on a microfluidic device can be described as

$$E_s + E_b = E_m \tag{2.70}$$

where E_X is the absolute value of the electric field strength in the sample (s), buffer (b), and mixing (m) channels. Figure 2.12 shows diffusive mixing of a sample with buffer. In this figure, the sample is combined with buffer at the tee intersection and mixed in the mixing channel. The potential at the tee intersec-

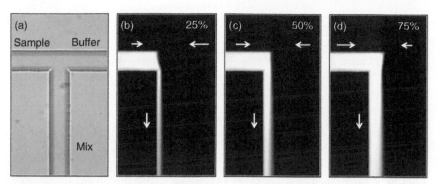

FIGURE 2.12 White light image of (a) a tee intersection and fluorescence images of mixing of (b) 25%, (c) 50%, and (d) 75% sample with buffer. Potentials were applied at the sample and buffer reservoirs with the mix reservoir grounded. Arrows depict the flow direction. (From Oak Ridge National Laboratory. With permission.)

tion is held constant and the potentials at the sample and buffer reservoirs are adjusted to achieve the 25% sample and 75% buffer in Figure 2.12(a), 50% sample and 50% buffer in Figure 2.12(b), and 75% sample and 25% buffer in Figure 2.12(c).

2.10 VALVING

Sample dispensing or valving on microfluidic devices is effected by controlling the material transport in two or more channels. The pinched,[32] double tee,[33] and gated[34] injection schemes are the most widely used in microfluidic separations. The pinched and double tee injection schemes dispense sample plugs with the shortest axial extent, but require the sample to be reloaded into the injection cross for each subsequent injection. The gated injection maintains the sample at the cross intersection but has a larger electrophoretic bias than the pinched or double tee schemes. These valving schemes function using electrophoretic, electroosmotic, and pressure-driven material transport alone or in combination.

2.10.1 Constant Volume (Pinched) Valve

Using simple circuit models, the basic valving functions can be understood for systems with homogeneous buffer and channel properties. For the pinched injection, the sample loading is described as a circuit with the following constraint:

$$E_b + E_s + E_a = E_w \qquad (2.71)$$

where E_X is the absolute value of the electric field strength in the buffer (b), sample (s), analysis (a), and waste (w) channels. The sample travels from the

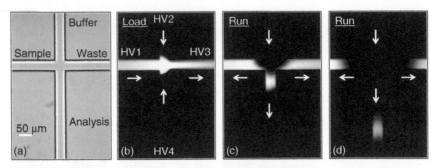

FIGURE 2.13 White light image of (a) a cross intersection (valve) and fluorescence images of (b) sample loading and (c, d) run modes for the pinched injection. Arrows depict the flow direction.

sample reservoir through the cross intersection to the sample waste reservoir. The sample is confined to the intersection by flows from the buffer and analysis channels. To dispense the sample, the voltages are reconfigured to produce the following condition:

$$E_b = E_s + E_a + E_{sw} \tag{2.72}$$

where the primary flow is from the buffer channel to the analysis channel with secondary flows from the buffer to sample and sample waste channels. To return to sample loading mode, the system is returned to the condition in Equation (2.71).

In Figure 2.13, the constant volume (pinched) valve is shown. The sample (the bright regions) is transported through the injection cross toward the waste reservoir. The potentials applied to the buffer and analysis reservoirs confine the sample in the cross intersection (Figure 2.13b). To dispense the sample onto the analysis column, the potentials are reconfigured as in Equation (2.72) so that the primary flow is from the buffer channel to the analysis channel. The potentials at the sample and waste reservoirs are chosen such that buffer flows into these channels to prevent bleeding of sample onto the analysis channel. In this scenario, the field strength in the buffer channel is greater than the field strengths in the sample, analysis, and waste channels. The arrows in the figure indicate direction of flow for the sample and buffer streams. This valving scheme dispenses the smallest axial extent sample plugs with the least amount of sample bias.

2.10.2 Variable Volume (Gated) Valve

With the gated valve, the run mode is described as a circuit with the following constraint:

$$E_b + E_s = E_a + E_w \tag{2.73}$$

where E_X is the absolute value of the electric field strength in the buffer (b), sample (s), analysis (a), and waste (w) channels. To prevent the sample from electrokinetically migrating from the cross intersection into the analysis channel in sample-loading mode, the following condition must be met:

$$E_s \leq E_w \quad \text{or} \quad E_b \geq E_a \tag{2.74}$$

The ratio of the sample to sample waste field strength needed to prevent the sample from diffusing into the analysis channel depends on both transit time of the sample through the cross intersection and the diffusion coefficient of the sample. To allow the sample to migrate into the analysis channel and therefore be dispensed, the criteria in Equation (2.74) must be reversed. The field strength in the sample channel must be greater than the field strength in the sample waste channel or the field strength in the buffer channel must be less than the field strength in the analysis channel:

$$E_s \geq E_w \quad \text{or} \quad E_b \leq E_a \tag{2.75}$$

To return to run mode, the system is returned to the condition in Equation (2.74).

In Figure 2.14, the variable volume (gated) valve is shown. The sample (the bright regions) is transported through the injection cross toward the waste reservoir. The potential applied to the buffer reservoir prevents sample transport into the analysis channel. To make an injection of the sample onto the analysis column, the potential at the buffer reservoir is lowered or removed by opening a high-voltage relay for a brief period of time (0.01 s or longer), and the sample migrates into the analysis column as in an electrokinetic injection. To break off the injection plug, the potential at the buffer reservoir is raised or

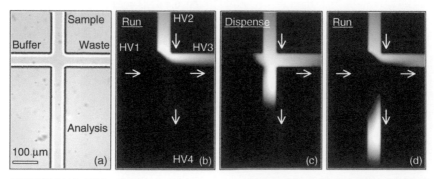

FIGURE 2.14 White light image of (a) a cross intersection (valve) and fluorescence images of (b, d) run and (c) dispense modes for the gated valve. The dispense time was 0.4 s. Arrows depict the flow direction. (Adapted from Jacobson SC, Ermakov SV, Ramsey JM. *Anal. Chem.* 1999, *71*, 3273–3276.)

reapplied. In this scenario, the electric field strength in the buffer channel is greater than the electric field strength in the analysis channel, and the electric field strength in the sample channel is less than the electric field strength in the waste channel. The arrows indicate direction of flow for the sample and buffer streams. This valve has a precision of better than 0.5% relative standard deviation (rsd) for delivering 20-pl aliquots of sample. Figure 2.15 shows the overlay of ten replicate injections each for dispensing times of 0.2, 0.4, and 0.8 s. As seen in this figure, the gated valve can deliver operator-selected volumes depending on the field strengths applied and dispensing times. The injection is biased by the relative electrophoretic mobilities of the analyte ions as with conventional electrokinetic injections. The gated valve also provides unidirectional fluid flow that can be advantageous for continuous flow conditions.

2.11 SCALING OF SEPARATIONS

With most assays, the analysis time is critical, and with microfabricated instrumentation, the reduced length scales can reduce the analysis time considerably. Taking an equation similar to Equation (2.14), where both pressure and voltage are applied across a channel or capillary and substituting in Equation (2.1), the total analysis time, t_{tot}, is

$$t_{tot} = L^2 \left(\frac{1}{\mu_{eo}V} + \frac{\eta}{\Delta P k_g} \right) \tag{2.76}$$

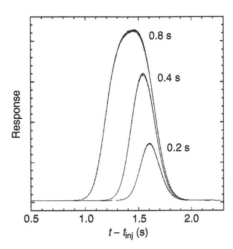

FIGURE 2.15 Overlay of 10 replicate sample profiles for injection times (t_{inj}) of 0.2, 0.4, and 0.8 s. The profiles were monitored 1.0 mm downstream of the valve in the analysis channel. (Adapted from Jacobson SC, Ermakov SV, Ramsey JM. *Anal. Chem.* 1999, *71*, 3273–3276.)

From this expression, the analysis time is proportional to the separation distance squared. Consequently, assuming the separation can be conducted appropriately in a shorter separation length, the analysis time can be dramatically reduced by shortening the separation length.

Also, the separation efficiency per unit length for microfluidic devices is comparable to other separation platforms. However, microfluidic devices have the advantage of being able to make injections having shorter axial extent using either the pinched or gated valve described above. The ability to introduce shorter sample plugs enables the separations to be performed faster using shorter separation lengths. Assuming a separation is limited by the injection plug length, i.e., other forms of dispersion such as diffusion and mass transfer are negligible, the minimum time, t_{sep}, and length, L_{sep}, to separate two components can be calculated.

To separate two components a distance ΔL depends on the relative velocity of the two components, Δu, and the separation time, t_{sep},

$$\Delta L = t_{sep}\Delta u \tag{2.77}$$

where Δu is the velocity difference between components 1 and 2. If the minimum distance to separate two components is the baseline width of the injected plug ($l_{inj} = 4\sigma$) the minimum time to separate two components can be calculated using Equations (2.37) and (2.77)

$$t_{sep} = \frac{R_s l_{inj}}{\Delta u} \tag{2.78}$$

The separation length, L_{sep}, corresponding to the separation time is

$$L_{sep} = t_{sep}u_1 \tag{2.79}$$

where u_1 is the velocity of the faster moving component 1.

In the example shown in Figure 2.16, the velocity of the first component is 5 mm/s, the velocity of the second component is 10% slower at 4.5 mm/s, the resolution, R_s, is 1.0, and the axial extent of the injected plug, l_{inj}, is assumed to be equal to the channel width. For an electrophoretic separation, using an electrokinetic mobility for the first component, μ_1, of 5×10^{-4} cm^2/V/s, and the electrokinetic mobility for the second component, μ_2, of 4.5×10^{-4} cm^2/V/s requires a separation field strength E_{sep} of 1000 V/cm. The applied potential, V_{app}, required to drive the separation is

$$V_{app} = E_{sep}L_{sep} \tag{2.80}$$

For a channel width of 1 μm corresponding to an injection length of 1 μm, the separation time is 2 ms, the separation length is 10 μm, and the applied

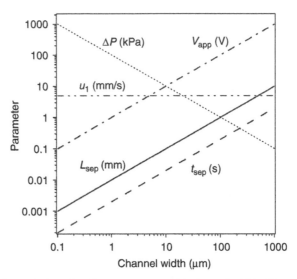

FIGURE 2.16 Variation of scaling parameters with channel width. The parameters are the applied pressure (ΔP), applied voltage (V_{app}), velocity of the first component (u_1), separation length (L_{sep}), and separation time (t_{sep}). See text for details.

potential required is only 1 V. Similarly, for a channel width of 10 μm again corresponding to an injection length of 10 μm, the separation time is 20 ms, the separation length is 100 μm, and the applied potential required is only 10 V. If the electric field strength is increased, the separation time decreases, the applied potential increases, but the separation length remains unchanged. Similarly, the same system can be used for liquid chromatographic separations. For example, components 1 and 2 can have capacity factors, k_1 and k_2, of 0.5 and 0.67, respectively.

Also shown in Figure 2.16 is the pressure required to transport the sample through the channel at the same linear velocity as the electrokinetic separation, e.g., 5 mm/s. Using the equation for a channel with rectangular cross section and an aspect ratio, width-to-depth, of 3, the pressure required is plotted as a function of the channel width. The pressure required to move the sample through the channel becomes exceedingly large and not feasible with many microfabricated devices. An alternative method to overcome these high pressures and to pump fluids through channels is shear driven flow and is described in Chapter 6.

Although this exercise is useful for demonstrating the advantage of reducing the channel dimensions, the primary drawback to decreasing the channel cross section to achieve smaller axial extent sample plugs is that every time the channel width is decreased by ten times, the number of molecules injected decreases by 10^3 times, placing increasing demands on the detection schemes.

REFERENCES

1. Bard, A.J.; Faulkner, L.R. *Electrochemical Methods*, John Wiley and Sons, New York, 1980.
2. Hunter, R.J. *Foundations of Colloid Science*, 2nd ed., Oxford University Press, Oxford, 2001.
3. White, F.M. *Viscous Fluid Flow*, 2nd ed., McGraw-Hill, New York, 1991.
4. Giddings, J.C. *Unified Separation Science*, John Wiley and Sons, New York, 1991.
5. McKnight, T.E.; Culbertson, C.T.; Jacobson, S.C.; Ramsey, J.M. *Anal. Chem.* 2001, *73*, 4045–4049.
6. Hunter, R.J. *Zeta Potential in Colloid Science*, Academic Press, Inc., San Diego, 1981.
7. Culbertson, C.T.; Jacobson, S.C.; Ramsey, J.M. *Talanta* 2002, *56*, 365–373.
8. Sternberg, J.C. In *Advances in Chromatography*, Giddings, J.C., Keller, R.A., Eds., Marcel-Dekker, New York, 1966, Vol. 2, pp. 205–270.
9. Alarie, J.P.; Jacobson, S.C.; Culbertson, C.T.; Ramsey, J.M. *Electrophoresis* 2000, *21*, 100–106.
10. Ermakov, S.V., Jacobson, S.C., Ramsey, J.M. *Anal. Chem.* 2000, *72*, 3512–3517.
11. Jacobson, S.C.; Culbertson, C.T.; Daler, J.E.; Ramsey, J.M. *Anal. Chem.* 1998, *70*, 3476–3480.
12. Culbertson, C.T.; Jacobson, S.C.; Ramsey, J.M. *Anal. Chem.* 1998, *70*, 3781–3789.
13. Rice, C.L.; Whitehead, R. *J. Phys. Chem.* 1965, *69*, 417–424.
14. Culbertson, C.T.; Ramsey, R.S.; Ramsey, J.M. *Anal. Chem.* 2000, *72*, 2285–2291.
15. Golay, M.J.E. *Gas Chromatography 1958*, Buttersworth, London, 1958.
16. Heiger, D.N. *High Performance Capillary Electrophoresis — An Introduction*, Hewlett-Packard, France, 1992.
17. Grushka, E.; McCormick, R.M.; Kirkland, J.J. *Anal. Chem.* 1989, *61*, 241–246.
18. Knox, J.H.; Grant, I.H. *Chromatographia* 1987, *24*, 135–143.
19. Peterson, N.J.; Nikolajsen, R.P.H.; Mogensen, K.B.; Kutter, J.P. *Electrophoresis* 2004, *25*, 253–269.
20. Ross, D.; Gaitan, M.; Locascio, L.E. *Anal. Chem.* 2001, *73*, 4117–4123.
21. Molho, J.I.; Herr, A.E.; Mosier, B.P.; Santiago, J.G.; Kenny, T.W.; Brennen, R.A.; Gordon, G.B.; Mohammadi, B. *Anal. Chem.* 2001, *73*, 1350–1360.
22. Griffiths, S.K.; Nilson, R.H. *Anal. Chem.* 2000, *72*, 5473–5482.
23. Culbertson, C.T.; Jacobson, S.C.; Ramsey, J.M. *Anal. Chem.* 2000, *72*, 5814–5819.
24. Paegel, B.M.; Hutt, L.D.; Simpson, P.C.; Mathies, R.A. *Anal. Chem.* 2000, *72*, 3030–3037.
25. Griffiths, S.K.; Nilson, R.H. *Anal. Chem.* 2001, *73*, 272–278.
26. Ramsey, J.D.; Jacobson, S.C.; Culbertson, C.T.; Ramsey, J.M. *Anal. Chem.* 2003, *75*, 3758–3764.
27. Griffiths, S.K.; Nilson, R.H. *Anal. Chem.* 2002, *74*, 2960–2967.
28. Dutta, D.; Leighton Jr., D.T. *Anal. Chem.* 2002, *74*, 1007–1016.
29. Giddings, J.C. *J. Chromatogr.* 1961, *5*, 46–60.
30. Kutter, J.P.; Jacobson, S.C.; Matsubara, N.; Ramsey, J.M. *Anal. Chem.* 1998, *70*, 3291–3297.

31. Kenndler, E. In *High Performance Capillary Electrophoresis: Theory, Techniques, and Applications*, 1st ed., Khaledi, M., Ed., Wiley-Interscience, New York, 1998, Vol. 146, pp. 25–76.
32. Jacobson, S.C.; Hergenröder, R.; Koutny, L.B.; Warmack, R.J.; Ramsey, J.M. *Anal. Chem.* 1994, *66*, 1107–1113.
33. Harrison, D.J.; Fluri, K.; Seiler, K.; Fan, Z.; Effenhauser, C.S.; Manz, A. *Science* 1993, *261*, 895–897.
34. Jacobson, S.C.; Koutny, L.B.; Hergenröder, R.; Moore Jr., A.W.; Ramsey, J.M. *Anal. Chem.* 1994, *66*, 3472–3476.

3 Microfabrication and Integration

Han J.G.E. Gardeniers
and Albert van den Berg

CONTENTS

3.1 INTRODUCTION

This chapter gives an overview of the techniques that are used to fabricate miniaturized chemical analysis and separation systems. The focus will be on the choice of the material(s) out of which the body of the separation system is build, the design rules for the use of the different micromachining techniques, and the specific characteristics (like the surface roughness) of the resulting three-dimensional (3D) structures.

Microstructures of growing complexity can be made in a number of different ways. In the field of integrated semiconductor circuit manufacture [1], which, as will be discussed below, formed the foundation for many of the methods that are presently used to fabricate separation chips, one often distinguishes between fabrication procedures that consist in sculpturing the complete microsystem out of a single piece of material, and methods that use micro-components fabricated in a variety of materials and assemble them to build a functional microsystem. The first method has adopted the name "monolithic" (from the Greek "mono" and "lithos," which mean "single" and "stone," respectively), although the original meaning of the word is not always correctly used in the field of integrated semiconductor circuit fabrication, where also processes like dopant diffusion (which alter the conductivity of semiconductors) or oxidation of that single piece of material, or even the deposition of a coating of a metal or a dielectric on the surface of the piece of material, are called monolithic.

It is worthwhile to digress a little bit on integrated circuit (IC) technology, one of the reasons already mentioned being that the field of micromachining started as a sort of "spin-off" of IC technology. The trends that can be observed in the history of IC development are also becoming apparent in the development of miniaturized chemical analysis systems. In particular, the successful development philosophy of IC technology is also leading the way for most of the microfabrication routes in the field of miniaturized separation systems. According to Fukuda and Menz [2], the following recipe of four steps is the key to the success of IC technology:

1. "Computer-aided design": design, optimization, and simulation of the circuit, in most cases including the corresponding fabrication process, is done on a computer; the design tools are so sophisticated that in the development of new circuits, expensive and time-consuming testing is hardly required.
2. Transfer by means of optical imaging (photolithography) of the designed circuit pattern to a substrate; patterns are transferred in series to a set of "masks" and subsequently imprinted in parallel on a substrate.
3. Batch fabrication: processes are applied that simultaneously treat the surface of a large number of substrates, or at least a large number of equal areas on one substrate, therewith minimizing the variation in process quality. The technological and metrological costs of a process step are thus distributed over thousands of components.
4. Linking a large number of identical components with high packing density to obtain a system with a new, high-quality function. Particularly in this step the power of the technology is manifested: due to large-scale integration unlimited opportunities are offered that are virtually impossible to achieve with other fabrication processes. The issue of integration will be discussed in more detail in Section 3.6.

Many of the processes that will be described in this chapter rely on the principles of above philosophy (which we will henceforth call "lithography-based" methods), mainly because such processes (with the exception of processes like hot embossing and injection molding that are used for polymer microfabrication) offer the best potential for large-volume production.

If one looks back to the history of microfabrication, it is observed that while IC technology came to maturity, new application fields were allocated for this versatile technology [3], and it turns out that the attempts to explore the use of microfabricated devices in the field of chemical analysis were very much inspired by the often appraised work of Terry and colleagues at Stanford University, as early as 1975, on a fully integrated gas chromatography system fabricated out of a 2-in. silicon wafer [4]. This system will be discussed in greater detail in Chapter 9.

It remains unclear why there was no direct follow-up on this work in the years thereafter, although a number of microfluidic devices were demonstrated that can be considered very useful components for miniaturized chemical analysis systems, like the piezoelectric micropumps developed first by Spencer and coworkers (fine-machined in stainless steel) [5] and later by van Lintel and colleagues (micromachined in silicon) [6] (for a recent comprehensive review on the history of micropumps the reader is referred to the SPIE paper by Woias [7]). The first liquid chromatography on a chip was reported in 1990 [8]. This particular chip was micromachined in silicon, using isotropic wet-chemical micromachining methods (described in Section 3.2). As a matter of fact, many miniaturized devices for use in chemistry were first demonstrated in silicon. The most important reason is the incredible wealth of silicon micromachining techniques that became available through the years, so that it is now possible to fabricate virtually any geometrical structure in silicon with very high precision. Other reasons to use silicon are the mechanical properties (hard, strong) and the possibility to passivate it with an inert insulating layer of silicon dioxide (which is very similar to fused silica, a material very well known to chemists).

It has often been noted that the Lab-on-a-Chip field, of which miniaturized separation systems form a very important part, has gained its fame because of the rise of the method of capillary electrophoresis (CE) that turned out to be so convenient to perform in chip format and was used to separate biologically relevant molecules like DNA fragments and amino acids. Due to the high electrical fields that are required in this method, the use of the semiconducting material silicon is not the best choice, and therefore microfluidic chips were fabricated out of glass substrates [9]. The next section will contain a few paragraphs describing the methods to sculpture glass. Details about how to perform CE on such chips are described in Chapter 8.

The methods used to micromachine glass are based on the batch fabrication philosophy described earlier, i.e., they include photolithography and etching steps. An important drawback of these processes is that they are only economically feasible if relatively large amounts of chips are fabricated, and in addition to that, the resulting chips become rather costly if their footprint is relatively large. The latter is the case for CE on a chip, where a certain minimum separation channel length is required to achieve reasonable separation efficiency (and as is discussed in Chapter 2, in order to achieve longer separation lengths it does not help to just fold the channel into a serpentine). On the basis of material costs, glass chips will be relatively expensive compared to chips fabricated out of polymeric materials [10]. The costs of typical polymer replication processes like hot embossing and injection molding are mainly determined by fabrication costs and lifetime of the mold. When replication methods that are well known from plastics industry are adjusted to a feature size in the range of a few micrometers, cheap disposable chips become feasible. In addition to this, the possibilities of chemical modification of polymers are often claimed to give the opportunity to manufacture chips with tailor-made surface properties. Looking into recent literature, it turns out that the latter possibility of polymer microfabrication is not yet extensively used, and in fact it is often more practical to just coat the chip inner walls with a polymer of the desired properties [11].

It should be made clear here that in addition to the above-mentioned methods of mass fabrication, a growing number of microfabrication tools have become available, based on a "fine-mechanical" approach, like laser ablation and electrochemical discharge drilling; see, e.g., Refs. [2, 12, 13]. Such methods are generally more expensive in large volume production than the other methods described above; nevertheless, they have been used to fabricate interesting demonstrators.

In the field of separation science, either for preparative or for analysis purposes, one may categorize all existing methods as either being based on differences in *affinity*, in *mobility*, or in *size*. Particularly for the latter separation principle the field of *nanotechnology* becomes important. Separation based on size (size exclusion and related methods) will benefit from the decreasing dimensions with improved dimensional tolerances as they have become available due to recent developments in nanotechnology.

3.2 THE MICROFABRICATION TOOLBOX

This section will give a comprehensive description of the techniques that have been reported for the microfabrication of separation systems. The text is divided according to the substrate material in which the basic structures for separation are engraved. Section 3.2.1 will focus on the important issue of surface roughness resulting from the micromachining process. This parameter has been claimed to be an essential factor for the quality of high-performance separation, in particular for electrophoresis, and we have collected information from the literature to investigate what the influence of surface roughness is on, e.g., analyte dispersion.

3.2.1 SILICON MICROMACHINING

An extensive overview of the silicon micromachining methods to be described below can be found in Refs. [14, 15]. Here we briefly summarize the principle of each method and describe what type of structures may be obtained with it.

An often-used method for micromachining of silicon is *anisotropic etching*. In this process, silicon is dissolved in a concentrated alkaline solution, e.g., 25 wt.% KOH in water at temperatures around 70°C. The method relies on the crystallographic properties of single-crystalline silicon substrates, i.e., during etching a structure develops that consists of the slowest etching crystal planes, the {1 1 1} planes. Depending on the selected crystallographic orientation of the wafer surface (defined by how the wafer was cut out of a larger piece of single-crystalline silicon), and the design of openings in the masking layer (usually a thin film of SiO_2 or Si_3N_4), several shapes can be fabricated with this method. If the substrate surface is of {1 0 0} orientation, V-grooves or grooves with a trapezoidal cross section are obtained; if the surface is {1 1 0}, grooves with parallel sidewalls result. Figure 3.1 shows typical examples of grooves etched in silicon. The anisotropic etching method is often used for the fabrication of membranes for pressure sensors. Such membranes may also serve as check valves and micropump membranes [16]. Microchannels with a trapezoidal cross section etched in a Si(1 0 0) substrate were applied in a liquid chromatography chip [17] as a separation column. This particular chip also had a split injector structure, a frit that is used to contain packing material with a particle size of 2 μm or more in the separation column, and an optical detector cell, which worked according to the principle shown in Figure 3.2(a).

A drawback of anisotropic etching is the limited design freedom. For example, in the fabrication of channels with bends the exact shape as defined by the mask is never obtained, because convex corners tend to develop toward a shape limited by the fastest etching planes, and concave structures will always end up with {1 1 1}-oriented surfaces. Corner compensation structures in the masking layer can help in this, but such compensation requires complex fine-tuning of the mask design and exact knowledge of the orientation dependence of etching rates. In addition to that, smoothly curved channel structures in single-crystalline silicon cannot be achieved with this method, while this is often desired in a separation system in order to avoid extensive plug dispersion.

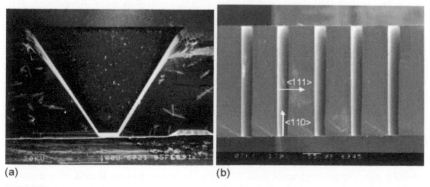

(a)   (b)

FIGURE 3.1 (a) Grooves etched in {1 0 0} Si. (Courtesy of M.J. de Boer, University of Twente.) (b) {1 1 0} Si.

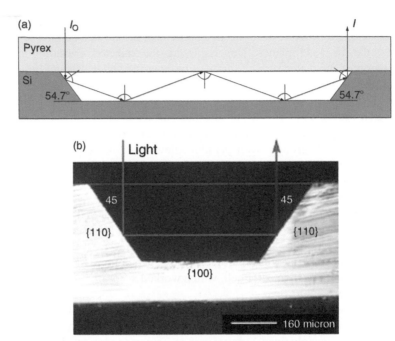

FIGURE 3.2 (a) Schematic of light beam propagation in an anisotropically etched detector cell. (b) Ten-millimeter-long detector cell obtained after the use of a special etchant solution and a special alignment of the mask structure with respect to the crystallographic directions in the silicon substrate.

The surface roughness of anisotropically etched channels can range from atomically smooth to relatively high, depending on the exact conditions of etching (temperature and composition of the etchant) and the care that is taken to align the mask patterns with the $\langle 1\ 1\ 0 \rangle$ crystal directions. Recent work has shown that even on perfectly aligned surfaces some roughness will develop, which is inherent to anisotropic etching. The reason for this is that etching on {1 1 1} surfaces only proceeds when nucleation of etch pits is initiated by crystal defects [18], and due to the fact that absolutely defect-free silicon is impossible to obtain, this inevitably leads to some surface height differences of the order of a few atomic steps or higher if the surfaces are contaminated by minute amounts of impurities [19, 20]. As far as we know, there were no studies of the effect of the surface roughness on the perform- ance of silicon-based separation microsystems. One of the reasons for this lack of information is that roughness is not an important parameter for the performance of the separation systems that have been fabricated in silicon, mainly gas chromatographs and to a much lesser extent liquid chromatog- raphy chips, with channel widths of several tens to hundreds of micrometers. It is relevant though for structures in the sub-micrometer range, like those used for hydrodynamic chromatography (HDC), described in Chapter 6, and other size-exclusion-based separation methods, which will be described in Section 3.7.

Although silicon is not transparent for ultraviolet (UV) light, it has turned out to be of interest for optical detectors for chromatography, due to the mirror-like surfaces (provided that the correct etching conditions are used) with well-defined geometry. Optical detection will be discussed in detail in Chapter 11; here, we will discuss only the fabrication of one particular type of detector structure by anisotropic etching.

The main problem with detection based on optical absorption in a micromachined device (and also in a capillary) is the limited optical path length that exists when one passes a light beam through the separation microchannel (or capillary). For absorption detection in capillary systems, therefore, special detector designs have been developed that intend to increase the optical path, like Z-shaped cells [21] or bubble cells [22].

In silicon-based detectors, increased optical paths can be achieved rather conveniently by using light reflection on the inclined silicon surfaces that result after anisotropic etching. Figure 3.2(a) gives an example for a structure etched in a Si(1 0 0) substrate [23]. The angle that the $\{1\ 1\ 1\}$ planes make with the top surface is 54.7°, so that it is possible to have a light beam perpendicular to the top surface reflected several times before it will leave the cell again via a second mirror-plane. The drawback of this method is substantial light intensity loss at every reflection, partially also because of the previously discussed unavoidable roughness of the silicon $\{1\ 1\ 1\}$ surfaces. Therefore, another approach was to use a 45° sidewall to reflect a vertically oriented light beam into the horizontal flow channel, which was done by creating Si$\{1\ 1\ 0\}$ planes, which make a 45° angle with the substrate surface of $\{0\ 0\ 1\}$ orientation, and can be obtained by a special etchant containing aqueous KOH solution mixed with isopropanol. In order to reduce the roughness of the $\{1\ 1\ 0\}$ sidewalls, the silicon surface was oxidized and the oxide subsequently removed by dissolution in a HF solution, the procedure being repeated several times, before finally a platinum layer was deposited on it to increase reflectivity of the surface even more [24]. Figure 3.2(b) shows a photograph of the cross section of the resulting detector cell.

If smooth surfaces are required, *isotropic wet-chemical etching* is the method of choice [25]. The etchants used for this generally consist of HF, an oxidant like HNO_3, and a solvent like water or acetic acid. The mechanism is claimed to be a combination of a silicon oxidation step followed by removal of the resulting silicon oxide by HF. The rate of this etching process is not limited by surface reactions, as is the case in the anisotropic method, but by mass transport in the solution. Due to the increased mass transport to and from extended features, under optimized conditions any protrusion from a nominally flat surface will be removed at a higher rate than the flat surface, and features are rounded off, so that surfaces become smooth. The channel shapes obtained with this method are tubular. Figure 3.3 shows a typical result.

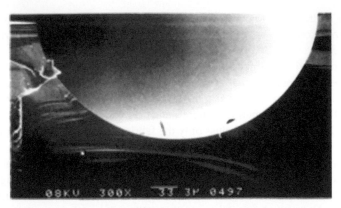

FIGURE 3.3 Cross section of isotropically etched channel in silicon; note that the width of the channel is almost twice its depth.

With this method the design freedom is also limited, because due to the isotropic nature of the process mask under-etching occurs, which increases the width of the channel to twice its depth. Nevertheless, this method has been used with success in some chromatographic microdevices, like in the original gas chromatograph of Terry et al. [4] and following gas chromatographs, to be described in Chapter 9. The method was also used to etch microchannels for a liquid chromatography chip [8].

One particular type of isotropic etching is a method that uses an electrical potential instead of the oxidant mentioned above. This anodic wet-chemical etching process gives essentially the same results as the previous method, except for conditions at relatively low anodic potentials, where the silicon does not dissolve completely but is turned into a porous layer [26], of which the pore size distribution is well defined and can be controlled accurately [27]. Many applications of such a porous material in microbiochemical analysis and microreactor devices can be imagined, and some interesting applications have already been realized. For example, it was reported that porous Si, used in an enzyme reactor, shows a 100-fold increased enzyme activity, compared to a nonporous reference [28]. Porous silicon can be modified in many ways; e.g., it can be oxidized to obtain porous silicon oxide [29]. In this way many different materials with a high surface-to-volume ratio can be manufactured, which can be used in gas chromatography, as molecular sieves (for microdialysis), as gas sensors, or as very effective catalysts in heterogeneous gas-phase microreactors. It was also used as a surface on which laser desorption can be performed without a matrix [30] for matrix-assisted laser desorption ionization mass spectrometry. For this particular application the morphological features of porous silicon are thought to provide a framework in which solvent and analyte molecules are retained, while the high thermal conductivity of silicon promotes efficient energy transfer from substrate to adsorbed analyte to allow desorption and ionization of the intact analyte.

Figure 3.4(a) gives a typical result of channels fabricated with a dry etching method, more specifically reactive ion etching (RIE), in this case in an SF_6–O_2-based inductively coupled plasma (ICP) process. This method allows the largest degree of freedom of structural design: the pattern defined in the mask can be projected directly into the silicon substrate, allowing sharp corners and channels with in-plane tapers [31]. The number of channels per unit area can be as high as the lithographic steps allow, which means that the method can also be used to make nanochannels, i.e., channels with a width below 100 nm (state-of-the-art optical lithography achieves feature sizes of ca. 0.15 μm, while with laser interference methods repetitive features smaller than 100 nm can be achieved [32]). One important property that should be noted here, however, is that the aspect ratio (i.e., ratio of depth to width), although high compared to other methods, is limited to typical values of 20 to 30.

An RIE process uses a gas discharge to create plasma that contains reactive species. These species, depending on their nature, either etch the surface chemically (radicals), physically (ions, which are accelerated toward the surface by the electrical bias that this surface has in a plasma), or in a combined physical–chemical mode, where ion bombardment via several different mechanisms stimulates chemical etching. Depending on the etching mode, isotropic or anisotropic (rather: directional) profiles develop, see Figure 3.4. An example of a more complex structure that was fabricated with this method in the directional mode, viz. a micromixer, is shown in Figure 3.5 [33].

The silicon micromachining methods described until now are all based on lithography. There are several reports in the literature describing direct-writing of structures in silicon, using lasers. The method consists of a

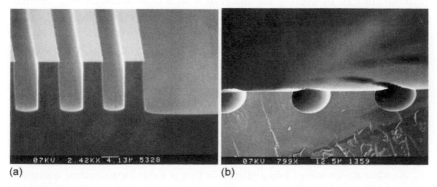

(a) (b)

FIGURE 3.4 (a) Deep trenches etched in silicon with an SF_6–O_2 plasma. (b) Reactive ion etching can also be tuned to give isotropic channel profiles.

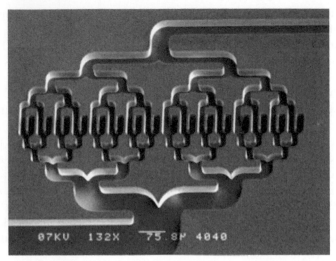

FIGURE 3.5 Photograph of a reactive ion etched micromixer in silicon. (Courtesy of C2V BV.)

laser-heated reaction of silicon in a chlorine atmosphere. Although these methods were originally claimed to be suitable for the fabrication of fluidic microstructures [34], it seems that lithography-based methods are preferred today.

Although the applications of silicon microchannels (or nanochannels, see below) are numerous, their use is limited for some of the important fields of current interest in chemical analysis. For example, in CE, high electrical fields along the length of a capillary (read: microchannel) are required, which is not possible in silicon because of its very high electrical conductivity. Some attempts were made to overcome the conductivity problems with silicon as a material for CE, by coating the material with a insulator layer like thermal oxide or low pressure chemical vapor deposition (LPCVD) silicon nitride [35, 36]. Such insulators are well known in IC technology, and if fabricated with IC-compatible deposition methods, have excellent electrical breakdown properties. An interesting demonstration can be found in the work of Mogensen et al., who used a 13-μm thick thermally grown silicon dioxide layer on a silicon substrate to isolate the electrolyte in the etched channels from the semiconducting substrate [37]. The breakdown voltage during operation of this chip was measured to be 10.6 kV, which is more than enough to perform electrophoresis. The chip also had integrated optical waveguides, connected to optical fibers, for absorption measurements in a 750-μm long U-shaped detection cell.

Another drawback of using silicon is that it exhibits no optical transparency for the visible or UV wavelength regime, frequently used for detection. For such applications a more suitable material, like glass, is highly desirable.

Glass has a much better stability against high electrical fields and is optically transparent for most of the visible wavelength range, and dependent on the exact material composition, also partially or completely transparent in the UV range. Glass micromachining will be discussed in the next section.

A particularly interesting strategy is to use silicon channels as a mold for the deposition of insulating materials [38, 39]. Figure 3.6 shows some examples, in which a combination of silicon etching, silicon dioxide deposition, and anodic bonding to a glass plate was applied, giving microtransparent insulating channels (μ-TICs). Earlier results have shown that these channels support electrical field strengths of up to 4 kV/cm and are well suited for CE [39]. Since these channels may have very thin walls and small diameters, completely new types of fluidic devices become feasible [40]. A similar replication technique, using micromachined silicon structures and hot filament chemical vapor deposition of polycrystalline diamond films, was used to create microchips with diamond tubes with cross-sectional areas of 4000 and 7000 μm². These channels were filled with continuous polymer beds and used for anion-exchange chromatography of proteins with results comparable to those obtained on fused silica capillaries and quartz microchips [41]. The

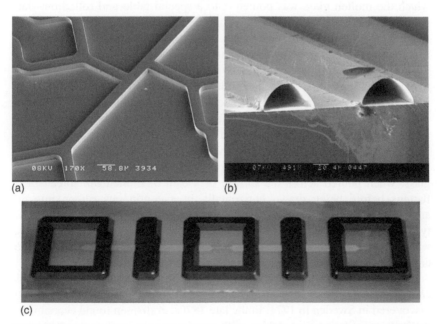

(a)

(b)

(c)

FIGURE 3.6 Examples of "μ-TICs": (a) silicon nitride channel based on a channel etched in Si by reactive ion etching; (b) silicon nitride replica of isotropically etched Si patterns, bonded to a glass plate; (c) separation channel with silicon mesas left over after etching, which serve as electrodes.

reasons to use diamond are the high thermal conductivity, excellent optical transparency, chemical inertness, and electrical insulation properties of the material.

3.2.2 GLASS MICROMACHINING

For many applications in solution chemistry, either for synthesis or analysis, glass is the preferred material, the most important reason being the familiarity of the chemist with the material. Glass is one of the oldest materials known and used by mankind [42]. Man-made glass objects date back to around 4000 BC, having been found in Egypt and Eastern Mesopotamia. These glasses have compositions very similar to those of modern soda lime silicate glasses, which are also employed in the microfluidic chips that are used for separation purposes today. A few events in the history of glass are worth mentioning. In the 11th century, German glass craftsmen developed a technique for the production of glass sheets by blowing a hollow glass sphere, swinging it vertically, and have gravity pull it into a cylindrical pod (measuring as much as 3 m long and up to 45 cm wide). While still hot, the ends of the pod were cut off and the resulting cylinder cut lengthways and laid flat. In 1688, in France, a new process was developed for the production of plate glass, in which the molten glass was poured onto a special table and rolled out flat. After cooling, the plate glass was ground on large round tables by means of rotating cast iron disks and increasingly fine abrasive sands, and then polished using felt disks. The result of this "plate pouring" process was flat glass with good optical transmission qualities. When coated on one side with a reflective, low melting metal, high-quality mirrors could be produced. A key figure in modern glass research was the German scientist Otto Schott (1851 to 1935), who used scientific methods to study the effects of chemical elements on the optical and thermal properties of glass. The float process developed by Britain's Pilkington Brothers Ltd., and introduced in 1959, combined the brilliant finish of sheet glass with the optical qualities of plate glass. Molten glass, when poured across the surface of a bath of molten tin, spreads and flattens before being drawn horizontally in a continuous ribbon into an annealing furnace. Borofloat® glass is now one of the most important glass types for the production of glass chips, not in the least because of its flatness, which is an important feature if glass plates have to be bonded in order to create a tight seal of microfluidic parts, as will be described later.

There are basically two ways to make precise patterns in glass; one is *chemical etching* where the surface is eroded by an acid. The method was first discovered in Sweden in 1771. In the late 1870s, craftsmen found economical methods of imitating acid etching with *sandblasting*, a method that "carves" structures into the glass by pressure blasting an abrasive at the glass surface. These two methods are nowadays still the most important techniques to make microstructures in glass.

In order to obtain the desired microstructures in glass, batch fabrication techniques similar to those used for silicon micromachining, based on photolithography, may be used. After the pattern is transferred lithographically into a resistant coating (a polymer layer, amorphous or polycrystalline silicon, or a metal like Cr), etching of glass is performed in solutions containing hydrogen fluoride, HF. With this process, it is possible to achieve features with a size down to several tens of nanometers. As was already noted for the isotropic etching of silicon, the width of any structure will be at least twice its width, so that deep trenches are not possible. Glass and fused silica removal rates are typically in the range 0.1 to 1 μm/min. The method is therefore particularly useful for shallow structures. Main drawbacks are the extreme health hazards and related safety precautions involved with HF, and the clean room facilities required for the photolithographic process, leading to relatively high investment costs for fabrication. A good overview of all the practical issues involved with isotropic etching of glass can be found in Ref. [43]. One of the examples of structures made with this method is a 50-cm long CE column on a Borofloat glass plate, which was used to sequence DNA with a performance similar to a fused silica capillary with similar cross section [44].

A cheaper and less critical process to obtain microstructures in glass was developed in the past few decades. This so-called *powder blasting* process, based on the original sandblasting method mentioned above but using much finer grains, was originally developed for flat panel displays [45] and micro-counter flow heat exchangers [46]. At present, it is also frequently used to fabricate fluidic structures for CE [47–49]. Glass-removal rates with this process, which consists of directing a jet of fine powder particles (particle size 9 or 30 μm) to a brittle material in order to erode that material locally, are typically between a few micrometers per minute and 1 mm/min. Thick elastomeric photosensitive coatings that can be photolithographically patterned are very suitable masks for this process, and structural features with a size of 50 μm and larger can be reliably manufactured, which is in the range of the dimensions that are of interest for chromatographic techniques. Figure 3.7 gives a typical example of structures obtained with this method. In addition to all this, powder blasting requires less stringent safety precautions and a less expensive infrastructure, so that this method may be a cost-effective alternative to HF etching. The main drawback of powder-blasted channels when compared to HF-etched fluidic channels is the much higher roughness of the surface of the channels (usually a few micrometers for powder blasting and below 50 nm for HF etching). We will discuss this issue in more detail in Section 3.3.

RIE is becoming a more common method now for etching of microstructures in glass. To recall, the basic features of an RIE process are that a plasma is created that contains reactive species, like ions that physically bombard the surface, and radicals that chemically etch it. Conventional plasma processing tools are essentially based on parallel-plate discharges [50], in which an

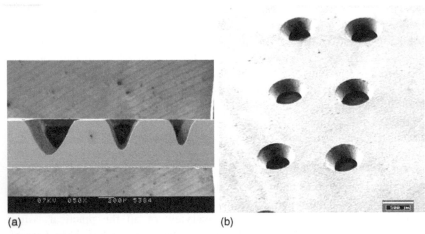

(a)   (b)

FIGURE 3.7 Typical structures in Borofloat glass, obtained by powder blasting: (a) microchannels. (Courtesy of H. Wensink, University of Twente.); (b) through-holes (Courtesy of Micronit Microfluidics BV.)

electric field between two plate electrodes is applied to generate and sustain a gas discharge. Plasmas thus generated are only partially ionized and have high neutral gas pressures. Because high ion fluxes toward the substrate to be etched are generally desired, high-plasma-density processing tools [51] with lower gas pressures are beginning to replace conventional plasma tools. Several different methods can be used to generate high-density plasmas, for example, oscillating radiofrequency (RF) electric fields can be used to generate and sustain a high-density plasma without using steady-state magnetic fields (ICP) or by using steady-state magnetic fields (known as a helicon plasma because helicon waves are induced along the magnetic field lines). Additionally, electron cyclotron resonance can be used to generate a high-density plasma. A neutral-loop discharge is a recently developed method in which a plasma is generated by RF fields along a closed magnetic neutral line [52]. The latter method has been used to create some very fine fused silica structures for DNA separation [53].

For a substantial etch rate it is necessary that the radicals form volatile species with constituents of the material to be etched, and exactly this may become a problem if glasses are to be etched. Since the most commonly used gases in RIE contain F, Cl, or O, it is of particular importance to have volatile fluorides, chlorides, or oxides, at the temperature of the surface that is exposed to the plasma. This surface temperature due to ion impact effectively is at least 100 to 200°C in conventional parallel-plate plasma reactors. It can be calculated on a thermodynamic basis that elements like Al, Ca, and Mg, which are some of the most important constituents of glasses, will not form volatile fluorides under such conditions (in fact, many other glass constituents that are present in lower concentrations, like Fe, Cu, Pb, and Ba, also do not). On the other hand, silicon and boron readily form volatile compounds in a plasma. It was found by some researchers that in the course of RIE of particular glass types, a residue became formed that increasingly slowed down the etching process with time, and resulted in extremely rough surfaces. These residues were not caused by redeposition of the mask material, which is a known factor causing increased roughness on etched surfaces [14], but consisted of nonvolatile compounds of the elements mentioned above.

It was found that physical etching conditions are required to remove the nonvolatile residues and therewith etch the glasses containing the previously mentioned elements. This implies that high substrate self-bias voltages and low pressures be used. Under such conditions, using a magnetically enhanced ICP RIE system, work was performed to etch Pyrex glass to achieve feed-through holes with a width of 50 μm in a 150-μm thick wafer, a process that took 10 h [54]. Another approach, used to etch microfluidic channels in borosilicate, uses high-flux Ar ion bombardment in addition to the etching gas SF_6 to remove the residues, achieving an etching rate of 0.6 μm/min and a surface roughness of less than 3 nm (without the addition of Ar the roughness was ca. 200 nm) [55].

One important type of glass that can easily be etched in a plasma is silicon dioxide, or rather fused silica. Silicon dioxide coatings are used frequently as insulating layers in IC technology or in planar optical waveguides; therefore, dedicated processes were developed to etch this material at a high rate and with excellent feature definition. These processes can directly be used to etch fused silica substrates. Figure 3.8 shows an example of microstructures in fused silica that were used to create injection slits in a HDC chip [56]. Work of particular interest was performed by Regnier and coworkers, who have fabricated a column with monolithic packing for liquid chromatography (LC), i.e., they used CHF_3-based RIE to create deep and narrow slits in a fused silica substrate in a regular pattern that acts as solid packing on which a stationary phase was bonded [57]. The monolithic pattern fabricated consisted of 5 μm by 5 μm square pillars of 10 μm height, defining channels 1.5 μm wide. Pictures of the resulting structures can be found in Chapter 7, which also contains more details about the operation and design of these LC chips. For chip formats, monolithic packing has some advantages over packing with particles, due to problems with fabricating frits in microchannels, the non-uniformity of packing at walls and corners of microchannels, and particularly the filling of channel networks to get the particles only at specific locations in the chip.

In the 1950s and 1960s, Corning extensively studied glasses that undergo a phase change upon illumination. One of the products that resulted was the so-called Fotoform® glass, but also the Schott product Foturan® and Hoya's PEG300 show more or less similar properties. These silicate glasses have a large concentration of lithium oxide, and small concentrations of silver and

FIGURE 3.8 Several trenches etched in fused silica, using a reactive ion etching process based on SF$_6$. (From Blom MT, On-chip Separation and Sensing Systems for Hydrodynamic Chromatography, PhD thesis, University of Twente, The Netherlands, 2002. With permission.)

cerium, which are responsible for the photosensitivity. If this type of glass is illuminated by UV light, the silver will be reduced to free silver atoms, which will coagulate when the glass is heated to a temperature of ca. 400°C, therewith forming silver particles of a few nanometers. When the glass is subsequently heated to 600°C, these silver particles will form nuclei on which the glass will crystallize into a material that dissolves at a much higher rate in hydrogen fluoride that the nonilluminated glass. By using lithographic masks in this way the glass may be patterned to achieve structures with almost vertical side-walls. The method has been used to make CE chips in Foturan, following the general procedure shown in Figure 3.9 [58], but with backside protection and a defined etching time so as to achieve a certain structure depth. Because the depth that is affected by UV illumination depends on the light intensity, it was also found possible to make structures with a continuously varying depth (i.e., with a slope), by using a special gray-tone lithographic mask that allowed a variation in light transmission over a certain area.

A drawback of the method of using photosensitive glass is the high surface roughness, 1 to 3 μm. Furthermore, during the curing process the changes in the material result in an increased surface roughness also in areas that are not etched, which complicates any subsequent bonding processes.

The methods described up to now were all based on photolithography and wafer-scale processing schemes. However, microstructures in glass have also been fabricated by conventional machining techniques. For example, 40-μm

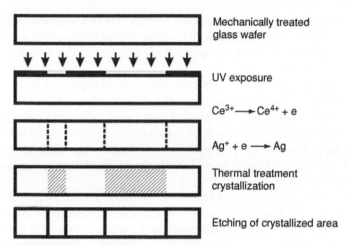

FIGURE 3.9 Procedure for fabricating structures in Foturan photosensitive glass. (Adapted from Becker H, Arundell M, Harnisch A, Hülsenberg D, *Sens. Actuators B*, 86, 2002, 271.)

wide (width accuracy 2.5 μm) high aspect ratio straight channels were made in glass using dicing methods [46]. Machining methods like mechanical or ultrasonic drilling are routinely used to make through-holes in glass plates with a diameter of 100 μm or more, e.g., spark-assisted etching (a method resembling spark erosion, but with the difference that in spark-assisted etching the workpiece is immersed in a concentrated alkaline solution) was used to make 200-μm holes in Pyrex glass [59]. Such methods are less suitable for the fabrication of microchannel structures with high accuracy. Excellent reviews of fine-machining methods, including, e.g., microelectric discharge machining, which can be used to make features with a size down to ca. 5 μm, can be found in Refs [12, 60, 61].

Although at first sight it would not seem to be a method useful for making structures in glass, lasers can be used to machine structures in glasses, even in fused silica with its good transparency in the UV range. This has become possible with the latest development in lasers, e.g., with a F_2-laser that operates at a wavelength of 157 nm, but it is also possible to use somewhat higher wavelength excimer lasers, e.g., at 193 or 248 nm, for glass types like Pyrex. In theory, very fine structures, even down to ca. 50 nm, are possible with such short-wavelength, short-pulsed lasers, in a direct-write process. However, due to problems associated with redeposition and crack formation, it is often more convenient to use the laser to illuminate a photoresist layer, which can be done with very high accuracy, and subsequently use the mask pattern to etch the glass with conventional etching methods.

One interesting way of doing this is by using computer-controlled laser writing to make a resist pattern with a well-defined tapered thickness, and use an RIE process that gives a photoresist etching rate that is almost similar to that of the glass, or at least has a well-characterized etching rate, to transfer the resist pattern taper into the glass. For example, in this way microlens arrays can be made in glass.

Microlenses are of great interest for miniaturized separation systems, and microlenses integrated on a microfluidic chip for chemical analysis were demonstrated recently [62]. In this particular work, the design is such that microlenses focus an excitation beam into a detection volume and collect the emitted light from fluorescent molecules. Successful detection of a 20-nM Cy5TM solution in a 50-μm deep microchannel with an excitation volume of ca. 250 pl was shown.

Many methods have been described to fabricate microlenses, in all kinds of materials. The simplest technique involves the formation of photoresist cylinders, heating the resist to a temperature that gives reflow. This will create a spherical surface, with a shape defined by surface tension properties. The restriction on this technique is that it can only be used to create spherical surfaces; it is not possible to adjust the profile of the lens to correct for optical aberrations. Finally, the lens profile is transferred into the substrate material, as described above, with an RIE process. An alternative technique is known as gray-tone lithography. This allows a better control of the lens profile, and also relies on a tapered resist profile. An extra possibility that exists is that the RIE selectivity of the substrate material to resist is tuned to another value than 1:1, to achieve a microlens that has a different height than the original photoresist dot [63]. Essential in these, or in all plasma processes that involve tapered resist profiles or resist patterns that withdraw slowly during etching, is the fact that the temperature during etching is controlled very well, to ensure that the resist will stay below its reflow temperature. Because the resist may heat up due to plasma interaction, this generally requires very efficient cooling, e.g., with helium on the backside of the substrate. For an overview on the fabrication of microlens structures, see Ref. [64].

One special method to create very small features in, e.g., glass is worth mentioning here: through-holes with a diameter of somewhat less than 1 μm, to be used for patch-clamping, were made by exposing a thinned area (80-μm thick) in glass with a gold ion beam of 2260 MeV and subsequently etch the glass from one side [65]. Heavy ions of such high energies produce so-called latent tracks characterized by a cylindrical damage zone of a few nanometers in diameter, which exhibit a much higher etching rate than the surrounding bulk material. Etching leads to a conically shaped groove along the track.

3.2.3 METAL MICROMACHINING

This chapter will not discuss metal micromachining in the same detail as the other materials, mainly because metals have hardly been used for separation systems, at least not in a micromachined fashion. A few examples exist of the use of stainless steel columns for certain chromatography applications like gas chromatography and liquid chromatography for petrochemicals and polymers. Nevertheless, the use of, e.g., steel has received an increasing amount of attention as an alternative substrate in microelectronic applications, and techniques to machine steel at wafer scale are becoming available. For example, discrete electronic devices fabricated from amorphous and polycrystalline silicon have already been successfully fabricated on steel substrates [66], and this progress in metal micromachining will eventually probably also find its way in the field of miniaturized chemical analysis and separation systems.

If micromachining of metal substrates, foils, or films is required, wet-chemical etchants may be used. As discussed above, most metals do not form volatile compounds in fluoride or chloride plasmas, and therefore cannot be etched unless under extreme conditions (higher temperatures and high ion flux). Some metals that can be etched in fluoride RIE are Ti, Ta, W, and Nb. For example, Ti RIE was used to create a well-defined texture on human implants [67]. Metals that are not etched (and therefore form good masks for RIE of silicon, polymers, or glass) are Al, Ag, Au, Co, Cr, Cu, NiFe, Pd, Pt, and Y [14]. Reference [68] gives a comprehensive list of wet-chemical etchants for most of the elemental metals. A remark that should be made with respect to etching is that most metals in their natural form (film, foil, or substrate) are polycrystalline and may have a certain texture (i.e., preferred crystallite orientation) depending on the method of manufacture. Both the presence of the polycrystal grain boundaries and the texture variations may give rise to variations in etching rates, even locally on a specimen, and therefore low reproducibility of pattern formation and relatively high surface roughness may be expected, in particular if the structural dimensions are close to the grain size.

A convenient method to micromachine metal substrates or foils that suffers less from these crystallite effects is laser ablation. By combining a stack of laser patterned metal foils in a lamination process it is possible to fabricate complex microfluidic structures, which are particularly attractive for microreactor applications. Heights of microchannels produced by this method are determined by the thickness of the foils, and typically range from 25 to 250 μm. The microchannel widths and lengths are determined by the patterned area. Using this assembly method, microchannels having a wide range of aspect ratios (height/width) can be produced. Typically, endplates having sufficient thickness to accommodate fluid connections are added to either end of the stack of patterned shims to complete the laminated device [69].

Finally, one may also build metal microstructures by using high aspect ratio photoresists and electroplating. One particular way to make high aspect ratio structures in a photoresist is via synchrotron x-rays, as is done in the so-called Lithografie, Galvanik, Abformung (LIGA) technique [70]. With conventional photolithography and specially developed resists like SU-8 one also achieves excellent pattern definition, at much lower costs than LIGA. SU-8 patterning will be discussed in some detail in the next section.

3.2.4 PLASTIC MICROFABRICATION

The word *plastic* comes from the Greek *plastikos*, which means pliant or pliable. Nowadays, the term *plastics* is used to define a group of natural or synthetic organic polymeric materials that can be formed or *molded* into products using heat and pressure. It also refers to materials that "set" or polymerize by chemical reaction or by evaporation of a solvent. Although most of the plastic materials in use today were developed after 1920, plastic substances have been employed since ancient times [71]. Early civilizations have applied natural plastic substances like gums, resins, and waxes for several purposes.

In polymer processing technology, a distinction is made between *thermoplastic* and *thermosetting* (or duraplastics) plastic materials, relating to their forming or molding properties. Thermoplastic materials become soft when heated and reharden on cooling without appreciable change of properties; thermosetting materials harden permanently after one application of heat and pressure. Thermosetting plastics, such as phenol-formaldehyde, cannot be remolded.

In some classifications, a third category is distinguished, viz., *elastomers* (or rubbers). Thermoplastics consist of linear or branched polymer chains, elastomers are weakly cross-linked polymers (that can be easily stretched to high extensions, but will adopt their original state when the stress is released; an example is polydimethylsiloxane, PDMS), and thermosets are heavily cross-linked polymers that are normally rigid, brittle, and intractable (e.g., Bakelite).

To make microstructures in plastics, one may use similar methods as the ones used for shaping silicon and glass, like bulk micromachining (i.e., "etch" the microstructure in a bulk piece of polymer) and surface micromachining (see Section 3.2.5). Polymers may be etched with excellent shape tolerance by lasers [72] or dry (plasma) etching methods (e.g., etching of parylene films in O_2 and CF_4, which has been found to produce straight sidewall profiles [73]). Wet-chemical etching of plastics similar to how it is used for glass and silicon is a virtually nonexisting method, except for the very well known *photolithographic* technique in which photosensitive polymer films or foils are structured by wet-chemical dissolution (i.e., development) of *illuminated* areas, for so-called *positive* photoresists, or dissolution of *nonilluminated* areas, for *negative* resists. Particularly, the negative photoresists that were developed in the last decade and that can be spin-coated or cast in relatively thick layers, like SU-8, are of importance to recent developments in microfluidic systems.

A special rapid prototyping method for polymer microstructures is *stereolithography*, which consists in "growing" a microstructure in the focal point of a laser beam directed into a primer solution [12]. A nice example is an electrophoretic chip in which an acrylic microfluidic channel was implemented directly on top of a photosensor array [74]. This integrated chip is able to follow the course of electrophoretic separations along the complete microchannel in real time. Another rapid prototyping method recently introduced, however, with a lower feature limit of ca. 250 μm, is *solid-object printing*. This is an inkjet printing technology that sprays layers of tiny droplets of waxlike material onto a surface, to create models that can either be used directly or as a mold for prototype microfluidic devices, e.g., of PDMS [75].

Most of the methods described up to now allow only two-dimensional (2D) fluidic networks. *3D microfluidic networks* were formed by construction of a large 3D scaffold array formed by depositing a paraffin-based ink in a criss-cross pattern of 16 consecutive layers using a computer-controlled ink-delivery nozzle. This scaffold was infiltrated with epoxy resin, the ink was removed after curing, and the channels were refilled with a photocurable epoxy resin. By selectively curing this resin by exposing it to UV light through a photomask and removing the uncured resin, a complex 3D network of channels was formed. The network was used as a mixer [76].

By far the most popular methods to fabricate plastic microstructures are *replication* techniques. All these methods use a *mold* to generate a structure. Replication of the mold can be performed in various ways, like *hot embossing* or *injection molding*. Molding techniques are known since ca. 1650. In those days, iron moulds were used to make buttons, combs, shoe buckles, and other products, mainly out of materials like natural keratin, a thermoplastic obtained from, e.g., horse and cow hooves, animal horn, and tortoise shell. Modern techniques to prepare plastics, like *extrusion*, are known since ca. 1840, while the introduction of high-volume production methods for pottery, porcelain, and chinaware domestic products during the 1830s had already influenced manufacturing technologies in *press molding* [71]. Rubber molding rapidly accelerated the use of higher temperature and pressure molding techniques, and influenced the establishment of the techniques as we know them today.

Early work on replicated microfluidic channels in plastic consisted in making an *imprint* of 50 to 75 μm outer diameter fused silica capillaries in 250-μm thick fluorinated ethylene–propylene copolymer foil. These microchannels were applied in a miniaturized flow injection analysis system [77]. The use of polymeric substrate materials for chip CE was proposed as early as 1990, by Soane and Soane [78], who focused on thermoplastics like polymethylmethacrylate (PMMA), and Ekström et al. [79], who studied elastomeric polymers such as PDMS, which is well suited for rapid prototyping of microfluidic devices, due to the ease with which micrometer-scale features from a *master* or mold can be transferred by a simple casting, curing, and parting process. Due to the simple and hence less expensive processing, plastic materials are nowadays very popular for microfluidic devices, and many different polymers have been applied [10], like PMMA [80], polycarbonate [81], PDMS [82], and polytetrafluoroethylene (PTFE) [83]. Polymeric materials are not only considered favorable because of the promise of lower production costs for higher volumes, which is of importance for disposables in medical applications, but are also used to obtain specific physical or chemical properties.

The *replication methods* most suitable for large-scale production of plastic microstructures can be divided into *injection* and *embossing or imprinting* methods. Basically, in *injection molding* the polymer is melted and then injected under high pressure into an evacuated cavity containing a precision master mold. During this process the cavity is maintained at a temperature close to the melting point of the polymer, to allow efficient fluid flow into all corners of the mold. Next, the cavity is cooled and the microstructured part ejected. For this process, thermoplastics of low viscosity at their melting point, and high mechanical strength at the demolding temperature, are preferred, which is ensured by polymers of average molecular weights between 40,000 and 100,000 Da [84]. The compact disk is in fact an excellent example of injection molding, although microfluidic structures generally require higher aspect ratios. A high aspect ratio mold, in particular for the very high aspect ratio structures fabricated with LIGA [70], constitutes a problem because of rapid heat exchange between molten polymer and mold, which leads to rising polymer viscosity and even complete solidification before all corners of the mold are filled. Therefore, in particular for LIGA parts, the mold insert is maintained at a temperature above the plastic melting point. An improved method of injection, called *injection compression*, uses a piston that presses the polymer into the mould insert while it cools down. This counterbalances shrinkage effects that otherwise would lead to slightly deformed microstructures.

In *hot embossing*, a mold and a planar polymer substrate are heated separately under vacuum to a temperature above the polymer's glass temperature. The mold is brought into contact with the substrate and embossed using a constant force, after which both are cooled to just below the glass temperature and pulled apart. Many different plastics may be hot-embossed, and a large number of microfluidic devices made with this method have been reported.

An essential tool in replication methods is the *mold*. The quality of surface features within the fabricated device is almost completely dependent on the quality and precision of the master template. To achieve the structural accuracy required for microfluidic devices this mold therefore has to be of high (surface) quality. Typically, the surface roughness should be below 100 nm [10], to ensure low friction forces on tool and polymer microstructure in the demolding step. This puts restrictions on the methods that can be used to machine the mold. The interface chemistry between mold and polymer is also of importance, low chemical and physical interface forces are required. Release agents used in conventional plastic replication methods are often not suitable for microfluidic devices since they may diffuse out of the polymer and contaminate the chemistry that is carried out in the microfluidic device, or may increase the autofluorescence of the polymer, which may deteriorate the signal-to-noise ratio of the optical detection processes performed on the device.

For injection molding, nickel masters are very suitable, showing no wear even after 10,000 molding cycles [84]. The best structural definition of the mold is obtained with the micromachining techniques described previously, i.e., LIGA, dry silicon etching, and optical lithography combined with electroforming. The main drawback of silicon as a mold is the limited lifetime. According to Ref. [10], polyamide and polyoxymethylene are among the best materials for micromolding.

Figure 3.10 gives an example of a microstructure imprinted in PDMS. Figure 3.10(a) shows the silicon mold and Figure 3.10(b) its replica. Imprinting was performed by casting a PDMS precursor on the mold and baking it in order to have it solidified. Afterwards, the PDMS structure is pulled from the mold.

As mentioned earlier, plastic microstructures may be made out of photosensitive polymers by conventional photolithographic techniques. The photoresists that are commonly used for lithographic purposes generally have a limited film thickness and would have to be deposited (by spin or dip coating) in a stack of layers to achieve the desired device dimensions, which is time-consuming and unpractical, if not unsuitable for high-resolution and smooth structures. To solve this problem, special resists have been developed that result in layers with a thickness ranging from 10 μm up to 1 mm. Most of these resists are of the negative type, i.e., illuminated areas cross-link and remain solid after development. One exception is the PMMA-based positive resist that is used in the previously mentioned LIGA process, which uses

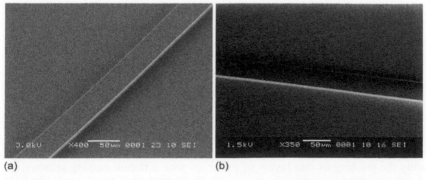

(a)   (b)

FIGURE 3.10 (a) Silicon mold fabricated by reactive ion etching. (b) PDMS replica of the mold. (Courtesy of R. Luttge, University of Twente, and H. Bouwes, Micronit Microfluidics BV.)

x-rays from a synchrotron to define the lithographic patterns [70]. This PMMA resist is mostly spun or cast from a solution, although laminated foils have also been applied [85]. The most popular negative resist is SU-8 [86], and ample examples exist in literature of microfluidic structures fabricated from this material.

An interesting method of using photopolymerizable materials (comparable to negative resists) is the one in which the precursor is flushed through a fluidic network (e.g., fabricated in glass) and is polymerized locally by a laser spot or by masked UV illumination. In this way, micropistons were fabricated that can be used for microvalves with a backpressure resistance of ca. 300 bars, which makes them suitable for high-performance liquid chromatography applications on a chip [87]. Similarly, photodefinable polyacrylamide gels, used as a sieving medium for DNA electrophoresis, were deposited inside a fluidic microchannel network in glass bonded to silicon or fused silica [88]. More complex structures were fabricated by Beebe and coworkers with a method named "microfluidic tectonics," which consists in the shaping of microchannels and components like hydrogel-based check valves within a cartridge by masked liquid-phase photopolymerization or laminar flow [89]. The latter makes good use of what is generally seen as a disadvantage of the laminar flow that is dictated by the small dimensions of microchannels: the slow, diffusion-controlled mixing in the laminar flow generates a spatial monomer concentration gradient along the width of a channel containing two liquids (one containing the monomer precursor), which after photopolymerization leads to a polymer with material properties slowly varying in the width direction of the channel [90].

Soft lithography is a relatively new high-resolution patterning method that uses the contact between a structured elastomer and a substrate to confine a surface reaction on the substrate. This method may be used to make fluidic microstructures, and was shown to be capable of making structures with a minimal feature size of 10 nm [91]. A variation on this method uses a PDMS mold that is pressed against a flat and solid substrate; the channels in the mold are filled with a polymer precursor, the polymer is let to solidify, and the mold is removed. This then leaves a replica of the mold on the substrate [92].

Table 3.1, adapted from Ref. [93], gives a comparative overview of a number of manufacturing techniques for plastic microstructures.

3.2.5 SACRIFICIAL LAYER ETCHING

Sacrificial layer etching is best known for its use in a fabrication procedure called surface micromachining. However, the method has a broader span than only the stacking of thin layers of materials with different etching selectivity.

Surface micromachining (for a review, see Ref. [94]) has become famous due to the tiny micromotors that have been presented in many different configurations throughout the literature. The method consists in the stacking

TABLE 3.1
Comparison of Techniques for Manufacturing Plastic Microstructures

Technology	Type[a]	Benefits	Drawbacks
E-beam lithography	P	Accuracy in 10- to 100-nm range	Long manufacturing times Thin structures Expensive equipment Restricted materials choice
UV lithography	P	Accuracy in μm range High aspect ratios Cycle times in seconds range Relatively simple equipment Suitable for mass production	Restricted materials choice Problematic stripping of exposed SU-8

X-ray lithography	P	Accuracy in sub-μm range
		Very high aspect ratios
		Thick structures, >1 mm
		Closely parallel walls
		Very smooth sidewalls
			Limited availability
			  synchrotrons
			Complex technology
			Restricted choice of
			  materials

Laser patterning	P	Accuracy in μm range Fast prototyping Large variety of structures	Serial manufacturing (time consuming) Limited surface quality

Stereolithography	P	Real 3D microstructures
		  possible
		Rapid prototyping
			Serial manufacturing
			  (time consuming)
			Very limited surface quality
			Restricted materials choice

Reactive injection
  molding, RIM
	R	Minute structures with
		  high aspect ratios are filled
			Long cycle times
			Complex equipment
			Restricted materials choice

Thermoplastic
  injection
  molding, TIM
	R	Short cycle times
		Wide range of materials
		Inexpensive in mass production
			Sophisticated equipment

Hot embossing	R	Wide range of materials
		Protective treatment of
		  material (low flow rate)
		Accurate positioning (μm
		  range) on prestructured
		  substrates
		Less equipment required than
		  in TIM
			Cycle times longer than
			  in TIM

[a]P = primary structuring, R = replication technique.

and patterning of layers of thin films, and afterwards selectively (by wet or dry etching techniques) removing some of the layers, the so-called *sacrificial layers*, therewith creating gaps between the unetched layers, see Figure 3.11.

The method is not limited to the conventionally used combination of poly-crystalline silicon as a structural layer and silicon dioxide or doped silicate glasses as sacrificial layers, but can be applied to any combination of materials that can be etched selectively with respect to one another. For example, metal microchannels using photoresist as a sacrificial layer were fabricated. The microchannels were meant to be used as a sample delivery system

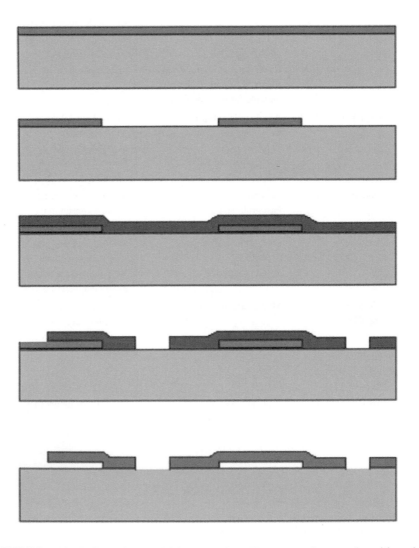

FIGURE 3.11 Surface micromachining procedure. From top to bottom: deposition of thin film (sacrificial layer) on surface, patterning of thin film, deposition of second film of different material, patterning of second film, removal of first film leaving cavities underneath second film.

consisting of microscale pipettes for highly parallel sample loading of pico-liter to milliliter volumes on a slab gel for electrophoresis [95].

A notorious problem in surface micromachining is stiction: in the drying step after sacrificial layer etching, the freestanding thin layer structures may stick to the surface underneath. Capillary forces during withdrawal of the liquid meniscus under these structures pull the microstructures together, or to the substrate surface underneath, and spontaneous bonding of the structure to its counterpart may occur if the surface contact area and the bonding forces are high enough (which often turns out to be the case). This effect is an example of the scaling laws in miniaturization: surface effects become more prominent when feature size goes down. Freeze drying is the most commonly used solution to the stiction problem, but other methods like reduction of the contact area by artificial asperities or intentionally increased roughness on the surface micromachined parts are also known to be effective [96].

A challenge in the case of surface micromachining is constructing long, sealed channel structures. Generally, the etching process in a very long and narrow tube will be diffusion limited, and the etchant will deplete quickly inside the channel since forced convection is not possible at that location. Due to the diffusion limitation, the release time will be proportional to the square of the channel length. For example, release of sacrificial oxide channels with lengths of 1 to 2 mm will require more than 200 min in concentrated HF solution [97]. During such long release times other films on the wafer may become significantly etched and the channel integrity cannot be maintained. To overcome the fundamental limitations of the conventional surface micro-machining technology based on polycrystalline silicon and (doped) silicon oxide, a technology based on photoresist as the sacrificial layer and conform-ally deposited poly-xylylene (parylene) (deposited by vapor deposition at room temperature in vacuum) was developed that could withstand very long release times. Figure 3.12 shows a scanning electron microscope (SEM) photograph of the cross section of such a plastic channel on a silicon sub-strate, which was used in a CE device [98]. The used parylene-C has been found to have a low background fluorescence.

The key property of sacrificial layer techniques is that modern film deposition processes lead to layers whose thicknesses can be very well controlled to the desired value, while also providing excellent thickness uniformity over relatively large areas. An example of a device in which this property was exploited is a HDC chip [99]. The principle of HDC and the reasons to perform this method on a chip are discussed in detail in Chapter 6. Essential is that the separation channel has a characteristic dimension (depth in this case) below 1 μm. A very small variation in the channel height is mandatory. With the aid of sacrificial layer techniques, using a thermal silicon dioxide layer on silicon (a layer that can be fabricated to a very accurately defined thickness) a height variation of less than 0.5% over the width of the channel was realized, while the surface roughness of the inner

FIGURE 3.12 SEM picture of cross section of 20-μm high, 200-μm wide parylene-C capillary on a silicon substrate constructed entirely from parylene-C and a sacrificial resist process. The parylene-C thickness is 8 μm. (From Webster JR, Monolithic Structures for Integrated Capillary Electrophoresis Systems, PhD thesis, University of Michigan, 1999. With permission.)

walls of the separation channel remained the same as the original roughness of the polished wafers, i.e., below 0.5 nm average roughness.

3.3   SURFACE PROPERTIES OF MICROMACHINED STRUCTURES

An influence of the inner surface roughness of capillaries or glass plates used for gel electrophoresis on separation efficiency was noted in the early stages of electrophoresis. In a publication on the effect of the flatness of capillary walls on electrophoretic resolution [100], it was concluded that the time-width increase of eluted zones shows a sixth-power dependence on fluctuations in capillary radius. For conventional electrophoresis on fused silica capillaries this effect is insignificant, the reported average roughness of 0.28 to 0.67 nm [101] on a nominal capillary diameter of 20 to 100 μm would lead to a dispersion increase of 0.04% at the most. However, the situation may be different on microfabricated channels. In studies reporting the use of powder-blasted glass chips for CE the effect of roughness on separation performance was noticed: Although Guijt et al. [102] and Schlautmann et al. [47] reported the successful application of powder-blasted microchips with integrated electrodes for zone electrophoresis and conductivity detection of alkali ions and organic acids, and found plate numbers for Li^+ as high as 28,700/m, Solignac

[48], on the other hand, concluded that powder-blasted chips are less attractive for electrophoretic separations because of large sample plug dispersion due to the roughness of the powder-blasted surface inside the channels. We will look into this issue in somewhat more detail here.

Blom et al. [103] investigated the effect of a nonuniform channel shape on the dispersion of a solute by electroosmotic flow (EOF), using microchannels with different types of artificial periodic variations of channel width along the length of a channel. Their results of the effective diffusion coefficient could be fitted with a mathematical model in which the Taylor–Aris dispersion in flow situations of very small Debye lengths (which is the case in EOF) was found to be described in terms of the Peclet number based on the transverse dimension of the channel (the Peclet number is given by $(v_{EOF}a)/D_{12}$, where v_{EOF} is the electroosmotic velocity and a is a characteristic channel dimension) [104] as follows:

$$D_{eff} = D_{12}(1 + \alpha_0 Pe^2) \qquad (3.1)$$

in which D_{12} is the molecular diffusivity and α_0 a parameter that depends on the Debye length. It was found experimentally [103] that for a 3-μm square periodic perturbation on a 100-μm wide channel the parameter α_0 takes the value 2.6×10^{-4}, from which it can be shown that for such a channel in a practical situation where the electroosmotic velocity is between 1 and 5 mm/sec, giving Peclet numbers between 50 and 250, the effective diffusivity will increase by a factor between 0.65 and 16.25. In other words, the time-width of eluted zones may increase by a factor of up to ca. 16 by channel width perturbation. The mentioned 3-μm variation on a nominally 100-μm wide channel is not uncommon for powder-blasted channels, as will be made clear below.

In an experimental study [49], Borofloat glass chips with channels either etched in a diluted aqueous HF solution or powder blasted with 9-μm diameter particles (for details about the powder-blasting process, see Ref. [47]) were compared. The channels were 90 to 110 μm wide and 20 to 50 μm deep. The surface roughness of these microchannels was measured to be in the range of 3 to 15 nm after HF etching and in the range of 1 to 5 μm after powder blasting. It is possible to reduce the roughness of powder-blasted surfaces to below 0.6 μm by a high-temperature treatment (1 h at 750°C), keeping the channel structure intact [105]; however, the surface will never become as smooth as after HF etching. The microchannels investigated in this study were subject to a heat treatment at 600°C for 2 h, which is part of the procedure for bonding glass substrates. This temperature is not high enough to reduce the surface roughness on the powder-blasted channel significantly [49]; however, it is thought that small cracks in the top layer of the glass, resulting from the mechanical impact of the powder particles, will be removed in this procedure.

The EOF is an indicator of the surface status in microchannels; further-more, it is an important parameter of which the exact value needs to be known in capillary electrophoretic separations. The mobility related to the EOF was measured with a fluorescent marker in the HF etched and powder-blasted channels, and found to be 7.6×10^{-4} and 7.0 to 7.3×10^{-4} cm^2/V/sec, re-spectively, at a pH of 9.2 to 9.3. This result differs from that obtained by Solignac et al. [48], who found that the electroosmotic mobility at pH 9.2 in a powder-blasted channel is about 50% lower than that in a HF etched channel, viz., 3.4 versus 7.8×10^{-4} cm^2/V/sec. These authors used a powder with a grain size of 30 μm and mention a surface roughness of 8 to 10 μm for their powder-blasted channel. It should also be noted that their chips were bonded at 100 to 200°C, which is for sure a much too low temperature to expect any annealing of possible surface cracks, and this contributed to their unexpect-edly low electroosmotic mobility.

The effect of roughness on EOF may be explained as follows: as a starting point, one should compare the surface roughness with the Debye layer thickness (the double-layer thickness). This thickness depends, among others, on wall material, pH, and ionic strength, and will be 10 to 30 nm for the situations considered here. It is within this layer close to the channel wall where the velocity gradient in EOF is located; outside of this layer the velocity is constant. Therefore, surface wall roughness can be expected to have an effect on EOF if that roughness exceeds the Debye length. This is not the case for HF-etched channels, where the roughness is only 3 to 15 nm, but for powder-blasted channels the surface roughness exceeds the Debye length by several orders of magnitude. As is explained in Ref. [49], the effect of surface roughness on EOF may be expressed as an effective electroosmotic mobility $\mu_{eo,eff}$:

$$\mu_{eo,eff} = \mu_{eo} \cos^2 \theta \tag{3.2}$$

where μ_{eo} is the electroosmotic mobility in a channel without significant wall roughness and θ the angle between the local surface vector (related to the exact surface roughness profile) and the applied electric field direction. Equation (3.2) was found to give an excellent fit to the experimental data: From the measured roughness profile for the powder-blasted channel an electroosmotic mobility change of 4% from the value in a smooth channel was calculated.

The separation efficiency of the different channels was also evaluated, and it was found that the highest plate numbers obtained for the HF-etched chip are seven to nine times those obtained for powder-blasted chips for rhodamine B and fluorescein, respectively. This indicates that the perform-ance of powder-blasted channels is indeed not as good as that of a HF-etched channel. Further analysis, based on a model for the relationship between plate height and applied electrical field [106] led to the conclusion that Joule heating plays only a minor role in the performance of the powder-blasted chips at higher electric field strengths. It was reported by others that heat transfer to the wall increases when the walls get rougher, see, e.g., Ref. [107] and references therein, but the effect is not very significant. On the other hand, it was experimentally demonstrated that on a chip heat exchange is more efficient than in capillaries with comparable inner diameter, particularly for rectangular channels with a relatively high surface-to-volume ratio. This led to the possibility to perform electrophoretic separations at fields up to 50 kV/cm without noticeable influence of Joule heating on dispersion [108]. An estimation for a fused silica chip with a channel of 10 μm by 50 μm and a wall thickness of 100 μm showed that fields up to 1.5 kV/cm could be applied in a buffer of 100 mM Tris–HCl (pH 8), while the temperature difference between buffer and surroundings stayed below 1° [109]; whereas when measured for a fused silica chip with a channel of 40 μm by 90 μm in which a 100 mM Tris/30 mM borate buffer with a resistivity of 1900 Ω cm was present, the average temperature change of the buffer was 1.48° at a field of 240 V/cm and 2.81° at 475 V/cm [110]. Such temperature changes will lead to a decrease in water viscosity of ca. 3 to 6%, which gives rise to a dispersion change of ca. 4%. The changes observed for the powder-blasted chips are much larger than this and therefore it seems unlikely that the lower plate numbers observed for powder-blasted channels are related to heat dissipation.

The influence of the effective separation channel length on separation efficiency was investigated for the powder-blasted and HF-etched channels, with fixed electric field. For all channels, the plate number increases linearly with an increase in effective length. It is therefore likely that the main source for dispersion in the powder-blasted chips is related to term B (in fact, the plate height is proportional to B [106]) given by

$$B = \frac{2D_{\text{eff}}}{\mu_{\text{eff}}} \tag{3.3}$$

We already discussed the effect of surface roughness on the effective electro-osmotic mobility and the effect of the Peclet number on the effective diffusivity. It is clear from Equation (3.1) that this will play an important role at higher voltages (where the electroosmotic velocity is higher), and this is most probably the main reason why the rough powder-blasted chips show an inferior performance than the HF chip. In addition to this, there are two other possible dispersion sources that may play a significant role in powder-blasted chip performance:

1. It was demonstrated by several authors, both mathematically [111] as well as experimentally [112], that the EOF velocity profile in channels with *radially* unequal ζ-potentials (as is the case here, one side is rough

     while the other is smooth) may deviate substantially from the well-known plug flow profile. The nonuniformity of EOF will lead to Taylor dispersion, as it is known for hydrodynamic flow.

2. Similarly, when the ζ-potential is nonuniform *axially*, pressure gradients arise, which give rise to Poisseuille flow with a parabolic velocity profile, which also gives rise to Taylor–Aris dispersion [113]. This mechanism may also apply for the rougher powder-blasted channels.

    In a study focusing on electrophoretic separation of low-density and high-density lipoproteins, of importance for cholesterol quantification in blood, the influence of the roughness of etched glass surfaces on separation efficiency was investigated [114]. The roughness measured by atomic force microscopy was 10.9 ± 1.6 or 2.4 ± 0.7 nm, depending on etching conditions. It was expected that *enhanced adsorption* of proteins on the inner surfaces of these chips may occur, compared to conventional capillaries, and that this would affect the performance of CE on the chips; however, no significant difference in either sample throughput or separation efficiency was observed. Some differences between the two chips were observed, but these were concluded to be due to the differences in channel layout, although the exact cause could not be identified.

    Significant surface roughness was also noted for microchannels fabricated in plastics. For example, Roberts and coworkers mention that especially for deeper (>20 μm) laser-ablated channels, the "rugosity" varied from 0.13 μm on polycarbonate and polystyrene surfaces to 0.27 μm on cellulose acetate and 0.40 μm on polyethylene terephtalate, while unablated surfaces had a rugosity of 0.01 μm [72]. (Other authors reported a roughness of 4.8 nm on the inner walls of excimer laser-ablated polycarbonate [115], which may be due to the use of a different laser and a different procedure of ablation.) These authors measured the EOF in the microchannels and noticed that the EOF increased in the range polycarbonate–polystyrene–cellulose acetate–poly (ethyleneterephthalate), which corresponds with the increased surface rugosity after ablation. They speculate that the increased surface area may have increased the charge density of the functional groups generated by the photo-ablative process, which would increase the ζ-potential (see also below) and thus the EOF. However, it may also be that the increased rugosity is the result of a material-specific increased absorption of UV energy during the ablation process, which would give rise to increased bond breaking and increased functionalization of the resulting surface. Whatever the exact reason may be, the result is that high-resolution electrophoresis on such chips is not possible.

    But, citing a famous sportsman: "Every drawback has its advantage" [116], so (artificial) surface roughness may also be exploited, for example, to enhance mixing in microchannels. The effect relates to Equation (3.1), i.e., the diffusivity can be enhanced by creating a perturbation on the transverse channel dimensions. Such perturbations, for the purpose of mixing, were created by soft lithography in PDMS [117] or by pulsed UV excimer laser in polycarbonate [118].

In addition to surface roughness, the *surface electrical properties*, conveniently expressed in the ζ-*potential*, are of great relevance to microfluidic devices applied for separation. It determines the magnitude of the EOF that is present in CE, but which is also used to establish pumping mechanisms for other types of separation. It is also relevant for adsorption of species, like protein molecules, on capillary walls. The ζ-potential depends mainly, in addition to the nature of the wall material, on the pH of the solution that is in contact with the wall surface. In this respect, the wall may be seen as being in an acid–base equilibrium with the solution, where the surface charge and therewith the ζ-potential is determined by the position of the equilibrium position. For example, the ζ-potential of fused silica was measured to vary from $+10\,mV$ at pH 2.0 to $-20\,mV$ at pH 6.0, with a zero value at pH 3.0 [119], while it was found to be zero for thin film SiO_2 at pH 2 [120]. Some other reported values of the pH at which ζ are zero are: 5.8 ± 0.1 for TiO_2 (rutile) and 9.1 ± 0.2 for Al_2O_3 (corundum) [121], and 3.1 for LPCVD Si_3N_4 [122]. The ζ-potential depends on surface treatment, e.g., for Si_3N_4 after a dip in aqueous HF the pH of zero ζ has risen from 3.1. to 4 [122], while for polymer surfaces a clear effect of rugosity due to the laser micromachining of the channel was found [72]. For more information on the theoretical aspects of ζ-potential and measurements thereof, we refer to Ref. [123].

3.4 WAFER BONDING AND LAMINATION

An important issue in the fabrication of microfabricated devices for separation is the sealing of the microfluidic capillary circuit that is formed by combining the two substrates. In most cases one of the substrates contains an etched or channel pattern engraved by other means, while the other substrate is an untreated (except perhaps polished or extensively cleaned) flat plate. Several sealing methods are known and summarized below. We will not discuss sealing methods based on glue. These are considered undesirable for fluidic chip sealing for several reasons: (i) it is difficult to dispense a uniformly thick material layer on exact positions along the periphery of an etched channel without leakage into the channel; (ii) most glues are permeable for gases and solvents and have low mechanical integrity; and (iii) glues may lead to contamination because they dissolve in the organic solvents in the channel during operation of the fluidic system.

We restrict ourselves here to sealing at the wafer scale, but in principle many of the described methods can also be performed on smaller pieces of material. An excellent comprehensive review of wafer bonding processes and the related fundamental principles can be found in a recent book by Tong and Gösele [124].

One may distinguish a number of bonding principles.

Direct bonding (without intermediate layer, also called *fusion bonding*). This process is used to bond, e.g., silicon to silicon, which leads to a monolithic structure, with the advantage of reduced stresses when the structure has to undergo temperature cycles. The bond is relatively strong at room temperature; however, excellent and durable bonding is obtained only after annealing at high temperatures, above 800°C for silicon [125] or at 1100°C for fused silica [126]. The most sophisticated example till today is a package consisting of six individually etched silicon wafers [127]. Thermal glass-to-glass direct bonding consists in heating both substrates to a temperature at which melting starts, or at least to a temperature at which the glass starts to soften, e.g., at 550°C for certain Borofloat glass types, and pressing the substrates together, by which a bond is formed. This is a relatively old and well-known method, and was used by Harrison et al. [9] in their classic paper describing CE on a chip. A drawback of the method is that leakage after bonding may occur when one of the substrates contained surface topography, like metal patterns used as drive or detector electrodes. Another issue is the deformation of the substrates when they are pressed together in a softened or partially molten state, by which the structural integrity of the fluidic circuit contained in one or both of the substrates will be affected.

It is not always necessary to heat the wafer pair to the melting or softening temperature, a lower temperature is sufficient if the roughness of the substrates is below 1 nm [128]. For a wafer pair consisting of silicon and Pyrex wafers, both with roughness below 1 nm, a curing temperature of 300°C was found sufficient to result in a bonded package that possesses adequate bonding strength for applications between 10 and 100 bar of hydrostatic pressure [129].

Anodic bonding (also called electrostatic bonding) mostly of a combination of silicon and Pyrex glass substrates. The earliest publication describing this method dates back to 1966 [130]. The method consists of applying a high electrical field across the wafer sandwich, at a temperature close to 450°C. Due to ion diffusion in the glass at these conditions, a space charge region forms at the surface of the glass wafers. This leads to a strong electrostatic attractive force between the two wafers. Thus, the gap between the wafers closes, and oxidation takes place at the interface, leading to a tight bond. The method also works with a thin glass layer in between two silicon wafers, see below. Also, the method can be used to bond only locally, if the areas that need to stay unbonded are covered with a pattern of a thin film of conductive material that does not become covered with an insulating oxide layer under the anodic bonding conditions, like chromium, so that no electrostatic attraction between the two substrates exists at those positions and no bond can form there [131].

Bonding with a thin intermediate layer. This can be done in a number of different ways:

1. Anodic bonding after deposition of a *thin metallic or semiconducting film* on one of two insulator substrates [132]. An alternative to this method is the use of an intermediate insulator layer like silicon nitride, which acts as a sodium diffusion barrier [133]. An advantage of these anodic bonding methods is that a roughness of several tenths of nanometers can be tolerated without a reduction in bonding quality. The drawback is the high electrical field that is required for the process, which in some cases will result in bonding of channel walls in unwanted locations [134]. *Direct* anodic bonding of two insulator substrates is also possible, optionally with a metal pattern in between [135]. This method comprises the evaporation of a thin layer of silicon oxide on, e.g., thin film circuitry, present on a substrate, and subsequent anodic bonding of a glass foil or substrate. This procedure is claimed to result in a hermetic seal because the bonding process presses the glass element on the metal pattern.

2. Bonding of two (glass) substrates through an *intermediate layer of a low-melting-point material*, or through an intermediate layer that solidifies from a solution during heat treatment. One example is the use of a sodium silicate spin-on-glass layer as an adhesive that solidifies at 90°C or after one night at room temperature [136]. This is probably the lowest temperature for a bonding process with intermediate layer reported in literature, and in fact lower temperatures would become impractical. The drawback of this method is that the layer during dispensing or during melting may destroy the structural integrity of the fluidic circuit, due to reflow of the material. The method is also known as *glass-frit bonding* [137], a process in which a glass layer (usually lead borates with a significant lead oxide content) is deposited or screen printed (which can be done locally) on one of the substrates after which the wafers are brought into contact at the melting temperature of the glass, which generally is below 600°C. Pressure is applied to keep the samples in intimate contact. Another special method is one in which a metal (gold, indium, aluminum, tin, and many alloys) layer is used in between two substrates, and where the metal layer is melted by inductive heating in a microwave. The metal layer is selectively heated to reach the eutectic temperature, without directly heating the other parts. Selective bonding may also be performed by using shielding layers to protect areas that should not be bonded. Silicon, quartz, certain ceramics, and plastics are transparent to microwaves so that most of the microwave energy applied is absorbed by the thin metal film and the bulk substrate materials are essentially left unaffected.

3. *Eutectic bonding* involves the deposition of intermediate metallic films. Usually two metals are used, one on each substrate, which exhibit a eutectic point in their two-component phase diagram which

point is located at the lowest melting temperature. An alloy is formed by solid–liquid interdiffusion at the metal contact interface, followed by solidification upon cooling. In the case of gold and silicon, this point is located at 363°C and corresponds to a eutectic composition of 2.85 wt.% Si. To accomplish the eutectic bond, the silicon surface should be made oxide free.

The demands on surface roughness are low for anodic bonding (asperities up to 1 μm are allowed). The use of a thin intermediate film may even allow higher surface steps; if thicker films are used they allow reflow over the steps. However, direct bonding requires a very smooth surface, with a roughness below a few nanometers. The details about the background principles of direct bonding can be found in a number of recent reviews ([124, 138, 139]; the paper by Haisma and Spierings contains an excellent historical review that brings bonding memories back to ca. 2000 BC; the paper by Plößl and Kräuter gives a comprehensive overview of methods for the evaluation of the quality of a wafer bond).

In brief, a good bond is achieved when surface conditions allow a contact area large enough for a sufficient change in surface energy. This implies that the key parameters in direct bonding are: the surface roughness on both substrates, the elasticity of the materials to be bonded, and the surface energy change that occurs during bonding. The fact that the elasticity plays a role here can be understood from the fact that asperities on the surface need to deform for the contact area to increase, and this has to occur also when no external pressure is applied and the bonding pressure arises from the surface energies on the surfaces that are in contact. Thus, soft materials are easier to bond. Furthermore, a surface with a roughness composed of asperities of small height or large pitch (or rather: wavelength) is easier to deform. A roughness of that kind can normally only be achieved with special polishing techniques [140].

The strict requirements on surface roughness can be exploited to bond only locally, for example, by covering the areas that need to stay unbonded with a pattern of a thin film with a higher roughness [141].

One important factor in all bonding techniques is the thermal budget: wafers may contain materials on the surface (e.g., metal or polymer thin films, or biological material) or in the bulk (e.g., doped regions in silicon) that are preferably not heated up for too long a time at a high temperature. More importantly, if two substrates each of a different material are bonded, thermal expansion differences may give rise to unwanted curvature, residual stresses, or, even worse, destruction of the package. Therefore, special surface treatments were investigated, which could help to increase the surface energy change during the direct bonding process, therewith reducing the need for high-temperature anneals. From the extensive literature on this topic and personal practical experience, it is seen that the surfaces to be bonded need to undergo hydration. Thus, most of the surface treatment methods try to achieve this. Examples are: oxygen plasma treatments or immersion in, e.g., boiling nitric acid or aqueous H_2O_2–H_2SO_4 mixtures [138]. Silicon, fused silica, or glass surfaces require more or less the same surface treatments. One special method of bonding, which can be performed at room temperature, is based on the treatment of SiO_2 surfaces in diluted aqueous HF [142]. The advantages of this method are clear: low thermal damage, low residual stress, and simple procedure (only a small pressure is required during bonding). Evaluation of the resulting wafer package led to the conclusion that a binding interlayer is formed between the substrates by solidification of dissolved silicon dioxide.

A topic already mentioned in previous sections is the *lamination* of metal [69] or polymer foils [72] to create a microfluidic channel network. Generally, lamination processes are of the direct type, i.e., a thermal process is used to bond the foils, and pressure is applied to enhance the contact area during bonding. For example, for metal foil lamination, bonding into a single solid piece is performed under vacuum using a high-temperature and high-pressure diffusion bonding process. Laminated parts are stacked into a high-temperature alloy clamping device to provide alignment and side support. An alloy endplate and a ram extension are used to transmit pressure from the hot press ram to the stacked laminate. Proven conditions to bond laminated stainless steel devices are 920°C and 300 atm. for 4 h. For laminates of different types of polymers, bonding conditions are, e.g., 125°C for ca. 3 sec on commercially available lamination equipment [72], a process that would even allow one to have biological material inside the microstructure during the bonding process without losing much of the activity.

Bonding of polymers to other polymers or to silicon or glass, in order to achieve sealed microfluidic structures, often requires surface treatments to enhance bonding. In the case of hydrophobic polymers, this is often done by making them hydrophilic, which may be achieved by wet-chemical treatments (e.g., treatment of PDMS in diluted HCl gives a good but reversible bond to PMMA [143]) or by oxygen plasma treatments (this gives a permanent bond between two PDMS pieces [144]). Other approaches are to apply a thin polyethylene terephthalate sheet with a heat-activated adhesive, which worked well for low fluid pressure applications with adhesive-compatible fluids [145], or thermal lamination with a 20 to 40 μm PET/PE film at temperatures around 100°C using standard industrial lamination apparatus [72].

3.5 COMPARISON OF MATERIALS

A few words need to be said about what material to use for what application. In principle, the previously described fabrication methods allow a large degree of freedom of geometrical structuring in virtually any material, the choice will mainly depend on the material properties that are required for the intended application. Properties of interest are electric, optical, mechanical, thermal, and possibly also magnetic. For most separation methods the surface

properties are of high importance, like the ζ-potential and the surface roughness, parameters that were discussed extensively in the previous section.
Surface properties of materials can in most cases be modified, by chemical
or plasma treatments, or by applying a coating, although the application of a
uniform well-adhering coating at the desired location in a microfluidic network is not at all trivial. Some of theses aspects will briefly be discussed in
Chapter 8.

Tables on all relevant properties of the materials that were discussed here
can be found in literature, we will not reproduce these data here. A number of
basic physical properties of molding polymer materials are listed in Ref. [10].

3.6   SYSTEM ASPECTS: INTEGRATION AND INTERFACING

One of the opportunities that are offered by microsystem technology is the
possibility of advanced integration. This can even go up to a stage where all
components of a complex and "intelligent" system are manufactured from
one piece of material, in so-called *monolithic integration*. Reasons to aim at
integration, be it monolithic or hybrid, can be one of the following:

1. *Reduction of signal quality loss.* This is obvious for modern day
   microelectronics: the tiny electronic signals would never reach their
   final destination if they would have to travel along conventional metal
   wires with their relatively large parasitic capacitance and resistance. In
   fluidic systems for chemical analysis, consisting of microchannels,
   resistances for flow can be high. Microfluidic systems require considerable pressure (or other means of propelling the fluids) to maintain an
   acceptable flow. Capacitance may exist in (flexible) tubing or dead
   volumes in fluidic connections, leading to the "loss" (i.e., dispersion)
   of chemical signals (i.e., chemical composition information). For such
   applications, monolithic integration may be beneficial, or even crucial.
   Similarly, for reasons of heat management, or protection against the
   environment, integration may be advantageous.
2. *Advanced miniaturization.* This needs no further debate, all the
   advantages of miniaturization have been extensively discussed in this
   chapter.
3. *Similarity of components.* For applications in high-throughput screening, a large amount of similar components operating in parallel on the
   same substrate may be preferred for reasons of dead volume, reduced
   footprint, and ease of fabrication. In this particular case, monolithic
   integration may offer the advantage of minimized variation in component properties. Also, for redundancy reasons, for example, in a
   portable system that is required to have a long lifetime, one may
   choose a larger amount of similar components that allows occasional
   drop-out without losing the overall performance of the system.

Three main concepts for integrating microfluidic components are known from the literature:

Vertical stacking of components [146], which was done in order to minimize the length of the connections between the various components.

Planar monolithic integration [16], which, similar to monolithic ultra large scale integration of microelectronic components, leads to the most advanced degree of miniaturization. The main advantage of this approach for fluidic circuits is that dead volumes and fluidic paths are reduced to the absolute minimum. The most important drawback, though, is that failure of one component may render the complete system useless. Either sufficient redundancy has to be contained in the monolithic system or the system design and the fabrication process have to be perfected to give high yields. The latter approach requires very high market volumes to compensate for the high development costs. For systems that do not fulfill these conditions, a hybrid solution to the integration issue is preferred.

Modular assembly [147], which consists in fixing the microfluidic components in a housing to form a robust module, and installing this module into a printed circuit board, a base plate with, on one side, all necessary electronic circuitry to control the system and perform data management, and, on the other side, a fluidic channel plate that defines the flow paths between the components.

Whereas a monolithic integrated system can be optimized in terms of connection length between components, and thus is particularly suited to reduce dead volume, the modular hybrid system is more flexible. The latter provides easy assembly and disassembly of components, either to replace a malfunctioning component or to upgrade a component to an improved version. Additionally, replacement of modules enables a change in the functionality of the system in order to perform a different function. For example, a specific reaction chamber could be replaced with one that is designed for a different temperature range, or an optical detector element could be replaced by a conductivity sensor.

An example of the monolithic as well as the modular approach is the development of a miniaturized flow injection analysis system for continuous measurement of the ammonia concentration in environmental water samples [148], based on a colorimetric method. The systems consist of micropumps, microheaters (in a reactor cell), connecting microchannels, and an optical absorption detector. Photographs of both systems are given in Figure 3.13. For details about the fabrication and performance of both systems we refer to Ref. [148].

If the question were asked: "How far are we really in terms of integration of microfluidic components ?" then the answer would have to be that we are now in the stage where microelectronics was some 40 years ago. Integration of

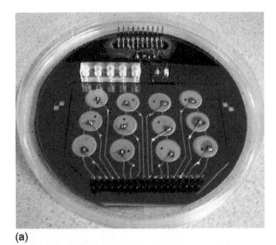

(a)

(b)

FIGURE 3.13 (a) Modular hybrid analysis system. (b) Same system integrated mono-lithically (silicon wafer size: 10 cm).

microfluidic components at this moment is for over 90% of the cases not done at all; that is, components are connected via conventional tubing and wiring, and most of the investigation and control of the "chemical signals" are done externally. Only those components that really need integration to achieve the desired performance are integrated, and these components are used for *thermal management* (heaters and temperature sensors), *optical* (integrated waveguides and optical detectors), *electrical, or electrochemical measurements* (the latter two with integrated electrodes), *sample preparation* (generally sample treatment steps can be categorized as follows: extraction or dissolution, concentration or dilution, separation or mixing, decomposition or binding, and derivatization; for each of these methods demonstrations on a chip exist, which are discussed in detail in Chapter 10), or *sample injection* (in order to establish a higher separation efficiency; also discussed in Chapters 2 and 8), all connected with microchannels that ensure low analyte dispersion. Nevertheless, there are ample descriptions of integrated components for detection in microchannels in literature, and we will give a small selection below. Only fabrication aspects of the detectors will be discussed, more details about detectors, including coupling of chips to, e.g., mass spectrometers, will be discussed in Chapter 11.

Most *electrical and electrochemical detectors* are based on noble metal electrodes that are positioned at the end of a separation channel. Such electrodes are deposited with well-known thin-film processes like evaporation, sputtering, electroplating, or chemical vapor deposition. The main issue with the deposition of such films in a microfluidic circuit is that sealing of the fluidic chip will have to occur over the electrode patterns, if the electrodes are *in* the microchannel (which, in principle, would give the best detection performance). The prevention of leakage is crucial for fluidic microsystems, since leakage will give rise to cross talk between adjacent fluidic conduits and leads to dead-volumes that give rise to memory effects and cross-contamination of subsequent sample injections. It was noted in the earliest papers on CE on a chip [9], describing glass chips that contained a pattern of platinum electrodes on the cover glass (bonded to a second glass plate with the etched channel), that such patterns caused difficulties in completely sealing the channels during bonding and required multiple bonding attempts, typically two to three times. Due to the long cooling cycles, this could take several days. A frequently pursued procedure to enhance sealing over metal patterns is one in which a recess is photolithographically defined and etched in one of the substrates, in which subsequently a metal pattern is disposed [47]. Such chips show hardly any leakage. It has to be mentioned that bonding problems usually only occur when relatively stiff materials are bonded over the metal patterns, like glass or silicon plates. For polymer foils, bonding over metallic patterns is less of an issue [149].

A serious problem with the placement of the electrodes inside a microchannel for the purposes of on-column detection in CE is the enormous drop of the potential across the channel. This may give rise to electrolysis at the electrodes, and possibly their destruction. Configurations are therefore often used in which the detector is placed outside the separation channel. The distance between the channel outlet and electrode affects the postcapillary band broadening, so that the channel–electrode spacing has to be kept to a minimum. Several configurations of this kind are described in Ref. [150]. The bonding problems described above do not occur for these configurations, since the electrode patterns are basically positioned outside the chip.

One very interesting way of integrating electrodes inside a microchannel is the one described by Kenis et al. [151]. These authors first used three-phase laminar flow to etch the central part of an Au stripe, so that two electrodes opposing one another result, and next used two-phase laminar flow to deposit a silver reference electrode in between these electrodes. The silver electrode develops at the interface between two phases containing components of electroless silver plating solution.

Examples of *optical detection elements* that were monolithically integrated with separation chips are optical cells that are illuminated with lasers or light emitting diodes [23, 24] (see also Figure 3.2), microlenses fabricated on the surface of the chip to focus light [62, 63, 152], planar waveguides that are positioned on both sides of a microchannel and are used to illuminate a part of the channel and collect the light opposite to the illuminating waveguide [37, 153], hollow waveguides that measure index changes as a function of concentration changes inside a hollow waveguide core [154], waveguides that sense via the evanescent field [155], and photodiodes integrated underneath a microchannel to collect fluorescent or transmitted light [156]. Fabrication of all these elements was performed with the micromachining and bonding methods described in this chapter.

One final issue with respect to integration is the fluidic interfacing of chips to external equipment. This is not at all a trivial problem; many different configurations have been described in literature and it can be stated that no method is really 100% satisfactory. Furthermore, a standard, which would be very welcome, has not been established yet. Figure 3.14 gives a selection of fluidic connections to chips.

3.7   SUMMARY AND OUTLOOK

It has been made clear in this chapter that a versatile microfabrication toolbox exists from which one can choose a method to sculpture micrometer-size structures in virtually any technical material that may be of use to separation microsystems. Also, a short overview was given of ways to integrate components in order to construct a system with a higher grade of functionality. In this section, we will discuss the latest developments in the field of fabrication of miniaturized fluidic systems, i.e., the fabrication of structures with at least one dimension below a micrometer.

The artificial structural dimensions developed in this *nanotechnology* area are getting closer to the size of single (bio)molecules, and it is foreseen that new physical–chemical phenomena would become available for (high-throughput) (bio)molecular characterization. Approaching the atomic scale, the behavior of liquids becomes inherently different, surface phenomena (liquid–gas and liquid–solid surface tensions, double-layer effects, etc.) start to dominate, and become important in the development of future liquid manipulation and (bio)chemical analysis concepts. The latest developments in spectroscopy, in particular, methods like mass spectrometry and laser-induced fluorescence, have led to nearly single-molecule sensitivity, which

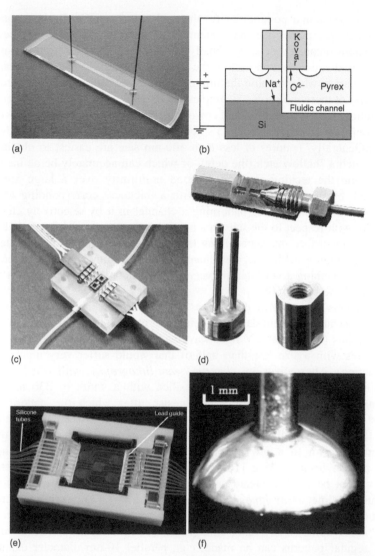

FIGURE 3.14 Examples of fluidic connections to a chip: (a) capillaries glued in access holes in a fused silica chip. (Courtesy of M.T Blom, MESA$^+$ Research Institute.) (b, d) Kovar pieces that can be machined easily, and can be bonded anodically to Pyrex glass. (From Blom MT, Chmela E, Gardeniers JGE, Berenschot JW, Elwenspoek M, Tijssen R, van den Berg A. *J. Micromech. Microeng.*, 11, 2001, 382. With permission.) (c) Connections made in a PMMA block. (Courtesy of R.E. Oosterbroek, MESA$^+$ Research to Institute.) (e) Socket for a micromixer array. (Reprinted from Yang Z, Maeda R, *Electrophoresis*, 23, 2002, 3474. With permission.) (f) Connection of a Kovar tube to silicon, using a molten glass frit. (Courtesy of S.M. Spearing, M.I.T. Cambridge, MA, from Harrison TS, London AP, Spearing SM, *Mater. Res. Soc. Symp.*, 657, 2001, EE6.5.1.)

leads to spectacular possibilities for the analysis of ever smaller amounts of biological specimens, like the contents of a single cell and (bio)molecular interactions on an atomic scale. Finally, when the characteristic dimension of a nanofluidic device approaches the size of molecules, novel separation methods arise, which exploit the limited mobility or configurational freedom (entropy) of these molecules. Examples of the latter can be found in a number of publications by Craighead and coworkers [157, 158], where gaps of 90 nm were defined by a shallow etch in silicon, which was then bonded to a Pyrex plate. Generally, features of less than 100-nm size are easiest to obtain by either such a shallow etch (the depth of which can accurately be defined by calibrating the etching rate) or, if good uniformity over a large area is required, by depositing a thin film with a thickness corresponding to the desired channel height and patterning a channel in it by selectively etching the film with respect to the substrate on which the film was deposited [159], or by using surface micromachining techniques. An example of the latter is shown in Figure 3.15. The nanochannel in this example was obtained by selectively etching a very thin polycrystalline silicon layer with respect to a silicon nitride film [160].

Channels with *lateral* dimensions below 100 nm are very difficult to make, because present-day conventional photolithography cannot yet achieve this (the lowest UV light wavelength obtainable is ca. 200 nm, and structures with a size less than half of that would suffer very much from diffraction). Alternatives are *electron-beam lithography*, with which it was possible to define channels in fused silica with a width of 330 nm (the e-beam was used to pattern a photoresist, after which the pattern was transferred into the fused silica by a fast-atom beam etching method [161]), which is still far from the limit of e-beam methods that allow structures down to a few nanometers, but with a very low throughput; *focused ion beam etching* [162], which is also a relatively slow method that cannot be used to create complete channel structures in a reasonable amount of time; *x-ray lithography* as used in LIGA [70], with a structural limit of nanometers but at the expense of high costs; or special optical tricks like *(holographic) interference lithography* [163]. With the latter, however, only regular patterns can be made, e.g., parallel 10-nm diameter channels with a titaniumoxide wall, obtained by overexposure during holographic resist patterning, covering the pattern with a TiO_x film, and subsequently removing the resist [164].

Relatively new methods to fabricate nanostructures are those classified under the name "soft lithography." For details about these methods we refer to the papers of Whitesides and Love, see, e.g., Ref. [165], which gives a low-level review of the essentials of the methods. Basically, the method exists in making an exact replica of a nanopatterned relief structure into a soft material like PDMS, and using that piece of material as a *stamp* to print the nanostructures on any substrate (which might even be curved). A derivative of these

FIGURE 3.15 V-groove channel etched in silicon with a concentrated KOH solution, connected with 100-nm deep surface-micromachined channels in silicon nitride.

methods was used to create channels in SiO_2 by using nanoimprint lithography and etching to create channels, and subsequently narrowing and sealing these channels with an asymmetric sputtering procedure to a width of 10 nm and a height of 50 nm [166].

An interesting approach to create nanostructures was one in which block copolymer self-assembly was combined with a long-range ordering method resembling graphoepitaxy [167]. One of the constituents of the block copolymer is resistant against an oxygen plasma, and remains on the surface so that it can be used as a mask to etch the underlying substrate (in this case SiO_2), e.g., by RIE. In this way, 20-nm wide SiO_2 pillars were obtained, which may be used, e.g., as a stationary phase in a chromatography method on a chip, as was done by Tezuka et al. [53] on similar SiO_2 nanopillars etched by a modified RIE method discussed in a previous section.

Finally, the ultimate way to achieve nanostructures would be to use a bottom-up method, i.e., by constructing a nanochannel starting from only atoms or molecules. Nature has given us ample examples of this concept; however, man has not yet reached the same level of nanofabrication "skills" to do the same, although it can be argued that carbon nanotubes are the first proof of our technical abilities in this direction. A first report on TEM studies of fluidic transport in nanotubes has recently appeared [168]. Whether such nanostructures will ever become valuable for separation systems remains to be seen.

REFERENCES

1. S.M. Sze, *VLSI Technology*, 2nd ed., McGraw-Hill, New York, 1988.
2. T. Fukuda and W. Menz, Eds., *Handbook of Sensors and Actuators, Volume 6, Micro Mechanical Systems, Principles and technology*, Elsevier, Amsterdam, 1998.
3. K.E. Petersen, *Proc. IEEE* 1982, *70*, 420–457.
4. S. Terry, J.H. Jerman, and J.B. Angell, *IEEE Trans Electron. Dev.* 1979, *ED–26*, 1880.
5. W.J. Spencer, W.T. Corbett, L.R. Dominguez, and B.D. Shafer, *IEEE Trans. Sonics Ultrason.* 1978, *SU–25*, 153–156.
6. H.T.G. van Lintel, F.C.M. van de Pol, and S. Bouwstra, *Sens. Actuators* 1988, *15*, 153–167.
7. P. Woias, *Proc. SPIE* 2001, *4560*, 39–52.
8. A. Manz, Y. Miyahara, J. Miura, Y. Watanabe, H. Miyagi, and K. Sato, *Sens. Actuators B* 1990, *1*, 249–255.
9. D.J. Harrison, A. Manz, Z.H. Fan, H. Ludi, and H.M. Widmer, *Anal. Chem.* 1992, *64*, 1926–1932.
10. H. Becker and C. Gärtner, *Electrophoresis* 2000, *21*, 12–26.
11. E.A.S. Doherty, R.J. Meagher, M.N. Albarghouthi, and A.E. Barron, *Electrophoresis* 2003, *24*, 34–54.
12. I. Fujimasa, *Micromachines. A New Era in Mechanical Engineering*, Oxford University Press, Oxford, 1996.
13. W. Menz and P. Bley, *Mikrosystemtechnik für Ingenieure*, VCH Verlagsgesellschaft GmbH, Weinheim, 1993.
14. M. Elwenspoek and H.V. Jansen, *Silicon Micromachining*, Cambridge University Press, Cambridge, U.K., 1998.
15. M.J. Madou, *Fundamentals of Microfabrication: The Science of Miniaturization*, 2nd ed., CRC Press, Boca Raton, 2002.
16. M.C. Elwenspoek, T.S.J. Lammerink, R. Miyake, and J.H.J. Fluitman, *J. Micromech. Microeng.* 1994, *4*, 227–245.
17. G. Ocvirk, E. Verpoorte, A. Manz, M. Grasserbauer, and H.M. Widmer, *Anal. Meth. Instrum.* 1995, *2*, 74–82.
18. A.J. Nijdam, J. van Suchtelen, J.W. Berenschot, J.G.E. Gardeniers, and M. Elwenspoek, *J. Crystal Growth* 1999, *198–199*, 430–434.
19. J.P. van der Eerden and H. Müller-Krumbhaar, *Phys. Rev. Lett.* 1986, *57*, 2431–2433.
20. A.J. Nijdam, E. van Veenendaal, H.M. Cuppen, J. van Suchtelen, M.L. Reed, J.G.E. Gardeniers, W.J.P. van Enckevort, E. Vlieg, and M. Elwenspoek, *J. Appl. Phys.* 2001, *89*, 4113–4123.
21. S.E. Moring, R.T. Reel, and R.E. van Soest, *Anal. Chem.* 1993, *65*, 3454–3459.
22. D.N. Heiger, P. Kaltenbach, and H-J. P. Sievert, *Electrophoresis* 1994, *15*, 1234–1247.
23. E. Verpoorte, A. Manz, H. Lüdi, A.E. Bruno, F. Maystre, B. Krattiger, H.M. Widmer, B.H. van der Schoot, and N.F. de Rooij, *Sens. Actuators B* 1992, *6*, 66–70.
24. R.M. Tiggelaar, T.T. Veenstra, R.G.P. Sanders, J.G.E. Gardeniers, M.C. Elwenspoek, and A. van den Berg, *Talanta* 2002, *56*, 331–339.

25. H. Robbins and B. Schwartz, *J. Electrochem. Soc.* 1959, *106*, 505.
26. R. Memming and G. Schwandt, *Surf. Sci.* 1966, *4*, 109–124.
27. L.T. Canham and A.J. Groszek, *J. Appl. Phys.* 1992, *72*, 1558–1565.
28. J. Drott, K. Lindstrm, L. Rosengren, and T. Laurell, *J. Micromech. Microeng.* 1997, *7*, 14–23.
29. K. Imai and H. Unno, *IEEE Trans. Electron Dev.* 1984, *ED-31*, 297.
30. J. Wei, J.M. Buriak, and G. Siuzdak, *Nature* 1999, *399*, 243–246.
31. M.J. de Boer, J.G.E. Gardeniers, H.V. Jansen, E. Smulders, M.-J. Gilde, G. Roelofs, J.N. Sasserath, and M. Elwenspoek, *J. Micro Electro Mech. Syst.* 2002, *11*, 385–401.
32. S.H. Zaidi and S.R.J. Brueck, *J. Vac. Sci. Technol. B* 1993, *11*, 658–666.
33. F.G. Bessoth, A.J. de Mello, and A. Manz, *Anal. Commun.* 1999, *36*, 213–215.
34. T.M. Bloomstein and D.J. Ehrlich, *J. Vac. Sci. Technol. B* 1992, *10*, 2671–2674.
35. D.J. Harrison and P.G. Glavina, *Sens. Actuators B* 1993, *10*, 107–116.
36. D.J. Laser, S. Yao, C.-H. Chen, J. Mikkelson, K. Goodson, J. Santiago, and T. Kenny, *Proc. 11th Int. Conf. Solid-State Sens. Act. (Transducers '01)*, Munich, Germany, June 10–14, 2001, pp. 920–923.
37. K.B. Mogensen, N.J. Petersen, J. Hübner, and J.P. Kutter, *Electrophoresis* 2001, *22*, 3930–3938.
38. R.W. Tjerkstra, M.J. de Boer, J.W. Berenschot, J.G.E. Gardeniers, M.C. Elwenspoek, and A. van den Berg, *Proc. IEEE Workshop Micro Electromech. Syst.*, Nagoya, Japan, January 26–30, 1997, pp. 147–152.
39. Y. Fintschenko, P. Fowler, V.L. Spiering, G.J. Burger, and A. van den Berg, *Proc. 3rd μTAS Workshop*, Banff, Canada, October 13–16, 1998, pp. 327–330.
40. R.B.M. Schasfoort, S. Schlautmann, J. Hendrikse, and A. van den Berg, *Science* 1999, *286*, 942–945.
41. H. Björkman, C. Ericson, S. Hjertén, and K. Hjort, *Sens. Actuators B* 2001, *79*, 71–77.
42. http://www.glassonline.com/history.html
43. P.C. Simpson, A.T. Woolley, and R.A. Mathies, *J. Biomed. Microdev.* 1998, *1*, 7–26.
44. C. Backhouse, M. Caamano, F. Oaks, E. Nordman, A. Carillo, B. Johnson, and S. Bay, *Electrophoresis* 2000, *21*, 150–156.
45. H.J. Lighthart, P.J. Slikkerveer, F.H. In't Veld, P.H.W. Swinkels, and M.H. Zonneveld, *Philips J. Res.* 1996, *50*, 475–499.
46. W.A. Little, *Rev. Sci. Instrum.* 1984, *55*, 661–680.
47. S. Schlautmann, H. Wensink, R. Schasfoort, M. Elwenspoek, and A van den Berg, *J. Micromech. Microeng.* 2001, *11*, 386–389.
48. D. Solignac, A. Sayah, S. Constantin, R. Freitag, and M.A.M. Gijs, *Sens. Actuators A* 2001, *92*, 388–393.
49. Q.-S. Pu, R. Luttge, J.G.E. Gardeniers, and A. van den Berg, *Electrophoresis* 2003, *24*, 162–171.
50. B.N. Chapman, *Glow Discharge Processes*, John Wiley & Sons, New York, 1980.
51. O.A. Popov, Ed., *High Density Plasma Sources*, Noyes Publications, Park Ridge, NJ, 1995.
52. T. Uchida, *Jpn. J. Appl. Phys.* 1994, *33*, L43.

53. Y. Tezuka, M. Ueda, Y. Baba, H. Nakanishi, T. Nishimoto, Y. Takamura, and Y. Horiike, *Micro Total Analysis Systems 2002*, Y. Baba et al. (Eds.), Kluwer Academic Publishers, Dordrecht, Vol. I, 2002, pp. 212–214.
54. X. Li, T. Abe, Y. Liu, and M. Esashi, *J. Microelectromech. Syst.* 2002, *11*, 625–629.
55. Y. Sugiyama, Y. Otsu, T. Ichiki, and Y. Horiike, *Micro Total Analysis Systems 2002*, Y. Baba et al. (Eds.), Kluwer Academic Publishers, Dordrecht, Vol. I, 2002, pp. 118–120.
56. M.T. Blom, On-Chip Separation and Sensing Systems for Hydrodynamic Chromatography, PhD thesis, University of Twente, The Netherlands, 2002.
57. B. He, N. Tait, and F. Regnier, *Anal. Chem.* 1998, *70*, 3790–3797.
58. H. Becker, M. Arundell, A. Harnisch, and D. Hülsenberg, *Sens. Actuators B* 2002, *86*, 271–279.
59. A. Daridon, V. Fascio, J. Lichtenberg, R. Wütrich, H. Langen, E. Verpoorte, and N.F. de Rooij, *Fresenius J. Anal. Chem.* 2001, *371*, 261–269.
60. T. Masuzawa, *Handbook of Sensors and Actuators, Volume 6, Micro Mechanical Systems, Principles and Technology*, T. Fukuda and W. Menz (Eds.), Elsevier, Amsterdam, 1998.
61. J. Franse, *Rep. Prog. Phys.* 1990, *53*, 1049–1094.
62. J.Ch. Roulet, R. Völkel, H.P. Herzig, E. Verpoorte, N.F. de Rooij, and R. Dändliker, *Opt. Eng.* 2001, *40*, 814–821.
63. Ph. Nussbaum and H.P. Herzig, *Opt. Eng.* 2001, *40*, 1412–1414.
64. D. Daly, *Microlens Arrays*, Taylor and Francis, London, 2001.
65. N. Fertig, Ch. Meyer, R.H. Blick, Ch. Trautmann, and J.C. Behrends, *Phys. Rev. E* 2001, *64*, 040901.
66. S. D. Theiss and S. Wagner, *IEEE Electron Device Lett.* 1996, *17*, 578–580.
67. E.T. den Braber, H.V. Jansen, M.J. de Boer, H.J.E. Croes, M. Elwenspoek, L.A. Ginsel, and J.A. Jansen, *J. Biomed. Mater. Res.* 1998, *40*, 425–433.
68. J.L. Vossen and W. Kern, *Thin Film Processes*, Academic Press, New York, 1978.
69. P.M. Martin, W.D. Bennett, D.J. Hammerstrom, J.W. Johnston, and D.W. Matson, *SPIE Conf. Proc.* 1997, *3224*, 258–265.
70. E.W. Becker, W. Ehrfeld, P. Hagmann, A. Maner, and D. Munchmeyer, *Microelectron. Eng.* 1986, *4*, 35–36.
71. http://www.nswpmitb.com.au/HistoryOfPlastics.html
72. M.A. Roberts, J.S. Rossier, P. Bercier, and H. Girault, *Anal. Chem.* 1997, *69*, 2035–2042.
73. R.D. Tacito and C. Steinbruchel, *J. Electrochem. Soc.* 1996, *143*, 1974–1978.
74. Y. Mizukami, D. Rajniak, and M. Nishimura, *Proc IEEE MEMS Conf.* 2000, 751–756.
75. J.C. McDonald, M.L. Chabinyc, S.J. Metallo, J.R. Anderson, A.D. Stroock, and G.M. Whitesides, *Anal. Chem.* 2002, *74*, 1537–1545.
76. D. Therriault, S.R. White, and J.A. Lewis, *Nat. Mat.* 2003, *2*, 265–271.
77. J. M. Hungerford, Theory, Chemical Kinetics, and Miniaturization in Flow Injection Analysis, PhD thesis, University of Washington, 1986.
78. D.S. Soane and Z.M. Soane, Method and Device for Moving Molecules by the Application of a Plurality of Electrical Fields, U.S. Patent 5,126,022, 1992.

79. B. Ekström, G. Jacobson, O. Ohman, and H. Sjodin, Microfluidic Structure and Process for its Manufacture, PCT 91/16966, 1991.
80. A.C. Henry, T.J. Tutt, M. Galloway, Y.Y. Davidson, C.S. McWhorter, S.A. Soper, and R.L. McCarley, *Anal. Chem.* 2000, *72*, 5331–5337.
81. J. Wen, Y.H. Lin, F. Xiang, D.W. Matson, H.R. Udseth, and R.D. Smith, *Electrophoresis* 2000, *21*, 191–197.
82. G. Ocvirk, M. Munroe, T. Tang, R. Oleschuk, K. Westra, and D.J. Harrison, *Electrophoresis* 2000, *21*, 107–115.
83. T. Katoh, N. Nishi, M. Fukagawa, H. Ueno, and S. Sugiyama, *Sens. Actuators A* 2001, *89*, 10–15.
84. R. Ruprecht, T. Hanemann, V. Piotter, and J. Haußelt, *Microsyst. Technol.* 1998, *5*, 44–48.
85. J.M. Hruby, S.K. Griffiths, L.A. Domeier, A.M. Morales, D.R. Boehme, M.A. Bankert, W.D. Bonivert, J.T.T. Hachman, D.M. Skala, and A. Ting, *Proc. SPIE Conf. Micromachining Microfab. Process Technol. V*, Santa Clara, September 1999, vol. 3874, pp. 32–43.
86. L.J. Guerin, M. Bossel, M. Demierre, S. Calmes, and P. Renaud, *Techn. Digest Int. Conf. Solid-State Sens. Act. (Transducers'97)*, pp. 1419–1421.
87. E.F. Hasselbrink, Jr., T.J. Shepodd, and J.E. Rehm, *Anal. Chem.* 2002, *74*, 4913–4918.
88. S.N. Brahmasandra, V.M. Ugaz, D.T. Burke, C.H. Mastrangelo, and M.A. Burns, *Electrophoresis* 2001, *22*, 300–311.
89. J. Moorthy and D.J. Beebe, *Anal. Chem.* 2003, *75*, 292A–301A.
90. Q. Yu, J.S. Moore, and D.J. Beebe, *Proc. Micro Total Analysis Systems 2002 Symp.*, Nara, Japan, November 3–7, 2002, pp. 712–714.
91. Y. Xia and G.M. Whitesides, *Angew. Chem. Int. Ed.* 1998, *37*, 550–575.
92. D. Juncker, H. Schmid, A. Bernard, I. Caelen, B. Michel, N. de Rooij, and E. Delamarche, *J. Micromech. Microeng.* 2001, *11*, 532–541.
93. P. Bley, *Proc. SPIE Conf. Micromachined Dev. Comp. V*, Santa Clara, September 1999, vol. 3876, pp. 172–184.
94. J.M. Bustillo, R.T. Howe, and R.S. Muller, *Proc. IEEE* 1998, *86*, 1552–1574.
95. A.B. Frazier, I. Papautsky, T.L. Edwards, and B.K. Gale, *Proc. Int. Symp. Mechatronics and Human Science (MHS '99)*, Nagoya, Japan, November 23–26, 1999.
96. C.H. Mastrangelo, *Proc. Mater. Res. Soc. Symp.* 2000, *605*, 105–116.
97. J. Liu, Y.C. Tai, J. Lee, K.C. Pong, Y. Zohar, and C.M. Ho, *Proc. IEEE Workshop Micro Electromech. Syst.* 1993, pp. 71–76.
98. J.R. Webster and C.H. Mastrangelo, *Proc. 1997 Int. Conf. Solid-State Sensors and Actuators (Transducers '97)*, 1997, pp. 503–506.
99. E. Chmela, R. Tijssen, M.T. Blom, J.G.E. Gardeniers, and A. van den Berg, *Anal. Chem.* 2002, *74*, 3470–3475.
100. G.W. Slater and P. Mayer, *Electrophoresis* 1995, *16*, 771–779.
101. R. Barberi, M. Giocondo, R. Bartolino, and P.G. Righetti, *Electrophoresis* 1995, *16*, 1445–1450.
102. R.M. Guijt, E. Baltussen, G. van der Steen, R.B.M. Schasfoort, S. Schlautmann, H.A.H. Billiet, J. Frank, G.W.K. van Dedem, and A. van den Berg, *Electrophoresis* 2001, *22*, 235–241.

103. M.T. Blom, E.F. Hasselbrink, H. Wensink, and A. van den Berg, *Proc. Micro Total Analysis Systems 2001*, Kluwer Academic Publishers, Dordrecht, 2001, pp. 615–616.

104. S.K. Griffiths and R.H. Nilson, *Anal. Chem.* 1999, *71*, 5522–5529.

105. H. Wensink, S. Schlautmann, M.H. Goedbloed, and M.C. Elwenspoek, *J. Micromech. Microeng.* 2002, *12*, 616–620.

106. J.P. Liu, V. Dolnik, Y.Z. Hsieh, and M. Novotny, *Anal. Chem.* 1992, *64*, 1328–1338.

107. S.G. Kandlikar, S. Joshi, and S. Tian, *Proc. 35th National Heat Transfer Conf.*, June 10–12, 2001, Anaheim, CA, pp. 1–10.

108. S.C. Jacobson, C.T. Culbertson, J.E. Daler, and J.M. Ramsey, *Anal. Chem.* 1998, *70*, 3476–3480.

109. S.V. Ermakov, S.C. Jacobson, and J.M. Ramsey, *Anal. Chem.* 1998, *70*, 4494–4504.

110. K. Swinney and D.J. Bornhop, *Electrophoresis* 2002, *23*, 613–620.

111. V.P. Andreev, S.G. Dubrovsky, and Y.V. Stepanov, *J. Microcol. Sep.* 1997, *9*, 443–450.

112. F. Bianchi, F. Wagner, P. Hoffmann, and H.H. Girault, *Anal. Chem.* 2001, *73*, 829–836.

113. A.E. Herr, J.I. Molho, J.G. Santiago, M.G. Mungal, T.W. Kenny, and M.G. Garguilo, *Anal. Chem.* 2000, *72*, 1053–1057.

114. B.H. Weiller, L. Ceriotti, T. Shibata, D. Rein, M.A. Roberts, J. Lichtenberg, J.B. German, N.F. de Rooij, and E. Verpoorte, *Anal. Chem.* 2002, *74*, 1702–1711.

115. S.-H. Chen, H.-N. Lin, and C.-R. Yang, *Rev. Sci. Instrum.* 2000, *71*, 3953–3954.

116. H. Davidse, *Je moet schieten, anders kun je niet scoren (en andere citaten van Johan Cruijff)*, BZZTôH, The Hague, ISBN: 9055015385.

117. A.D. Stroock, S.K.W. Dertinger, A. Ajdari,I. Mezic', H.A. Stone, and G.M. Whitesides, *Science* 2002, *295*, 647–651.

118. T.L. Johnson, D. Ross, and L.E. Locascio, *Anal. Chem.* 2002, *74*, 45–51.

119. S. Koch, P. Woias, L.K. Meixner, S. Drost, and H. Wolf, *Biosens. Bioelectron.* 1999, *14*, 417–425.

120. L.J. Bousse, S. Mostarshed, B. van der Schoot, N.F. de Rooij, P. Gimmel, and W. Göpel, *J. Colloid Interface Sci.* 1991, *147*, 22–32.

121. G.A. Parks, *Chem. Rev.* 1965, *65*, 177–198.

122. L.J. Bousse, S. Mostarshed, and D. Hafeman, *Sens. Actuators B* 1992, *10*, 67–71.

123. R.J. Hunter, *Zeta Potential in Colloid Science. Principles and Applications*, Academic Press, New York, 1981.

124. Q.Y. Tong and U. Gösele, *Semiconductor Wafer Bonding*, Wiley-Interscience, New York, 1999.

125. M.A. Schmidt, *Proc. IEEE* 1998, *86*, 1575–1585.

126. S.C. Jacobson, A.W. Moore, and J.M. Ramsey, *Anal. Chem.* 1995, *67*, 2059–2063.

127. A.P. London, A.A. Ayón, A.H. Epstein, S.M. Spearing, T. Harrison, Y. Peles, and J.L. Kerrebrock, *Sens. Actuators A* 2001, *92*, 351–357.

128. F. Pigeon, B. Biasse, and M. Zussy, *Electron. Lett.* 1995, *31*, 792–793.

129. M.T. Blom, N.R. Tas, G. Pandraud, E. Chmela, J.G.E. Gardeniers, R. Tijssen, M. Elwenspoek, and A. van den Berg, *J. Micro Electro Mech. Syst.* 2001, *10*, 158–164.

130. D.I. Pomerantz, Anodic Bonding, U.S. Patent 3,397,278, filed October 3, 1966.

131. T.T. Veenstra, J.W. Berenschot, J.G.E. Gardeniers, R.G.P. Sanders, M.C. Elwenspoek, and A. van den Berg, *J. Electrochem. Soc.* 2001, *148*, G68–G72.
132. H. Wohltjen and J.F. Giuliani, Method for Bonding Insulator to Insulator, U.S. Patent 4,452,624, filed December 21, 1982.
133. A. Berthold, L. Nicola, P.M. Sarro, and M.J. Vellekoop, *Sens. Actuators A* 2000, *82*, 224–228.
134. J.A. Plaza, J. Esteve, and E. Lora-Tamayo, *Sens. Actuators A* 1997, *60*, 176–180.
135. D.I. Pomerantz, Bonding an Insulator to an Insulator, U.S. Patent 3,506,424, filed May 3, 1967.
136. H.Y. Wang, R.S. Foote, S.C. Jacobson, J.H. Schneibel, and J.M. Ramsey, *Sens. Actuators B* 1997, *45*, 199–207.
137. S.A. Audet and K.M. Edenfeld, *Proc. Int. Conf. Solid-State Sens. Act. (Transducers '97)*, Chicago, IL, 1997, pp. 287–289.
138. A. Plößl and G. Kräuter, *Mater. Sci. Eng. R* 1999, *25*, 1–88.
139. J. Haisma and G.A.C.M. Spierings, *Mater. Sci. Eng. R* 2002, *37*, 1–60.
140. C. Gui, M. Elwenspoek, N.R. Tas, and J.G.E. Gardeniers, *J. Appl. Phys.* 1999, *85*, 7448–7454.
141. C. Gui, R.E. Oosterbroek, J.W. Berenschot, S. Schlautmann, T.S. J. Lammerink, A. van den Berg, and M.C. Elwenspoek, *J. Electrochem. Soc.* 2001, *148*, G225–G228.
142. H. Nakanishi, T. Nishimoto, R. Nakamura, A. Yotsumoto, T. Yoshida, and S. Shoji, *Sens. Actuators A* 2000, *79*, 237–244.
143. D.C. Duffy, H.L. Gillis, J. Lin, N.F. Sheppard, Jr., and G.J. Kellogg, *Anal. Chem.* 1999, *71*, 4669–4678.
144. D.C. Duffy, J.C. McDonald, D.J.A. Schueller, and G.M. Whitesides, *Anal. Chem.* 1998, *70*, 4974–4984.
145. P.M. Martin, D.W. Matson, W.D. Bennett, Y. Lin, and D.J. Hammerstrom, *J. Vac. Sci. Technol. A* 1999, *17*, 2264–2269.
146. B.H. van der Schoot, S. Jeanneret, A. van den Berg, and N.F. de Rooij, *Sens. Actuators B* 1993, *15*, 211–213.
147. J. Wissink, A. Prak, M. Wehrmeijer, and R. Mateman, *Proc. VDE World Micro Technol. Congr.* 2000, Vol. 2, pp. 51–56.
148. R.M. Tiggelaar, T.T. Veenstra, R.G.P. Sanders, J.W. Berenschot, M.C. Elwenspoek, A. van den Berg, J.G.E. Gardeniers, A. Prak, R. Mateman, and J.M. Wissink, *Sens. Actuators B* 2003, *92*, 25–36.
149. B. Graβ, A. Neyer, M. Jöhnck, D. Siepe, F. Eisenbeiβ, G. Weber, and R. Hergenröder, *Sens. Actuators B* 2001, *72*, 249–258.
150. J. Wang, *Talanta* 2002, *56*, 223–231.
151. P.J.A. Kenis, R.F. Ismagilov, and G.M. Whitesides, *Science* 1999, *285*, 83–85.
152. M.-H. Wu and G.M. Whitesides, *Adv. Mater.* 2002, *14*, 1502–1506.
153. K.B. Mogensen, P. Friis, J. Hübner, N. Petersen, A.M. Jørgensen, P. Telleman, and J.P. Kutter, *Opt. Lett.* 2001, *26*, 716–718.
154. B.G. Splawn and F.E. Lytle, *Anal. Bioanal. Chem.* 2002, *373*, 519–525.
155. G. Pandraud, T.M. Koster, C. Gui, M. Dijkstra, A. van den Berg, and P.V. Lambeck, *Sens. Actuators A* 2000, *85*, 158–162.
156. J.R. Webster, M.A. Burns, D.T. Burke, and C. H. Mastrangelo, *Anal. Chem.* 2001, *73*, 1622–1626.
157. J. Han and H.G. Craighead, *Science* 2000, *288*, 1026–1029.

158. J. Han, S.W. Turner, and H.G. Craighead, *Phys. Rev. Lett.* 1999, *83*, 1688–1691.

159. M.T. Blom, E. Chmela, J.G.E. Gardeniers, R. Tijssen, M. Elwenspoek, and A. van den Berg, *Sens. Actuators B* 2002, *82*, 111–116.

160. J.W. Berenschot, N.R. Tas, T.S.J. Lammerink, M. Elwenspoek, and A. van den Berg, *J. Micromech. Microeng.* 2002, *12*, 621–624.

161. A. Hibara, T. Saito, H.-B. Kim, M. Tokeshi, T. Ooi, M. Nakao, and T. Kitamori, *Anal. Chem.* 2002, *74*, 6170–6176.

162. D.F. Moore, J.H. Daniel, and J.F. Walker, *Microelec. J.* 1997, *28*, 465–473.

163. C.J.M. van Rijn, W. Nijdam, S. Kuiper, G.J. Veldhuis, H.A.G.M. van Wolferen, and M.C. Elwenspoek, *J. Micromech. Microeng.* 1999, *9*, 170–172.

164. J. Scarminio, E.L. Rigon, L. Cescato, and A. Gorenstein, *J. Electrochem. Soc.* 2003, *150*, H17–H20.

165. G.M. Whitesides and J.C. Love, *Sci. Am.* 2001, *September*, 39–47.

166. H. Cao, Z. Yu, J. Wang, J.O. Tegenfeldt, R.H. Austin, E. Chen, W. Wu, and S.Y. Chou, *Appl. Phys. Lett.* 2002, *81*, 174–176.

167. J.Y. Cheng, C.A. Ross, E.L. Thomas, H.I. Smith, and G.J. Vancso, *Appl. Phys. Lett.* 2002, *81*, 3657–3659.

168. C.M. Megaridis, A.G. Yazicioglu, J.A. Libera, and Y. Gogotsi, *Phys. Fluids* 2002, *14*, L5–L8.

4 New Tools: Scalar Imaging, Velocimetry, and Simulation

Ernest F. Hasselbrink, Jr. and Sneha Madhavan-Reese

CONTENTS

Fanne sperientia e poi regola. ("Make experiments and then the rule.") 51r
(c. 1499)

Leonardo da Vinci, *Madrid Codex I*

4.1   INTRODUCTION

The symbiosis between experiment and theory is well known; likewise, the advancement of chip-based analytical instruments relies in part on new tools for diagnosing them. As microfabrication offers us unprecedented control over the geometry and surface chemistry of our devices, transport phenomena that were previously negligible can become dominant, and thus, new techniques for observation and theory evaluation have recently been developed. This chapter focuses on such techniques for diagnosing the performance of microfluidic systems: novel fluorescence imaging techniques, microparticle imaging velocimetry (μ-PIV), and numerical simulation.

4.2   NOVEL FLUORESCENCE IMAGING TECHNIQUES

4.2.1   FLOW VISUALIZATION USING CAGED-DYE OR BLEACHED-DYE METHODS

Standard epifluorescence imaging usually provides too little contrast to provide information about the details of solute transport through components with fine geometric features. This is because diffusion and dispersion smooth the dye interface as it makes its way through the microfluidic device to the region of interest. Even if the dye is transported to the region of interest by a flat velocity profile (closely approximated by electroosmotic flow [EOF] in a straight channel with thin Debye layers), the dye front diffuses to a thickness $\delta \sim (2Dt)^{1/2}$, where D is the binary diffusion coefficient and t is time. This is about 140 μm after 10 sec for a dye molecule, and this is a best-case scenario (D is replaced by a much larger effective diffusion coefficient [1, 2] if the dye is introduced by pressure-driven flow). Thus, for investigating transport through geometries smaller than 100 μm, one needs a method for injecting contrast, a very short distance upstream of the region of interest.

This need is satisfied using caged-dye or bleached-dye imaging methods, wherein, one creates spatial contrast in the fluorescence image optically, typically using a focused sheet of light from a pulsed UV source. In the case of caged-dye imaging, the dye is derivatized with a moiety that quenches fluorescence; however, the moiety can be photocleaved by UV radiation. The dyes have been developed extensively in recent years for cell biology research applications [3–7], and a wide variety of dyes are now available [4]. Their utility for visualization of fluid motion in microchannels was introduced by Paul et al. [8]; a similar technique developed for large-scale flows is due to Koochesfahani et al. [9, 10]. The essence of the method is simple: shaping the uncaging beam into a thin sheet oriented across the channel creates a sliver of photoactive dye, the transport of which can be observed in time. An example of an image sequence obtained using this method is shown in Figure 4.1. This figure shows an interesting artifact of caged dyes: they contain impurities which can be electrophoretically separated during visualization. Figure 4.1 shows a faint blue band that leads the bright yellow band; a very faint trailing band is also visible. The dye is nominally zwitterionic (caged Rhomamine-110, Molecular Probes), but the leading and trailing bands are positive- and negative-charged impurities in the dye, which separate from the zwitterionic dye (yellow band) electrophoretically.

The caged-dye images shown in Figure 4.1 were obtained using the optical setup shown in Figure 4.2. A ring of 24 blue LEDs (3 mW, Nichia, Japan) arranged on a custom circuit board (courtesy E.B. Cummings, Sandia National Labs) provides excellent uniformity of fluorescence illumination. LEDs may be pulsed using a delay generator (Stanford Research Systems Model DG-535) and a power transistor connected to a variable power supply. The major drawback to LED illumination is that they emit over a fairly wide band (450–510 nm), so rejection of elastic scattering is not as efficient with typical filters. A pulsed Nd:YAG laser (Minilite II, Continuum Laser Corp.) is used as the pulsed UV source. The UV beam can be focused to a thin sheet directly using a cylindrical singlet lens, but the thinnest sheets can be obtained by illuminating a thin slit with an unfocused beam, and then focusing the resulting diffraction pattern using a spherical lens (these are usually superior

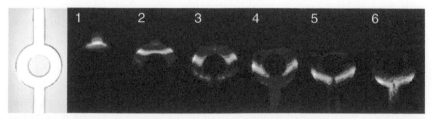

FIGURE 4.1 (See color insert following page 208) First frame: a microfluidic channel (50 μm wide × 10 μm deep) with a circular feature, with flow inlet and outlet at the top and bottom, respectively. Numbered frames: caged dye is uncaged just above the entrance to the circular feature, using a 10-μm-wide beam, achieved using a Fourier-transformed diffraction optic. Flow is electroosmotically driven flow at an electric field of approximately 400 V/cm; fluid is 50:50 acetonitrile–phosphate buffer pH 7.6. In frame 3, a positively charged impurity is visible, which electrophoretically separates from the zwitterionic dye.

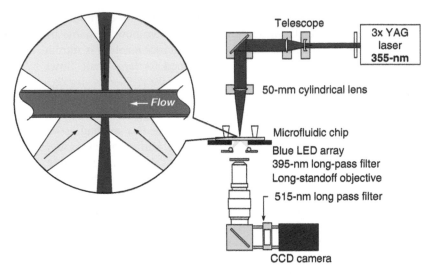

FIGURE 4.2 (See color insert) Caged-dye imaging optical arrangement. Synchronization details not shown for clarity. A Nd:YAG laser beam provides a 5-nsec pulse of 355-nm light, which is shaped into a planar sheet via focusing a diffraction pattern from an adjustable slit using a cylindrical lens. The laser sheet strikes the channel, in which caged dye is uniformly distributed, and the dye is locally photoactivated. Continuous or pulsed illumination from a blue light source (e.g., Ar$^+$ laser, blue LED, Hg lamp) permits observation of the convection of the dye through the channel. Telescope and cylindrical lens for a 355-nm beam may be replaced with a slit and spherical lens for the narrowest possible uncaging beam.

to cylindrical lenses, in terms of aberrations). In the focus plane, this results in an intensity profile that is the Fourier transform of the slit function (an Airy function); hence, the longer the slit, the thinner the sheet. One must be careful with the UV laser beam to avoid damaging microscope objectives; even a 3-mJ pulsed beam focused to a point can damage optics and even the chip itself.

Many possible variations on fluorescence illumination (e.g., using a typical fluorescence microscope) are possible. Paul et al. [8] used a CW blue diode laser, coupled with a polarizer and Faraday rotator, in order to gate the illumination for short periods (a few milliseconds) to avoid blurring due to fluid motion. The same method with an Ar$^+$-ion laser also works quite well (this is the illumination source used in the images in Figure 4.1). Laser illumination provides much more signal; however, use of coherent light can result in diffraction patterns in the excitation beam due to any particles on the surface of the chip. Continuous illumination from a standard mercury lamp is adequate for many situations, as demonstrated in Ref. [11].

Bleached-dye imaging [12] is a somewhat simpler method that uses almost the exact same setup as caged dye imaging. The signal obtained is essentially the inverse of the signal obtained with caged dye: the channel is filled with actively fluorescing dye, which is selectively bleached by UV light, creating a *lack* of signal that provides contrast for the observation of fluid motion. Bleached-dye imaging is somewhat cheaper than caged-dye imaging, since it avoids the relatively high cost of caged dyes. The method is decidedly inferior in terms of sensitivity (cf. [12]), because the contrast must be detected as a small difference from a large signal, and shot noise is proportional to the square root of the total number of photons incident on the detector (see Section 4.2.4 of this chapter for further details). Even for well-illuminated pixels, say 10^4 photons per pixel, in locations where the signal is bleached by only 10%, the signal-to-noise is at best $(0.1 \times 10^4/10^2) = 10$. In contrast, caged-dye imaging of only 10^3 photons per pixel gives a best-case signal-to-noise of $(10^3/10^{3/2}) = 31$.

Caged-/bleached-dye imaging is complementary to PIV (which is discussed later) in two ways: (1) it provides different *information* — it directly images scalar transport, whereas PIV obtains velocity fields and (2) it can be used in situations where PIV cannot — specifically, when introducing particles would disrupt the device function or otherwise be unacceptably invasive.

4.2.2 MIXING QUANTIFICATION

4.2.2.1 Mixing Measurement Theory and the Mixture Fraction, $\xi(x,y,z)$

Mixing of reagents is a prerequisite for chemical reaction, and so it is of central importance for precolumn reactions (e.g., fluorophore labeling) as well as numerous other microchip applications. To achieve the reaction $A + B \longrightarrow C$ in a microsystem, for example, one introduces certain concentrations of A and B at a wye or tee intersection as shown in Figure 4.3(a), whereupon product C is produced via mixing and chemical reaction in the mixer section. If reaction kinetics are fast compared to mixing (the high Dahmköhler number limit, where the Dahmköhler number Da is the ratio of reaction rate to mixing rate), the flux of product from the mixer is limited only by the degree of mixing that occurs within it. As shown in Figure 4.3(b), presuming large Da and equal diffusivities of all species involved, the concentrations [A], [B], and [C] can all be deduced if one knows the stoichiometry of the reaction, the concentrations of A and B in the inlet streams, and the value of a single spatial variable known as the mixture fraction $\xi = x_1/(x_1+x_2)$. Here, x_1 and x_2 indicate the volume fractions (in an infinitesimal fluid element) of fluid at a given point in space that originated in streams 1 and 2, respectively.

Since inlet concentrations and reaction stoichiometry are typically known, all one needs then is to know the spatial distribution of ξ, for example, as shown at locations 3 and 4 in Figure 4.3(a). The distribution of ξ in space depends strongly on the fluid mechanics of the individual mixer and the Peclet number $Pe = u_c d/D$ of the reactants, where u_c and d are a character-

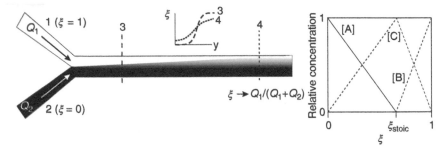

FIGURE 4.3 (Left) Two streams of fluid are introduced at a wye intersection, and mix asymptotically toward a uniform distribution $\xi(y) = \xi_m$. Spatial distribution of ξ is shown in the plot, through cross sections 3 and 4. (Right) Mapping of the concentrations of reactants A and B, and product C, to the single variable ξ, for the chemical reaction $A + B \longrightarrow C$. Depending on the initial concentrations of A and B in fluid streams 1 and 2, ξ_{stoich} may not be equal to 0.5, but for the symmetric mixer shown, $\xi_m = 0.5$.

istic speed and lengthscale of the system. For all mixers, the goal is to homogenize ξ to a uniform value of ξ_m, where $\xi_m = Q_1/(Q_1+Q_2)$, the ratio of the flowrate of stream 1 to the total flowrate. This occurs asymptotically in time and space, however, and in the meantime the ξ distribution in space can be quite complicated depending on the mixer.

4.2.2.2   Complications for Experimental Methods

In the simplified case in Figure 4.3(a), it appears that the ξ field can be measured by simply putting fluorescent dye into stream 1, and observing its rate of dilution with stream 2 with an epifluorescence microscope. However, this method has a hidden peril that becomes obvious when the mixer creates significant three-dimensional (3D) motion: $\xi(x,y,z)$ is a function of all three spatial variables, and epifluorescence imaging typically has poor resolution in the axial direction. Thus, the span of concentration gradients can be smaller than the resolution limit of the imaging system; while this would not appear to be serious for the case depicted in Figure 4.3(a), the situation is much worse for vertically stratified layers (relative to the microscope objective), because this is the direction in which resolution is most limited. For example, in Figure 4.3(a), a microscope objective oriented perpendicular to this page will correctly observe homogenization of the two streams, but one oriented, say, from top to bottom in the plane of the page will observe no difference in dye signal between stations 3 and 4 due to line-of-sight integration of the dye signal. The situation is depicted graphically in Figure 4.4. Often the axial resolution of an epifluorescence microscope is in the order of 10 μm or more, which is in the order of the depth of many microfluidic channels. In such a situation, layers of unmixed fluid are indistinguishable from mixed fluid. For this reason, epifluorescence imaging of a passive dye always overestimates the amount of mixing

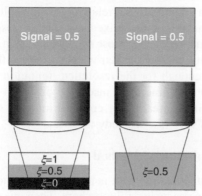

FIGURE 4.4 Fluid (bottom) is imaged through an epifluorescent microscope objective (middle), providing an intensity image (top). (Left) The example of three vertically stratified lamina of fluid is shown, in which the fluid is only partially mixed. Due to line-of-sight integration (for an objective with axial resolution larger than the channel depth), the signal obtained is the same as the signal obtained for perfectly mixed fluid streams (right).

that occurs, and in most good mixers, it overestimates the mixing by a significant amount. Indeed, mixers that induce helical streamlines often show apparent "unmixing," which happens when an upstream vertically stratified dye distribution is misinterpreted as mixed fluid, until it rotates to reveal the umixed layers as it moves downstream.

The problem exists even for two pressure-driven streams introduced side by side in a simple, straight channel — Ismagilov et al. [13] showed that there are concentration gradients in the vertical direction even for this simple case, due to the nonuniform velocity profile created by pressure-driven flow. Furthermore, since mixing in straight channels is quite slow,[1] a number of workers have developed various means for achieving rapid mixing [14–20], which invariably create strong 3D flow patterns. The straightforward but expensive way to conquer this difficulty is to use a commercially available 3D confocal scanning laser fluorescence microscope (as in Ref. [14]) to image

---

[1] Almost paradoxically, in microchannels, rapid mixing can be difficult to achieve in a small area without sustaining very large pressure drops. For example, to mix two reactant streams in a simple channel via pure diffusion, so that the concentration of product is some arbitrary fraction of its ultimate value, the two streams must be in contact for a time of about $t \sim d^2/D$, where d is the channel diameter. Assuming a square channel cross-section, the flowrate through the channel is $Q = Ud^2$, where U is the average fluid velocity. The length of channel required for mixing is then approximately $L = Ut = Ud^2/D = Q/D$; so, to mix two reagents with relatively high-diffusion coefficients, $D = 10^{-9}$ m^2/sec at a rate of 1 μl/sec (10^{-9} m^3/sec) requires a channel of length 1 m. This is typically an undesirable expense of microchip real estate; furthermore, the pressure required to force fluid through a given channel increases linearly with its length, so this can be a practical issue.

the dye. However, in addition to the relatively large capital expense, there is still some overestimation of the mixedness due to resolution limitations. These are probably negligible for most situations, but in the case of mixing of very low-diffusivity, high-molecular weight proteins, the diffusion length-scale may be small enough for the resolution limitation to matter (e.g., for $D \sim 10^{-11}\,\mathrm{m^2/sec}$, contact time of 70 msec, the diffusion lengthscale will be only ~1 μm). In addition, present-day confocal imaging systems may not be able to keep up with very fast unsteady mixers.

4.2.2.3 The "Unmixedness" Method

Ideally, the way around this problem is to use a direct indicator of mixing: a fast chemical reaction producing a fluorophore. Then, we could simply obtain fluorescence images of the product of a fast fluorogenic reaction (wherein the signal would be proportional to the product). However, most fluorogenic labeling reactions are relatively slow (in the order of 1 sec) compared to microfluidic convective timescales.

The method presented here, represented in Figure 4.5, measures the "unmixedness" of two fluid streams; this method was originally developed to solve similar resolution problems in turbulent flows [21–23]. The idea is to create a fluorescence signal which, at a point in space, has a normal-

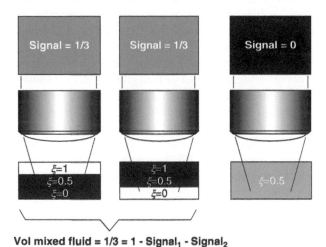

$$\text{Vol mixed fluid} = 1/3 = 1 - \text{Signal}_1 - \text{Signal}_2$$

FIGURE 4.5 The "unmixedness" method. Left: a binary dye signal is obtained in which if $\xi > 0.51$, unity signal is obtained, and if $\xi < 0.51$, zero signal is obtained. Thus, the signal is proportional to the volume of fluid with $\xi > 0.5$ ($= 1/3$). A second experiment, in which the signal is unity for $\xi > 0.49$, will yield a relative signal of 2/3. The fraction of "mixed" fluid ($0.49 < \xi < 0.51$) is equal to Signal($\xi > 0.49$) – Signal($\xi > 0.51$) = 2/3 – 1/3 = 1/3, the correct value. This is clearly distinguished from uniformly mixed fluid (right), in which case the mixed fraction is Signal($\xi > 0.49$) – Signal($\xi > 0.51$) = 1 – 0 = 1.

ized fluorescence yield of unity or zero, depending on whether ξ is greater than, or less than, a cutoff mixture fraction, ξ_c. The fraction of fluid with mixture fraction between these two "cutoff" values is simply $x(\xi_{c1} < \xi < \xi_{c2}) = x(\xi > \xi_{c1}) - x(\xi > \xi_{c2})$.

Several variations on this theme are possible. For example, one may construct the distribution $x(\xi)$ to increasing accuracy with an increasing number of experiments at various ξ_c. Alternatively, if one simply desires to measure the distance downstream at which, say, all the fluid has a mixture fraction $\xi_m - \varepsilon < \xi < \xi_m + \varepsilon$, then only two experiments need to be performed, at $\xi_c = \xi_m \pm \varepsilon$. It is also possible to save reagent preparation time by conducting "flip" experiments where one measures $x(\xi > \xi_c)$ and $x(\xi < 1 - \xi_c)$ with the same reagent mixtures. In this case $x(1 - \xi_c < \xi < \xi_c) = 1 - x(\xi > \xi_c) - x(\xi < 1 - \xi_c)$.

The fluorescence yield can be made to be nearly binary with mixture fraction by exploiting the steepness of the acid–base titration curve near the equivalence point, in conjunction with the pH-dependence of fluorescence yield for many dyes. Sodium fluorescein, for example, has a fluorescence yield, which drops about 2 orders of magnitude across 3.5 pH units (from 7.5 to 4). Thus, as the acid–base titration reaction occurs during mixing, with base and dye solution introduced into stream 1, the fluorescence signal corresponds to $x(\xi > \xi_c)$. For example, Figure 4.6(a) shows the titration curve for strong acid and strong base at $1\,mM$ concentrations, in the presence of $50\,\mu m$ fluorescein (note that it is *very* important to account for the pK_as of the dye in these unbuffered solutions). Due to fluorescein's pH-dependence fluorescent yield (Figure 4.6b), the fluorescence signal is nearly binary, equal to unity when $\xi > 0.7$.

Some care must be taken in applying this technique. First of all, calibration of the fluorescence yield curves is essential before employing this

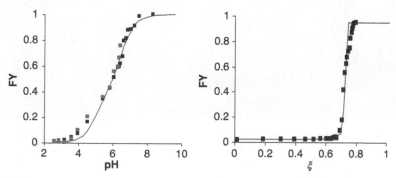

FIGURE 4.6 (Left) FY versus pH and (right) FY versus mixture fraction, for a case in which fluid 1 is a strong base solution and fluid 2 is a strong acid solution of slightly higher concentration.

method. Dyes typically have significant contaminants, and so calculated fluorescence yield curves are not always very accurate. Also, since dyes buffer near their pK_as, it is necessary to use relatively high-strength acid and base solutions (at least $1\,mM$ for $50\,\mu M$ dye) to maintain the steepness of the fluorescence yield curves. Buffer solutions (pH 4 and 8, for instance) would allow the use of much higher ionic strength than the dye concentration, while not creating extreme pH values. Regardless of whether buffered or unbuffered solutions are used, CO_2 dissolution over time can titrate the working solutions, so they must always be used immediately after the calibration curves are obtained.

This method [24] has been demonstrated on a helical micromixer modeled after that presented by Stroock et al. [14], laser-machined in silicon using an Nd:YAG laser. Sample images are presented in Figure 4.7, demonstrating rapid extinction of fluorescence in the case of $\xi_c = 0.8$, but that fluorescence persists further downstream as ξ_c approaches ξ_m (0.5 in this case). Stream 1 was KOH (pH ~3) while stream 2 was HCl (pH ~11), and both solutions contained $5\,\mu M$ fluorescein.

In closing this section, we would like to mention that a promising alternative to this method might be the fluo-3 Ca^{2+} [25] indicator available from Molecular Probes (Eugene, OR); this, however, has significant nonlinearity (about 30%) over even just one decade of Ca^{2+} concentration (other fluo-4 and fluo-5 variants are significantly worse in this regard). Careful

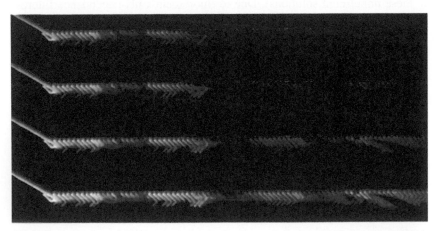

FIGURE 4.7 Sample "unmixed fluid" measurements for (from top to bottom) $\xi_c = 0.8, 0.7, 0.6$, and 0.5 for a herringbone mixer. Mixing of nearly all the $\xi = 1$ fluid to below 0.8 occurs within about nine channel widths, while mixing of all fluid below $\xi_c = 0.7$ requires about 18 channel widths' distance; these results appear to be in rough agreement with the results obtained via confocal laser scanning microscopy for this design by Stroock et al. [14]. Mixing of all fluid below $\xi_c = 0.5$ never occurs, since this can only occur asymptotically.

calibration and interpretation of the image data might make this uncertainty acceptable for numerous applications, however, and the speed with which data could be obtained is significantly higher, because multiple experiments in the "unmixedness" method could be condensed into a single experiment. A complete characterization of the reaction kinetics, however, has yet to be performed and is needed before its utility as a mixedness marker can be ensured.

4.2.3 TEMPERATURE IMAGING

Temperature imaging diagnostics are of interest because Joule heating often imposes a practical limit on electrophoresis systems. Furthermore, in micro-chip systems that integrate multiple components, nonuniform channel dimensions and conductivity may exist, resulting in localized Joule heating, and the temperature distribution can have significant transients, so a rapid, 2D imaging method is desirable. Infrared imaging can be quite accurate, but infrared cameras are typically very expensive; nuclear magnetic resonance imaging methods [26, 27] are likewise expensive and cumbersome for these applications; thermochromic liquid crystals [28] are inexpensive and accurate, but they are usually large compared to microchannels. Methods have also been developed for gas-phase flows based on NO, OH, and acetone spectroscopy [29–32], and for solid–liquid interfaces [33], but these are not very useful for liquid-phase flows.

However, a new technique, which employs standard epifluorescence microscopy and results in 2D temperature measurements (spot-size averaged) with better than $1°C$ accuracy, has recently been developed by Ross et al. [34]. The method has two incarnations, both of which exploit the temperature dependence of the fluorescence yield of rhodamine B, as shown in Figure 4.8. The first method is to obtain two epifluorescence images: the first image is obtained with no heating applied, and the second while the device is at operating temperature. Denoting these background-subtracted signal images as $S_{ij}(T_{ref})$ and $S_{ij}(T)$, respectively, one simply calculates $S_{ij}(T)/S_{ij}(T_{ref})$ and uses the mapping in Figure 4.8 to deduce the temperature at pixel (i,j). Transients can likewise be observed if a suitable frame-rate camera is used. By taking a ratio of fluorescence signal, nonuniform illumination and channel depth are normalized out of the measurement, leading to much better accuracy. Some care and planning must be taken so that the imaged region is not moved between these two experiments.

Improved accuracy, at the expense of more detailed calibration, can be obtained by taking the ratio of signals from two dyes that are both distributed uniformly throughout the system. A ratio of fluorescence signal from rhodamine B and carboxyfluorescein (which is relatively insensitive to temperature) is similarly monotonic, owing to the relative insensitivity of carboxyfluorescein to temperature. In this case, one simply calculates the pixelwise ratio of the

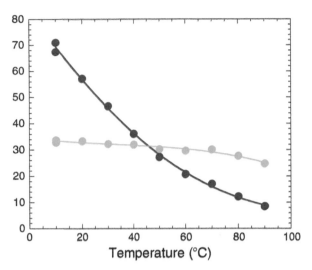

FIGURE 4.8 Curves of rhodamine B (black) and carboxyfluorescein (gray) fluorescence yield versus temperature. Vertical axis is linear, but has arbitrary (relative fluorescence) units. Data courtesy David Ross (NIST).

signal obtained from these two fluorophores, $S_{ij,\mathrm{RB}}(T)/S_{ij,\mathrm{CF}}(T)$, obtained with a single two-color image. Again, this method automatically normalizes out variations in channel depth, but one must first also measure the ratio of signals at a known reference temperature $S_{ij,\mathrm{RB}}(T_{\mathrm{ref}})/S_{ij,\mathrm{CF}}(T_{\mathrm{ref}})$ in order to account for dye concentration. Formulating a new calibration curve for each dye solution is also recommended, and this may require some effort. However, once the calibration curve is established the experiment can proceed quickly, and the results are accurate. Figure 4.9 shows a demonstration image obtained in a channel with a linear temperature gradient established along its length, showing that "common-mode noise" due to channel imperfections and dust are normalized out by this ratiometric method. However, it is emphasized that the method still has imaging resolution spot-size limitations; if appreciable temperature differences occur over lengthscales comparable to the spot size, there can even be errors in the "average" temperature within this spot owing to the curvature of the fluorescence yield curves shown in Figure 4.8.

4.2.4 QUANTITATIVE IMAGING CONSIDERATIONS

4.2.4.1 Linearity, Background Signal, and Signal-to-Noise Ratio

Charge-coupled devices (CCDs) typically have a linear response, and thus in principle can be very good quantitative detectors. Usually, the image intensity "II" at pixel (i,j) is written as the sum $II(i,j) = G(i,j)[S(i,j) + B(i,j) + N(i,j)]$,

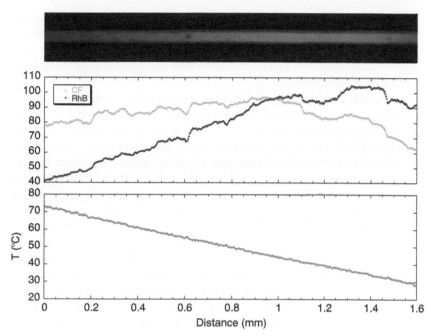

FIGURE 4.9 (See color insert) Top: a color fluorescence image of a channel with a linear temperature gradient. Middle: fluorescence signals from carboxyfluorescein and rhodamine B. Bottom: temperature profile, deduced from the ratio of these signals. Note the rejection of common-mode noise from the raw images due to the use of a ratiometric technique. (Image courtesy David Ross [NIST].)

where S represents the desired signal, B represents any undesired background (e.g., scattering from surfaces), N represents noise, and G is the camera or CCD response ("gain") at each pixel. As long as each of these factors are properly accounted for, one can obtain $S(i,j)$ ($\Pi(i,j)/G(i,j) - B(i,j)$. Often $G(i,j)$ is assumed to be some constant, but there may be significant spatial variation of illumination intensity, or of collection efficiency (especially with high numerical aperture (NA) objectives). Depending on the demands of a particular quantitative method, a suitable means for correcting for spatially varying response may be required. For example, one strategy for establishing G is to image a uniform-depth channel, filled with a dye solution of uniform concentration. It may also be necessary to subtract background signal, although removing nonresonant scattering is always inherently problematic.

The total noise, $N(i,j)$, is typically expressed in terms of photoelectrons, and can be expressed as the sum-of-squares from various independent noise sources, for example, $N^2 = N_{shot}^2 + N_{therm}^2 + N_{read}^2$. Here, three noise sources are considered:

- Shot noise (N_{shot}): This is unavoidable noise due to the statistical–mechanical nature of photon emission from molecules. The number of noise photons is the square root of the number of total photons emitted, so signal-to-noise can never be better than the square root of the number of photons collected.
- Thermal noise (N_{therm}): The camera sensor collects a certain number of random long-wavelength thermal photons and phonons due to similar statistical–mechanical processes. This noise is uncorrelated with the signal, and can be reduced by cooling the sensor array. Since CCD sensors integrate incident photons, the effect is not only one of random fluctuations, but of increased background levels, and is quite pronounced for long exposure times. Thus, most cameras intended for long-exposure astronomy are cooled, but video-rate cameras are only rarely cooled. This component is often negligible for short exposure times, but can grow significantly. An easy way to measure the effect is to keep the camera shutter closed or place a cap over the aperture, taking images with successively longer exposure times, and plotting the root-mean-square fluctuations in the image versus exposure time. In this experiment, readout noise should be constant and shot noise will be nearly zero, so the component of the noise dependent on exposure time is likely to be thermal noise.
- Readout noise (N_{read}): Pixels on CCD cameras are read out in a bucket-brigade fashion, where accumulated charge is passed from pixel to pixel (or along masked lines on a chip) until it reaches an A/D converter. The noise associated with A/D conversion is typically proportional to the square root of the A/D conversion frequency. Most scientific CCD manufacturers allow the user to set the readout speed, and so setting the speed to as low a speed as practical or as needed will minimize this source of noise. Values are about $10e^-$ for very good scientific grade cameras.

Using this approach, the signal-to-noise ratio (SNR) of an image can be estimated. For example, if peak signal collected by the optics at a pixel is about 2000 photons, the CCD quantum efficiency is 0.5, thermal noise is negligible, and readout noise is $10e^-$, one can expect to obtain an SNR of $(2000 \times 0.5)/(2000 \times 0.5 + 100)^{1/2} = 30$.

4.2.4.2 Intensifiers

Image intensifiers allow the detection of low-level light signals, but the detection limit comes at a significant price. First of all, there is a loss of resolution, due to the use of a microchannel plate (MCP). The microchannel plate also places a limited dynamic range, due to the limited capacity of the multicathode plate to source electrons for signal amplification. Last but not least, there is a significant decrease in the collection efficiency of the system.

Typically, one suffers >80% losses of the incident photoelectrons due to photon–electron conversion at the cathode; there is significant gain in the MCP, but additional noise as well, which scales with the square root of the gain. Thus, typically an intensified CCD (ICCD) is the best option only if a transient event of very low-level emission must be observed.

As an example, assume an average of 500 photons will be collected by the objective and imaged onto a set of pixels. Using an ICCD camera, one expects about 100 photons to make it through the cathode, with about $10e^-$ of shot noise (due also to shot noise from the MCP); for a camera of $10e^-$ total readout and thermal noise, the SNR will be about $100/(10^2 + 10^2)^{1/2} = 7$. For comparison, a backside-thinned cooled CCD with slow readout will have a quantum efficiency of >50%, so about 250 photons make it to the CCD array, with $250^{1/2} = 16e^-$ shot-noise photoelectrons and $10e^-$ of readout noise. SNR is then $250/(100 + 250)^{1/2} = 13.3$, almost twice as good. Only at extremely low signal levels, where the nonamplified signal begins to fall below the bit noise from the A/D converter, thermal noise, or background signal, does the intensified system compete.

4.2.4.3 Photobleaching

Because 2D images are obtained, the number of photons incident on each pixel may be quite low. One might be tempted in many cases to increase the intensity of illumination to compensate, but doing so often leads to fluorescence photobleaching; fluorescein is particularly susceptible to this. A simple test is to obtain several images over time with nonflowing dye, and observing whether the signal decreases with time. The test is quite simple but is often overlooked. Some improvement may be possible by degassing solutions — oxygen is the usual culprit, and eliminating it from the solution as much as possible may help somewhat. Even with this precaution, many dyes have a finite number of photons one can hope to get from it, at least over short timescales. Significantly, depleting the dye as it moves through an image (say, in a flowing system) will significantly contaminate the results.

4.3 PARTICLE IMAGING VELOCIMETRY

4.3.1 PRINCIPLE

PIV allows the experimentalist to measure the in-plane components of the fluid velocity field within a quasi-2D "slice" through a flow by acquiring images of small particles in the flow at two successive instants in time, and estimating their displacements. PIV was developed in the 1980s primarily for probing macroscale unsteady fluid flows [35–39]. Early workers used high-resolution film cameras, but the realization that highly resolved images of individual particles were unnecessary led to the development and rapid growth in popularity of digital PIV, wherein images are obtained on comparatively low-resolution CCD sensors (typically 1000×1000 pixels, compared to the 3400×2500 or better resolution of many films). Key improvements in image processing and experimental methods, as well as the development of a market for interline frame transfer cameras tailored for this application, have resulted in what can now be purchased as a user-friendly system. A comprehensive introduction is given in review papers by Willert [40] and Westerweel [41], and in the comprehensive book by Raffel et al. [42]. Adaptation of the method to microfluidic systems is detailed in Refs. [43–47].

PIV is closely related to particle tracking velocimetry (PTV). In PTV, particles are introduced into the flow at very low density, and a pair of digital images of the particles is acquired in rapid succession. The particles move a short distance, say, less than 10 pixels, between image frames. Since the particle density is very low, an individual particle may be identified, the change in its position $\delta x = (\delta x, \delta y)$ over a short time δt may be measured from the two images, and thus the particle velocity calculated as $u_p = \delta x / \delta t$. If the particle is very small, and the flow steady enough, the particle will follow the flow, and so it can be hoped that the particle velocity is equal to the fluid velocity (more discussion on this subject will follow later). Thus, the fluid velocity field can be determined, but the downside to this method is that the resolution is poor since the particle number density is necessarily low for the image processing method to work.

PIV differs from PTV primarily in that a much higher density of particles is used, as shown in Figure 4.10(a). It is computationally extremely expensive and difficult to track individual particles under these conditions, but instead one obtains a *statistical* measure of the average particle displacement within a small (typically square) subwindow of the image, called an interrogation region (IR), as illustrated in Figure 4.10(b). This is accomplished by cross-correlating the IR from the first image of particles (say, an $M \times N$ pixel subimage; typically, $M = N \cong 32$ with the corresponding IR in the second image [48]. Denoting the signal in the two subimages at pixel (m,n) as $S_1(m,n)$ and $S_2(m,n)$, respectively, the correlation C_{ij} is

$$C(i,j) = \sum_{m=1}^{M} \sum_{n=1}^{N} S_1(m, n) S_2(m - i, n - j)$$

The correlation at pixel (i, j) is the result of shifting $S_2(m,n)$ by (i, j) (e.g., up i rows and over j columns, multiplying it pixelwise by $S_1(m,n)$, and adding up all the resulting products).

When (i, j) is very close to the mean particle displacement (in pixels, to integer accuracy), all the high-intensity signal regions multiply and add to a much larger value than otherwise, leading to a prominent peak in the correlation signal, as shown in Figure 4.10(c). This peak signal corresponds to the

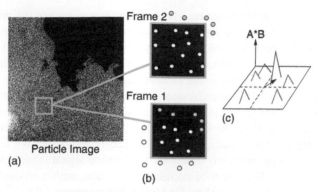

FIGURE 4.10 (a) A particle image. This is taken from a large-scale, unsteady turbulent flame, with nonuniform particle density, and the flow is generally upward and to the right. (b) A cartoon of an IR, and the particles in or near the IR at two instances in time. In frame 1, some particles below and to the left of the IR have moved into the IR by the time frame 2 is exposed, and likewise, some particles in the IR in frame 1 have left the IR by frame 2. These are shown in a gray shading. However, particle displacement from frame 1 to frame 2 is small enough so that a significant fraction of the particles (shown in white) are in the IR during both exposures. These particles contribute to a positive correlation (c) of these exposures. The peak signal in the cross-correlation plane corresponds to the velocity vector. Other peaks typically correspond to random noise, but sometimes they can be the result of a poor imaging technique.

best estimate (to integer accuracy) of the mean particle displacement. Most PIV image processing codes use a Fourier transform method to compute the cross-correlation. The advantage of this method is a significant increase in computational speed: presuming an $N \times N$ IR, the direct calculation of the correlation requires $O(N^4)$ operations, while the Fourier transform method requires $O(4N^2 \log 2N)$ operations. For $N = 32$, the computational cost is approximately 50 times less by the transform method.

Subpixel accuracy is obtained by fitting the intensities near the peak to an analytical curve. The method employed in many PIV codes is a simple Gaussian three-point fit to the signal at, and adjacent to, the peak center at (i_{peak}, j_{peak}):

$$\Delta i_{frac} = \ln[C(i_{peak} + 1, j_{peak})] - \ln[C(i_{peak} - 1, j_{peak})] - 2\ln[C(i_{peak}, j_{peak})]$$
$$\Delta j_{frac} = \ln[C(i_{peak}, j_{peak} + 1)] - \ln[C(i_{peak}, j_{peak} - 1)] - 2\ln[C(i_{peak}, j_{peak})]$$

This estimator is computationally inexpensive and is highly accurate. Interestingly, the results are more accurate if the mean value of the correlation is subtracted from the correlation [41]. Although this mean value is small, it results in the small possibility that $C(i, j)$ may have negative values adjacent to the peak, and these cannot be handled robustly by the above estimator due to the natural logarithm function. The problem is worst when the correlation peak is narrow, and when displacements are near 0.5 pixels (when most of the signal is divided between two adjacent pixels, leaving little signal in the third pixel). In these cases, the fractional displacement might be estimated using a centroid fit or other estimator that is more robust. For example, in a freeware code called "StanPIV" written by one of us (E.F.H.), a centroid estimator (the same formula as above, without taking the natural logarithm of the signals) is used in rare cases when the Gaussian estimator fails due to negative values (approximately 0.5% of the vectors or so). The results also have significant error for larger displacements unless the correlation signal is multiplied by a "tent" function, called a loss-of-correlation function, which accounts for the lower expected correlation values for higher displacements [49].

All of this image processing results in a single vector, representing the velocity in the vicinity of this $M \times N$ pixel region. A field of vectors can be obtained by repeating this algorithm with an array of IRs throughout the image. Typically, the IRs are allowed to overlap by 50% (motivated by Nyquist sampling criteria). Thus, a 1024×1024 pixel image sampled with 32×32 subwindows at 50% overlap provides a 63×63 vector field.

Due to the statistical nature of the cross-correlation method, occasionally, the peak of the cross-correlation occurs at a spurious location. For example, if the largest correlation peak is located at $(+15, +15)$, but the true displacement corresponds to a smaller correlation peak at $(+5.0, 0.0)$, some automated means for detecting the error is needed. In order to deal with this, a number of "filters" and other means for eliminating erroneous vectors have been developed [50–53]. The simplest of these is a range filter. For example, if one knows that particle displacements in the x-direction should all be in the range 0 to 10 pixels, and y-direction displacements should be in the range -5 to 5 pixels (e.g., in a flow that was largely proceeding from left to right with no backward flow expected), vectors outside this range are rejected. An obvious extension of the $(\delta x, \delta y)$ range filter is the (r, θ) range filter, which can eliminate vectors which point in an obviously incorrect direction or are unusually long or short. Median filters, as another example, typically look at the velocity components in the 3×3 grid neighborhood of a given vector, and eliminate the vector if it is not within some tolerance of the median u and v components. A variation on this filter ensures that the vector has at least a specified number of neighbors with velocity vectors within some tolerance of the vector in question. Yet another filter, based on the idea that neighboring vectors should probably not differ by more than 1 pixel or so, multiplies the correlation function obtained by the one from the adjacent IR, thus giving a strong weighting factor that is very likely to identify the correct peak in the first place. Some algorithms replace the rejected vector with a vector interpolated from "valid" neighboring vectors. Better algorithms seek to find the largest peak that satisfies the filtering criteria and replace the spurious vector with that measured value.

4.3.2 IMAGE PROCESSING AND ACCURACY

Like all measurement techniques, PIV has errors that can be classified as either "bias errors" (systematic errors) or "random errors"; typical PIV software is now able to reduce bias errors to less than 0.01 pixels, at least for ideal images. Random errors, however, tend to be on the order of 0.1 pixels even for very good images, as shown in Figure 4.11. For this reason, particle displacement is set (by the time delay between images) to be about 5 to 10 pixels, to limit the uncertainty to 1 to 2%.

These guidelines are by no means exact, and sophisticated image processing can reduce uncertainty considerably as compared to the generic PIV algorithm described above. For example, one would like, if possible, to reduce the IR size so that greater resolution could be obtained. However, there are two important effects that lead to accuracy tradeoffs when attempting to do so. (1) Larger particle displacements (relative to the IR size) lead to a larger percentage of particles leaving and entering the IR, as shown in Figure 4.9(b), and these lead to loss of correlation. This not only increases the uncertainty but also increases the chance to get a completely erroneous vector. For this reason, it has been suggested as an experimental "rule of thumb" that the displacement should not exceed one third of the IR size. (2) The uncertainty scales with the number of particles common to both IRs to the $-1/2$ power, as illustrated from the StanPIV shareware code in Figure 4.10, and thus the uncertainty is inversely proportional to the IR side length.

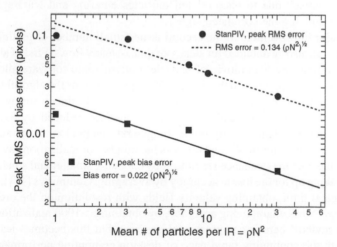

FIGURE 4.11 Bias and RMS (random) errors in PIV. Values were calculated by a particular algorithm (StanPIV, Hasselbrink, 1999) against simulated "ideal" particle images with a range of known displacements. As in this plot, it is generally true with PIV algorithms that (1) systematic bias errors can be kept much smaller than random errors (a percentage or so of a pixel) and (2) statistical errors such as these show a power-law scaling ($-1/2$ power) with the number of correlated particles in the IR.

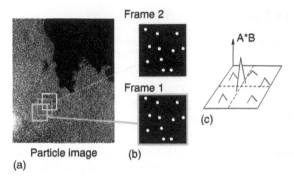

Particle image   (b)
(a)

FIGURE 4.12 PIV algorithm using an iterative procedure. An estimate of the displacement is first obtained using the algorithm shown in Figure 4.9, and now the IRs are offset from each other as shown in (a), so that the particles are ideally all captured in both frames, as shown in (b). The correlation peak is now near the center as shown in Figure 4.11.

A major improvement in the first issue (loss of correlation) can be realized with an iterative method. A good estimate of the velocity field is obtained first using the generic algorithm outlined above, and then a second pass is made using IRs which have been offset, as shown in Figure 4.12, to account for the particle displacement. In this way, the particles are almost all correlated, and the cross-correlation peak is located very near the center. This reduces the likelihood of a spurious vector compared to the generic algorithm, by reducing the "noise" due to uncorrelated particles entering and leaving the IR while maximizing the correlated signal.

A major improvement in the second issue (small number of particles in a small IR) can also be obtained for steady or quasi-steady flows. First of all, if the user has a time-resolved movie of particle motion, some software allows for "adaptive timestepping," so that in very low-speed regions the algorithm skips frames until the total displacement is at least a few pixels, thus ensuring a lower relative error. Secondly, random uncertainty can be reduced by averaging numerous independent velocity fields. In general, the random uncertainty can be reduced by a factor of $N^{1/2}$, where N is the number of realizations averaged. In principle, one may enhance resolution by choosing smaller and smaller IRs, while making up for the loss in accuracy by averaging. Cummings [47] take this to its logical limit, obtaining velocity fields with resolution on the order of a *single pixel* while maintaining accuracy, by obtaining >1000 realizations. The obvious tradeoff here is computational expense, but this becomes less of an issue with the continuing rapid pace of desktop computing performance improvements. Figure 4.13 shows an example of a particle image and the flowfield resulting from a 2000-frame movie at video resolution. That is, the velocity field and the image are both at the same resolution. The image processing techniques for achieving this type of resolution are too detailed for this introduction, but it is noted that some commercially available codes

FIGURE 4.13 (See color insert) A PIV particle image and the resulting high-resolution map of velocity magnitude for steady EOF over an array of circular posts. A banded grayscale table is used to show the variations in velocity, which is very slow at the leading and trailing edges (left and right edges in the image) of each post, and fastest at the sides (top and bottom edges in the image) of each post. Flow is from left to right, but the velocity magnitude $(u^2 + v^2)^{1/2}$ is symmetric about each post. High resolution is made possible from a statistical technique that allows for grid refinement in steady flows. (Images courtesy by E.B. Cummings, Sandia National Labs.)

now incorporate this feature in their μ-PIV systems. The technique is also unique for its use of dye-filled liposomes, rather than commercial nanoparticles, as flow tracers.

As a final note, while image processing techniques have received much focus in the literature, the most important determinant of PIV accuracy is experimental technique. For example, a mistake often made by researchers new to the technique is to attempt to obtain very sharp-looking particle images that are saturated. Since subpixel PIV accuracy depends on fitting a Gaussian function to the correlation peak, and this peak will not be a Gaussian function if the particle images are not Gaussian — for example, if they are so bright as to saturate the CCD, this leads to large errors. Particle images should appear as diffuse gray spots, not sharp white dots, for this reason. Another easy pitfall is to allow signal from something other than the particles to appear in the IR. The most important example of this is light scattering from the sides of the microfluidic channel walls. In this case, the correlation function is not only operating on the particle images, but on signal from a nonmoving channel wall, which can bias the velocity measurement significantly. Even the use of fluorescent beads and stringent filtering is not enough near channel walls sometimes, because fluorescent light emitted from a bead that is very near a wall will scatter from the surface, and this cannot be removed by optical filtering. The only way to be sure in this case is to check the correlations themselves for signs of contamination — if the brightest peaks in the correlation appear as elongated lines rather than compact spots, the correlation is corrupted.

In addition to the wall scattering problem, there are a few other experimental considerations that are unique to, or more dominant in, microfluidics experiments. The first is that, in the case of flows in the presence of E-fields, the particles employed may have nonzero zeta potential, in which case they will have their own electrophoretic velocity superposed on the bulk fluid velocity. This presents a problem for interpreting the velocity field, especially in mixed pressure-driven and EOFs. It could also present a problem in that the particles may no longer be considered nonintrusive, since the electromagnetic forces acting on them impart a net force on the fluid due to viscosity. Globally, the effect is small — one can determine the global effect on the flow by comparing the surface area of all the particles in a channel to the total surface area of the channel itself (assuming similar zeta potentials of the particles and wall, and relatively large particles, the total charge will be proportional to the zeta potential and the surface area for a given electrolyte solution). However, even for 100-μm channels, and dense particle loadings (100-nm-diameter spheres, 1 μm^{-3}), the total particle surface area is only 2% of the channel area, so typically these forces do not affect the global flow. Another potential problem, however, is dielectrophoresis. Cummings (see Ref. [54, chapter 8]) shows that sharp corners create high electric field gradients which lead to dielectrophoretic forces that can not only modify particles trajectories, but can even trap them. For these reasons, some estimate of dielectrophoretic forces should be carried out, and one should be very cautious about PIV results in flows where high electric field gradients are present. In such cases, caged-dye imaging of a zwitterionic dye offers an alternative method that may be worth considering.

4.3.3 PIV HARDWARE SELECTION

The equipment required for conducting μ-PIV is relatively minimal: a suitable camera, a pulsed light source, and a microscope objective. There are several companies offering complete PIV systems, usually with several options for these basic components. In selecting these components, one must take into consideration the speed of the flow that will be imaged and the magnification that will be used. When selecting a pulsed light source, the important issues are adequate illumination, a short enough pulse time so that there will be no streaking of the particle images, and a short enough time *between* pulses so that the particles displace no more than about 10 pixels. These requirements are directly coupled to the CCD chip size (typically, about 1 cm), the objective magnification, and the flowspeed. For fast flows imaged at high magnification, one can easily find that the time delay between image realizations is shorter than can be obtained with a standard (e.g., 30 fps, interleaved) video CCD camera. Special interline frame transfer CCD cameras (sometimes called "cross-correlation" cameras by PIV vendors) are capable of taking two completely separate images with time delays as short as 1 μsec or even less. These image pairs may typically be obtained at 30-msec time intervals. Recently, CMOS imaging chips have become available and are incorporated into cameras capable of very high frame rates (>10,000 fps). These devices are completely programmable (one may select only to read out a subset of the image), which adds significant flexibility. Not surprisingly, these cameras are quite expensive (about US$50,000 at the present time, compared with US$10,000 for an interline frame transfer camera, and about US$1500 for a good monochrome video camera), but they are expected to become a replacement for the CCD in many applications, and if this happens they should become much less expensive in the near future.

4.3.4 PIV RESULTS AND INTERPRETATION

A typical PIV vector field can be displayed as a vector field or as streamlines (which are defined as lines everywhere parallel to the velocity field). Most PIV software will do this; but often one wants to obtain more quantitative information, for example, the flowrate through a certain conduit in a network. In this case, it is necessary to extract the velocities in a region of interest for further processing, or obtain a specialized software package for carrying out such calculations and displaying the results. For example, Tecplot^{TM} (Amtec Engineering, Bellevue, WA) is a popular package in the computational fluid dynamics community; of course, custom codes for data processing and display may be written in any programming language, or a general mathematics package such as MATLAB. For example, the flowrate Q through a surface is simply the integral $\int \mathbf{u} \cdot d\mathbf{A}$, which reduces to an integral along a contour in a 2D flow (flowrate per unit depth is obtained). When one has very thin depth-of-field compared to the channel width, and multiple planes of 2D data can be obtained as described previously, the area integral may be evaluated more exactly. Other quantities of interest such as the vorticity (out-of-plane-component) and various strain rates may also be calculated from numerical derivatives of the velocity field.

4.4 NUMERICAL SIMULATION

Numerical simulation of fluid flow (also known as computational fluid dynamics [CFD]) is a well-developed field of much greater scope than can be covered in this chapter. Fortunately, commercial CFD software exists, and the entire point of such a tool is that it can be used with accuracy, without understanding all the arcane details of how the computational engine works. Commercial software exists now that can simulate high Mach number turbulent flows as well as steady creeping (low Reynolds number) flows in microchannels with reasonable accuracy, if the problem is given the appropriate physical models, boundary conditions (BCs), grid resolution, and solution method. The bad news is that these codes are often so general, and offer so many options, that the user still needs a reasonable degree of sophistication to use the tool effectively. The goals of this section are therefore: (1) to provide some perspective on the use of simulation as an engineering tool, (2) provide a primer on the typical physical models and BCs that are specifically needed in microfluidics applications, and (3) to endow the reader with an appropriate level of skepticism on the validity of the results obtained. More depth for those desiring to write their own CFD codes can be found in numerous texts on the subject of numerical methods (e.g., Refs. [55–57]); a recent text by Karniadakis and Beskok [54] also describes methods for simulations for gas-phase microflows (as are frequently encountered in traditional MEMS devices) in detail.

4.4.1 TO SIMULATE, OR NOT TO SIMULATE?

Choosing how to spend one's time and money is an investment decision, and so one must assess costs against expected benefits. Briefly, these are:

- *Costs*: Site licenses for commercial CFD packages cost on the order of US$10,000 to US$30,000 depending on how much training, support, and functionality is desired. But a bigger expense is the investment in trained personnel to run the software and extract results from it. This has a sunk cost as well — obviously people have a wide range of abilities, but one can expect an individual with a strong technical background but no formal training in fluid mechanics to spend at least a full month or two just learning to become productive with the software, depending on the amount of training they receive. If one has no trained staff, and a

CFD calculation is simply a one-time effort on a particular device, it may cost less to take advantage of the consulting service that most CFD software companies also offer, in which they will set up and perform simulations for a fee. And, of course, one can always seek a collaborator who has experience in performing simulations.

- *Benefits*: CFD is ultimately a supplement for experiment, not a replacement (especially if absolute numbers, not trends, are important). In the archetypical design process, many ideas are narrowed down to a few options, which are reduced to one after semidetailed designs have been completed and performance estimated. CFD is particular useful in this latter phase, especially when prototype construction is expensive or time-consuming. In later phases of device development, CFD solutions require experimental validation of some form or another. On the other hand, when problems or unexpected behavior arise, CFD can offer a level of detailed information about the computed solution that is rarely possible to obtain via experiment, and this level of detail can give the insight needed to solve a microchip transport problem.

Like any other calculation, the old adage applies: "garbage in, garbage out." The benefits of simulation are only possible when the simulation has been posed properly. A lesser caveat is that CFD codes, due to uncertainties in properties, numerical error, and imperfect domain modeling, are usually not more accurate than within 5 to 10% when compared with experimental results, unless the physical models involved are quite simple. However, running a series of CFD simulations while varying an independent parameter will usually give accurate trends in a dependent variable, if not the correct absolute value.

In short, numerical simulation is a net benefit approach when it is expected that (1) valid results will be obtained, (2) the results will lead to improved physical insight or engineering design, and (3) the results would have been significantly more expensive to obtain via experiment. These criteria seem axiomatic, but the remainder of this chapter is devoted to the fundamentals of simulation, which will hopefully also convey when and how it is possible to reap these benefits.

4.4.2 Fundamentals: Flow Equations, BCs, Domain Selection

4.4.2.1 Equations: Pressure-Driven Flow in Channels with Walls of Zero Surface Charge Density

Equations for pressure-driven fluid flow are the conservation of mass, balance of momentum, and conservation of energy. In differential form, these are:

$$\frac{\partial \rho}{\partial t} + \vec{u} \cdot \nabla \rho + \rho \nabla \cdot \vec{u} = 0 \quad \text{(mass)} \tag{4.1}$$

$$\rho\left(\frac{\partial \vec{u}}{\partial t} + \vec{u} \cdot \nabla \vec{u}\right) = -\nabla P + \nabla \cdot \boldsymbol{\tau} + \vec{f} \quad \text{(momentum)} \quad (4.2)$$

$$\rho\left(\frac{\partial(e + u^2/2)}{\partial t} + \vec{u} \cdot \nabla(e + u^2/2)\right)$$

$$= -P\left(\nabla \cdot \vec{u}\right) - \nabla \cdot \vec{q} + \boldsymbol{\tau}{:}\nabla \vec{u} \quad \text{(energy)} \quad (4.3)$$

The reader with training in a field other than fluid mechanics may at first be daunted by these equations.[2] But to use CFD software, the user need only determine whether sufficient information has been provided to the software to adequately solve them, and so this is the approach we will follow from here.

The first thing we need to check is whether the number of equations is equal to the number of unknowns. The momentum equation is really three equations, one for each component of the momentum vector. For pressure-driven flows in channels with zero wall surface charge, the usual remaining body force is gravity $\vec{f} = \rho\vec{g}$, which is known (and can be neglected in flows not driven by gravity). We then have five equations in 15 variables or parameters (the three velocity components $\vec{u}$, pressure P, internal energy e, the fluid mass density ρ, the three heat flux vector components $\vec{q}$, and the six components of the symmetric stress tensor $\boldsymbol{\tau}$), and so clearly we need more equations. This can be reduced to a closed set using the following simplifications and constitutive relations:

- *Constant mass density or perfect gas law*: In the case of liquids, it is usually possible to say $\rho = $ constant; that is, the fluid can be considered *incompressible*, unless pressures exceed several hundred megapascals (although experience tells us that a pressurized syringe plunger often springs back when the pressure is released, this comes from air bubbles in the pressurized system, not from compressibility of the liquid). In the case of compressible (gas) flows, density can be usually calculated from a perfect gas relationship $P = \rho RT$, but of course now R and T may be functions of space (since R is a function of gas composition).
- *Newtonian fluids*: Most solvents (water, alcohols, DMSO, and acetonitrile) have viscosity that is nearly independent of shear rate, making them *Newtonian* fluids by definition. In these cases, for liquids, the shear stress tensor can be reduced to a single (known) nonzero com-

---

[2] Here ρ is fluid density, $\vec{u}$ is the fluid velocity vector, P is pressure, $\boldsymbol{\tau}$ is the stress tensor, f is a body force (see text), e is internal energy, and $\vec{q}$ is a local heat flux vector. Fluid mechanics texts normally use μ for dynamic viscosity, but in this chapter we denote dynamic viscosity with η, and mobility with μ.

ponent, which we call the dynamic viscosity (typically μ in the fluid mechanics literature, η in the chemistry literature — here we represent it with η). In gases, a "bulk" viscosity may be important, but it is usually incorporated as a factor into η. Note that in both cases, viscosity depends strongly on temperature!

- *Gradient transport*: The heat flux vector is usually modeled as proportional to the temperature gradient ($\vec{q} = k\nabla T$), where k is the local thermal conductivity (units typically W/m K). This eliminates three unknowns in favor of one new unknown (T), and a known coefficient of thermal conductivity (denoted k in most handbooks and heat transfer texts). Typically, conductivity varies appreciably with temperature, so curve-fits to data (found in handbooks) can be useful in nonisothermal cases.

- *Thermodynamics*: Internal energy e can also be written as a simple function of temperature using an approximate relation. Over the liquid range of temperatures for most solvents, $e = cT + e_{ref}$ is usually sufficiently accurate.

The result of these simplifications is that we now have five equations in five unknowns ($\vec{u}$, P, T), which can be solved in principle, given well-posed BCs and a well-resolved computational domain. Problems that are more complicated (e.g., pressure-driven flow of a non-Newtonian gel) require additional information (e.g., a rheological model for the fluid). Relationships for temperature-dependent transport properties can usually be entered into commercial software, and dedicated CFD codes often include a database of these properties or coefficients for curve-fits to tabulated data. In cases where thermal gradients may be neglected, one can typically tell the solver to ignore the energy equation completely, leaving only four equations with ($\vec{u}$, P) as the four unknowns.

4.4.2.2 Domain and BC Selection: Pressure-Driven Flow in Channels with Walls of Zero Surface Charge Density

In order to reduce computational time, it is preferred to choose the simulation domain to be the smallest domain that satisfies two criteria: (1) the domain captures the flow phenomenon of interest and (2) accurately known BCs can be applied at the boundaries. As with all partial differential equations, it is possible to overspecify or underspecify BCs. Modern-day CFD software does its best to simplify the procedure by applying good default BCs for certain types of boundaries (e.g., a no-slip BC for velocity at a solid wall) and in attempting to catch over- and underconstrained situations while the BCs are being posed. Nonetheless, some simple rules of thumb can make the problem setup go much quicker.

On solid walls, the no-slip BC $\vec{u} = 0$ normally applies; this condition may be replaced with a partial slip condition for low-density gas flows (see Ref. [53] for details), or a constant shear stress BC at a fluid–fluid interface. At gas–liquid interfaces, the viscosity difference is so large that a better choice is a zero shear stress BC at the liquid side of the interface, and $\vec{u}_{gas} = \vec{u}_{fluid}$ is for the gas side of the interface. However, various CFD packages have various preferred methods for dealing with such interfaces.

At inlet and outlet boundaries, usually some sort of convective assumption is being made, for example, that velocity derivatives are zero and the second derivatives of pressure are zero in the direction normal to the surface. It is important to note that since inflow–outflow BCs assume no streamwise changes in the flow, significant error will result if, in reality, there are sudden changes in the velocity field where the BC is applied. Figure 4.14 shows examples of a poorly chosen and a well-chosen domain for the same flow problem. Clearly, in Figure 4.14, the sudden expansion just outside the left boundary makes the "smooth outflow" BC $\partial u/\partial x = 0$ and $\partial v/\partial x = 0$ a very poor approximation. The solution is to move the boundary a few channel widths to the left, which results in a much more accurate solution.

4.4.2.3 Equations: Flow in Channels with Walls of Nonzero Surface Charge Density

There are many cases where a surface potential does exist, and this has some bearing on the overall flow. The obvious case is EOF in a glass channel; another less obvious possibility is the "electroviscous effect" which arises during pressure-driven flow of low-conductivity fluid through a channel with a surface charge (a significant streaming potential can develop, creating an opposing EOF which reduces the overall flowrate).

To account for such phenomena, an additional term must be added to Equation (4.2) in the presence of an applied electric field, namely, $\rho_c \vec{E}$, where ρ_c is the electric charge density and $\vec{E}$ is the applied electric field. This term accounts for Coulombic forces, and is responsible for EOF.

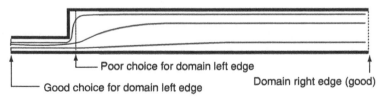

Poor choice for domain left edge

Good choice for domain left edge

Domain right edge (good)

FIGURE 4.14 Good and bad choices for the edge of a domain. Edges near sudden changes in geometry lead to a poor approximation for inlet–outlet BCs, since they usually assume mild changes in the streamline direction.

Although this term is dominant in the electric double layer, this term is negligible in flows of classical engineering interest, so it does not usually appear in classic texts on fluid mechanics, or in CFD software that is not tailored for microfluidics. $\vec{E}$ is obtained from $\vec{E} = -\nabla\phi$ after solving for the potential field ϕ arising from applied voltages; the potential field ϕ is obtained as the solution of $\nabla(\sigma)\nabla(\phi) = 0$, taking σ to be the fluid's electrical conductivity. In a typical computational domain, ϕ requires one BC at each point on the boundary. The BC for nonconducting surfaces is the zero-flux BC $\nabla\phi \cdot \hat{n} = 0$, where $\hat{n}$ is the outward normal from the surface. At a conductor, and usually at an inlet–outlet boundary, ϕ is given a specified value. Again, as shown in the previous section, to simulate flow through a detailed region of interest without including long lengths of channels leading to electrode reservoirs (which becomes computationally expensive), a reasonably accurate approximation is to truncate the channels to no shorter than about 10 channel diameters, and apply fixed-potential BCs across the inlets and outlets, matching the electric field that one plans to impose in the real device.

The charge density ρ_c can, in principle, be obtained from direct solution of the Poisson–Boltzmann equation

$$\nabla^2 \psi = -\frac{\rho_c}{\varepsilon} = -\sum_i \frac{n_i z_i e}{\varepsilon} \exp\left(\frac{z_i e \psi}{kT}\right) \tag{4.4}$$

where the summation is over all charged species. Here n is the local number density, z is the valency, e is the electron charge, k is the Boltzmann constant, T is the temperature, and ψ is the *naturally* occurring potential. The BC on ψ is usually expressed as the "zeta" potential, ζ, which is the potential at a plane very near the wall (approximately, the Stern plane; see also Chapter 2). However, in the case of thin Debye layers it is often not necessary to solve this equation directly; indeed, it becomes quite difficult to do so with accuracy in a reasonable amount of time. Simulating problems with widely disparate length scales is inherently difficult, and requires a very nonuniform grid, with many gridpoints near the wall, as compared to the center of the channel, to solve efficiently.

When the Debye layer is much thinner than the smallest channel dimension, a reasonable approximation is to replace the no-slip BC with the velocity at the outer edge of the Debye layer. Thus, $\vec{u} = \mu_{eo}\vec{E}$ on the boundary, where the electroosmotic mobility $\mu_{eo} = \varepsilon\zeta/\eta$. Ignoring the finite thickness of the double layer with this approximation leads to relative errors in the mass flux, which are (to first order) equal to λ_D/h for flat plates and $2\lambda_D/a$ for round capillaries, where h is the half-height of the channel and a is the capillary radius. Thus, there is less than 0.4% error in the mass flux for a channel of 10-μm smallest dimension and electrolyte of 1 mM ionic strength (10-nm Debye thickness).

4.4.3 Caveats

4.4.3.1 Physics Models

CFD packages tailored for microfluidics do allow for the thin Debye layer approximation described above. However, there is an important element of physics that some CFD packages do *not* capture: the zeta potential, ζ, can be strongly influenced by the local pH and ionic strength of the electrolyte solution in contact with the surface. As demonstrated in Ref. [58], these can be responsible for dominant dispersive phenomena in capillary tubes; without accounting for this effect, chromatographic peak widths can be grossly underpredicted. Silica, for example, has a ζ which becomes increasingly negative between pH 3 and 8, and a logarithmic dependence on ionic strength also exists [59, 60]. Thus, the EOF of fluid with pH 4 is about one fifth the speed of one at pH 8, and so if two slugs of such electrolytes are in series within a channel, a pressure-driven flow is created due to one slug attempting to move faster than another. (Notably, one of the most popular CFD packages in the microfluidics community does *not* calculate a local μ_{eo} as of the March 2002 release.)

Other poor predictions can be made when the user allows the software to solve equations that are overly restrictive. As an example, many packages by default assume steady laminar flow. In very low Reynolds number flow ($Re \ll 1$), the nonlinear $\vec{u} \cdot \nabla \vec{u}$ term (term 2) in the momentum equation may be ignored, leading to significant improvements in the stability and speed of numerical solution. The Reynolds number $Re = \rho u_c d / \eta$, where u_c is a characteristic velocity such as the mass-flux averaged velocity, and d is a characteristic length scale such as a channel diameter. Re is a dimensionless ratio of inertia forces to viscous forces. However, the assumption that $Re \ll 1$ is not always the case in all microflows: a 1-m/sec flow of water through a 100-μm channel (which can easily be achieved with syringe pressure) has $Re \cong 100$. While unlikely to be turbulent, in complex geometries such a flow could certainly have some interesting features that would not be correctly predicted.

Particle transport (e.g., flow of a heavily laden flow) is another area where one should be particularly skeptical of results. Particle-laden flow is generally considered a very complicated problem that is difficult for CFD to capture with accuracy, due to the fact that models must be employed to keep a reasonable grid size — the flow in the vicinity of each particle is not calculated in detail. Also, particle motion is a relatively complex physics problem, since they rotate near walls and they are often subject to significant dielectrophoretic, electrokinetic, and acoustic forces.

Last but not least, problems involving free surfaces are particularly difficult. Progress in numerical methods for such systems in recent years (e.g., front tracking and volume-of-fluid techniques) has been significant, but there is still a practical difficulty in analytical devices, which often have to handle fluids containing micellar surfactants. These can cause drastic changes in surface tension coefficient even when small amounts of surfactants are present, and spatial variations in the concentration can cause a number of interesting convective phenomena to occur.

4.4.3.2 Discretization and Convergence

Setting up a grid that is "resolved" is always a major issue in any CFD calculation. CFD works by estimating the governing differential equations with discretized versions of these equations, so the estimation of the equations usually gets better as the grid gets finer. A good rule of thumb is that a well-resolved grid has about 10 to 20 gridpoints across any feature where relatively large differences in velocity occur. For example, in a fully developed laminar pressure-driven flow in a round capillary, a grid with 10 to 20 points in the radial direction (between the wall and the center of the capillary) will lead to a fairly accurate solution; an example of square cross-sectioned channels is shown in Figure 4.15.

The primary difficulty is that physics often allows important phenomena to occur over very short lengthscales compared to the domain of interest. The obvious example is EOF in a 10-μm channel with Debye layers on the order of 10-nm thickness. If limited to a uniform grid, the requirement to resolve the Debye layer with 20 gridpoints means that 20,000 gridpoints are required across the channel. Grid difficulties can also arise due to geometries that have sharp corners and complex geometries. In such cases, there are three options: (1) a nonuniform grid (most CFD software has some means for

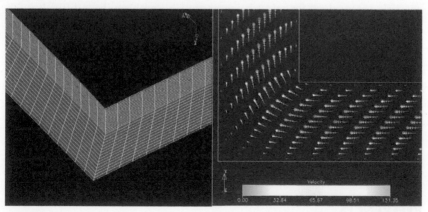

FIGURE 4.15 (See color insert) Image of a discretized corner of a microfluidic channel (left), showing the automatically generated computational grid. Refining the grid very close to the sharp bend may be needed to achieve accurate solution. The solution for the velocity field (vectors colored by their magnitude) of this coarse grid is shown at the right.

automatically generating a nonuniform grid, as well as for locally "tweaking" grids according to the user's intuition — do not underestimate your intuitive ability to improve on the software's grid), (2) a model for the BC that accounts for the phenomena without having to simulate it directly (as for EOF, described previously), or (3) to accept the high computational cost.

Regardless of how one deals with grid generation, one must always test for *grid independence* of the solution. The tried-and-true (and necessary) technique for grid independence is to run the simulation a second time with a grid that is twice as fine. If the results of interest are unchanged (or only very slightly changed), one can be at least somewhat sure that the resolution is adequate; otherwise, the resolution must be doubled again, and so on until the solution is found to be independent of the grid. This test is important enough so that doubling the grid resolution is usually a built-in feature of commercial CFD software. However, the test gets computationally expensive quickly: one can expect about 2^N more computation time for each doubling of grid resolution, where N is the number of spatial dimensions. Note that in the solution of time-dependent problems, one must *also* check for timestep independence in a similar manner (cutting the timestep in half, and resimulating the problem), which again doubles the computational time. Although these tests are computationally expensive, they are necessary to achieve confidence in the result.

It is possible to have a simulation which does not converge, for example, the residual oscillates from iteration to iteration. Often this is an indication that the simulation has inadequate grid resolution, incorrect BC, or a poor assumption in the physical model. If the grid is resolved and BCs appear to be realistic, one should also revisit assumptions and make sure the solver is appropriate to the problem. Certain solvers are more stable for incompressible flow than others; for example, some solvers for unsteady incompressible problems have difficulty with pressure, because acoustic waves move at infinite speeds through truly incompressible media. Alternative solvers allow for an artificial (but small) amount of compressibility to stabilize the solution for pressure. Also, some steady-flow solvers do not converge well if the flow would normally be turbulent or otherwise unsteady. Always check the Reynolds number, based on the narrowest channel dimension — if Re (= fluid density $\times$ maximum velocity $\times$ smallest dimension, divided by dynamic viscosity) is greater than 10, there is a chance the flow might be unstable. Van Dyke [61] has compiled an excellent and inexpensive photo-gallery of fluid flow, which may help to enhance one's intuition about various flow patterns and instabilities that can arise in flows at various Reynolds numbers.

REFERENCES

1. G.I. Taylor, *Proc. R. Soc. Lond. A* 219 (1953) 186.
2. R. Aris, *Proc. R. Soc. Lond. A* 235 (1956) 67.
3. T.J. Mitchison, K.E. Sawin, J.A. Theriot, K. Gee, A. Mallavarapu, *Methods Enzymol.* 291 (1998) 63.

4. R.P. Haugland, *Handbook of Fluorescent Probes and Research Products*, Molecular Probes, Inc., Eugene, OR, 2002.
5. H. Kasai, N. Takahashi, *Philos. Trans. R. Soc. Lond. Ser. B Biol. Sci.* 354 (1999) 331.
6. G.D. Housley, N.P. Raybould, P.R. Thorne, *Hear. Res.* 119 (1998) 1.
7. S.A. Stricker, *Dev. Biol.* 170 (1995) 496.
8. P.H. Paul, M.G. Garguilo, D.J. Rakestraw, *Anal. Chem.* 70 (1998) 2459.
9. C.P. Gendrich, M.M. Koochesfahani, *Exp. Fluids* 22 (1996) 67.
10. C.P. Gendrich, M.M. Koochesfahani, D.G. Nocera, *Exp. Fluids* 23 (1997) 361.
11. D. Ross, T.J. Johnson, L.E. Locascio, *Anal. Chem.* 73 (2001) 2509.
12. B.P. Mosier, J.I. Molho, J.G. Santiago, *Exp. Fluids* 33 (2002) 545.
13. R.F. Ismagilov, A.D. Stroock, P.J.A. Kenis, G. Whitesides, H.A. Stone, *Appl. Phys. Lett.* 76 (2000) 2376.
14. A.D. Stroock, S.K.W. Dertinger, A. Ajdari, I. Mezic, H.A. Stone, G.M. Whitesides, *Science* 295 (2002) 647.
15. Z. Yang, S. Matsumoto, H. Goto, M. Matsumoto, R. Maeda, *Sens. Actuators A Phys.* 93 (2001) 266.
16. N. Schwesinger, T. Frank, H. Wurmus, *J. Micromech. Microeng.* 6 (1996) 99.
17. M.H. Oddy, J.G. Santiago, J.C. Mikkelsen, *Anal. Chem.* 73 (2001) 5822.
18. P. Hinsmann, J. Frank, P. Svasek, M. Harasek, B. Lendl, *Lab Chip* 1 (2001) 16.
19. R.H. Liu, J.N. Yang, M.Z. Pindera, M. Athavale, P. Grodzinski, *Lab Chip* 2 (2002) 151.
20. R.H. Liu, M.A. Stremler, K.V. Sharp, M.G. Olsen, J.G. Santiago, R.J. Adrian, H. Aref, D.J. Beebe, *J. Microelectromech. Syst.* 9 (2000) 190.
21. M.M. Koochesfahani, P.E. Dimotakis, *J. Fluid Mech.* 170 (1986) 83.
22. P.S. Karasso, M.G. Mungal, *J. Fluid Mech.* 323 (1996) 23.
23. P.S. Karasso, M.G. Mungal, *J. Fluid Mech.* 334 (1997) 381.
24. S. Madhavan-Reese, D. Lim, J. Mazumder, E.F. Hasselbrink, in Y. Baba, S. Shoji, A.v.d. Berg (Editors), *Proceedings of Micro Total Analysis Systems*, Kluwer Academic, Nara, Japan, 2002, p. 900.
25. A. Minta, J.P.Y. Kao, R.Y. Tsien, *J. Biol. Chem.* 264, 14, (1989) 8171–8178.
26. K. Kuroda, A.H. Chung, K. Hynynen, F.A. Jolesz, *J. Magn. Reson. Imaging* 8 (1998) 175.
27. J. Delannoy, C.N. Chen, R. Turner, R.L. Levin, D. Lebihan, *Magn. Reson. Med.* 19 (1991) 333.
28. C.R. Smith, D.R. Sabatino, T.J. Praisner, *Exp. Fluids* 30 (2001) 190.
29. J.L. Palmer, R.K. Hanson, *Appl. Opt.* 35 (1996) 485.
30. M.C. Thurber, F. Grisch, R.K. Hanson, *Opt. Lett.* 22 (1997) 251.
31. M.C. Thurber, F. Grisch, B.J. Kirby, M. Votsmeier, R.K. Hanson, *Appl. Opt.* 37 (1998) 4963.
32. M.C. Thurber, R.K. Hanson, *Exp. Fluids* 30 (2001) 93.
33. C.H. Fan, J.P. Longtin, *Exp. Thermal Fluid Sci.* 23 (2000) 1.
34. D. Ross, M. Gaitan, L.E. Locascio, *Anal. Chem.* 73 (2001) 4117.
35. R.J. Adrian, *Appl. Opt.* 23 (1984) 1690.
36. R.J. Adrian, *Opt. Lasers Eng.* 9 (1988) 211.
37. R.J. Adrian, *Opt. Lasers Eng.* 9 (1988) 317.
38. R.J. Adrian, *Meas. Sci. Technol.* 8 (1997) 1393.
39. R.J. Adrian, *Exp. Fluids* 29 (2000) S1.

40. C.E. Willert, M. Gharib, *Exp. Fluids* 10 (1991) 181.

41. J. Westerweel, *Meas. Sci. Technol.* 8 (1997) 1379.

42. M. Raffel, C. Willert, J. Kompenhans, *Particle Image Velocimetry: A Practical Guide*, Springer, Berlin, 1998.

43. J.G. Santiago, S.T. Wereley, C.D. Meinhart, D.J. Beebe, R.J. Adrian, *Exp. Fluids* 25 (1998) 316.

44. C.D. Meinhart, S.T. Wereley, J.G. Santiago, *Exp. Fluids* 27 (1999) 414.

45. C.D. Meinhart, H.S. Zhang, *J. Microelectromech. Syst.* 9 (2000) 67.

46. S.T. Wereley, L. Gui, C.D. Meinhart, *AIAA J.* 40 (2002) 1047.

47. E.B. Cummings, *Exp. Fluids* 29 (2000) S42.

48. R.D. Keane, R.J. Adrian, *Appl. Sci. Res.* 49 (1992) 191.

49. J. Westerweel, D. Dabiri, M. Gharib, *Exp. Fluids* 23 (1997) 20.

50. J. Westerweel, *Exp. Fluids* 16 (1994) 236.

51. X. Song, F. Yamamoto, M. Iguchi, Y. Murai, *Exp. Fluids* 26 (1999) 371.

52. J. Nogueira, A. Lecuona, P.A. Rodriguez, *Meas. Sci. Technol.* 8 (1997) 1493.

53. D.P. Hart, *Exp. Fluids* 29 (2000) 13.

54. G.E. Karniadakis, A. Beskok, *Microflows: Fundamentals and Simulation*, Springer-Verlag, New York, 2002.

55. S.V. Patankar, *Numerical Heat Transfer and Fluid Flow*, McGraw-Hill, New York, 1980.

56. C.M. Bender, S.A. Orszag, *Advanced Mathematical Methods for Scientists and Engineers*, McGraw-Hill, New York, 1978.

57. E.S. Oran, *Numerical Simulation of Reactive Flow*, Cambridge University Press, Cambridge, 2001.

58. A.E. Herr, J.I. Molho, J.G. Santiago, M.G. Mungal, T.W. Kenny, M.G. Garguilo, *Anal. Chem.* 72 (2000) 1053.

59. P.J. Scales, F. Grieser, T.W. Healy, L.R. White, D.Y.C. Chan, *Langmuir* 8 (1992) 965.

60. I. Gusev, C. Horvath, *J. Chromatogr. A* 948 (2002) 203.

61. V. Dyke, *An Album of Fluid Motion*, Parabolic Press, Stanford, CA, 1982.

5 Transport Modes: Realizations and Practical Considerations

Susanne R. Wallenborg, Per Andersson, and Gunnar Thorsén

CONTENTS

5.1 GENERAL INTRODUCTION

Moving liquids has always been a challenge for mankind. Modern solutions for controlling liquid movement use pumping technology that is constantly being refined, with new pump designs frequently being introduced into the market. Pumps with flow rates typically in milliliters per minute have dominated in the areas of chemical analysis and separation. Efforts to miniaturize liquid transport systems began more widely in the early 1980s. The first efforts were primarily directed from a cost and environmental perspective toward reducing the consumption of organic solvents. However, such miniaturized pumping systems, typically microliters per minute down to nanoliters per minute, did not gain wide acceptance in the high-performance liquid chromatography (HPLC) market. A new, practical way to pump and direct the flow of nanoliter volumes was published in *Science* in 1993 and gained much attention. Suddenly it was possible to run an entire system using less volume than the volumetric error on a conventional system. Manz and Harrison described a planar, microfabricated glass chip platform that provided high separation efficiencies by controlling an electrical field and using electroosmotic flow (EOF) to pump liquids. The significant impact of this publication is demonstrated by the fact that the technique has dominated the micrototal analysis system (μ-TAS) field during the last decade. However, alternative concepts are being developed with increasing pace and most of these alternatives circumvent transport challenges that have not yet been solved by the EOF approach. We describe some of these challenges in this chapter. In addition, we describe some approaches such as the very promising technological approach to generate high-pressure flows on the chip. We also describe some of our own contributions using centrifugal force to drive separations in parallel as well as some other transport modes that are still in the research field. We have focused on transport modes that have found an application in the field of on-chip separations. However, some new and promising concepts, not yet mature for separation applications, have also been included.

5.2   EOF-BASED LIQUID TRANSPORT

5.2.1   Introduction to EOF

EOF has played a crucial role in the development and maturation of microanalytical systems. A vast majority of the earliest chip-based separation applications used EOF to transport fluid within the microfabricated structures.[1] So why was EOF chosen as the main driving force? First, the fluid transport mechanism involved in EOF is based on an applied electric field and the surface characteristic of the microchannel itself. Voltage switching can alter the fluid flow direction, thus eliminating the need for moving parts such as mechanical valves. Second, there is no need for pressure connections between the chip and an external source such as a pump. Basically, a high voltage power supply and two or more electrodes are all that is needed to generate a flow of fluid within the structures, so EOF readily lends itself to applications within the μ-TAS area.

Another reason for the popularity of EOF for chip-based separations is the separation efficiency that can be achieved. While the flow profile of a pressurized flow in a laminar flow region has a parabolic profile, EOF is characterized by a plug-shaped flow profile. This difference in flow profile reduces longitudinal analyte spreading, making high-resolution separations possible in relatively short channels. This is important as chip-based systems originally were regarded as devices capable of achieving high separation power in a short time and on a small footprint area. The difference in flow profile between EOF and pressure-driven fluid transport was visualized in work by Paul et al.[2] Figure 5.1 shows the flow profiles for pressure-driven and EOF flow, recorded using a technique involving a photoactivated fluorescent dye. The dye is activated close to the point of detection, resulting in a sample plug shape with minimum sample dispersion.

Flow profiles in plastic microchannels have been studied in a similar way as described above.[3] Microchannels having both the channels and the cover material made of poly(dimethylsiloxane) (PDMS) generated a similar plug-like flow profile as shown in Figure 5.1 for a fused silica capillary.

5.2.2 EOF IN MICROCHANNELS

In order to take full advantage of the integrated nature of interconnected channels on a μ-TAS device it is essential to be able to control the fluid flow. EOF has the advantage of "direct control" and a fast response time. As described in more detail in Chapter 2, EOF is generated by applying high

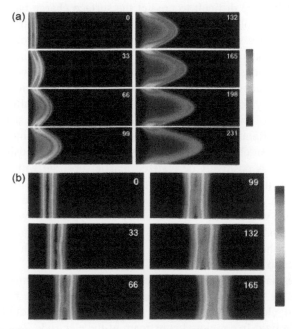

FIGURE 5.1 (See color insert following page 208) Images of (a) pressure-driven flow through an open 100-μm-i.d. fused-silica capillary using a caged fluorescein dextran dye (pressure differential of 5 cm of H_2O per 60 cm of column length, viewed region 100 μm by 200 μm) and (b) electrokinetically driven flow through an open 75-μm-i.d. fused-silica capillary using a caged rhodamine dye (applied field of 200 V/cm, viewed region 75 μm by 188 μm). The frames are numbered in milliseconds as measured from the uncaging event.

voltage along a capillary or channel. A flow of fluid is formed as the electric field interacts with the mobile ions present in the electric double layer adjacent to the charged channel surface. These mobile ions move toward the electrode with the opposite charge, dragging with them the bulk fluid. It is thus easy to understand that the nature of the surface in the channel as well as the magnitude of the applied electric field are key elements in EOF. One of the first attempts to achieve controlled flow was to apply active voltage control to all reservoirs on the chip.[4] To reduce the number of power supplies required in such a setup, the channel itself is designed to give a suitable voltage division.[5]

Since EOF is a "surface-driven pump," another approach to control the flow is to control the surface of the channel. A large number of channel coatings have been presented in the literature. Some have been utilized to reverse the EOF in glass or fused silica chips while others have been used to minimize the differences in surface charge of the chip substrates. The latter is especially common when devices have been made from polymeric material.[6,7] In glass or quartz, where the bulk material is well characterized, the surface charge and thus EOF may be predicted. In plastic materials, on the other hand, it is more difficult to predict the surface charge based solely on the structure bulk polymer since additives are commonly introduced in the manufacturing process.[7] In these materials it is thus more important to have a reliable and reproducible surface chemistry in order to control the EOF. Surface modifications can be used to generate different flow directions in different intersecting channels and even different flow directions within the same channel. Polyelectrolyte multilayers have been used to alter the surface charge and control the direction of flow in polystyrene and acrylic microfluidic devices.[8] Relatively complex flow patterns were obtained by derivatization of different channels within a single device with oppositely charged polyelectrolytes. A positively derivatized plastic substrate with a negatively charged lid was used to achieve top–bottom opposite flows, while derivatization of the two sides of a plastic microchannel with oppositely charged polyelectrolytes was used to achieve side-by-side opposite flows.[8] In addition to chemical modifications, the EOF can be altered by applying external voltages along the channels.[9,10] Figure 5.2 shows an approach using embedded electrodes to alter the EOF.[9]

Electrodes have also been incorporated into the microchannel by patterning them on the cover material. By applying a high voltage between electrodes placed close to each other in the microchannel, liquid flow was generated in a field-free region downstream of the electrodes (see the section on electroosmotic pumping below).[10]

EOF is changed permanently by chemical modification of a surface. Applying an external voltage may, from this perspective, be viewed as a temporary form of EOF control, that is, there are no permanent changes made to the microchannel so the EOF control can be activated as required. In some instances it may be desirable to have both temporary and spatial control of

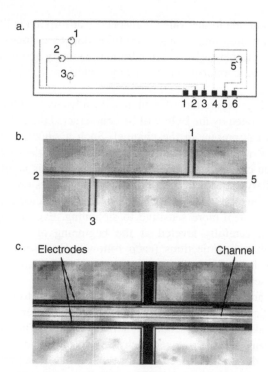

FIGURE 5.2 (a) Schematic of the CE microdevice equipped with external voltage electrodes: 1, injector waste port; 2, separation port inlet; 3, sample inlet port/injector port; 4, external voltage electrode; 5, separation waste port; 6, external voltage electrode. Total size of the structure: 2.54 cm × 7.62 cm. (b) Micrograph of the injection (150 pl) zone within the separation channel. The sample is injected from reservoir 3 to reservoir 1 and separated from reservoir 2 to reservoir 5 using applied voltage fields. (c) Micrograph of external voltage electrodes on the CE microdevice. Electrodes are placed 50 μm away from the surface of the channel (30 μm wide × 10 μm deep).[9]

the EOF within a channel, using the temporary control to switch the EOF on and off. A number of promising approaches, although not used for separation applications, have been presented in the literature. An approach was to apply a thin layer of UV-sensitive TiO_2 to the channels and used light modulation to change the zeta potential of the surface.[11] With this approach the magnitude of the change in fluid velocity was 7 to 242 mm/sec. The largest effect was seen at a buffer pH of 4.7 (higher or lower pH values resulted in smaller changes).[11]

5.2.3 EFFECTS OF GOING SMALL

A large surface-to-volume ratio within the microchannel is an inherent characteristic of a chip-based separation system and increases the risk of nonspecific adsorption to the walls. Nonspecific adsorption is likely to cause unwanted sample loss and changes in EOF due to changes in the zeta-potential of the channel surface.[12] In addition, capillary action and meniscus effects will be the dominant forces in any system with a large surface-to-volume ratio.

In terms of high separation efficiencies, the advantage of the plug-shaped flow profile generated by the EOF will be counteracted if a parabolic flow (in either direction) is present in the channel. Such a situation may occur by siphoning, for example, if the fluid in the inlet and outlet reservoirs is not leveled or if the chip is not positioned horizontally. This is, of course, also true for a capillary system but the smaller liquid volumes used in chip-based separation and the more complex network of interconnected channels on the chip make this system more sensitive to siphoning. Even though the fluids may have been carefully leveled at the beginning of an experiment or analysis, less than ten injections (each followed by a separation taking a few minutes or less) may be enough to create significantly different liquid levels in the reservoirs on the chip.

Separation anomalies in EOF-driven separations due to pressure effects have been studied in detail by Crabtree et al.,[13] who observed a significant flow of fluid also in the absence of a high voltage. The effects of this unwanted pressure-driven flow were a drifting baseline and increased migration times for a dye that was used as model substance. The pressure-induced flow rates were of the same order of magnitude as the generated EOF flow (ca. 0.5 mm/sec) and 10 to 20 times higher than would be predicted from siphoning alone. It was suggested that meniscus effects (i.e., capillary action) made a significant contribution to the pressure-induced flow observed in the microchannel. Similar separation anomalies were also observed when performing consecutive injections of explosive compounds.[14] In this case indirect laser-induced fluorescence was used for detection and the separation was carried out by micellar electrokinetic chromatography. An increase in migration time and a drifting baseline were observed after more than seven consecutive injections. The conclusion made from the observed results was that electrolysis, occurring in the electrode reservoirs, caused significant pH changes in the separation buffer, thus altering the velocity of the EOF. The authors suggested the use of divided reservoirs, as shown in Figure 5.3, where the fluids going into the channels are never in direct contact with the high-voltage electrode. A "sacrificial buffer" and a plug of porous glass were used to protect the separation buffers and the sample.

Effects originating from buffer depletion or liquid evaporation and meniscus effects such as capillary action will be more pronounced in chip-based systems than in larger-scale systems (e.g., capillary-based systems). A capillary electrophoresis (CE) system utilizes liquid volumes in the order of $= 1$ ml. The liquid volume in a chip-reservoir is commonly $= 100 \, \mu l$. Precautions also have to be taken not to cause sample degradation by electrochemical reactions

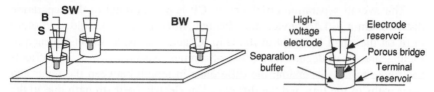

FIGURE 5.3 Schematic and expanded view of the separation chip including the divided reservoirs and the porous elements. Reservoirs; buffer (B), buffer waste (BW), sample (S), sample waste (SW).[14]

in the sample reservoir. This effect will be more pronounced for small sample volumes or when a countervoltage is applied to the sample reservoir during the separation step. Degradation may occur as a result of direct electrochemical reactions of the sample or by an indirect effect caused by pH changes in the sample reservoir.

5.2.4 SAMPLE DISPERSION

The generation of high separation efficiencies (i.e., minimal band broadening) on a small footprint device such as a microchip relies on minimal sample dispersion within the system. Sample dispersion may originate from a number of sources including unwanted pressure flow (as discussed above), dispersion at the point of injection or during the separation, or at the point of detection. This section will discuss a number of reports that have studied different causes of band broadening on microchip devices in which EOF has been used as the driving mechanism (see also Chapter 2 for more on [sample] dispersion). The total sample dispersion in the system will be the sum of all the dispersions caused by the different components. An argument that has been used to show the advantages of a chip-based system is the potential to create zero dead volume connections between channels. Thus, using a microchip, coupling devices such as those used in capillary-based systems are replaced by an integrated channel network. Nevertheless, the individual sources of dispersion have to be considered also in a microchip separation system.

In the early days of separations on a chip, injection was studied in detail since a crucial part of injection is the generation of small, discrete injection plugs. It was found that active voltage control in all the intersecting channels was necessary to create discrete sample plugs and to avoid leakage of sample during the separation.[4,15,16] A pinched injection mode was introduced[4] to create well-defined sample plugs. A widely used approach, the method includes countervoltages applied at the separation channel inlet and waste during sample loading and at the sample inlet and waste reservoir during the separation. This results in well-defined sample plugs during sample loading and a small back-flow into the injection channels during the separation.

The overall separation efficiency in CE is proportional to the separation distance. On a μ-TAS device, the channels often have to be folded in order to fit on the small footprint of the chip. Channel turns, as a consequence of folding, are a well-documented cause of sample dispersion.[4,17–20] Two effects contribute to the dispersion: the difference in length between the inner and outer paths of the turn and the difference in electric field strength due to the dissimilarity in path length. Figure 5.4 illustrates an analyte band traveling through a turn. In systems utilizing a pressure-driven flow the effect of a turn will be less significant as only the difference in travel distance will cause dispersion.

A number of approaches to minimize sample dispersion in folded channels have been presented. Most of these include various modified turn geometries.[17–19] For molecules with sufficiently small diffusion rates, an additional turn (in the opposite direction to the first one) can be included in the channel design.[17] However, for this solution to be effective, the second turn must be introduced shortly after the first turn, that is, before the analyte molecules have had time to diffuse across the channel. Another approach is to selectively increase the flow rate in the outer radius of the turn, thus creating comparable velocities in the inner and outer parts of the turn.[20] This approach was shown in a poly(methyl methacrylate) (PMMA) channel, where treatment with a UV excimer laser below the ablation threshold was used to increase the surface charge and thus the zeta potential of the outer channel wall. An

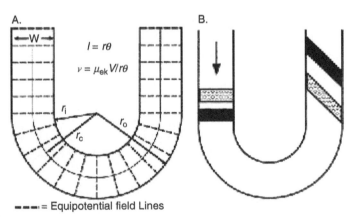

A.

$l = r\theta$

$v = \mu_{ek} V/r\theta$

r_i

r_o

r_c

B.

━ ━ ━ = Equipotential field Lines

FIGURE 5.4 (A) The distance (l) that an analyte molecule travels around a turn and the field strength experienced in a turn depend on the molecule's radial position. (B) When the diffusion time across the channel is slow compared to the transit time around the turn (t_D/t_t large), excess dispersion is introduced by the turn as the molecules situated along the inside of the turn move more quickly through the turn than those along the outside. This results in the parallelogram analyte band shape seen. The molecules in these distorted bands will diffuse and redistribute themselves across the channel width, restoring the rectangular shape of the band further down the separation channel.[17]

increase in EOF of 4% was observed due to this treatment. Compared to a nontreated 90° turn, the laser-modified turn resulted in an average decrease in band broadening of ca. 40%.

In some chips the channel and cover material are not the same. This is especially true for plastic microchannels where a cover plate made of glass is commonly used. Due to the differences in EOF velocity, this design can generate quite severe sample dispersion. When sample dispersion was compared in a chip made entirely of PDMS and a chip where the channels were made of PMMA and the cover material was PDMS, it was found that the latter generated a flow profile deviating from the expected plug flow.[3]

5.2.5 EOF in Open Channels and in Packed Beds

EOF can be used as a pumping mechanism in both open channels and channels packed with beads. In packed channels and in channels with a very small cross section the thickness of the double layer has to be considered, as a double layer overlap could reduce and even eliminate EOF.[21] In a system where the channel dimensions approach the thickness of the double layer, the flow rate will decrease nonlinearly with the cross-sectional area.[22]

Capillary electrochromatography is the common terminology for EOF-driven separations using a chromatographic stationary phase (i.e., the separation is not dependent only on electrophoretic mobility). The stationary phase may be coated on the channel walls or be present in a bulk material packed or polymerized *in situ* in the channels.[23–25] For chip-based applications several issues are raised when packing beads into the channels: the mechanism of the packing (pressure or electrokinetic packing), how to create a restriction (frit) to hold the particles in place, and how to get uniform packing in the channels (especially in curved channels). An alternative is to utilize *in situ* polymerization or a continuous bed. Such an approach was reported by, for example, Ericson et al.,[23] who showed comparable van Deemter curves for a capillary and a microchannel of the same effective length, the same channel–capillary cross-sectional area, and filled with the same continuous bed material.

5.2.6 Electrokinetic Pumping

EOF has been used not only as a direct transport mechanism, but also to create pressure and thus a pressure-driven flow. The use of EOF to create pressurized flow was discussed more than three decades ago.[26,27] Lately, two strategies have been shown: (a) using EOF created in packed beds to generate high pressures and thus create a fluid flow and (b) using EOF generated in open channels[28] to facilitate a fluid flow. The latter approach is not capable of generating high pressures. Paul et al.[29] have shown the use of packed capillaries to get pressures of ≥10 psi/V. The technology is also utilized by the company Eksigent (www.eksigent.com), who use electrokinetic flow control in a variety of application areas. The critical issue in the use of EOF to create a pressure-driven flow is to minimize the loss of fluid flow due to back-flow through the pump itself. In the case of a packed capillary the back-pressure can be made sufficiently high by using densely packed particles or nonporous material. In a similar pump, using open channels, the back-pressure can be maximized by narrowing the channels. This was demonstrated by Lazar et al.[28] using an array consisting of 1 to 100 narrow (2.5 μm wide, 1 to 6 μm deep), open channels. Figure 5.5 shows the design of this pump.

Using the design described in Figure 5.5, flow rates of 10 to 400 nl/min and pressures up to 80 psi could be generated. In addition, eluent gradients were created by simultaneously using two pumps. In this first application the pump was used to feed an electrospray ionization mass spectrometer with peptide samples.[28] A similar approach to an open-channel electroosmotic pump has been described from a theoretical point of view.[30] This pump design may have a great potential, but is a complex system requiring careful calculations prior to the actual chip design and manufacturing.

As mentioned above, electrodes placed in direct contact with fluid in only a section of the microchannel can be used to generate fluid flow in a field-free region.[10] Electrodes patterned on a glass cover plate in combination with microchannels cast in PDMS demonstrate this approach. The channels are fabricated in PDMS, a gas-permeable polymer, to minimize the accumulation of gas bubbles generated at the electrodes. However, care must be taken not to

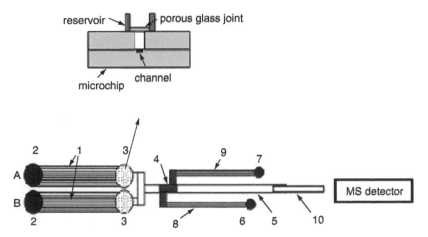

FIGURE 5.5 Diagram of the microfabricated electroosmotic pumping system: (1) open channel electroosmotic pump, (2) micropump inlet reservoir, (3) micropump outlet reservoir, (4) double-T sample injection element, (5) channel for sample infusion or separation, (6) sample inlet reservoir, (7) sample waste reservoir, (8) channels for sample inlet, (9) channels for sample outlet, and (10) ESI emitter. The inset shows an expanded view of the micropump outlet reservoir containing the porous glass disk.[28]

apply too high voltages. The generated current must be kept low enough so that the diffusion rate of the electrolytic gaseous products through the PDMS is faster than the formation of the bubbles. Application of too high voltages may result in macroscopic bubbles, affecting fluid flow by disrupting the electrical contact between the electrodes or changing the hydrodynamic conditions in the channel. The linear velocity in the field-free region was measured as approximately 0.03 to 0.07 mm/sec for three different chips. Although a relatively large chip-to-chip and run-to-run variation was observed, the results clearly showed the potential to generate fluid flow by the use of electrodes placed inside the microchannels.[10]

Surface coating of selected channels may also be used to obtain electro-osmotic pumping.[31] The design of such a chip is shown in Figure 5.6. A viscous polymer was selectively coated in the ground channel, resulting in a dramatic reduction in EOF compared to the EOF in the separation channel. This difference creates an overpressure at the intersection and thus fluid is pushed out into the field-free region.

5.2.7 Combining EOF and Hydrodynamic Flow

Since EOF and pressure-driven flow both have their advantages and disadvantages; a combination of the two is an interesting option for μ-TAS applications. This concept has been investigated by, for example, Caliper Life Science (www.caliperls.com) (California, USA). Hydrodynamic flow is

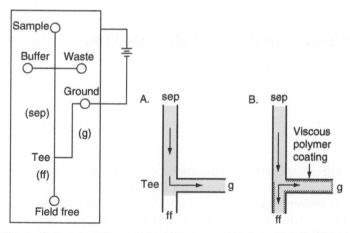

FIGURE 5.6 Schematic of the chip used for ion transport experiments. The potential is applied between the sample and ground reservoirs. The separation channel (sep), the ground channel (g), and the free-field or floating channel (ff) are indicated on the diagram. The enlarged drawings show the schematic of the fluid flows through the "tee" intersections at which the electroosmotic velocity in the ground arm (g) is the same (A) or less than (B) that found in the separation (sep) channel arm.[31]

used to transport sample plugs to the separation and reaction channels, where an electrophoretic separation is performed based on differences in electrophoretic mobility. This methodology prevents sample bias as found when using certain electrokinetic injection methods.[32] The combination of pressure and EOF driven flow is very attractive as the injection can be made with a minimum of sample discrimination and the separation can be made in the most efficient way. However, to be able to generate high-quality results, it is important to understand well the electroosmotic and hydrodynamic fluid movements within the microchannel network. A computational model is recommended (not to say necessary) in order to simulate and visualize the flows generated within the channels and intersections.

5.2.8 SUMMARY

In summary, EOF-driven flow is still the most widely used mode of transport in chip separations. However, as flow rate is dependent on the surface of the microchannels as well as the nature of the liquid being pumped, linear velocities must be carefully monitored and internal standards used to account for any changes. Using EOF to create a pressure-driven flow is an interesting technique for the future. As yet, this method has not been used for separation purposes on chip, but rather to generate a pressure-driven flow for separation in capillary format or to transport a liquid to a specific detector (e.g., electrospray MS). A combination of EOF and pressure-driven flows may also be a future option as this approach takes advantage of the best features of both techniques.

5.3 PRESSURE-DRIVEN LIQUID TRANSPORT

5.3.1 INTRODUCTON

As yet, very few pressure-driven separation systems have been developed for planar microstructures. This may, in part, be due to the dominant role of electroosmosis-driven separation systems or other reasons such as the following:

1. Little or no gain in performance compared to the capillary tubing approaches developed during the last 20 years
2. The challenge of low dead volume, high-pressure couplings to the chip
3. The challenge of introducing a stationary phase into the chip
4. The challenge of generating a continuous liquid elution gradient to the column.

The first pressurized μ-TAS system presented was the Stanford gas chromatograph in the mid-1970s (which is described in Chapter 9 of this book). In this section, we will focus on liquid-phase separations that have become increasingly important during the past decades, driven by the rapid and strong development in the life sciences.

5.3.2 COUPLING AN EXTERNAL PUMP TO THE CHIP

Ocvirk et al.[33] were one of the first to demonstrate a packed liquid chromatography bed in a chip. A packed small-bore column placed against a frit was used with optical detection. However, one obstacle with on-chip high-pressure separation is that the injection of sample into the high-pressure zone requires valving from the low-pressure section to the high-pressure column. O'Neill et al.[34] have developed an injection system able to deliver picoliter sample plugs for miniaturized liquid chromatography.

Open tubular separation where the stationary phase is present as a thin film on the wall of the tubing was first demonstrated on a chip by Jacobson et al.[35] Although this approach was driven by EOF it showed that the low sample capacity leading to high sensitivity demands of the detection unit was not solved. Until now, there have been few arguments as to why a chip approach should outperform the separation system based on capillary tubing. However, if multiple processes could be integrated on the chip, advantages might be demonstrated. Work leading in this direction has been published by the Harrison group. Large-volume samples have been captured on a micro-column incorporated on a chip.[36] The chip also included a separation channel and an interface to electrospray MS.[37] There are only a limited number of publications in this field to date and we believe that gains in performance, throughput, and integration must be shown in order to expand the application.

5.3.3 CENTRIFUGAL FORCE-DRIVEN SEPARATIONS

The growing "omics" areas such as genomics and proteomics place a heavy demand on obtaining more information from each analysis. Although products are available commercially (www.gyros.com), there are few publications in the field of miniaturized separations using centrifugal force.[38–40] We have therefore included some details of our own published and unpublished work in this section. In our work, we have addressed throughput issues by developing a parallel centrifugal analysis platform in a compact disk (CD) format. Figure 5.7 shows the first commercially available CD, released in May 2002. Digests from 96 protein samples are added to the CD, concentrated on 10-nl RPC columns, desalted, eluted, and crystallized on the CD for subsequent MALDI-MS analysis. The gold layer is present for mass spectrometric purposes.

Conventional centrifugal chromatography was developed decades ago and used for semipreparative purposes, but has not been widely used. One reason could be that the potential for multiple separations run in parallel was never demonstrated in a practical way. A miniaturized centrifugal analyzer for parallel sample analysis was developed in 1973 by Scott and Burtis.[41] Liquid in the 10- to 100-μl range was added close to the center of a CD-sized

FIGURE 5.7 Gyrolab MALDI SP1 CD from Gyros AB.

plastic disk and processed outward by spinning the disk. Similar technology has been used by the company Abaxis who, in the 1990s, successfully developed a diagnostic disk that used centrifugal force for liquid transport (www.abaxis.com). Kellogg and Madou extended the microliter concept by the use of geometrically defined microfabricated burst valves.[42,43] Centrifugal microfluidic technology suited to process discrete nanoliter volume as defined by Burn's group,[45] was first presented in 2000 in the work led by Andersson at Amersham Pharmacia Biotech, Sweden.[44] Work has continued under the company name of Gyros AB, Sweden. Discrete nanoliter volume metering combined with centrifugal force results in a fluid transport system that has several advantages:

1. Miniaturization benefits: Reduced sample and reagent consumption, a high number of parallel structures on one CD.
2. Simple automation: Processing different functional steps often only requires a revolutions-per-minute (rpm) shift.
3. Easy parallel sample processing since no connections to the disk are necessary.
4. On-disk liquid volume definition eliminates the need for precise sample transfer, a challenge in the nanoliter range, and so less complex instrumentation is required.
5. Spinning the disk over a stationary detector simplifies and speeds up the optical detection of reactions in parallel columns.

A potential drawback can be that it is not easy to generate high pressures.

To obtain high-resolution separations, strategies such as the use of smoother solvent gradients, smaller particles, longer columns, monolithic columns, or multidimensional chromatography are all viable. However, many applications in life sciences require only basic sample clean-up and limited separation power for which the parallel centrifugal separation system is already well suited.

5.3.4 FLOW GENERATION BY CENTRIFUGAL FORCE

The theory of flow generation using centrifugal force was described by Kellogg et al.[42] Figure 5.8 shows the basic principle for flow generation when moving a liquid from chamber 1 (r_0) to chamber 2 (r_1).

For a liquid flowing through a channel from a reservoir, the average velocity of the liquid (U) and its volumetric flow rate (Q) depend on the rheological properties of the liquid, the size, location, and configuration of the channels, and the rate of rotation, through Equations (5.1) and (5.2) (see further below), where ρ and η are the density and viscosity of the liquid, respectively, A is the cross-sectional area of the channel, d_H is the hydraulic diameter of the channel (defined as $4A/P$, where P is the perimeter of the channel), L is the length of the channel, ω is the angular velocity, Δr is the average distance of the liquid in the channels from the center of the disk, and $\bar{r}$ is the radial extent of the fluid subject to centrifugal force. It is convenient to define Δr and $\bar{r}$ in terms of r_0, r_1, and H, the head of the liquid in the reservoir that feeds the channel, such that $\bar{r} = (r_1 + (r_0 - H))/2$ and $\Delta r = r_1 - (r_0 - H)$. The geometric parameters are shown schematically in Figure 5.8:

$$U = d_H^2 \rho \omega^2 \bar{r} \Delta r / 32 \eta L \tag{5.1}$$

$$Q = UA = A d_H^2 \rho \omega^2 \bar{r} \Delta r / 32 \eta L \tag{5.2}$$

The flow generation principle is valid also for micrometer (nanoliter) systems. However, when working with dimensions suited for nanoliter volumes, performance is affected by scaling laws. For example, to introduce the liquid to the nanoliter-sized channels on the CD, capillary action must be the preferred means to fill the channels. A pressure connection could be introduced to push or pull the liquid through the device, but this would require hundreds of connections to the CD. We therefore use oxygen plasma to make the surface of the CD (made from polycarbonate or Zeonor) hydrophilic. Capillary action drives the liquid under the lid into the channel system. To

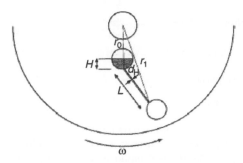

FIGURE 5.8 Geometric parameters for flow rate consideration in centrifugal fluidics.

prevent the liquid from filling the entire channel system we again alter the surface by introducing hydrophobic breaks. To pressurize the stopped liquid, we alter the local gravity field through spinning the CD. The combination of g-force (or gravity) and surface forces used in this concept is illustrated in a paper from Wente in 1968 called "The Size of Man."[46] The slopes of gravity-dependent and surface-dependent forces (called molecular forces) are plotted in a force vs. size graph. There is a point of size corresponding to 1 mm at which surface forces start to dominate.

One way to alter this situation and allow gravitational force to dominate even micrometer-sized events is to change the local gravity (e.g., by spinning a CD) until gravity becomes the dominant force. We used this strategy to develop a chromatography system. We use the dominant surface forces to load and restrict the movement while the CD is at rest. By spinning the disk we perform functions within the generated liquid transport system such as volume definiton, injection, and elution. Figure 5.9 shows an example of a parallel chromatography system that can be incorporated on a CD. A commercially available version has 112 of these systems in one CD (www.gyros.com). Samples are loaded by capillary action, filling up the channel system until the liquid reaches the hydrophobic breaks.

In order to define the volume that is to be applied onto the column, the disk is rotated to generate enough pressure to overcome the resistance of the hydrophobic break placed furthest out on the disk radius (Figure 5.9). The volume in the volume definition chamber is maintained since it has a shorter liquid height and so generates a lower pressure at the second hydrophobic break. A second rotation at higher rpm passes the defined volume onto the

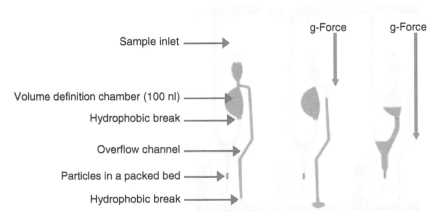

FIGURE 5.9 Exact volume definition, ranging from 5 nl to 5 μl, takes place within a CD channel system. The overflow channel is activated at a low spinning frequency, removing excess liquid. The defined liquid is moved on through a packed bed by increasing the spinning frequency.

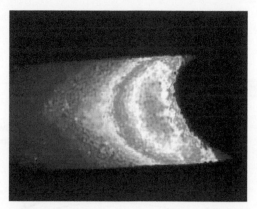

FIGURE 5.10 (See color insert) Close-up of a chromatography column in a CD channel. Flow direction is from right to left. Picture is from a fluorescence microscope.

column. The volume definition process in the example shown in Figure 5.9 defined 100-nl plugs. In spite of imprecise liquid transfer to the CD, the measured precision of the within-CD volume definition is typically better than 1.0% in standard deviation.

The columns are formed by volume definition of a slurry of 15-μm chromatography beads. The beads are packed into columns against a shallow (10 μm) part of the channel. In the resulting chromatography system, a sample concentration effect of 100 times can easily be achieved as seen in the microscope image in Figure 5.10, where 200-nl sample solutions have been captured in the first 2 nl of the column bed.

In this system, detection is performed by fluorescence on-column. Additional information about the performance of the separation system can be derived from the sample elution profile on the column. The relatively short columns often used in the CD (0.3–5 nm) limits the separation power in isocratic elution mode. To improve the separation power, gradient elution mode is frequently used.

To generate a continuous gradient in parallel systems and microstructures, we developed a flow-through dispenser system that feeds the parallel separation columns simultaneously with a continuous solvent gradient.[40] Figure 5.11 shows the system setup.

A chromatographic pump generates a gradient through the flow-through dispenser. Each time the CD completes a revolution, a controller triggers a train of pulses to actuate the dispenser. Thus, each droplet formed is dispensed to an individual CD microstructure and, by using a triggered train of pulses, several structures are fed quasi-simultaneously. The flow that enters the structures is controlled by balancing the angular velocity of the CD with the triggering frequency (e.g., one trigger every two revolutions).

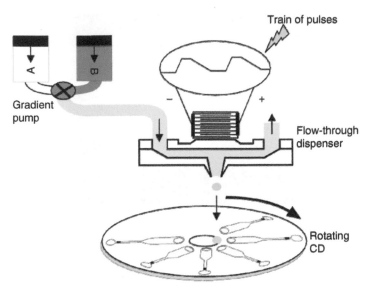

FIGURE 5.11 Solvent gradients are formed in parallel inside the CD by dispensing a solvent gradient using a flow-through dispenser triggered by the position on a rotating CD.

The gradient profile coming from the dispenser is shown in Figure 5.12 for water–acetonitrile (ACN).

Using this parallel feeding system, potentially 100 gradient separations can be achieved. However, only one example of a single gradient separation has been shown to this date.[40] An example of a stepwise gradient separ-

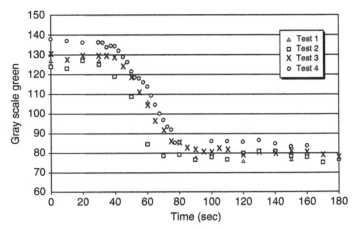

FIGURE 5.12 Four sequential gradient profiles coming out of the dispenser measured by an indicator added to the water phase.

| 17.5% | 20% | 22.5% | 25% | 27.5% |

FIGURE 5.13 (See color insert) Stepwise elution of Cy5-labeled angiotensin I from Cy5 dye. Pictures show the early stage of the elution as free Cy5 dye separates from Cy5-angiotensin I. The eluent flows from right to left in the pictures. ACN concentrations are given as a percentage under the respective elution.

ation is shown in Figure 5.13. Here, incremental changes of 2.5% ACN lead to separation between Cy5 and the Cy5-labeled peptide. Each addition of 200 nl by the dispenser is equal to 10-column volumes of isocratic elution.

5.3.5 SUMMARY

Overall, parallel, miniaturized, centrifugal chromatography is useful for applications where limited separation power is required and a high throughput is desirable. Future developments hold the promise of high-throughput gradient separations and better resolution using monolithic columns with a low-pressure drop.

5.4 OTHER MEANS OF LIQUID TRANSPORT

Other methods for generation of liquid flow in microfluidic systems include capillary forces,[47] electromagnetic forces,[48,49] evaporation,[50] gravity,[51] and shear-driven flow.[52,53] The number of applications where these different methods have been used for separation purposes, and especially high-performance separations is, however, limited.

Electromagnetic forces can be used to pump liquids as long as the liquids are at least slightly conductive. The flow is generated due to forces acting upon a liquid placed in a channel where two sides of the channel are patterned with electrodes, or consist of electrodes, and the channel is placed in a magnetic field. When a voltage is applied to the electrodes, an electrical current will flow from one electrode to the other resulting in the generation of Lorentz forces directed perpendicular to the plane formed by the magnetic field flux and the electrical current (Figure 5.14). Coupled AC electric and magnetic fields can be used to reduce hydrolysis.[48] High linear flow rates have been shown for saline solutions using this so-called magnetohydrodynamic pumping. The pressure generated by the hydrodynamic pump depends on the intensity of the magnetic field and the voltage applied across the electrodes as well as the length of the channel used for pumping. Long, meandering channels may thus be used to create substantial pressures.[49]

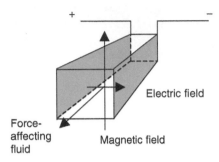

FIGURE 5.14 Principle of magnetohydrodynamic pumping: when a channel with two opposing sides acting as electrodes is placed in a magnetic field, a force is generated that can act as a driving force to propel liquid through the channel.

Shear-driven flow has been proposed for the propulsion of liquid in microchannel separation systems.[52,53] The concept is based on the movement of liquid in a channel situated between a stationary wall and a moving wall (Figure 5.15). Drag forces will generate a flow in the channel having a mean velocity equal to half that of the moving wall. There is no pressure drop along the length of the channel making it possible to use shear-driven flow in long channels with very small diameters or cross-sectional areas. Such channels would be beneficial in open tubular chromatographic systems. Channels with dimensions of 20 mm × 4 mm × 0.8 μm have been studied for potential use in chromatographic separations. The possibility of using a wide, but thin, channel may also facilitate low detection limits using the whole width of the channel as the detection cell length. The moving barrier does not need to be pressed against the stationary barrier with great force as there is no pressure generated (see also Chapter 6 for more on this subject).

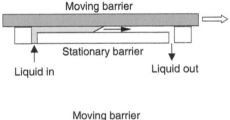

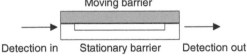

FIGURE 5.15 Principle of shear-driven flow: one side of the channel consists of a moving substrate that propels liquid through the channel by viscous drag. The width of the channel is best exploited to create a sufficient path length for absorbance detection.

The lack of moving, mechanical parts or a high electric field strength is a great advantage for microfluidic systems relying on capillary forces for liquid propulsion. There are significant advantages if the analyses are being performed in remote locations or places where high-tech equipment is difficult to maintain. It is, however, difficult to achieve the pressure needed for most high-efficiency separation techniques. Capillary force-driven systems aimed at the separation of particulate material from whole blood by passing the liquid phase through a filter or membrane have been developed for semi-quantitative assays of glucose or high-density lipoprotein cholesterol (Ref. [47] and references cited therein).

Evaporation of a liquid can also be used to force a liquid to move through a microfluidic channel system.[50] A membrane evaporation pump has been developed for the generation of low, but constant, flow rates. The authors suggest that the pump may be used to support implantable sensors or micro-dialysis systems, but the possibility of using this pumping method for (high performance) separation systems is not discussed. High salt content or crude samples may lead to clogging of the air–liquid interface limiting the reliability and robustness of the technique.

Another passive pumping method is the use of gravity to generate a liquid flow. Diffusion-based separation systems of the T-sensor type have been shown in which a crude sample containing red blood cells is allowed to flow alongside a reagent solution and a solution containing a known reference sample.[51] Diffusion of sample molecules from the sample and the reference solution into the reagent solution generates zones where measurement of the relative concentrations of the analyte of interest can be determined. The effect of forces influenced by, for instance, surface tension will increase rapidly when dimensions are reduced as the gravitational force scales with the volume of the liquid. This may result in a lower practical volumetric limit for the use of gravity as a propulsion method. For more on the T-sensor-type approach, see also Chapter 7.

REFERENCES

1. Dolnik, V.; Liu, S.; Jovanovich, S. *Electrophoresis* 2000, *21*, 41–54.
2. Paul, P. H.; Garguilo, M. G.; Rakestraw, D. J. *Anal. Chem.* 1998, *70*, 2459–2467.
3. Ross, D.; Johnson, T. J.; Locascio, L. E. *Anal. Chem.* 2001, *73*, 2509–2515.
4. Jacobson, S. C.; Hergenroder, R.; Koutny, L. B.; Warmack, R. J.; Ramsey, J. M. *Anal. Chem.* 1994, *66*, 1107–1113.
5. Jacobson, S. C.; Ermakov, S. V.; Ramsey, J. M. *Anal. Chem.* 1999, *71*, 3273–3276.
6. Barker, S. L. R.; Tarlov, M. J.; Canavan, H.; Hickman, J. J.; Locascio, L. E. *Anal. Chem.* 2000, *72*, 4899–4903.
7. Locascio, L. E.; Perso, C. E.; Lee, C. S. *J. Chromatogr. A* 1999, *857*, 275–284.

8. Barker, S. L. R.; Tarlov, M. J.; Ross, D.; Gaitan, M.; Locascio, L. E. *Anal. Chem.* 2000, *72*, 5925–5929.

9. Polson, N. A.; Hayes, M. A. *Anal. Chem.* 2000, *72*, 1088–1092.

10. McKnight, T. E.; Culbertson, C. T.; Jacobson, S. C.; Ramsey, J. M. *Anal. Chem.* 2001, *73*, 4045–4049.

11. Moorthy, J.; Khoury, C.; Moore, J. S.; Beebe, D. J. *Sens. Actuators B* 2001, *75*, 223–229.

12. Ghosal, S. *Anal. Chem.* 2002, *74*, 771–775.

13. Crabtree, H. J.; Cheong, E. C. S.; Tilroy, D. A.; Backhouse, C. J. *Anal. Chem.* 2001, *73*, 4079–4086.

14. Wallenborg, S. R.; Bailey, C. G.; Paul, P. H. *Micro Total Analysis Systems 2000, Proceedings of the mTAS Symposium, 4th, Enschede, Netherlands, May 14–18, 2000*, 2000, 355–358.

15. Fan, Z. H.; Harrison, D. J. *Anal. Chem.* 1994, *66*, 177–184.

16. Jacobson, S. C.; Culbertson, C. T.; Daler, J. E.; Ramsey, J. M. *Anal. Chem.* 1998, *70*, 3476–3480.

17. Culbertson, C. T.; Jacobson, S. C.; Ramsey, J. M. *Anal. Chem.* 1998, *70*, 3781–3789.

18. Griffiths, S. K.; Nilson, R. H. *Anal. Chem.* 2001, *73*, 272–278.

19. Paegel, B. M.; Hutt, L. D.; Simpson, P. C.; Mathies, R. A. *Anal. Chem.* 2000, *72*, 3030–3037.

20. Johnson, T. J.; Ross, D.; Gaitan, M.; Locascio, L. E. *Anal. Chem.* 2001, *73*, 3656–3661.

21. Cikalo, M. G.; Bartle K. D.; Myers, P. *J. Chromatogr. A* 1999, *836*, 35–51.

22. Jacobson, S. C.; Alarie, J. P.; Ramsey, J. M. *Micro Total Analysis Systems 2001, Proceedings mTAS 2001 Symposium, 5th, Monterey, CA, United States, October 21–25, 2001*, 2001, 57–59.

23. Ericson, C.; Holm, J.; Ericson, T.; Hjertén, S. *Anal. Chem.* 2000, *72*, 81–87.

24. Ngola, S. M.; Fintschenko, Y.; Choi, W.; Shepodd, T. J. *Anal. Chem.* 2001, *73*, 849–856.

25. Singh, A. K.; Throckmorton, D. J.; Shepodd, T. J. *Micro Total Analysis Systems 2001, Proceedings mTAS 2001 Symposium, 5th, Monterey, CA, United States, October 21–25, 2001*, 2001, 649–651.

26. Rice, C. L.; Whitehead, R. J. *J. Phys. Chem.* 1965, *69*, 4017–4024.

27. Pretorius, V.; Hopkins, B. J.; Schieke, J. D. *J. Chromatogr.* 1974, *99*, 23–30.

28. Lazar, I. M.; Karger, B. L. *Anal. Chem.* 2002, *74*, 6259–6268.

29. Paul, P. H.; Arnold, D. W.; Neyer, D. W.; Smith, K. B. *Micro Total Analysis Systems 2000, Proceedings of the mTAS Symposium, 4th, Enschede, Netherlands, May 14–18, 2000*, 2000, 583–590.

30. Morf, W. E.; Guenat, O. T.; de Rooij, N. F. *Sens. Actuators B* 2001, *72*, 266–272.

31. Culbertson, C. T.; Ramsey, R. S.; Ramsey, J. M. *Anal. Chem.* 2000, *72*, 2285–2291.

32. Kerby, M. B.; Spaid, M.; Wu, S.; Parce, J. W.; Chien, R.-L. *Anal. Chem.* 2002, *74*, 5175–5183.

33. Ocvirk, G.; Verpoorte, E.; Manz, A.; Grasserbauer, M.; Widmer, H. M. *Anal. Methods Instrum.* 1995, *2*, 74–82.

34. O'Neill, A. P.; O'Brien, P.; Alderman, J.; Hoffmann, D.; McEnery, M.; Murrihy, J.; Glennon, J. D. *J. Chromatogr.* 2001, *924*, 259–263.

35. Jacobson, S. C.; Hergenroder, R.; Koutny, L. B.; Ramsey, J. M. *Anal. Chem.* 1994, *66*, 2369–2373.

36. Oleschuk, R. D.; Shultz-Lockyear, L. L.; Ning, Y.; Harrison, D. J. *Anal. Chem.* 2000, *72*, 585–590.

37. Li, J.; LeRiche, T.; Tremblay, T.-L.; Wang, C.; Bonneil, E.; Harrison, D. J.; Thibault, P. *Mol. Cell. Proteomics* 2002, *1*, 157–168.

38. Palm, A.; Wallenborg, S. R.; Gustafsson, M.; Hedstrom, A.; Togan-Tekin, E.; Andersson, P. *Micro Total Analysis Systems 2001, Proceedings mTAS 2001 Symposium, 5th, Monterey, CA, United States, October 21–25, 2001*, 2001, 216–218.

39. Gustafsson, M.; Hirschberg, D.; Palmberg, C. H. J.; Bergman, T. *Anal. Chem.* 2004, *76*, 345–350.

40. Jesson, G.; Andersson, P. *Micro Total Analysis Systems 2003, Proceedings mTAS 2003 Symposium, 7th, Squaw Valley, CA, United States, October 5–9, 2003*, 2003.

41. Scott, C. S.; Burtis, C. A. *Anal. Chem.* 1973, *45*, 327A–340A.

42. Duffy, D. C.; Gillis, H. L.; Lin, J.; Sheppard, N. F.; Kellogg, G. J. *Anal. Chem.* 1999, *71*, 4669–4678.

43. Madou, M. J.; Lua, Y.; Laib, S.; Kohb, C. G.; Leeb, L. J.; Wennerc, B. R. *Sens. Actuators A* 2001, *91*, 301–306.

44. Ekstrand, G.; Holmquist, C.; Orlefors, A. E.; Hellman, B.; Larsson, A.; Andersson, P. *Micro Total Analysis Systems 2000, Proceedings of the mTAS Symposium, 4th, Enschede, Netherlands, May 14–18, 2000*, 2000, 311–314.

45. Burns, M. A.; Johnson, B. N.; Brahmasandra, S. N.; Handique, K.; Webster, J. R.; Krishnan, M.; Sammarco, T. S.; Man, P. N.; Jones, D.; Heldsinger, D.; Mastrangelo, C. H.; Burke, D. T. *Science* 1998, *282 (5388)*, 484–487.

46. Went, F. W. *Am. Sci.* 1968, *56*, 400–413.

47. Cunningham, D. D. *Anal. Chim. Acta* 2001, *429*, 1–18.

48. Lemoff, A. V.; Lee A. P. *Micro Total Analysis Systems 2000, Proceedings of the mTAS Symposium, 4th, Enschede, Netherlands, May 14–18, 2000*, 2000, 571–574.

49. Zhong, J. Y.; Mingqiang, B.; Haim, H. B. *Sens. Actuators A* 2002, *96*, 59–66.

50. Effenhauser, C. S. H., Krämer, P. *Micro Total Analysis Systems 2001, Proceedings mTAS 2001 Symposium, 5th, Monterey, CA, United States, October 21–25, 2001*, 2001, 397–398.

51. Weigl, B. H. B. R.; Schulte, T; Williams, C. *Micro Total Analysis Systems 2000, Proceedings of the mTAS Symposium, 4th, Enschede, Netherlands, May 14–18, 2000*, 2000, 299–302.

52. Desmet, G. B.; Gino, V. *Anal. Chem.* 2000, *72*, 2160–2165.

53. Desmet, G. V.; Nico, B.; Gino, V. *Micro Total Analysis Systems 2000, Proceedings of the mTAS Symposium, 4th, Enschede, Netherlands, May 14–18, 2000*, 2000, 599–602.

6 Pressure-Driven Separation Methods on a Chip

Gert Desmet, Emil Chmela,
and Robert Tijssen

CONTENTS

6.1 INTRODUCTION

Compared to the large number of electrically driven (ED) applications, the number of pressure-driven (PD) on-chip separation systems is very small. For newcomers in the field, this might come as a surprise, because in the macroscopic world, the application of a pressure gradient (mostly by means of a mechanical pumping system) is *the* method of choice for the propulsion of fluid flows. The lack of success is even more intriguing when considering that PD flows have a number of important advantages over ED flows for the conduction of chemical analysis: (i) the flow velocity in a PD system can be controlled much more accurately, as it is independent of the pH, electrolyte concentration, wall surface material, adsorption of large molecules onto the wall, composition of the sample matrix, etc.; (ii) the range of applicable solvents is much broader [1], allowing better use of solvent selection as an additional means to increase the separation selectivity and to enhance the detection sensitivity; (iii) when desiring to use an electrical detection method, there is no interference between the force field needed for the flow propulsion and the electrical field needed for the detection. Another advantage of PD flows is that they offer a broader choice for the selection of the substrate material for the etching of the microfluidic channels, whereas for ED flows it is imperative that the channel substrate material should have a very low electrical conductivity; for example, excluding the use of silicon-etched channels. This is very unfortunate, because, due to its widespread use in the microelectronics industry, and due to its crystalline structure, the micro-machining procedures for silicon are much more advanced than, for example, for fused silica, which is, together with a number of polymeric materials, the most suitable substrate for the generation of a large, stable electroosmotic force (EOF).

Considering all these theoretical advantages, it should be obvious that the lack of success of PD microanalytical systems is due to the inlet pressure limitation. This limitation already hinders the performance of analytical PD systems in the macroworld [2], and the practical and mechanical problems related to the small scale of the microfluidic channels only magnify this problem. Using off-chip pumps, the mechanical sealing and the inlet and outlet tube connection problems of PD microfluidic systems currently limit the maximally allowable inlet pressure to 10 to 20 bar, which is more than an order of magnitude smaller than in the traditional, macroworld liquid chroma-tography (LC) applications. Considering on-chip pumps, the situation is even much worse. Despite the research efforts of the past, the pressure heads that can be delivered by on-chip pumps are still disappointingly small (<0.5 bar), and the leakage problems seem to be too difficult to overcome. The on-chip pumps, furthermore, require complicated multilevel production processes, and are very prone to mechanical wear and damage by particles of dust and contaminants in the fluid. Considering the ease with which the electrodes needed to control ED flows can be integrated onto the surface of the chips, the absence of any moving parts or pressure gradient, and also considering that the presence of an elec-trical field induces an additional separation effect for the separation of charged species, as is so nicely exploited in capillary electrophoresis (CE) and capillary electrochromatography (CEC), the small number of practical PD applications becomes fully comprehensible.

Hence, it is also not surprising to find that the single PD flow applications with a potential for immediate commercial success currently have to be sought in a few specific separation methods for which the parabolic flow profile, often cited as one of the major disadvantages of PD flows, is absolutely indispensable for the induction of the separation influence. The two main analytical separation methods making explicit use of the parabolic profile of a PD flow are field flow fractionation (FFF) and hydrodynamic chromatography (HDC). Both these can be classified as "noninteractive" separation methods, i.e., they do not rely on the selective exchange with a second phase (cf. the use of a stationary phase in "true" liquid [LC] or gas chromatography [GC]). Whereas miniaturization is only favorable for a few types of FFF [3, 4], the reduction of scale is highly advantageous for HDC (see Section 6.4.1). In addition to these types of chromatographic methods, mainly focusing on the size separation and classification of macromolecular particles (large proteins, large molecular-weight polymers, cells, etc.) and hence in a sense complementary to "true" LC, there also remains a continuous interest in the development of on-chip LC application [5]. The latter is still considered to be the most versatile and selective of all analytical separation methods. Finding solutions to overcome all the practical problems related to high-pressure on-chip applications remains an important research area. An overview of the recent advances in the field of on-chip LC is given in Section 6.4.2.

Apart from the above-mentioned applications, the use of PD flows in lab-on-a-chip systems remains limited to a number of separation methods involving only low flow velocities. In this case, cheap and simple syringe pumps can be used and the sealing problem is not critical. Examples of such types of low-pressure separations can be found in the field of flow-through biosensors for the conduction of immuno- and enzymatic assays and for the measurement of binding kinetics (see Section 6.4.3). One of the advantages of PD flows in this case is that, apart from the less-expensive auxiliary equipment, they circumvent a possible interference between the electrical field needed to create an EOF and the binding or recognition process.

6.2 FUNDAMENTAL CONSIDERATIONS

Since microscale separations inevitably involve the use of thin capillaries and microchannels, the flow conditions are without any doubt purely laminar. In the present contribution, the emphasis will be on three basic flow types: the flow through a packed bed of spheres, the flow through an open-tubular circular capillary, and the flow through an open-tubular microchannel with flat-rectangular cross section (Figure 6.1). For packed beds of individual particles, as well as for monolithic packings, the exact cross-sectional shape of the enveloping channel wall is totally irrelevant, because all the important performance characteristics (pressure drop, separation resolution, etc.) are

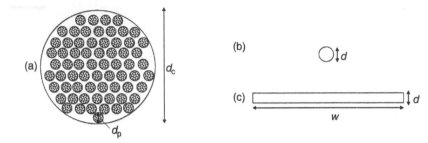

FIGURE 6.1 Overview of typical column cross sections most frequently used in microanalytical separations: (a) the packed-bed of spheres; (b) the circular capillary; and (c) the microchannel with flat-rectangular cross section.

determined by the particles themselves and not by the surrounding column or channel wall (Figure 6.1a). The cylindrical capillary (Figure 6.1b) represents, of course, the type of channels used in capillary tube systems (open-tubular LC, CE, etc.), and the flat-rectangular open-tubular channel (Figure 6.1c) represents most closely the type of channels typically used in the lab-on-chip field, although it is for most practical cases only an over-simplification of the reality. In practice, the side walls of the channels are very often curved, due to the specific nature of the applied etching processes. In some cases, even semispherical channels are used. An excellent overview of the influence of the exact shape of the side walls and the overall cross-sectional shape of the channels on the observed band broadening is given by Dutta and Leighton [6]. In the present contribution, we will mainly focus on the idealized flat-rectangular channel shape with $d \ll w$, as this strongly simplifies the theoretical calculations.

6.2.1 GENERAL PROPERTIES OF PRESSURE-DRIVEN FLOWS

Perhaps the most eye-catching difference between a PD and an ED flow is the shape of the velocity profile: parabolic for PD and perfectly flat (except for the double-layer region) for ED flows.

Solving the Navier–Stokes equations [7] in a cylindrical coordinate system for a laminar, unidirectional, and incompressible flow:

$$\frac{\eta}{r} \frac{d}{dr} \left(r \frac{du}{dr} \right) = \frac{dP}{dx} \tag{6.1}$$

it is found that the velocity profile in an open-tubular cylindrical capillary with radius R is given by

$$u(r) = u_{max} \left(1 - \frac{r^2}{R^2} \right) \tag{6.2}$$

where

$$u_{max} = \frac{\Delta P d^2}{12 \eta L_{tot}} \tag{6.3}$$

Equation (6.2) clearly represents a parabolic variation (in the r-direction) of the axial velocity profile. Integrating Equation (6.2) across the entire cross-sectional surface shows that the mean velocity in a cylindrical capillary is exactly equal to one-half of the maximum of the parabolic distribution:

$$u_{mean} = \frac{\int_0^R u 2\pi r \, dr}{\pi R^2} = \frac{u_{max}}{2} \tag{6.4}$$

For a channel with a flat-rectangular cross section, the expression for the velocity field can relatively easily be found by solving an analogous heat transfer problem (uniform heat generation in a flat-rectangular rod with the walls kept at $T = 0$ [8]). The resulting solution is anyway much more complicated than the cylindrical capillary case given in Equation (6.2):

$$u(z,y) = \frac{\Delta P}{2\mu L} \left(\frac{d^2}{4} - y^2 \right) - \frac{4 \Delta P d^2}{\mu L \pi^3}$$

$$\times \sum_{n \text{ odd}}^{+\infty} \frac{1}{n^3} \sin \left(\frac{n\pi}{d} \left(y + \frac{d}{2} \right) \right) \frac{\cosh(n\pi z/d)}{\cosh(n\pi w/2d)} \tag{6.5}$$

A closer inspection of Equation (6.5) shows that the velocity profile consists of a broad central region wherein the mean fluid velocity is equal to the flow velocity that would have been obtained in a hypothetical infinitely wide channel without side-walls, but with the same thickness d. In the regions near the side walls, there is a small region (with a width of order d) where the velocity sharply decreases from this infinite channel value in the central region to $u = 0$ at the side-wall surface (Figure 6.2a). In the radial direction (Figure 6.2b), the flow in the central bulk region displays the same parabolic profile as given by the expressions in Equation (6.2). The only difference with the cylindrical tube case is that now the mean velocity amounts up to two-thirds of the parabolic maximum:

$$u_{mean} = \frac{\int_0^R u \, dr}{R} = \frac{2}{3} u_{max} \tag{6.6}$$

Averaging Equation (6.5) across the y-dimension for a z-value in the central bulk region allows to show that:

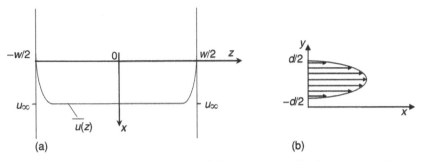

FIGURE 6.2 Schematic representation of the velocity profiles in the (a) (x,z)-plane (top view) and (b) in the (x,y)-plane (side view) of a channel with flat-rectangular cross section.

$$u_{\max} = \frac{\Delta P d^2}{8\eta L_{\text{tot}}} \tag{6.7}$$

This value is also slightly different from that for the cylindrical tube case.

Repeating the above analysis for systems with a different cross-sectional shape, and defining an apparent hydraulic diameter as

$$d_{\text{H}} = \frac{4 \times \text{cross-sectional area}}{\text{perimeter of cross section}} \tag{6.8}$$

it is found that the following expression for the mean fluid velocity appears to be generally valid for all types of channel cross sections:

$$u_{\text{mean}} = \frac{\Delta P d_{\text{H}}^2}{\psi \eta L_{\text{tot}}} \tag{6.9}$$

The only difference between the different cross-sectional shapes is the value of the proportionality constant ψ. For a cylindrical tube with diameter d it is found that $\psi = 32$ $(d_{\text{H}} = d)$, whereas for a channel with a flat-rectangular cross section (thickness $d \ll$ width w) it is found that $\psi = 48$ $(d_{\text{H}} = 2d)$. Equation (6.9) reveals one of the problematic scaling features of PD flows: when attempting to reduce the characteristic channel thickness d_{H} to reduce the radial mass transfer distances, which is one of the essential aims of the microfluidics and the lab-on-a-chip fields, the fluid velocity decreases according to the second power of d. This implies that from a given degree of miniaturization on, the increased mass transfer rates can no longer be translated into an increase of the overall speed of the process. This is further illustrated in Section 6.4.2 (cf. Figure 6.19–Figure 6.21).

The origin of the parabolic shape of the velocity profile of PD flows can be understood from the fact that the stationary channel walls exert their flow-arresting influence to the adjacent fluid layers through the viscous effect that is present in every fluid. Fluid elements in the center of the channel are least influenced by this flow-retarding influence, such that the flow velocity reaches its maximum at this position. This situation is, of course, completely different from the situation in the ED case, where instead of retarding the flow, the side walls (the positive counterions adsorbed to the side walls to be more precise), in fact, constitute the momentum source. Since the central region of the channel is free of flow-retarding objects, all fluid layers are dragged along without any retardation and a perfectly flat velocity profile is established.

The difference between a PD and an ED flow has been very elegantly illustrated by Paul et al. [9], using an ultraviolet (UV) laser writing a pattern in the flow by uncaging a fluorescent dye to define a sharply delimited color tracer plug. As can be noted from Figure 6.3, the PD flow

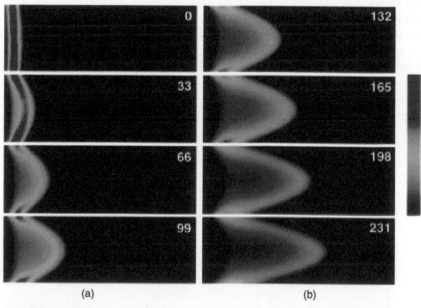

(a)   (b)

FIGURE 6.3 Difference between a pressure-driven and an electrically driven flow visualized using an uncaging fluorescence dye to mark the flow pattern. (Photographs taken from Paul, P. H.; Garguilo, M. G.; Rakestraw, D. J. *Anal. Chem.* 1998, *70*, 2459. Copyright 1998 American Chemical Society. With permission.) (a) Images of pressure-driven flow through an open 100-μm i.d. fused silica capillary (viewed region 100 by 200 μm). (b) Images of electrically driven flow through an open 75-μm i.d. fused silica capillary (viewed region 75 by 188 μm). The timescale indication is in milliseconds. The other experimental conditions can be found in Ref. [9].

exhibits a perfectly symmetric parabolic flow profile, whereas under ED conditions the tracer plug remains perfectly straight. When considering Figure 6.3(a), it should, however, be noted that the presented flow images are somewhat misleading. The fact that the dye molecules exactly represent the shape of the parabolic flow profile is only due to the very short time-scale on which the photographs have been taken. For larger times, and especially for cases with a small channel width and a rapid molecular diffusion, the parabolic shape is "destroyed" by the molecular diffusion. The latter forces the dye molecules to spread in the radial direction (from regions with a high concentration [original tracer band] to regions with a small concentration). This process continues until after a given lateral equilibration time (see further on, Equation (6.12)) all radial concentration gradients have vanished (Figure 6.4). Since the axial distance is usually orders of magnitude larger, the elimination of the axial concentration gradient occurs on a much longer timescale than the elimination of the radial concentration gradient.

In packed-bed systems, the flow is still determined by the Navier–Stokes equations, and it is hence obvious to expect that the local velocity profiles are also of a parabolic nature. This is, however, difficult to confirm experimentally. Pressure-drop measurements, on the other hand, are much easier to conduct, and they reveal the same square dependence of the pressure drop on the characteristic flow-through pore diameter as found for the open-tubular PD systems (cf. Equation (6.9)). For a packed bed, the elaboration of Equation (6.8) shows that the characteristic flow-through pore diameter is given by [10]:

$$d_{\mathrm{H}} = \frac{4\varepsilon}{(1-\varepsilon)a} = \frac{4\varepsilon}{(1-\varepsilon)} \frac{d_{\mathrm{p}}}{6} \tag{6.10}$$

where a is the specific surface area of a sphere ($a = 6/d_{\mathrm{p}}$). Usually, the pressure drop through a packed bed of spheres is described by the well-established Kozeny–Carman equation [11]:

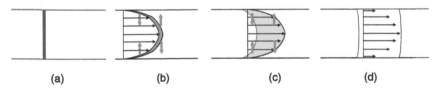

(a)   (b)   (c)   (d)

FIGURE 6.4 Schematic representation of the evolution of a rectangular tracer band (a) injected in a pressure-driven flow, first rapidly assuming the shape of the parabolic velocity profile (b), but then gradually losing this original shape (c, d) due to the effect of the molecular diffusion. The velocity profile itself, of course, remains parabolic.

$$u_m = \frac{1}{180} \frac{\varepsilon^2}{(1-\varepsilon)^2} \frac{d_p^2 \Delta p}{\eta L} \tag{6.11}$$

where u_m is the mean interstitial velocity, and showing that indeed $\Delta P \sim d_H^2$. Filling in the appropriate value for ε (i.e., $\varepsilon = 0.4$), and comparing Equation (6.11) with Equation (6.9), we find that a packed bed of particles with diameter d_p has a flow resistance that is about ten times larger than the flow resistance in an open-tubular channel with diameter or thickness $d = d_p$. The ratio can even become significantly larger than 10 when the particles have a broad distribution around their mean size [11].

6.2.2 GENERAL SEPARATION PERFORMANCE MEASURES

Adopting a short-cut analysis outlined by Poppe [2, 12, 13], each analytical separation process can be represented as a process involving one or more radial diffusive equilibration steps. Defining τ_{rad} as the mean time needed for the analyte molecules to exchange once with a second phase, and noting that in a laminar flow each mass transfer process in a direction perpendicular to the mean flow field has to occur by molecular diffusion, it can generally be stated from the basic law of molecular diffusion that:

$$\tau_{rad} \approx \frac{l^2}{D} \tag{6.12}$$

where l is the characteristic radial distance to be traveled by the molecules. For a packed bed of spheres, l is usually taken as the particle diameter d_p, for an open-tubular system with negligible porous solid-phase thickness, the straightforward measure for l is the hydraulic channel diameter d_H (cf. Equation (6.8)), whereas for an open-tubular system for which the channel wall is covered with a thick porous solid-phase zone, l should best be taken as the diffusion rate weighed sum of the distances to be traveled in the liquid and in the solid phase zone [14]. In Equation (6.12), the value of the exact numerical constant has been omitted, because its value depends on the specific geometry of the problem (e.g., 1/30 for a sphere with diameter d_p and 2/3 for a stagnant fluid layer with thickness d).

For a separation process involving a single adsorption or binding step, such as, for example, in an immuno- or enzymatic assay or in a flow-through biosensor, it is obvious that the residence time in the system should be selected such that:

$$t_R \geq \tau_{rad} \approx \frac{l^2}{D} \tag{6.13}$$

Equation (6.13), showing that the analysis time can be decreased according to l^2 ($= d^2$, when d is the diameter of the flow-through cell), provides a strong argument for the miniaturization of biosensor and bioassay systems. It should, however, be noted that Equation (6.13) does not account for the finite time needed for the adsorption or reaction step. The total analysis time can, of course, never be decreased below the latter time.

For a chromatographic process, where the quality of the separation increases with the number N of equilibration steps between the mobile phase and the stationary phase, it is obvious to find that the total time is proportional to the number of equilibration steps N:

$$t_R \approx N\tau_{rad} \tag{6.14}$$

where N can also be interpreted as [12]

$$N = \left(\frac{t_R}{\sigma_t}\right)^2 \tag{6.15}$$

In Equation (6.15), σ_t represents the standard deviation of the chromatographic peaks. Even for a separation method like HDC, where no exchange with a stationary phase has to occur, the performance still remains determined by the radial diffusion rate. The quality of a HDC separation namely depends on the rate with which the analyte molecules can sample all the different streamlines of the flow (see Section 6.4.1).

Although the expressions in Equations (6.12)–(6.14) are by no means complete, they provide an excellent basis to understand the performance dynamics of all possible analytical separation systems. Equation (6.14), for example, directly shows that the speed of a chromatographic separation is fully determined by the radial diffusion time. In addition, since combining Equations (6.14) and (6.15) directly yields an expression for the width of the chromatographic peak:

$$\sigma_t \approx N^{1/2}\tau_{rad} \tag{6.16}$$

it can readily be assessed that the quality of the separations (i.e., the achievable separation resolution) is also directly proportional to the radial diffusion time. The smaller τ_{rad}, the narrower, and hence the better separated the chromatographic peaks become. Traditionally, the achievable resolution of a separation system is also very often described in terms of a theoretical plate height H. It can easily be understood [12] that the relation between τ_{rad} and H is given by

$$H \approx u\tau_{rad} \tag{6.17}$$

Equation (6.17) now directly shows that the smaller τ_{rad}, the smaller H, and hence the larger the number of theoretical plates that can be achieved in a column with given length L.

The explicit need for short radial diffusion times revealed by Equations (6.12)–(6.17) precisely forms the rationale behind the use of miniaturized systems, as already outlined elsewhere in this book.

6.2.3 INFLUENCE OF THE FLOW PROFILES ON THE SEPARATION PERFORMANCE

To accurately assess the influence of the velocity profile on band broadening in a given flow system, one has to actually solve the complete convection–diffusion mass balance [10]:

$$\frac{\partial c}{\partial t} = D\nabla^2 c - \mathbf{u}\nabla c \tag{6.18}$$

For axisymmetric flow systems, Aris [15] has established a general solution to the long time limit behavior of Equation (6.18). He showed that, for any type of cross-sectional geometry, the degree of axial dispersion can always be expressed as

$$D_{\text{ax}} = D_{\text{mol}} + \kappa \frac{u_{\text{m}}^2 d_{\text{H}}^2}{D_{\text{mol}}} \tag{6.19}$$

where κ is a dimensionless, geometry-specific parameter. Complete outlines of the calculation method for κ can, for example, be found in Refs. [8, 15–17].

Rewriting Equation (6.19) in terms of a theoretical plate height, which is more customary in the field of analytical separations, it is found that

$$H \equiv \frac{2D_{\text{ax}}}{u_{\text{m}}} = \frac{2D_{\text{mol}}}{u_{\text{m}}} + 2\kappa \frac{u_{\text{m}} d_{\text{H}}^2}{D_{\text{mol}}} \tag{6.20}$$

Equation (6.20) shows that, if d_{H} is sufficiently small, as is the case in a very thin channel, or when D_{mol} is large (cf. the gas-phase separations considered in Chapter 9), the contribution of the second term, representing the band broadening caused by the presence of a velocity gradient, becomes negligibly small. Physically this implies that, when the time τ_{rad} needed for the radial diffusion can be made sufficiently short (cf. Equation (6.12)), the band broadening effect of the velocity gradient can be wiped out to a very large extent because under these conditions the analyte molecules "hop" very rapidly from streamline to streamline, and hence very rapidly assume the same mean velocity.

To assess the influence of the velocity field upon the degree of band broadening in a laminar flow system, it is instructive to compare the solution to Equation (6.18) for three totally different flow types (Figure 6.5): a perfect plug flow, representative for the ED case; a parabolic flow, representative for

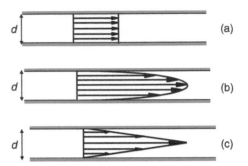

FIGURE 6.5 Schematic representation of three different velocity profiles: (a) plug flow; (b) parabolic flow; and (c) a (hypothetical) axisymmetric linear flow.

the PD case; and a flow with a linear velocity profile. The latter is a purely imaginary flow type, but has been added here as an additional point of reference. All three flow types are assumed to be established in a channel with a flat-rectangular cross section (i.e., with width w much larger than the thickness d). Table 6.1 gives the κ-value for the three different flow types depicted in Figure 6.5. For each flow type, two values are given: one for the peak broadening in a pure flow system, and one for the band broadening in a chromatographic system wherein a diffusive exchange with a stationary phase (with retention factor k') is taking place.

Using the κ-values from Table 6.1, Figure 6.6 shows that, in the case of a chromatographic exchange process, the κ-curves for the different flow types lie relatively close to each other, i.e., the influence of the exact shape of the velocity profile on the peak broadening is only of secondary importance. Looking in closer detail, it is even found that for $k' > 2$, the range wherein most chromatographic separations are conducted, the flow with the flat flow profile (representing the ED flow case) surprisingly leads to a larger peak broadening than the parabolic and the linear flow types. This certainly is in

TABLE 6.1
Overview of κ-Values to Be Used in Equations (6.19)–(6.20) for the Flow Types Shown in Figure 6.5

	κ (Unretained Solute)	κ (Retained Solute)
Plug flow	$\kappa = 0$	$\kappa = \dfrac{k^2}{6(1+k)^2}$
Parabolic flow	$\kappa = \dfrac{1}{210}$	$\kappa = \dfrac{1 + 9k + 25.5k^2}{210(1+k)^2}$
Axisymmetric linear flow	$\kappa = \dfrac{1}{120}$	$\kappa = \dfrac{1 + 7k + 16k^2}{120(1+k)^2}$

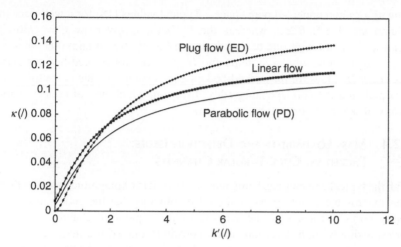

FIGURE 6.6 Comparison of the variation of κ with the retention coefficient k' for the three different flow types shown in Figure 6.5.

contrast to what is generally intuitively believed, but it is a consequence of the fact that the linear and the parabolic flows create a steeper radial concentration gradient than the flat flow profile, and this speeds up the radial mass transfer process.

In some cases, where the stationary phase has a considerable thickness, or has a very small diffusion rate, an additional term, representing the slow stationary phase mass transfer, has to be added to Equation (6.20):

$$H = \frac{2D}{u_{\mathrm{m}}} + 2\kappa \frac{u_{\mathrm{m}} d_{\mathrm{H}}^2}{D} + \frac{2}{3} \frac{k'}{(1+k')^2} \frac{u_{\mathrm{m}} d_{\mathrm{f}}^2}{D_{\mathrm{s}}} \tag{6.21}$$

Considering the packed bed of spheres, it is found that the expression for H is still related to the radial diffusion time. An exact expression for H, however, does not exist, partly due to the complex geometry of the pore space. A number of empirical expressions therefore coexist in literature. One of the most frequently used expressions is the so-called Knox equation, usually written in its dimensionless form [18]:

$$h = A\nu^{1/3} + \frac{B}{\nu} + C\nu \frac{D_{\mathrm{m}}}{D_{\mathrm{s}}} \tag{6.22}$$

where $h = H/d_{\mathrm{p}}$ and $\nu = u d_{\mathrm{p}}/D_{\mathrm{m}}$. In Equation (6.22), the A term is composed out of two contributions, one stemming from the finite duration of the mass transfer in the mobile phase (the second term in Equation (6.21)), and one (the so-called eddy-diffusion term) stemming from the additional longitudinal mixing experienced by the analytes when flowing through the tortuous network of interconnected pore spaces. Typical values [18] for a well-packed column are $A \cong 1$, $B \cong 2$, whereas the C factor is given by $C = 1/30 \cdot k'/(7.1 + k')^2$. As a rough rule of thumb, it can be said that minimal H value is $H_{min} = d$ for open-tubular systems and $H_{min} = 2d_p$ for packed-bed systems. The larger H values for the packed-bed case are essentially due to the tortuous flow path and the corresponding eddy diffusion, which are, of course, completely absent in an open-tubular system.

6.2.4 MASS LOADABILITY AND DETECTION ISSUES: PACKED VS. OPEN-TUBULAR CHANNELS

With the typical channel size being one or two orders of magnitude smaller than their macroscale counterparts, it should be obvious that the poor mass loadability (which scales with the volume of the separation channel) and the correspondingly high detection limits constitute one of the major Achilles' heels of microscale separation systems. The single exception wherein a miniaturized system may yield an improved detectability is when the amount of sample is limited to a few nanoliters or less, such as is the case in forensic and in some types of clinical analysis. In this case, the miniaturized column helps to avoid the dilution following upon the injection into a too wide column [19].

To properly discuss the influence of the column miniaturization, often a distinction is made between mass-sensitive and concentration-sensitive detection methods. For mass-sensitive detection methods (e.g., all fluorescence-based schemes), the loss in response signal that is caused by the reduction of the amount of sample molecules that can be injected in a miniaturized system is pretty obvious. For concentration-sensitive detection methods such as UV absorption (still the workhorse in virtually all liquid-phase fractionation techniques), on the other hand, one is easily tempted to think that, provided the optical path length remains constant, a miniaturization of the system would not affect the detectability. However, even in UV absorption, the amount of absorbed light is in the end also simply proportional to the number of molecules irradiated, i.e., it is not only the path length that determines the signal-to-noise ratio, also the area of the illuminated surface starts to play a dominant role when the dimensions of the detection cell are too drastically shrunk [12]. Hence, even for so-called concentration-sensitive detection methods, the volumetric and the mass loadability eventually become the detection sensitivity determining parameter.

For packed-bed systems, where the mass loadability is directly proportional to the internal (for porous particles) or the external (for full particles) surface area of the packing, the mass loadability is directly proportional to the cross-sectional column area. Decreasing then, for example, the diameter of the column from the standard 4-mm format used in high-performance liquid chromatography (HPLC) to the 50-μm narrow bore capillaries typically used in micro-HPLC or CEC, it can easily be calculated that the latter system can only be loaded with 6400 times less molecules. This already gives a good qualitative view on the huge loss in concentration detectability following a miniaturization of the separation equipment. For open-tubular applications, where only the enveloping wall is available for an adsorptive or a reactive exchange, such that the mass loadability is automatically orders of magnitude smaller than in the packed-bed case, the two most straightforward channel formats are the circular and the flat-rectangular channel (Figure 6.1b–c). It can easily be verified that for a given radial distance d (which determines the mass transfer rate between the fluid and the wall; see Equation (6.12)), the flat-rectangular channel with width w and thickness d yields a mass loadability which is $2/\pi(w/d)$ times larger than a circular capillary with diameter d. From the perspective of mass loadability, the flat-rectangular channel certainly is the ideal microfluidic channel format: the small channel thickness d guarantees a rapid mass transfer, whereas the large width w yields an increased mass loadability and optical path length.

6.2.5 Additional Peak Broadening Sources in Pressure-Driven Flows through Flat-Rectangular Channels

Channels with a flat-rectangular cross section, however, display a number of hidden sources of additional axial dispersion or peak broadening. When using them for separations for which the peak broadening is critical, as is, for example, the case for all chromatographic separations, these effects should certainly be taken into account. For ED flows, the effects discussed below do not occur, at least not to the same dramatic extent.

The first effect that has to be taken into account is the so-called side-wall effect: when considering ever wider channels, one would intuitively expect that the amount of axial dispersion would converge to the value expected for the axial dispersion in a flow between two infinitely wide parallel plates (i.e., without side walls). This is, however, not the case, and it has been demonstrated both theoretically [8, 16] and practically [20] that, when letting $w/d \to \infty$, the effective axial dispersion in a flat-rectangular channel converges to a value that is significantly larger, a factor of 7.95 to be exact. In other words, it is found that, despite of the fact that their effect on the velocity field decreases according to d/w, the presence of the side walls leads to a persistent, disproportional strong contribution to the overall axial dispersion (cf. the value of κ in Equations (6.19) and (6.20)):

$$\kappa = 1/210 \quad \text{flow between infinite plates} \tag{6.23a}$$

$$\kappa = 7.95/210 \quad \text{flow in flat-rectangular channel with } w/d \to \infty \tag{6.23b}$$

Using a chromatographic analog, the origin of this unexpectedly strong phenomenon can be understood as follows [17]. The thin boundary layers (relative thickness $\approx d/w$) near the side walls (cf. Figure 6.2a) can be considered as stagnant fluid zones, acting as a chromatographic stationary phase exchanging mass with the plug flow filling up the major part of the channel. As the liquid in the bulk and in the stagnant side layers has the same composition, the chromatographic distribution coefficient between both "phases" is simply equal to unity. This implies that the stagnant boundary layers near the side walls lead to a chromatographic retention with a retention coefficient equal to $k' \cong d/w$. Although this value tends to zero when $w/d \to \infty$, it should be noted that, according to Equations (6.19)–(6.20), and using the κ-expression for a retained component in a plug flow given in Table 6.1, this decreasing k'^2 value has to be multiplied by the (increasing) square of the diffusional distance, which, for the case of diffusion in the z-direction, scales according to w. For the plug flow under consideration this yields (A is a geometric constant):

$$\lim_{w/d\to\infty} \kappa_{\text{sw}} = A\frac{k'^2}{1+k'^2}u\frac{w^2}{D_\text{m}} \cong Ak'^2 u\frac{w^2}{D_\text{m}} = A\left(\frac{d}{w}\right)^2 u\frac{w^2}{D_\text{m}} = Au\frac{d^2}{D_\text{m}} \quad (6.24)$$

Equation (6.24) now directly shows that the additional peak broadening stemming from the flow-retarding action of the two side walls does not vanish when $w/d \to \infty$, but reaches a given constant value that is of the same order as the other dispersion terms (all of order d). An exact calculation then shows that $A = 6.95/210$ [16]. It can also be shown that roughly for all cases for which $w/d > 10$, the total plate height of a chromatographic flow through a flat-rectangular channel is simply found by adding the κ_{sw}-contribution given in Equation (6.24) to the other dispersion terms already accounted for in Equation (6.21):

$$H = \frac{2D}{u_\text{m}} + 2(\kappa + \kappa_{\text{sw}})\frac{u_\text{m}d_\text{H}^2}{D} + \frac{2}{3}\frac{k'}{(1+k')^2}\frac{u_\text{m}d_\text{f}^2}{D_\text{s}} \quad (6.25)$$

The calculation of the side-wall effect has very recently been extended by Poppe for the cases where the side walls of the channels are also covered with the stationary phase [21].

On the practical side, Dutta and Leighton [6] have recently developed an ingenious manufacturing method to overcome the side-wall effect (Figure 6.7a). Making numerical tracer dispersion calculations, they showed that by devising a channel that is slightly thicker than the average thickness near the side walls, hence allowing for a locally larger flow rate and allowing compensating for the flow-retarding action of the stagnant side walls, it is indeed possible to virtually completely eliminate the κ_{sw}-term in Equation (6.25). They also proposed a two-step etching procedure with which the

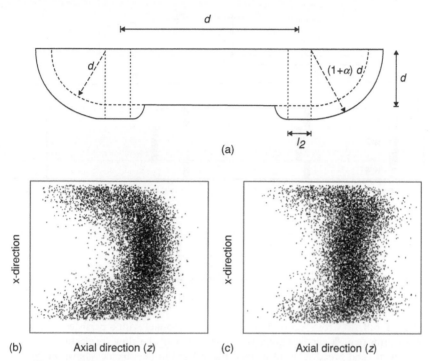

FIGURE 6.7 (a) Double-etched channel profile yielding a reduction of the side-wall dispersion effect. (b, c) Difference in band broadening between a channel with a perfectly rectangular cross section and a channel with a cross-sectional profile as shown in (a). The results were obtained using a random walk numerical simulation method, coupled to a numerical solver to yield the velocity profile. (Photographs taken from Dutta, D.; Leighton, D. T. *Anal. Chem.* 2001, *73*, 504. Copyright 2001 American Chemical Society. With permission.)

desired cross-sectional channel shape can be obtained. Figure 6.7(b) and (c) confirms that the double-etched channel indeed produces much less tailing than a single etched channel. They also showed that for the case of $w/d \gg 1$ (case not represented here), their special channel design allows to completely eliminate the side-wall effect.

An effect that in practice turns out to be even much stronger than the side-wall effect is the additional peak broadening stemming from small but persistent variations in the channel height. The strong impact of this effect can be understood as follows. Consider a channel with a curved top wall as depicted in Figure 6.8(a), and assume that this cross-sectional shape persists over the entire length of the channel. To a first approximation, the curved top wall can be represented by the dashed line system also depicted in Figure 6.8(a), dividing the channel into a central portion with thickness d_2 and two side wall regions with depth d_1.

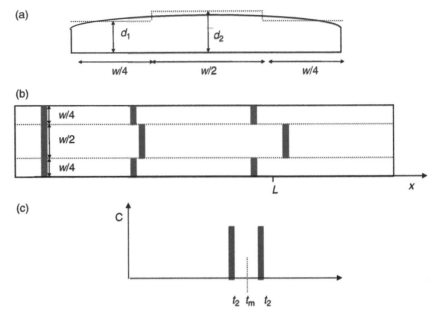

FIGURE 6.8 (a) Cross-sectional view of a microchannel with a persistent manufacturing error (e.g., a curved top-wall). (b) Top view of the flow (neglecting all other sources of band broadening) of a small tracer band in the channel shown in (a). (c) Recorded tracer signal at a detector located at $x = L$.

Denoting the mean channel thickness as

$$d_m = \frac{d_1 + d_2}{2} \tag{6.26a}$$

it is straightforward to define a dimensionless channel height variability factor x such that:

$$d_1 = d_m(1 - x) \tag{6.26b}$$

$$d_2 = d_m(1 + x) \tag{6.26c}$$

With this notation, and with Poiseuille's pressure drop law (put $\psi = 12$ in Equation (6.9)), the mean velocity in regions 1 and 2 can easily be calculated to be given by

$$u_1 = \frac{\Delta P d_1^2}{12\eta L} = \frac{\Delta P d_m^2 (1 - x)^2}{12\eta L} = u_m(1 - x)^2 \tag{6.27a}$$

and

$$u_2 = u_\mathrm{m}(1 + x)^2 \qquad (6.27\mathrm{b})$$

Neglecting the diffusive equilibration between both regions (an assumption becoming more and more valid with increasing w), it can now easily be shown that the tracer in side regions and the tracer in the central region will reach a detector positioned at L in, respectively, t_1 and t_2 seconds, with:

$$t_1 = \frac{L}{u_1} = \frac{t_\mathrm{m}}{(1 - x)^2} \qquad (6.28\mathrm{a})$$

and

$$t_2 = \frac{L}{u_2} = \frac{t_\mathrm{m}}{(1 + x)^2} \qquad (6.28\mathrm{b})$$

If there would be no other sources of band broadening, the response measured at the detector at L would hence consist of two ideal Dirac pulses (Figure 6.8b and c). Calculating the variance $\sigma_{\Delta d}^2$ of such a signal, we have

$$\sigma_{\Delta d}^2 = \int_{-\infty}^{+\infty} t^2 C(t) \, dt - t_\mathrm{m}^2$$

$$\sigma_{\Delta d}^2 = \frac{1}{2} \int_{-\infty}^{+\infty} t^2 \delta(t - t_1) \, dt + \frac{1}{2} \int_{-\infty}^{+\infty} t^2 \delta(t - t_2) \, dt - t_\mathrm{m}^2 \qquad (6.29)$$

with δ being the well-known Dirac function ($\delta(t) = 1$ for $t = 0$ and $\delta(t) = 0$ for all $t \neq 0$). Equation (6.29) yields

$$\sigma_{\Delta d}^2 = \frac{1}{2}(t_1^2 + t_2^2) - t_\mathrm{m}^2 \qquad (6.30\mathrm{a})$$

$$\Leftrightarrow \frac{\sigma_{\Delta d}^2}{t_\mathrm{m}^2} = \frac{\frac{1}{2} t_\mathrm{m}^2 \left(\frac{1}{(1-x)^4} + \frac{1}{(1+x)^4} \right)}{t_\mathrm{m}^2} - 1 \qquad (6.30\mathrm{b})$$

Expanding now $(7.1 - x)^4$ and $(7.1 + x)^4$ and considering only the first- and second-order terms, Equation (6.30b) becomes:

$$\frac{\sigma_{\Delta d}^2}{t_\mathrm{m}^2} = \frac{\frac{1}{2}(1 - 4x + 6x^2 - \cdots + 1 + 4x + 6x^2 + \cdots)}{(1 - x)^4 (1 + x)^4} - 1 \qquad (6.30\mathrm{c})$$

$$\frac{\sigma_{\Delta d}^2}{t_\mathrm{m}^2} \simeq \frac{1 + 6x^2}{(1 - x)^4 (1 + x)^4} - 1$$

which for small x can be further approximated as

$$\frac{\sigma_{\Delta d}^2}{t_m^2} = \frac{1}{N_{\Delta d}} \cong 6x^2 \tag{6.31}$$

Noting now that the effective plate number is the reciprocal of the sum of the reciprocal of the individual band broadening phenomena [13]:

$$\frac{1}{N_{eff}} = \frac{1}{N_{theo}} + \frac{1}{N_{\Delta d}} \tag{6.32}$$

where N_{theo} is the number of plates that would have been achieved in a channel without channel height variation, and where $N\Delta d$ is the number of plates corresponding to the channel height variability effect, Equation (6.31) directly shows that, for example, for a small difference in channel height of $x = 0.01$ ($= 1\%$ deviation from the mean channel thickness), the maximal achievable plate number in the column is only about 1666 (even if N_{theo} would be infinitely large). That is, if one would consider a channel that is sufficiently long to yield more plates, the effective plate number cannot exceed $N_{eff} = 1666$. For $x = 0.05$, still a small relative deviation from the ideal rectangular shape, the maximum number of plates is only $N_{eff} = 66$. This extreme sensitivity to small deviations in channel height readily explains the lack of successful open-tubular on-chip LC separations presented in the literature.

6.2.6 ALTERNATIVES TO PD FLOWS

As already mentioned, the variable channel thickness effect is much less pronounced for ED flows, because there the flow rate is independent of the channel thickness, whereas in a PD system the flow automatically tries to follow the path with the smallest resistance. Recently, alternative flow driving principles, based on the use of centrifugal or shear forces [22, 23], have been proposed (cf. Chapter 5). For shear-driven flows through straight-running rectangular channels, the flow rate is always exactly to one-half of the moving wall velocity, and is hence also not affected by channel defects as the one shown in Figure 6.8(a). Figure 6.9 shows a series of images of a shear-driven tracer plug flow. It can clearly be noted that the tracer plug remains perfectly straight. As shear-driven flows offer all the advantages of a PD flow (free selection of mobile and stationary phase composition), without being impeded by the pressure-drop limitation stemming from Poiseuille's pressure-drop law and by the large sensitivity to small deviations from a perfectly rectangular cross section, it should be obvious that, for the conduction of rapid open-tubular LC separations, shear-driven flows provide a promising alternative to PD flows [23]. In fact, shear-driven flow

FIGURE 6.9 Image sequence of a shear-driven flow of a coumarin dye tracer through an ultrathin channel with depth = 0.1 μm and width = 800 μm, corresponding to an aspect ratio of 8000. The velocity of the sample plug was 5 mm/sec.

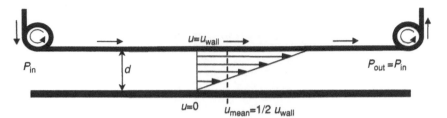

FIGURE 6.10 Operating principle of shear-driven microchannel flow.

separations can be considered as taking place *inside* a pump (Figure 6.10), instead of in a capillary coupled *behind* a pump, as is in the traditional PD mode.

6.3 PRACTICAL ASPECTS

6.3.1 PUMPS AND VALVES

When discussing the practical limitations of the currently employed pumping systems, a distinction has to be made between on-chip and off-chip pumping systems. The electrokinetic induction of pressure heads will not be considered here, as it is treated elsewhere in this book (Section 5.2.6).

Despite intense research efforts in the field of on-chip pumping, originally initiated by the work of van Lintel et al. [24] in 1988, it turns out to be really difficult to obtain pressure heads exceeding 0.1 to 0.3 bar. The most successful on-chip pumping systems are of the membrane type, using piezoelectric, thermopneumatic, or electrostatic forces for their actuation. Actuation principles based on the use of bimetallic systems or on the shape-memory effect of some TiNi alloys have been considered as well. An extensive overview of the work in this field and the employed fabrication techniques can, for example, be found in Ref. [25]. The fabrication of robust on-chip valving systems, which are absolutely indispensable in the complex microfluidic networks needed for on-chip combinatorial chemistry or whole-cell analysis, also turned out to be extremely difficult. A variety of cantilever and diaphragm valves, mostly machined in silicon and metallic substrates, has been proposed and tested. The back-leakages (mostly due to the mechanical stiffness of the employed materials) and the limited operating pressure (maximally of the order of a few bar), however, seem to form an insurmountable impediment for their use in robust, high-performance separation systems. In the last few years, the use of more soft materials (mainly elastomeric polymers and hydrogels) has been proposed to reduce the leakage problems. With these materials, radically new actuation principles could be devised as well. One such method is based on the autonomous swelling properties of hydrogels [26]; another recent method relies on the application of an external pressure to squeeze and deform tangentially running elastomeric PDMS channels with a well-defined contact area [27]. Exploring a variety of formulations and photopolymerization methods, Hasselbrink et al. [28] recently reported on a series of *in situ* fabricated flow control systems that are capable of operating under much larger pressures (up to 30 bar) than can be achieved with the traditional, hard material valving systems. An excellent overview of the research on microfluidic valve systems can be found in Ref. [29].

Comparing the small pressure heads achievable with the current on-chip pumping systems to the high-pressure heads (order of 100 bar and more) required to conduct high-performance separations, and considering the complexity of the manufacturing schemes and the fact that on-chip pumping systems are extremely sensitive to clogging and mechanical wear, it is not surprising to find that most of the currently developed PD applications are based on off-chip pumping schemes, using the robust and high-performance pumping and valving systems developed for the conduction of macroscale separations. With this approach, however, a new problem turns up. Interfacing the tubing used in the off-chip circuit with the on-chip microchannels appears to be much more difficult than originally assumed. The originally employed gluing procedures proved to be very tedious and lead to very poorly sealed junctions. Better performances were obtained using, for example, vertically etched sleeve couplers [30], PEEK interfaces [31]. A twin bonded, silicon-etched horizontal tube connector has been proposed as well [32].

Surveying all current chromatographic microscale chromatographic on-chip separations, it can, however, be concluded that, despite all the efforts in the field, the maximal operating pressure of on-chip microanalysis systems is maximally of the order of 10 bar [33–36]. This is still more than one order of magnitude smaller than the pressure heads used to conduct high-performance macroscale separations.

6.3.2 Packing Methods, Monolithic Packings, and Micromachined Packings

Although most microfluidic systems have an open-channel architecture, it cannot be denied that for a large number of important applications, such as chromatography, enzyme assays, flow-through immunoassays, solid-phase extraction, etc., the loadability of such systems is very low (cf. Section 6.2.4). A straightforward solution, allowing for a dramatic increase of the surface-to-volume ratio, is to incorporate a packing of porous particles into the microchannels. Similar to the tube connection problem discussed in Section 6.4.1, this however turns out to be easier said than done.

The first paper reporting on the huge practical obstacles one is confronted with when trying to pack small (i.e., microsized) beads into the narrow microfluidic channels is the classic work of Ocvirk et al. [36]. The task of arranging the particles in a uniform and densely packed manner appears to be especially impossible when the channel is not straight, but curved into a serpentine configuration [5]. Apart from the difficulties encountered in filling up the narrow channels, for which a variety of empirically developed procedures, sometimes even referred to as "black magic" [37], exist, another important issue is the need to retain the beads inside the channels after being packed. Traditionally, the beads are retained inside the channels by means of *in situ* sintered retaining frits. This approach is, however, generally considered as being too cumbersome and too irreproducible. Recently, two promising approaches (both based on the use of micromachining methods) have been proposed to overcome the problems related to the traditional method of *in situ* frit generation. One approach relies on the use of two micromachined weirs enclosing a packing chamber that is being filled via a third, hooked bead filling channel [38]; the second approach uses a tapered channel outlet section, and relies on the so-called cobble stone or key stone effect [39] to immobilize the bed without the need for a retaining frit or weir. Attempts such as the one cited above are important, because they open the road toward the use of the extremely broad variety of commercially available HPLC beads for on-chip applications, hence, omitting the need for coating the chips and allowing for a fully optimized selection of the stationary phase [38].

Recently, two radically new approaches to overcome the packing problems and the huge flow resistance of the traditional packed bed of spheres have been proposed and successfully demonstrated. One of these alternatives is the use of so-called monolithic or continuous beds, where a stationary phase structure (mostly a polymeric one) is formed *in situ* by UV-induced polymerization of a monomer mixture pumped into the microchannels [5, 40–42]. The thus formed bed can be considered as a polymer monolithic structure composed of covalently linked 0.1 to 0.4 µm particles. The total porosity of the polymer monolithic beds can often be predicted from the composition of the initial monomer mixture (often including so-called porogen agents [43]). The ratio of void to total channel volume can be made as large as 60%, i.e., significantly larger than the typical packed-bed porosity of around 40%. Accordingly, the through-pore/skeleton size of such columns can be greater than in the packed-bed situation, resulting in shorter stationary phase diffusion lengths and in a larger bed permeability K [44]. The latter is especially advantageous for on-chip applications, because, since K is defined as

$$K = \frac{u\eta L}{\Delta P} \qquad (6.33)$$

a large bed permeability simply reflects the fact that a given required fluid velocity can be realized with a smaller inlet pressure. The technique of monolithic polymeric column packings has now also been applied in an inventive manner by Svec and co-workers to form an integrated reaction and separation system for on-chip tryptic digest assays [45].

A second alternative to the packed bed of particles has been provided by Regnier and coworkers [46, 47], and is based on the use of the same micro-lithographic machining techniques used in the electronics industry and capable of fabricating arrays of sub-micrometer structures over large surfaces. With these so-called COllocated MOnolith Support Structures (COMOSS), much more uniform and reproducible chromatographic beds are obtained (Figure 6.11). As it can be inferred [18] that a large part of the eddy-diffusion contribution to the theoretical plate height in packed-bed HPLC stems from nonuniformities in the packing, the potential advantage of these structures is obvious. The silicon-machined chromatographic packings can either be used directly or after replication in a suitable polymer (e.g., PDMS). The latter option is, of course, much more cost-effective [48, 49]. Recently, an alternative design for the layout of the COMOSS beds was proposed by Knox [50]. Taking advantage of the fact that the use of lithographic etching techniques also allows to optimize the shape of the particles, he suggested developing beds consisting of more elongated structures (Figure 6.12). The advantage of these structures is again to be found in a reduction of the flow resistance of the beds, making them more compatible with the small inlet pressures on-chip.

6.3.3 INJECTION AND DETECTION ISSUES

The design of the detection and injection sections is most critical for chromatographic applications. Similar to ED systems, the overall band broadening in PD chromatography is the sum of the contributions coming from the band broadening in the column, and also from that in the injection and detection sections [2]:

$$H_{tot} = H_{col} + H_{inj} + H_{det} \qquad (6.34)$$

Considering the injection of a rectangular plug with width w_{inj} and considering an on-column detection method monitoring a rectangular volume with width w_{det}, the H_{det} and H_{inj} values are, respectively, given by [51]

$$H_{inj} = \frac{w_{inj}^2}{12} \frac{1}{L_{tot}} \quad \text{and} \quad H_{det} = \frac{w_{det}^2}{12} \frac{1}{L_{tot}} \qquad (6.35)$$

Considering a typical column length of $L = 10$ cm, and knowing that as a rule of thumb it can be said that $H_{col,min} = d$ for open-tubular systems and $H_{col,min} = 2d_p$ for packed-bed systems, and adopting the arbitrary rule that the contribution to H_{tot} coming from the injection and detection band broadening should be smaller than 10%, it can immediately be seen from Equation

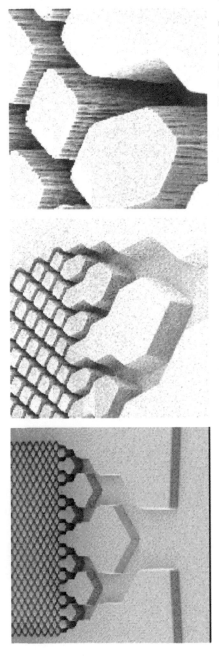

FIGURE 6.11 Scanning electron micrographs of an *in situ* micromachined column for on-chip LC. (Reproduced from He, B.; Tait, N.; Regnier, F. E. *Anal. Chem.* 1998, 70, 3790. Copyright 1998 American Chemical Society. With permission.)

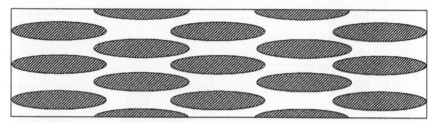

FIGURE 6.12 Alternative design for micromachined packings for on-chip LC proposed in Ref. [50] to minimize the pressure drop. (Drawing adapted from Knox, J. *J. Chromatogr. A* 2002, *960*, 7.)

(6.35) that $w_{inj} < 350$ μm and $w_{det} < 350$ μm for an open-tubular system with $d = 1$ μm.

Considering first the injection problem, it should hence be obvious that a "good" injection system for on-chip LC should be able to deliver sample plugs with a well-defined and reproducible length of the order of 200 to 300 μm. With the current state of the art, this is not a very difficult problem. The main bottleneck for the injection in PD systems is that the integration of mechanical components such as a microvalve or a plunger into a chip often makes the device too complex and expensive. Given the required degree of miniaturization, such systems would also be very prone to mechanical wear and damage. As a consequence, most often off-chip valves and injection plungers are used. The injected material is then introduced into the separation channel using either so-called cross injector [52, 53] where the sample delivery channel intersects the separation channel at one position, or a twin-T injector [54, 55] where both channels share a common, short length. This length defines the injection volume. The main bottleneck with such an off-chip injection scheme is that it is extremely difficult to avoid a continuous sample leakage (after the actual injection) coming from the injection channels, a problem that is similar to the tailing problem in the earlier electrokinetic injection schemes [56]. The recently proposed twin T-injection technique [55, 57] helps to overcome this problem. Figure 6.13(a) shows a schematic representation of the sequence of the events in a twin T-injection for PD chip separations. The injection can be realized in three steps, using a combination of external flow actuation and internal chip resistances as depicted in Figure 6.13(a). The valves C and B are always switched simultaneously. In the first step, fresh sample is introduced directly from the syringe through valve B into collector C and thus deposited in a well-defined volume within the separation channel. The collector C is made as a larger groove of volume of a few microliters, having a low resistance in order to ease its purge. During the next step, pressure is applied via the main channel inlet (filled with the normal mobile-phase liquid) to push the sample in the direction of the separation channel. As can be noted from Figure 6.13(a),

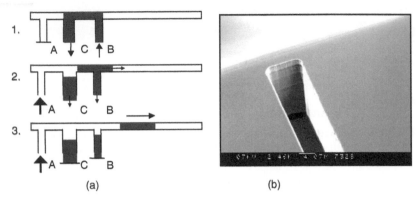

(a)   (b)

FIGURE 6.13 (a) Longitudinal sketch of a twin-T injector for PD separation. Sample is filled in between the second and third slits (7.1). After switching the valves, the carrier liquid flow is set on, sweeping the sample plug out into the separation channel (7.2). Partial back-leak into the slits, controlled by the valve and resistance R prevents sample tailing (7.3). (b) Bird's eye view of one of the injection slits.

the sample is also pushed back in the auxiliary channels. In the third step, these auxiliary channels are then closed off in order to prevent the loss of the mobile-phase liquid. In order to avoid complex fabrication procedures, the delivery slits are made as open grooves in the bottom of the channel, entering it from the same side (Figure 6.13b). With such transversal conduits, uniform filling can only be achieved if their flow resistance is much smaller than that of the shallow channels. In this way, pressure gradients and corresponding flow nonuniformities across the channel width are prevented.

Solving the problem of defining narrow sample plugs in the separation channel alone is however not sufficient. The problem of the large volumes of the connection tubing (obliging to send much more sample to the waste than to the separation channel itself) and the difficulty in sealing of the coupling between the chip and the connection tubing remains a critical issue. Another problem is that the junction interfaces usually comprise a sharp bend or a sudden diameter reduction and are hence highly prone to clogging. One elegant solution, providing the possibility for *in situ* cleaning of the connection tubing, while still allowing the use of operating pressures up to about 10 bar is the use of a stainless steal clamp (Figure 6.14) to which stainless steel connection tubing is soldered using a silver-based alloy. The clamp–chip interface is sealed by chemically resistant Kalrez O-rings. Another recently proposed connection solution is the use of v-grooves wherein the connection capillaries can be fixed [35].

Concerning the on-chip detection, the problems of the small available detection volumes and the short optical path lengths, currently forming the major bottleneck for the commercial application of microscale and lab-on-a-chip separation methods, are not different for PD than for ED systems,

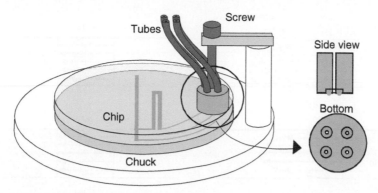

FIGURE 6.14 Mechanical clamp connection system for connection to off-chip injection and pumping systems.

and will therefore not be treated here. More on detection issues can be found in Chapter 11.

6.4 OVERVIEW OF APPLICATIONS

6.4.1 HDC and FFF on a Chip

In the field of analytical chemistry HDC takes a special position, because HDC is the only type of chromatographic separation for which no ED variant can be conceived. This is due to the fact that HDC is based on the presence of a parabolic velocity gradient, which is an exclusive feature of PD flows (Figure 6.15a). As depicted in Figure 6.15(b), the flat velocity profile of ED flows does not induce any HDC separation effect. The main application of HDC is the size separation of larger molecules, i.e., polymers and macromolecules (DNAs, proteins, and other biopolymers, but also colloidal particulates, vesicles, cells, etc.), which currently is one of the main issues in the field of analytical chemistry. For the separation of charged species, the most popular methods up to date are, of course, CE (e.g., for proteins) or capillary gel electrophoresis (for nucleic acids). For the separation of uncharged species (e.g., synthetic polymers), several methods are in use, mainly based on steric exclusion effects, viz., size exclusion chromatography (SEC) [58] and HDC [59, 60]. Both methods are in the family of HPLC techniques, where solvent propulsion takes place with applied pressure drops up to 400 bar. The advantage of HDC over SEC is that, whereas the latter is based on the exchange of the molecules with a porous matrix, the former does not involve any mass transfer to a second matrix and is therefore acclaimed for its faster separation speed and for its reduced column fouling.

As it was originally developed to be performed in packed columns with nonporous particles, HDC however finds its limitation in the large pressure

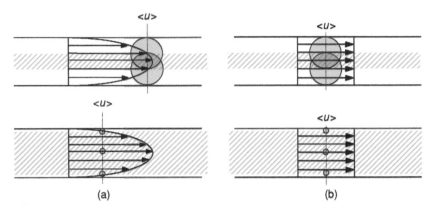

FIGURE 6.15 (a) Schematic drawing (longitudinal cross section of an open-tubular channel) of the hydrodynamic size separation effect in a pressure-driven flow: larger analytes cannot sample the regions with the low fluid velocities near the channel wall and therefore move faster (the shaded areas represent the region of the channel that is accessible to the center of mass of analytes with a given size). (b) Schematic drawing showing that the flat profile of an ED flow cannot induce a difference in migration velocity between large and small molecules.

drops that are inherent to packed-bed flows, and which prohibits the use of extremely small packing particles <1 μm. This implies that packed column HDC can essentially only be performed with relatively large (up to 100 μm in diameter) packing particles, and can hence only be used satisfactorily to separate molecules situated at the higher end of the molecular mass scale [59, 60]. To enable separations of smaller molecules, the obvious solution is to switch from a packed column format to an open-channel or tubular column format, because the latter has a much smaller flow resistance factor. In addition, open-channel columns have perfectly predictable separation characteristics, whereas the packed column approach suffers from poor column-to-column reproducibility due to slight differences in packing material (e.g., packing particle shape and packing particle size distribution). Still, even with the open-channel variant, the problem of the fluid velocity limitation and the connection to external pumping systems remains, not to speak of the detection problems arising from the tiny cross sections used in open-tubular capillaries [59].

All this implies that the ideal column format for HDC is the open-tubular channel with flat-rectangular cross section. In HDC, the retention behavior has been described by several models, the simplest taking into account only geometrical effects. The more extended models by DiMarzio and Guttman [61] and Brenner and Gaydos [62] can be unified for all HDC geometries to obtain the calibration relationship between residence time and size:

$$\tau = (1 + B\lambda - C\lambda^2)^{-1} \tag{6.36}$$

where τ is the relative retention time and λ the relative size of the analyte with respect to the channel. These quantities are defined as $\tau = t_R/t_0$ (where t_0 is the time of elution of a small compound) and $\lambda = d_a/d$ if d_a is a properly chosen analyte diameter (for a hard sphere d_a is its geometrical diameter) and d the diameter of a capillary or the thickness of a flat channel. Constants B and C are both geometry dependent, and constant C is model dependent. Constant B, together with a certain minimum value of C, represents directly the effective "cutoff" of the parabolic flow profile for an analyte at the inner boundary of the wall-adjacent layer from which this analyte is excluded and thus reflects the main retention mechanism of HDC. Flat-rectangular channels can be accurately described by plan-parallel plates for the case of retention, resulting in $B = 1$. In circular cylindrical and polygonal cross sections $B = 2$ [63]. The constant C accounts also for the fact that large particles slightly lag the virtual local velocity of the fluid in their center of mass. This slip velocity is caused by the curvature of the flow profile and also by the rotation–translation coupling as the particles are forced to rotate by a force couple stemming from different fluid velocities at opposite sides. DiMarzio and Guttman [61] found for plan-parallel plates $C = (7.1/2) + (7.3/4)\gamma$, γ being a function of the particle shape and density distribution ($\gamma = 2\pi/27$ for free draining polymer coils and $\gamma = 2/3$ for impermeable hard spheres). Thus, the HDC calibration equation (6.40) becomes

$$\tau = (1 + \lambda - 0.68\lambda^2)^{-1} \quad \text{[flat channel; polymers]} \tag{6.37a}$$

$$\tau = (1 + \lambda - \lambda^2)^{-1} \quad \text{[flat channel; particles]} \tag{6.37b}$$

However, those authors neglect the hydrodynamic interaction of a sphere moving close to a wall, which increases the value of the C constant substantially as was pointed out by Brenner and Gaydos [62]. They present a corrected approximation for a cylindrical tube but no explicit result for a planar geometry. Furthermore, the above formulae do not reflect radial forces such as the hydrodynamic "tubular pinch" effect [64, 65], which occurs at higher velocities, and electrostatic and electrokinetic lift forces acting upon charged particles in mainly aqueous solutions [66]. Those forces change the uniformity of the concentration profile of the particles over the channel cross section, thus favoring certain velocities and influencing the retention. With all the effects included, the retention curve especially for colloids is more complicated than Equation (6.36), as was described by Tijssen [64]. This is also apparent from experimental results of Small [67] and interpreted by Ploehn [65], the latter presenting trendlines with several inflex points.

Neglecting these secondary effects, and considering only Equation (6.37b), the dashed line in Figure 6.16 shows that the retention time for

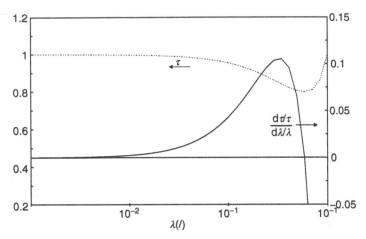

FIGURE 6.16 Variation of the retention time (τ) and the relative separation resolution $(d\tau/\tau)/(d\lambda/\lambda)$ as a function of the relative ratio (λ) of analyte diameter to channel diameter for HDC in a channel with a flat-rectangular cross section.

HDC in a flat-rectangular channel varies between $\tau = 0.8$ (for $\lambda = 0.5$) and $\tau = 1$ (for $\lambda = 1$ and for $\lambda \rightarrow 0$). This implies that the particles elute relatively short after each other and that relatively long channels (i.e., channels yielding 100,000 plates or more) are needed to obtain a sufficient separation resolution. Figure 6.16 also shows that for $\lambda > 0.5$ a reversal of the elution order occurs. This is due to the fact that above $\lambda = 0.5$, the interactions with the wall become important. Since the larger analytes are impeded more strongly by the successive contacts with the (stationary) channel walls, they can no longer elute faster than the smaller analytes. Defining a dimensionless separation sensitivity parameter $(d\tau/\tau)/(d\lambda/\lambda)$, it can easily be understood from the full line in Figure 6.16 that the highest separation resolution is obtained in the range between roughly $\lambda = 0.02$ and $\lambda = 0.4$.

The other parameters affecting the separation efficiency of a HDC separation in a channel with a flat-rectangular cross section, i.e., the degree of peak broadening and the effective pressure-drop, are, respectively, given by Equations (6.9) and (6.25).

Given that the thickness of the channel should not be larger than 1 μm in order to be able to effectively separate polymer fractions and proteins, and given the need for a sufficient optical path length (>500 to 1000 μm) needed for on-column detection, the aspect ratio of the channels preferably should be of the order of 1000 or more. Since the single viable option to fabricate such channels is via the use of microfabrication methods [68], one automatically comes to the HDC-on-a-chip concept. Very recently, Chmela et al. [33] have delivered the first-proof-of-principle results for this HDC-on-a-chip concept. Using a prototype setup as shown in Figure 6.17, and using flat-rectangular

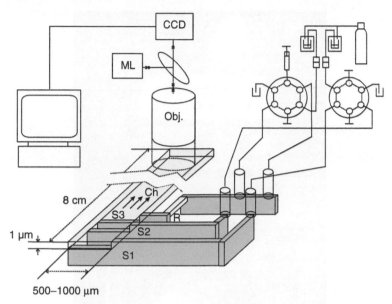

FIGURE 6.17 Schematic view of the prototype experimental set-up used for the HDC-on-a-chip experiments. (Drawing taken from Chmela, E.; Blom, M. T.; Gardeniers, J. G. E.; van den Berg, A.; Tijssen, R. *Anal. Chem.* 2002, *74*, 3470. Copyright 2002 American Chemical Society. With permission.) The HDC chip (not to scale) contains a separation channel (Ch), a resistance channel (R), and inlet and outlet slits for sample (S2, S3) and carrier liquid (S1). The chip is connected to a valve flow-actuation system, solvent being pressed in from gas-pressurized vessels. An imaging microscope is used to view the separations.

HDC channels (8-cm long, 1-mm wide, and 1-μm deep) fabricated by glass and silicon microtechnology, they showed the separation of a mixture of four differently sized polystyrene nanobeads (respectively, 26, 44, 110, and 180 nm in diameter) in less than 3 min (Figure 6.18). The separation of biopolymers has been demonstrated as well. A mixture of fluorescently labeled bovine serum albumin (BSA, estimated diameter $\cong$ 8 nm), bovine eye alpha-crystalin (estimated size $\cong$ 9 nm $\times$ 18 nm), and a t_0-marker (7.4(5)-carboxyfluorescein) were separated in about 1 min (operating pressure = 8 bar). The gain in speed as compared to the slow SEC and packed HDC processes traditionally used is entirely due to the chip format and the enhanced separation kinetics offered by the smaller dimensions.

Presently, the HDC-on-chip separations are however still strongly compromised by the variable channel thickness effect described in Section 6.2.5. Solving this problem should either come from a better materials choice or from the use of smaller channels. The latter option has already been explored experimentally, and it was indeed found that by narrowing the width of the

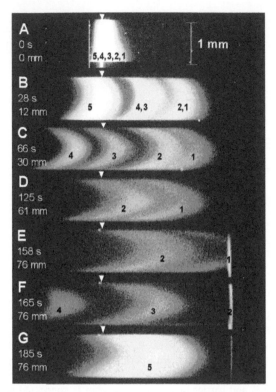

FIGURE 6.18 Separation on the HDC-chip (top view) of fluorescein (0.05 mg/ml) (5), fluorescently labeled anionic dextran 10 kDa (7.4) as the marker, and fluorescent polystyrene particles of 26 nm (7.3), 44 nm (7.2), and 110 nm diameter (7.1); concentration 0.25 to 0.5 mg/ml in 10 mM phosphate buffer (pH 7.0); working pressure 4 bar. Flow is from the left to the right. The channel is 1-mm wide and 1-μm high. After the mixture injection (A), fluorescein is separated first (B), further downstream also the dextran and the particles are resolved (C–F). Distance (arrow pointer) and the time passed since injection are indicated on the left. At the end (E, F, G) some of the zones are eluting already into the outlet slit, visible as a bright line on the right. (Photographs taken from Chmela, E.; Blom, M. T.; Gardeniers, J. G. E.; van den Berg, A.; Tijssen, R. *Anal. Chem.* 2002, *74*, 3470. Copyright 2002 American Chemical Society. With permission.)

channels from, for example, 1000 to 500 μm, a significant reduction of the variable channel thickness effect was obtained [57].

Field flow fractionation, invented by Calvin Giddings in 1966, is another well-established separation method making explicit use of the presence of a velocity gradient to induce a size separation of macromolecules and larger particles and is also usually conducted channels with a flat-rectangular cross section. As has very recently been discussed in a very elegant and condensed manner by Gale et al. [3], a reduction of scale is, however, counterproductive for most of the different members of the FFF family.

The miniaturization of FFF is therefore not discussed here, and we simply refer the interested reader to the work of Gale et al. They showed that the single types of FFF for which a miniaturization yields a number of important advantages are electrical field-FFF (EFFF) and thermal field-FFF (TFFF). They also demonstrated the practical feasibility of on-chip EFFF [4] and TFFF [69].

6.4.2 HPLC ON A CHIP

Liquid-phase chromatography, be it in its packed-bed variant (the so-called HPLC) or in its open-tubular variant, probably is the most versatile of all analytical separation methods. The classic paper on HPLC-on-a-chip undoubtedly is from Ocvirk et al. [36], who published in 1995 the first results of a prototype LC system, wherein a split injector, a packed small-bore column, a retaining frit, and an optical detector cell were all integrated onto the surface of a silicon chip. Demonstrating the advantages of a partially integrated approach (pumping occurred off-chip), they showed that the critical extra-column dead volume could be limited to a value below 2.5 nl, i.e., about 200 times smaller than the separation column volume of 0.5 µl. The resulting separation performance was, however, relatively disappointing: the separation of fluorescein and acridine orange yielded only a maximum of 200 theoretical plates in about 3 min.

The following theoretical analysis readily allows to understand that, apart from a number of experimental error sources that were undoubtedly present, there is also a fundamental problem underlying this poor performance. With the theoretical plate height being fully determined by the particle size and the applied flow velocity (cf. Equation (6.22)), and with the number of plates N needed to achieve a given separation quality being a given design criterion, the length of the separation channel is given by

$$L = NH \qquad (6.38)$$

Noting that H for a packed bed of spheres is given by Equation (6.22), and that the pressure drop is determined by Equation (6.11), it can easily be verified that Equations (6.38), (6.22), and (6.11) can only be satisfied simultaneously for one given value of the velocity. Calculating this value numerically, and using this value to calculate the analysis time, which is well known [13] to be given by

$$t_R = \frac{L}{u}(1 + k) \qquad (6.39)$$

we can establish a plot yielding the minimal analysis time needed to achieve a given separation resolution as a function of the particle diameter (Figure 6.19).

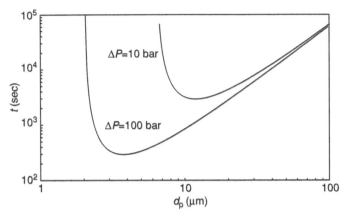

FIGURE 6.19 Variation of analysis time t_R versus the particle diameter d_p for a packed-bed LC separation requiring $N = 11,400$ plates ($k' = 3$), and for two different values of the pressure drop: $\Delta P = 20$ bar and $\Delta P = 100$ bar (other parameters: $D = 1 \times 10^{-9}$ m²/sec, $D_s = 5 \times 10^{-10}$ m²/sec, $\mu = 1 \times 10^{-3}$ kg/(m sec), and $\psi = 1000$).

In Figure 6.19, two different values of the available pressure drop are considered: one ($\Delta P = 10$ bar) corresponding to the maximally achievable value with the current state of the art, and another ($\Delta P = 100$ bar) corresponding to a highly desired target value. The difference between the $\Delta P = 10$ bar curve and the $\Delta P = 100$ bar curve immediately shows that the current on-chip are pressure-drop limited. The $\Delta P = 10$ bar curve also reveals that the packing material used in the Ocvirk paper ($d_p = 5$ μm) was too small. With a diameter of say 10 μm, much better results would have been obtained. This result appears counterintuitive at first, and it emphasizes the dramatic consequences of the current inlet pressure limitation for on-chip LC.

Figure 6.19 clearly shows that, in the large particle diameter range, a decrease of the particle diameter yields a clear decrease of the analysis time. This simply reflects the fact that the mass transfer in a smaller particle occurs faster than in a larger particle (cf. the τ_{rad}-argumentation in Section 6.2.2). Figure 6.19, however, also shows that this kinetic potential cannot be exploited *ad infinitum*. For the presently considered case, i.e., for $N \cong 11.400$, it is found that, for the typical on-chip $\Delta P_{max} = 10$ bar case, decreasing d_p below $d_p \cong 10$ μm leads to a dramatic increase of the required analysis time. This can be understood from the fact that in this range the increased pressure drop resulting from the decreased particle size no longer allows to operate the packed bed at the optimal mobile-phase velocity, leading to an increase of H and the required column length, which in turn leads to an additional flow resistance, implicating that even smaller flow velocities have to be used. This mechanism eventually leads to the existence of a lower particle diameter boundary below which the desired separation resolution can no longer be achieved within the given available pressure-drop limitation. Figure 6.19 also shows that a tenfold increase of the available inlet pressure (which is totally inconceivable with the current state-of-the-art PD microfluidic systems) would lead to a tenfold decrease of the minimal analysis time This finding can in fact be generalized to the following general rule, stating that for a PD LC system, the minimal analysis time needed to achieve a given number of plates is inversely proportional to the available pressure drop:

$$t_{R,\,min} \sim 1/\Delta P \tag{6.40}$$

Equation (6.40) constitutes a clear incentive for the development of novel methods to increase the inlet pressure for on-chip LC systems.

Given that the available inlet pressure is clearly a limiting factor for the performance of a chromatographic system, and considering that the flow resistance of an open-tubular channel is roughly 10 to 20 times smaller than for the packed case, it is straightforward to assume that much faster separations can be obtained with open-tubular channels. This is clearly illustrated in Figure 6.20, comparing the required analysis time for an open-tubular and a packed-bed on-chip LC system. The data in Figure 6.21 were obtained by starting from the plate height value for the open-tubular system given by Equation (6.21). Rewriting this expression as

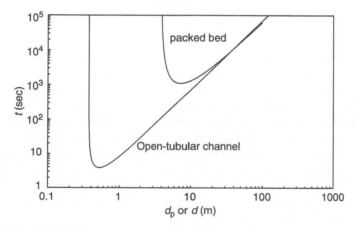

FIGURE 6.20 Comparison between a packed-bed and an open-tubular PD flow for an LC separation requiring $N = 11,400$ plates ($k' = 3$) and with $\Delta P = 20$ bar. The other parameters are identical to that in Figure 6.19.

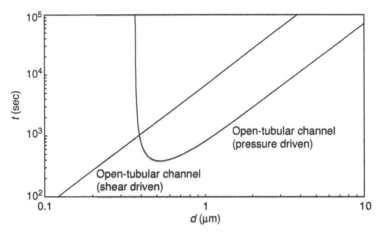

FIGURE 6.21 Comparison between an open-tubular PD flow and an open-tubular shear-driven flow for an LC separation requiring $N = 11,400$ plates ($k' = 3$ and with $\Delta P = 20$ bar for the PD flow). The other parameters are identical to that in Figure 6.19.

$$H = \frac{2D}{u_{\mathrm{m}}} + 2u_{\mathrm{m}}C \tag{6.41}$$

and combining this expression with Equations (6.38) and (6.11), it can easily be shown that the achievable velocity under the influence of a given available pressure drop is given by

$$u_{\mathrm{m}} = \sqrt{\frac{1}{2C}\left(\frac{\Delta P d^2}{12\mu N} - 2D\right)} \tag{6.42}$$

Equation (6.42) again shows that for each value of d, only one value of u_{m} exists for which the desired number of plates N can be achieved. Equation (6.42) also directly points at the existence of a minimal required channel thickness (determined by the condition that the expression under the square root sign in the right-hand side of Equation (6.42) should be ≥ 0). This value of d_{min} corresponds to the value for which the (t_{R}, d) curve for the open-tubular capillary shown in Figure 6.20 reaches its vertical asymptote. Comparing the packed and the open-tubular systems, it should be noted that, although the open-tubular channel data have been corrected for the side-wall effect described in Section 6.2.5, the resolution loss caused by a systematic deviation from a perfect flat-rectangular shape has been neglected.

To further illustrate the fundamental limitation of a PD chromatographic system, Figure 6.21 compares an open-tubular PD flow with an open-tubular shear-driven flow (cf. Section 6.2.6). As can be noted, the advantage of the operation without a pressure drop in the shear-driven mode can be large. The fact that with a shear-driven flow the velocity does not need to be reduced when the channel thickness is decreased clearly allows to fully exploit the kinetic advantage of nanometric thin channels. As can be seen from Figure 6.21, this possibility obviously opens the road to analysis times that are orders of magnitude smaller than the theoretical HPLC limit. Unfortunately, this also implies that, in order to realize this otherwise inaccessible separation potential, channels with a very small mass loadability have to be used. The realization of the full potential of the SDC concept will depend on future developments in the sensitivity of the detection equipment.

6.4.3 LOW-PRESSURE IMMUNOASSAYS AND ENZYMATIC ASSAYS + FLOW-THROUGH BIOSENSORS

Apart from the above-mentioned high-pressure applications for which the practical connection and sealing problems still form a major bottleneck, much attention has in the recent years also been paid to low-pressure applications, such as immunoassays and enzymatic assay, wherein relatively small fluid velocities and thick channels are required, such that simple, low-pressure syringe pumps can be used. Most of the current research efforts in this field are aiming at improvements in detection sensitivity, considering either improved optical and electrical detector designs [70], or by pursuing an increased mass loadability. For the latter, the fact that PD flows allow the use of silicon substrates provides an important advantage over ED, because silicon can easily be processed to a porous state, yielding a high surface area matrix comparable to any other solid-state matrices [71]. Recent applications of silicon microchips with enhanced loadability are an enzyme immunoassay with immobilized antibodies [71] and an enzymatic microreactor for the tryptic digest of β-casein [72]. Packed microchannels can of course also be used to increase the detection sensitivity of immunoassays [38]. Recently, a novel cancer diagnosis chip based on an integrated bead-bed immunoassay has been presented [73]. Good overviews of all recently developed on-chip immunoassays and flow-through biosensors are given in Refs. [74–76].

Another original and promising PD flow application is that of the 3D DNA hybridization microarray developed by Gene Logic Inc. [77]. In this concept (Figure 6.22), the hybridization fluid is pumped through the thinnest dimension of a glass slide perforated by an array of microchannels (typically 10 μm in diameter or less). The advantage of this approach is that the surface available for hybridization is dramatically increased compared to the conventional hybridization microarray systems using only the top glass surface. Another advantage is that the diffusional distances are decreased significantly. Overall, improvements in detection sensitivity of over 44 times that of the conventional 2D approach have been reported.

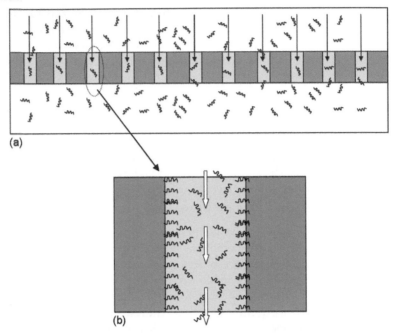

(a)

(b)

FIGURE 6.22 Schematic representation of the pressure-driven flow-through DNA hybridization system developed by Gene Logic Inc. [77]. (a) Global overview, showing the two fluid chambers from which the DNA sample is continuously pumped back and forth through the microchannels running across the glass slide. (b) Detailed view of a microchannel (not to scale) showing the hybridization of matching sample strands to the target probes immobilized along the wall of the microchannels.

6.5 CONCLUSIONS AND OUTLOOK

PD flows display a number of theoretical advantages over ED flows. Practical tube connection and sealing problems, however, impede the development of high-performance, industrially operable PD microfluidic analysis systems. The future of PD flows in the field of microanalytical systems, therefore, depends largely on improvements on the *hardware* side: the fabrication of flat-rectangular microchannels with a less than 1% tolerance on the lateral channel depth, better tube connection and sealing procedures allowing the use of inlet pressures above 100 bar, and the development of materials that are less prone to materials deformation stresses. For the conduction of complex reaction and separation networks on a single-chip surface, and for the development of portable PD microanalysis instruments for in-the-field analysis (another key promise of the lab-on-a-chip concept), the performance of the current on-chip valving and pumping systems is also far too insufficient.

Waiting for these revolutions to occur, the use of PD flows in microanalytical separations will remain limited to a number of (important) niche applications such as on-chip HDC for the rapid size separation of macromolecules and some types of flow-through biosensors. Desperation is however a bad counselor in this rapidly growing field of lab-on-a-chip systems, where novel, inventive solutions are brought out nearly every day.

REFERENCES

1. Bartle, K. D.; Myers, P. *J. Chromatogr. A* 2001, *916*, 3–23.
2. Poppe, H. *J. Chromatogr. Library* 1992, *51A*, 151–225.
3. Gale, B. K.; Caldwell, K. D.; Frazier, A. B. *Anal. Chem.* 2001, *73*, 2345–2352
4. Gale, B. K.; Caldwell, K. D.; Frazier, A. B. *Anal. Chem.* 2002, *74*, 1024–1030.
5. Ericson, C.; Holm, J.; Ericson, T.; Hjertén, S. *Anal. Chem.* 2000, *72*, 81–87.
6. Dutta, D.; Leighton, D. T. *Anal. Chem.* 2001, *73*, 504–513.
7. Incropera, F. P.; De Witt, D. P. *Fundamentals of Heat and Mass Transfer*, 2nd Ed.; John Wiley & Sons, New York, 1990.
8. Cifuentes, A.; Poppe, H. *Chromatographia* 1994, *255*, 391–404.
9. Paul, P. H.; Garguilo, M. G.; Rakestraw, D. J. *Anal. Chem.* 1998, *70*, 2459–2467.
10. Wilkes, J. O. *Fluid Mechanics for Chemical Engineers*; Prentice-Hall, London, 1999.
11. Coulson, J. M.; Richardson, J. F. *Chemical Engineering, Volume 2 — Particle Technology and Separation Processes*; Pergamon Press, Oxford, 1991.
12. Poppe, H. *Analusis* 1994, *22*, 22–24.
13. Poppe, H. *J. Chromatogr. A* 1997, *778*, 3–21.
14. Ruthven, D. *Principles of Adsorption and Adsorption Processes*; Wiley-Interscience, New York, 1984.
15. Aris, R. *Proc. Roy. Soc. A* 1959, *252*, 538–550.
16. Golay, J. E. *J. Chromatogr.* 1981, *216*, 1–8.
17. Desmet, G.; Baron G. V. *J. Chromatogr. A* 2002, *946*, 51–58.
18. Knox, J. H. *J. Chromatogr. A* 1999, *831*, 3–15.
19. Vissers, J. P. C. *J. Chromatogr. A* 1999, *856*, 117–143.
20. Martin, M.; Jurado-Baizaval, J.-L.; Guiochon, G. *Chromatographia* 1982, *16*, 98.
21. Poppe, H. *J. Chromatogr. A* 2002, *948*, 3–17.
22. Desmet, G.; Baron G. V. *J. Chromatogr. A* 1999, *855*, 57–70.
23. Desmet, G.; Vervoort, N.; Clicq, D.; Gzil, P.; Huau, A.; Baron, G. V. *J. Chromatogr. A* 2002, *948*, 19–34.
24. van Lintel, H. T. G.; van de Pol, F. C. M.; Bouwstra, S. *Sens. Actuators* 1988, *15*, 153–167.
25. Koch, M.; Evans, A.; Brunschweiler, A. *Microfluidic Technology and Applications*; Research Studies Press Ltd., Baldock, UK, 2000.
26. Beebe, D. J.; Moore, J. S.; Bauer, J. M.; Yu, Q.; Liu, R. H.; Devadoss, C.; Jo, B. H. *Nature* 2000, *404*, 588.
27. Ismagilov, R. F.; Rosmarin, D.; Kenis, P. J. A.; Chiu, D. T.; Zhang, W.; Stone, H. A.; Whitesides, G. M. *Anal. Chem.* 2001, *73*, 4682–4687.
28. Hasselbrink, E. F.; Shepodd, T. J.; Rehm, J. E. *Anal. Chem.* 2002, *74*, 4913–4918.
29. Reyes, D. R.; Iossifidis, D.; Auroux, P.-A.; Manz, A. *Anal. Chem.* 2002, *74*, 2623–2636.

30. Mourlas, N. J.; Jaeggi, D.; Flannery, A. F.; Gray, B. L.; van Drieënhuizen, B. P.;
    Storment, C. W.; Maluf, N. I.; Kovacs, G. T. A. In *Proceedings of the μTAS '98
    Workshop*, Banff, Canada; Kluwer Academic Publishers, Dordrecht, 1998;
    pp. 27–30.
31. Nittis, V.; Fortt, R.; Legge, C. H.; De Mello, A. J. *Lab-on-a-Chip* 2001, *1*,
    148–152.
32. Gonzalez, C.; Collins, S. D.; Smith, R. L. *Sens. Actuators B* 1998, *49*, 40–45.
33. Chmela, E.; Blom, M. T.; Gardeniers, J. G. E.; van den Berg, A.; Tijssen, R. *Anal.
    Chem.* 2002, *74*, 3470–3478.
34. McEnery, M.; Tan, A.; Alderman, J.; Patterson, J.; O'Mathuna, S. C.; Glennon,
    J. D. *Analyst* 2000, *125*, 25–27.
35. O'Neill, A. P.; O'Brien, P.; Alderman, J.; Hoffman, D.; McEnery, M.; Murrihy, J.;
    Glennon, J. D. *J. Chromatogr. A* 2001, *924*, 259–263.
36. Ocvirk, G.; Verpoorte, E.; Manz, A.; Grasserbauer, M.; Widmer, M. *Anal. Meth.
    Instrum.* 1995, *2*, 74–82.
37. Colòn, L. A.; Maloney, T. D.; Fermier, A. M. *J. Chromatogr. A* 2000, *887*, 43–53.
38. Oleschuk, R. D.; Schultz-Lockyear, L.L.; Ning, Y.; Harrison, D. J. *Anal. Chem.*
    2000, *72*, 585–590.
39. Ceriotti, L.; de Rooij, N. F.; Verpoorte, E. *Anal. Chem.* 2002, *74*, 639–647.
40. Hjertén, S.; Liao, J. L.; Zhang, R. *J. Chromatogr.* 1989, *473*, 273.
41. Liao, J. L.; Chen, N.; Ericson, C.; Hjertén, S. *Anal. Chem.* 1996, *68*, 3468.
42. Wang, Q. C.; Svec, F.; Fréchet, J. M. J. *J. Chromatogr. A* 1994, *669*, 230–235.
43. Viklund, C.; Ponten, E.; Glad, B.; Irgum, K. *Chem. Mater.* 1997, *9*, 463–471.
44. Tanaka, N.; Kobayashi, H.; Nakanishi K.; Minakuchi, H.; Ishizuka, N. *Anal.
    Chem.* 2001, *73*, 420–429.
45. Peterson, D. S.; Rohr, T.; Svec, F.; Frechet, J. M. J. *Anal. Chem.* 2002, *74*,
    4081–4088.
46. Regnier, F. E. *J. High Resol. Chromatogr.* 2000, *23*, 19–26.
47. He, B.; Tait, N.; Regnier, F. E. *Anal. Chem.* 1998, *70*, 3790–3797.
48. Slentz, B. E.; Penner, N. A.; Regnier, F. *J. Chromatogr. A* 2002, *948*, 225–233.
49. Slentz, B. E.; Penner, N. A.; Regnier, F. *Electrophoresis* 2001, *22*, 3736–3743
50. Knox, J. *J. Chromatogr. A* 2002, *960*, 7–18.
51. Liang, Z.; Chiem, N.; Ocvirk, G.; Tang, T.; Fluri, K; Harrison, J. *Anal. Chem.*
    1996, *68*, 1040–1046.
52. Fan, Z. H.; Harrison, D. J. *Anal. Chem.* 1994, *66*, 177.
53. Desphande, M.; Greiner, K. B.; West, J.; Gilbert, J. R.; Bousse L.; Minalla, A. In
    Proc. ÌTAS 2000, Kluwer Academic Publishers, Dordrecht, The Netherlands;
    2000, pp. 339–343.
54. Effenhauser, S. C.; Manz, A.; Widmer, H. M. *Anal. Chem.* 1993, *65*, 2284.
55. Ruzicka, J.; Hansen, E. H. *Anal. Chim. Acta* 1984, *161*, 1.
56. Polson, N. A.; Hayes, M. A. *Anal. Chem.* 2001, *73*, 312A–319A.
57. Chmela, E., PhD Thesis, University of Amsterdam, The Netherlands, 2002.
58. Chi-san Wu (Ed.), *Handbook of Size Exclusion Chromatography*, Chromato-
    graphic Science Series, Vol. 69; Marcel Dekker, Inc., New York, 1995, p. 453.
59. Bos, J.; Tijssen, R. Hydrodynamic chromatography of polymers (Chapter 4). In
    Chromatography in the Petroleum Industry.; Adlard, E. R. (Ed.); *J. Chromatogr.
    Library Series* 1995, *56*, 95–126.
60. Venema, E.; Kraak, J. C.; Poppe, H.; Tijssen, R. *Chromatographia* 1998, *48*, 347.

61. DiMarzio, E. A.; Guttman, C. M. *Macromolecules* 1970, *3*, 131–146.
62. Brenner, H.; Gaydos, L. J. *J. Colloid. Interface Sci.* 1977, *58*, 312–356.
63. Unpublished theoretical study in cooperation with Dr H. Hoefsloot (University of Amsterdam, The Netherlands).
64. Tijssen, R. In *Theoretical Advances in Chromatography and Related Separation Techniques*; Dondi, F.; Guiochon, G. (Eds.); Kluwer Academic Publishers, Dordrecht, The Netherlands; 1992, pp. 397–441.
65. Ploehn, H. J. *Int. J. Multiphase Flow* 1987, *13*, 773–784.
66. Hollingsworth, A. D.; Silebi, C. A. In *Particle Size Distributions III, Assessment and Characterization*; Provder, Th. (Ed.); ACS Publishing, Washington, DC; 1998, pp. 244–265.
67. Small, H. *J. Colloid. Interface Sci.* 1974, *48*, 147–161.
68. Blom, M.; Chmela, E.; Gardeniers, J. G. E.; Tijssen, R.; Elwenspoek, M.; van den Berg, A. *Sens. Actuators B* 2002, *82*, 111–116.
69. Edwards, T. L.; Gale, K.; Frazier, A. B. *Anal. Chem.* 2001, *74*, 1211–1216.
70. Luppa, P. B.; Sokoll, L. J.; Chan, D. W. *Clin. Chim. Acta* 2001, *314*, 1–26.
71. Yavkovleva, J.; Davidsson, R.; Lobanova, A.; Bengtsson, M.; Eremin, S.; Laurell, T.; Emneus, J. *Anal. Chem.* 2002, *74*, 2994–3004.
72. Bengtsson, M.; Ekström, S.; Marko-Varga, G.; Laurell, T. *Talanta* 2002, *56*, 341–353.
73. Sato, K.; Tokeshi, M.; Kimura, H.; Kitamori, T. *Anal. Chem.* 2001, *73*, 1213–1218.
74. Verpoorte, E., *Electrophoresis* 2002, *23*, 677–712.
75. Lichtenberg, J.; de Rooij, N. F.; Verpoorte, E. *Talanta* 2002, *56*, 233–266.
76. Khandurina, J.; Guttman, A. *J. Chromatogr. A* 2002, *943*, 159–183.
77. Benoit, V.; Steel, A.; Torres, M.; Yu, Y.-Y.; Yang, H.; Cooper, J. *Anal. Chem.* 2001, *73*, 2412–2420.

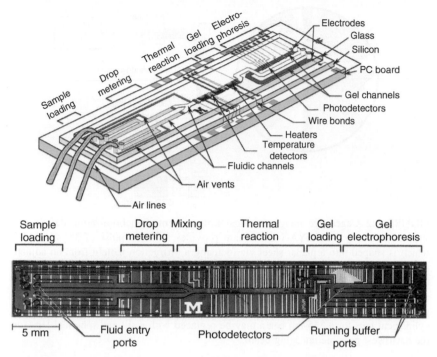

FIGURE 1.1 Integrated silicon-based device for DNA analysis. (From Burns MA, Johnson BN, Brahmasandra SN, Handique K, Webster JR, Krishnan M, Sammarco TS, Man FP, Jones D, Heldsinger D, Namasivayam V, Mastrangelo CH, Burke DT. *Science* 1998, *282*, 484–487. With permission.)

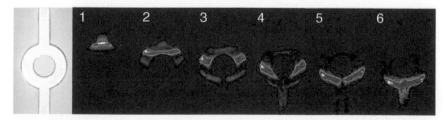

FIGURE 4.1 First frame: a microfluidic channel (50 μm wide × 10 μm deep) with a circular feature, with flow inlet and outlet at the top and bottom, respectively. Numbered frames: caged dye is uncaged just above the entrance to the circular feature, using a 10-μm-wide beam, achieved using a Fourier-transformed diffraction optic. Flow is electroosmotically driven flow at an electric field of approximately 400 V/cm; fluid is 50:50 acetonitrile–phosphate buffer pH 7.6. In frame 3, a leading blue band is visible; this is a positively charged impurity which electrophoretically separates from the zwitterionic dye (yellow band).

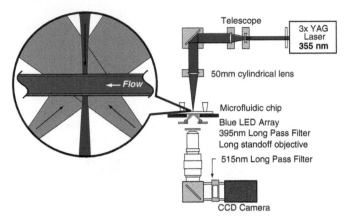

FIGURE 4.2 Caged-dye imaging optical arrangement. Synchronization details not shown for clarity. A Nd:YAG laser beam provides a 5-nsec pulse of 355-nm light, which is shaped into a planar sheet via focusing a diffraction pattern from an adjustable slit using a cylindrical lens. The laser sheet strikes the channel, in which caged dye is uniformly distributed, and the dye is locally photoactivated. Continuous or pulsed illumination from a blue light source (e.g., Ar$^+$ laser, blue LED, Hg lamp) permits observation of the convection of the dye through the channel. Telescope and cylindrical lens for a 355-nm beam may be replaced with a slit and spherical lens for the narrowest possible uncaging beam.

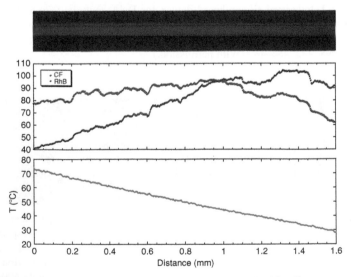

FIGURE 4.9 Top: a color fluorescence image of a channel with a linear temperature gradient. Middle: fluorescence signals from carboxyfluorescein and rhodamine B. Bottom: temperature profile, deduced from the ratio of these signals. Note the rejection of common-mode noise from the raw images due to the use of a ratiometric technique. (Image courtesy David Ross [NIST].)

FIGURE 4.13 A PIV particle image and the resulting high-resolution map of velocity magnitude for steady EOF over an array of circular posts. A banded grayscale table is used to show the variations in velocity, which is very slow at the leading and trailing edge (left and right edge in the image) of each post, and fastest at the sides (top and bottom edge in the image) of each post. Flow is from left to right, but the velocity magnitude $(u^2 + v^2)^{1/2}$ is symmetric about each post. High resolution is made possible from a statistical technique that allows for grid refinement in steady flows. (Images courtesy E.B. Cummings, Sandia National Labs.)

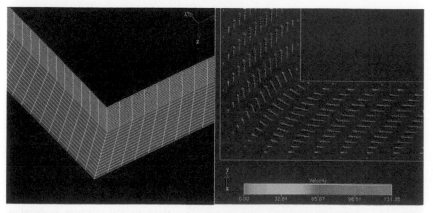

FIGURE 4.15 Image of a discretized corner of a microfluidic channel (left), showing the automatically generated computational grid. Refining the grid very close to the sharp bend may be needed to achieve accurate solution. The solution for the velocity field (vectors colored by their magnitude) of this coarse grid is shown at the right.

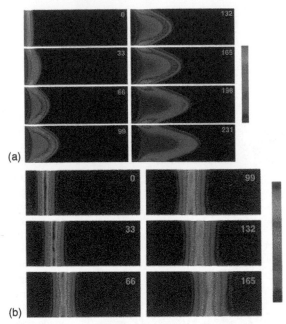

(a)

(b)

FIGURE 5.1 Images of (a) pressure-driven flow through an open 100-μm-i.d. fused-silica capillary using a caged fluorescein dextran dye (pressure differential of 5 cm of H₂O per 60 cm of column length, viewed region 100 μm by 200 μm) and (b) electro-kinetically driven flow through an open 75-μm-i.d. fused-silica capillary using a caged rhodamine dye (applied field of 200 V/cm, viewed region 75 μm by 188 μm). The frames are numbered in milliseconds as measured from the uncaging event.

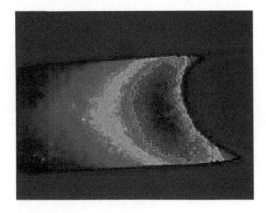

FIGURE 5.10 Close-up of a chromatography column in a CD channel. Flow direction is from right to left. Picture is from a fluorescence microscope.

17.5%   20%   22.5%   25%   27.5%

FIGURE 5.13 Stepwise elution of Cy5-labeled angiotensin I from Cy5 dye. Pictures show the early stage of the elution as free Cy5 dye separates from Cy5-angiotensin I. The eluent flows from right to left in the pictures. ACN concentrations are given as a percentage under the respective elution.

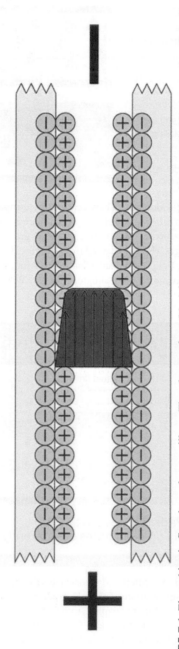

FIGURE 7.1 Electrokinetic flow in a glass capillary. The glass surface (blue) has a net negative charge at the glass–water interface. Positive counter-ions from the glass (pink) are tightly bound but undergo electrophoresis along the device surface in response to the applied potential. The resulting velocity profile (yellow) is nearly uniform.

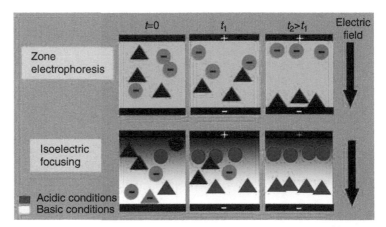

FIGURE 7.12 Comparison between zone electrophoresis (ZE) and isoelectric focusing (IEF). The two different species of particles are represented by triangles and circles. The pH gradient created in IEF is represented by a gradient of color between the electrodes in the lower images. (Figure adapted from Cabrera, C., Microfluidic Electrochemical Flow Cells: Design, Fabrication, and Characterization. Doctoral dissertation in Bioengineering, 2002, University of Washington: Seattle, WA.)

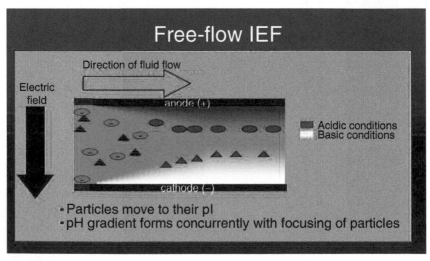

FIGURE 7.13 Schematic of transverse IEF implemented in a microfluidic device. Note the self-generation of the pH gradient as OH^- is produced at the cathode and H^+ at the anode. (Figure adapted from Cabrera, C., Microfluidic Electrochemical Flow Cells: Design, Fabrication, and Characterization. Doctoral dissertation in Bioengineering, 2002, University of Washington: Seattle, WA.)

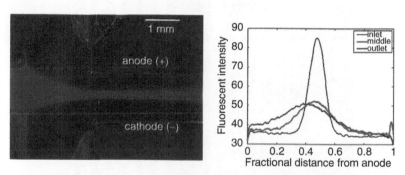

FIGURE 7.17 At left is an image of three solutions entering the channel between the two electrodes in a zone electrophoresis experiment. The central stream is entering from the left. At right is a set of fluorescence intensity profiles taken across the channel at a different point. Note the effects of both electrophoresis (to the left) and diffusion, which broadens the band with time. (Figure adapted from Cabrera, C., *Microfluidic Electrochemical Flow Cells: Design, Fabrication, and Characterization*. Doctoral dissertation in Bioengineering, 2002, University of Washington: Seattle, WA.)

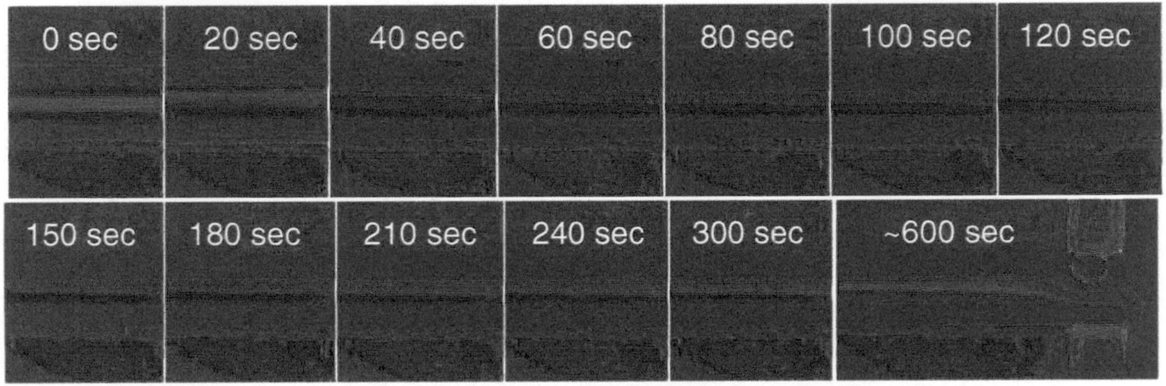

FIGURE 7.18 Images of the effect of time of application of voltage on the continuous separation of bovine serum albumin (BSA) (orange) and lectin (green) fluorescent conjugates over time in a flow channel. The proteins were pumped into the channel together (at left). Fluorescent images near the exit were made at the times shown after the application of 2.0 V. Note that separation is most pronounced about 60 sec after application of the voltage, followed by progressive loss of resolution and, perhaps, loss of fluorescence from the BSA. Note that reversing the voltage in this experiment immediately restores the "lost" fluorescence, implying that the BSA fluorophores were only reversibly quenched, not chemically altered. (From Macounová, K., C.R. Cabrera, and P. Yager, *Analytical Chemistry*, 2001, 73(7): 1627–1633. With permission.)

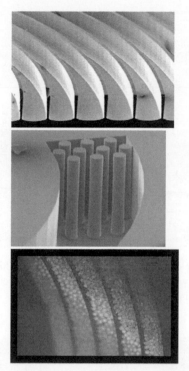

FIGURE 9.12 Microfabricated packed column; cross section is 300 μm by 300 μm; posts are used as "gate" material to retain packing; packing shown is a small diameter porous polymer (70 μm, Hayesep A).

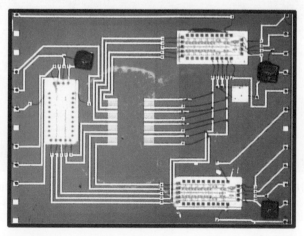

FIGURE 9.14 Surface acoustic wave array detector showing three prominent application-specific integrated circuits for radio frequency excitation and phase detection. Four delay line SAW devices shown in center.

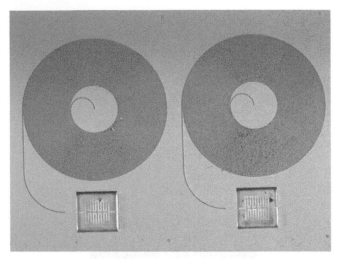

FIGURE 9.17 Prototype chip integration of micromachined high aspect ratio column and membrane device (two devices shown). Membrane is used as solid-phase micro-extraction device and serves as GC injector.

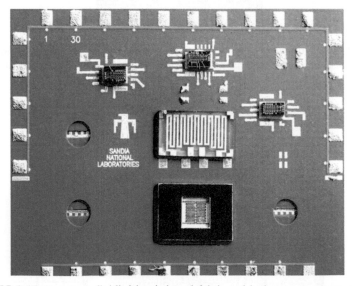

FIGURE 9.18 Prototype fluidic/circuit board fabricated in low-temperature co-fired ceramic. Membrane preconcentrator and SAW array attached in pick and place scheme. Fluidic vias are circular features.

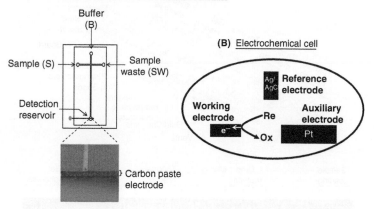

FIGURE 11.17 (A) A typical microchip CE configuration for amperometric detection with end-channel detection. (B) A three-electrode electrochemical cell which contains a reference, auxiliary, and working electrode.

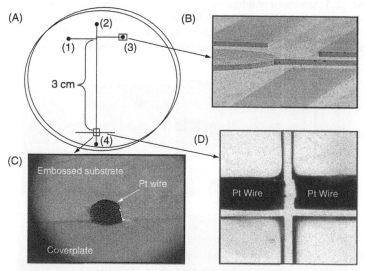

FIGURE 11.21 (A) Topographical layout of the PMMA-based microchip with an integrated conductivity detector. (1) sample reservoir, (2) buffer reservoir, (3) waste reservoir, and (4) receiving reservoir. (B) SEM of the Ni electroform embossing die used to make the chip. (C) Optical micrograph of PMMA microchip that was assembled with a coverplate and electrodes and then cut down the center of the fluidic channel. (D) Optical micrograph of the conductivity detector (T-cell, electrode gap ~20 μm) integrated to the PMMA microfluidic device. (Reprinted from Galloway, M.; Styjewski, W.; Henry, A.; Ford, S. M.; Llopis, S.; McCarley, R. L.; Soper, S. A. *Anal. Chem.* 2002, *74*, 2407–2415. With permission.)

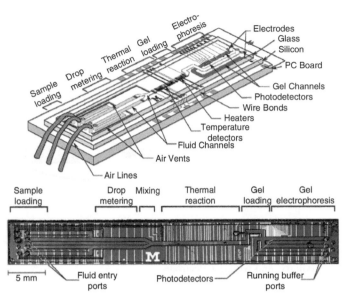

FIGURE 12.5 Schematic of integrated device with two liquid samples and electrophoresis gel present. (From Burns, M.A.; Johnson, B.N.; Brahmasandra, S.N.; Hhandique, K.; Webster, J.R.; Krishnan, M.; Sammarco, T.S.; Man, P.M.; Jones, D.; Heldsinger, D.; Mastrangelo, C.H.; Burke, D.T., *Science* 1998, 282, 484. With permission.)

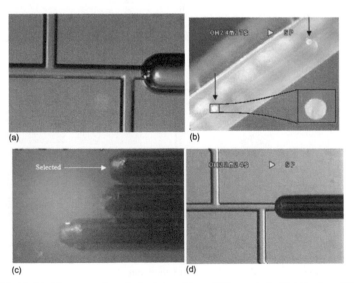

FIGURE 12.11 Pictures of real devices: (a) a 400-μm twin-T injector; (b) cross section profile of a round channel; (c) ground capillary tips; and (d) a twin-T injector incorporated with a capillary.

7 Transverse Transport in Microsystems: Theory and Applications

Paul Yager, Catherine Cabrera, and Andrew Kamholz

CONTENTS

7.1   MOTIVATIONS

While most of the progress made in microfluidics in the last decade has, for very good reasons, been devoted to enhancing the performance of capillary electrophoresis on a chip, there have been very good reasons to pursue problems that have been both intellectually and physically orthogonal to the mainstream. This chapter will discuss issues related to the transport of matter *across* flow lines. One strong motivator for this work has been extending the utility of microfluidic processes beyond the clean homogeneous samples necessary for good chromatographic separations. "Raw" samples, particularly biological fluids, contain a wide range of proteins and larger particles, most of which are present at concentrations not under experimental control. This variability in composition, with concomitant variability in fluid conductivity, pH, and concentration of surface-active compounds, renders such samples incompatible with electrokinetic pumping — the cornerstone of microfluidic capillary electrophoresis. Such samples, therefore, dictate the use of pressure-driven flow.

The current chapter will dwell on issues related to the transport of matter across flow lines in such pressure-driven devices, some of which may have multiple input and output streams. The driving force for transfer of solutes between parallel streams may be simple diffusion down a concentration gradient, or may involve application of an external force, such as gravitation or electric fields. Unlike capillary electrophoresis and other chromatographic techniques, separation relies on spatial, not temporal, differences in elution. This difference facilitates adaptation of these transverse separation techniques to continuous-flow applications. While the basic operating principles of these "transverse devices" are well known, there are important (nonintuitive) secondary effects caused by the presence of flow perpendicular to the relevant fields. As a consequence, application of one-dimensional solutions to the distribution of mass is generally inadequate to describe the operation of the devices when flow is present. Because such devices operate best in circumstances where samples are large (compared to the volume of the channels in which they flow), the devices are most suitable to "continuous" operations. They can be useful as components of a variety of practical instruments or quantification of the concentrations of analytes and fractionation of complex samples. The first applications of these devices will probably be in point-of-care diagnostics and detection of chemical and biological warfare agents, although one can foresee many others.

What follows is a brief description of some of the fundamentals required for understanding transverse transport under pressure-driven flow, and of examples of how such transport can be put to useful work in microfluidic devices for research and clinical diagnostics. This discussion will be limited to Newtonian fluids, and generally to Reynolds numbers sufficiently low that turbulence is not a factor.

7.2 FUNDAMENTALS OF PRESSURE-DRIVEN FLOW IN MICROCHANNELS

Nearly all microfluidic devices incorporate fluid flow for their operation. In some cases, this flow is limited to the introduction of a sample, and measurements are made after a binding or incubation step. In other cases, however, flow is continuous and measurements are performed at some point of detection downstream. The method of fluid actuation is usually one of two types: pressure-driven flow or electrokinetic pumping. In the former case, flow is driven by an applied pressure, usually produced by a pressurized gas source, a hydrostatic column of fluid, or the positive displacement of an actuator such as a syringe. By contrast, electrokinetic flow (also called electroosmotic flow) is induced by electrophoresis of solvent molecules in a potential gradient. It is covered extensively in other chapters in this book.

Many microfluidic devices use pressure-driven flow. In the earliest days of the microfluidics renaissance of the 1990s, the majority of investigators actuated flow using glass tubing attached to the backside of devices with epoxy. These open-air reservoirs allowed users to pipette columns of liquid reagents onto individual devices. In this scenario, the flow rate, determined entirely by the hydrostatic pressure and fluidic resistance, necessarily changed continuously as the reservoir levels changed with flow. Therefore, this method of fluid actuation was most useful for early test-phase device, as opposed to quantitative assays.

The need for controlled, constant flow rates turned most users of pressure-driven microfluidic device to syringe pumps. These devices, utilizing positive displacement, can provide a steady, predetermined flow rate against the typical resistance of a microfluidic network. The adoption of syringe pumps also incorporated a number of useful features, such as computer control, programmability, valving, and sequencing, as well as allowing substantially more complex fluidic manipulations.

In addition to syringe pumps, a number of methods are used to generate pressure driven flow. One still-common method is the use of hydrostatic pressure, generated with columns of liquid. Converse to the old style use of small columns, modern hydrostatic pressure sources use tall columns with large volumes (100 ml or more), such that the volumes transported during device operation (less than 1 ml) are small enough to have negligible effect on the pressure gradient driving flow. Some laboratories use peristaltic pumps, avoiding the high cost of syringe pumps while sacrificing some measure of steadiness of flow. Gas pressure can also be effectively used, both in the form of a pressurized gas source for pushing fluid, as well as a vacuum source for pulling fluid.

7.2.1 VELOCITY PROFILES IN MICROSYSTEMS

The majority of microfluidic devices that do not use pressure-driven flow instead actuate fluid motion through electrokinetic pumping. Based on traditional capillary electrophoresis, the use of an applied voltage to induce fluid transport was first adapted to microsystems in the mid-1990s. Utilizing a high-voltage power supply, an aqueous phase is mobilized utilizing loosely associated counter-ions from the vessel material, such as glass (Figure 7.1).

The most widely utilized feature of electrokinetic pumping is its nearly uniform (or blunt) velocity profile. This allows reagents to be transported as plugs through the channels. In addition, molecules or particles in the aqueous phase react to the applied electric field in accordance with their respective electrophoretic mobilities. Therefore, each entity with a different electrophoretic mobility will have a unique velocity through the system, after considering the superimposed electrokinetic velocity of the carrier fluid and electrophoretic velocity of the individual substance.

In pressure-driven flow systems, the velocity profile always has substantial nonuniform character. In a round capillary using pressure-driven flow, the fully developed velocity profile is a paraboloid, with the maximum velocity in the center of the channel and a zero-velocity no-slip condition imposed at the inner wall of the capillary. This flow, termed Poiseuille, has an average fluid velocity that is one half of the maximal velocity.

In microfluidics, there are many instances of pressure-driven channels with aspect ratios much greater than 1. This is due in part to the nature of popular fabrication methods, whereby channels are often 100 μm to several millimeters wide, while usually less than 100 μm in etch depth (or thickness of photoresist, as the case may be). Where the aspect ratio of the channel dimensions is greater than about 4, the resulting pressure-driven velocity

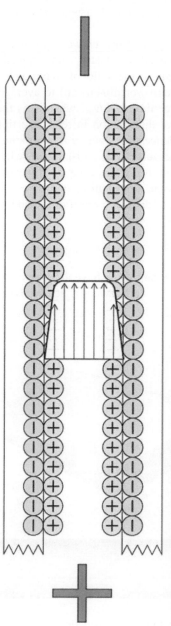

FIGURE 7.1 (See color insert following page 208) Electrokinetic flow in a glass capillary. The glass surface has a net negative charge at the glass–water interface. Positive counter-ions from the glass are tightly bound but undergo electrophoresis along the device surface in response to the applied potential. The resulting velocity profile is nearly uniform.

profile has parabolic character across the narrower dimension, and is unchanging across the majority of the wider dimension. The exact fully developed velocity profile for a rectangular duct with aspect ratio of 22 is shown in Figure 7.2.

Approximating the geometry as flow between infinite parallel plates (valid for aspect ratios of about 20 or greater, and neglecting behavior at the edges of the wider dimension), the ratio of maximum velocity (v_{max}, occurs at the midplane of the narrower dimension) to average velocity (v_{av}) is 1.5 [1]. The average residence time is L/v_{av}, where L is the length of the rectangular duct. The theoretical distribution in residence time ranges from two-thirds of the average residence time (where the velocity is v_{max}) up to infinity (at the walls, where the velocity is zero). There will therefore be a corresponding distribution in the residence time of any particle or analyte contained in the fluid. This time distribution depends on many factors, including the size, concentration, diffusion coefficient, and spatial distribution of the species.

7.2.2 LAMINAR FLOW

Because microfluidic systems by definition have at least one very small (less than 1 mm) dimension, the flow in such devices is nearly always laminar. The

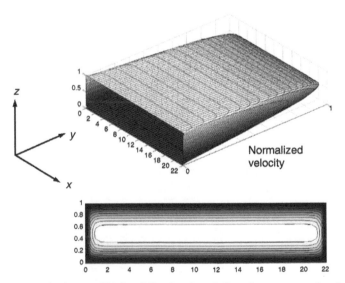

FIGURE 7.2 Velocity profile for fully developed flow in a rectangular duct with aspect ratio of 22 (top). The bottom image shows a contour plot of the same data, with maximal velocity in white and minimal velocity in black. Note that the flow is parabolic across the narrower dimension.

most common metric used to characterize the type of flow is the Reynolds number (*Re*), given by

$$Re = \frac{\rho v l}{\eta} \qquad (7.1)$$

where ρ is the fluid density, v the characteristic velocity, l is a characteristic length dimension, and η the fluid viscosity. The exact parameter selections for v (choosing between maximum velocity or average velocity) and for l (choosing among channel width, depth, and length) are not critical as *Re* is only meant as a broad indicator. However, when comparing among different microfluidic devices, it is important to consider these selections, particularly of l, as vastly different definitions may affect the ability to directly compare *Re* values. It is noteworthy that many authors, when appropriate, use hydraulic radius (four times the ratio of cross-sectional area to cross-sectional perimeter) as the length scale.

The Reynolds number, a nondimensional parameter, is commonly interpreted as the ratio of inertial forces to viscous forces. A larger *Re* is indicative of dominance of fluid momentum, promoting flow separation and recirculation, and even turbulence with fast enough flows. A smaller *Re* is conversely indicative of dominance of viscous interactions, where flows generally do not have enough energy to separate. The transition between laminar flow and turbulent flow occurs generally around a *Re* of 1000; most microfluidic systems have *Re* on the order of 1. Consequently, the flow is not only laminar, but is usually in a regime of laminar flow referred to creeping flow, where viscous interactions completely dominate the behavior.[1] There are occasional microfluidic devices with higher *Re*, where some near-turbulent characteristics are seen; these devices usually have somewhat larger device dimensions or very fast flow rates (~ml/min) and *Re* substantially larger than 1. These systems are the exception and represent no more than 1% of the total population of microfluidic devices.

Laminar flow presents a number of major advantages and disadvantages. The most significant advantage is the predictability of device behavior. Since flow will be laminar for almost any reasonable flow rate, the designer can predict with high accuracy *a priori* how the given device will function. Designers also benefit from not having to worry about whether particular bends, corners, or bifurcations will induce flow separation or recirculation. Microfluidics also offers unique ability to handle very small samples by avoiding the irregularity of turbulence. The largest disadvantage is being unable to use turbulence when desired, most notably in mixing. While there

---

[1] For excellent visualized examples of turbulent and laminar flow, see *An Album of Fluid Motion* by Martin van Dyke, Parabolic Press, 1982.

are a number of microsystems that use particular geometries or flow schemes to minimize mixing time by diffusion, it is still the case that very few truly microfluidic systems can mix reagents any faster than can be achieved at the benchtop scale. Microsystems are also, by their nature, unable to process large volumetric flow rates. Although many groups offer the promise of multi-plexed or scaled systems, very few examples of such schemes have evolved, meaning that many of the novel measurements and manipulations made in microdevices are as of yet unfeasible for realistic operations.

7.2.3 BASIC RELATIONSHIPS OF LAMINAR FLOW

Many types of flow systems can be characterized by the well-known Navier–Stokes equations, which relate many aspects of fluid flow, including kinetic energy distribution, pressure effects, viscous forces, and gravitational forces [2]. There are many examples of simplifications of these equations for particular flow systems. One common simplification is the assumption that water is an incompressible fluid, which is often nearly true to the extent that compression effects can be effectively neglected. Because viscous forces dominate microfluidic systems, none of the time-dependent behavior characteristic of turbulent flows is present. Furthermore, in systems that avoid compliant elements, such as large flexible storage loops, there is often nearly no time dependence, and systems can reach steady flow within a few seconds of initiation.

To describe a laminar microfluidic system, it is useful to begin with the equation of continuity for incompressible flow [2]:

$$\frac{\partial c_N}{\partial t} + \left(v_x \frac{\partial c_N}{\partial x} + v_y \frac{\partial c_N}{\partial y} + v_z \frac{\partial c_N}{\partial z} \right) = D_N \left(\frac{\partial^2 c_N}{\partial x^2} + \frac{\partial^2 c_N}{\partial y^2} + \frac{\partial^2 c_N}{\partial z^2} \right) + X_N \quad (7.2)$$

The coordinate axes are as in Figure 7.2, c_N is the concentration of species N, v is velocity, D_N is diffusion coefficient of species N, and X_N is the production or elimination of species N by chemical reaction. A number of simplifications can be used for a microfluidic geometry, particularly if only a single straight main channel is considered. Modeling of the behavior in flow inlets and curved channels is more easily achieved by numerical methods [3] than by analytical means. If continuous flow inputs of fixed concentration are assumed (instead of bolus), this allows assumption that the flow is steady state, making the time-dependent term in Equation (7.2) zero. Furthermore, if the flow development is neglected, all velocities in the x- and y-directions are zero. Also, because convection in this system is the dominant form of transport in the z-direction, diffusion in this dimension need not be considered. The resulting simplified equation is

$$v_z \frac{\partial c_N}{\partial z} = D_N \left(\frac{\partial^2 c_N}{\partial x^2} + \frac{\partial^2 c_N}{\partial y^2} \right) + X_N \quad (7.3)$$

In this simplified form, the rate of change of concentration while traversing the channel in the z-direction changes as molecules diffuse down x- and y-concentration gradients and are consumed or produced by chemical reaction. The rates of change are scaled inversely by the local velocity. This confirms what intuition reveals: near the walls of the device, where the velocity is slowest, individual molecules have a greater opportunity to diffuse or react because their residence time is high. Similarly, molecules near the center of the device will tend to experience the least diffusion and reaction since their residence time in the device is the shortest.

In some geometries, the diffusion in the y-direction need not be explicitly considered. This is true when the size of the y-dimension is substantially smaller than the x-dimension, and the average residence time in the device is sufficiently long that molecules diffuse, on average, further than the y-dimension. The above equation can therefore be further simplified by dropping the concentration gradient term in the y-direction, as this is assumed to be zero. A complete discussion of the proper use of different forms of the analytical solution has been made elsewhere [3].

7.2.4 OTHER FORCES THAT MAY MOVE SOLUTES AND SUSPENDED PARTICLES

There are forces that move solutes across flow lines other than diffusion down concentration gradients. These can be of varying importance depending on the nature of the device and the flow rates used. However, they cannot be ignored. Their importance varies with the nature of the fluid being processed.

7.2.4.1 Gravitation

Gravity is the most serious perturbant on a flow system. It can influence both solutions and suspended particles. If two or more solutions of different densities are placed in a microfluidic channel, even though the dimensions of the channel may be extremely small, there is a tendency for the solutions to seek to stack in order of their density, independent of how they are introduced into the system. Therefore solutions that are introduced into a microfluidic channel side-by-side may be found stacked one atop the other by the time they exit the channel. Clearly, this could result in unexpected outputs from the channel exit ports. This sort of "macroscopic" rearrangement of flow is very sensitive to initial conditions and minor perturbations in the system. Obviously if the solutions are miscible and consist of highly diffusive molecules, the streams may completely interdiffuse before they exit the channel, but at any time before complete interdiffusion of the slowest-diffusing component, there will be a problem.

Even if the solutions that enter a microchannel are similar in density, particles in those solutions that are large enough to sediment a significant fraction of the channel dimensions during the transit of the solutions through the channel will be in the wrong locations by the time they exit the channel. This has many implications. First, the particles will no longer be traversing the channel at the mean flow rate — they will be moving either faster or slower depending on their (changing) distribution among the flow lines. Furthermore, if those particles are introduced as a bolus into the microchannel, the dispersion of that bolus will not be that predicted in a gravity-free simulation. In the worst case, they can all sink to the lowest wall in the channel, and there travel at extremely low rates through the channel. If the particle density is lower than the solution in which it is suspended, positive buoyancy with float the particles to the top of the channel. This can be used to one's advantage, of course, if the aim is to remove such particles, or to bring them to an interface between two streams, for example. In general, for particles consisting of reasonable densities, sedimentation can be ignored for particles much smaller than 1 μm.

These problems may be avoided by proper layering of the solutions, or working in a geometry in which gravity has no effect on motions perpendicular to the flow direction.

7.2.4.2   Magnetic (Magnetic Field Applied Transverse to Flow Direction)

If forces other than gravity can be applied to particles, they can be pulled to arbitrary locations in the channel by altering the direction of the applied field (when using gravity, the device itself must be moved to change the position of the particles). The most common approach has been to pull magnetic particles to one side of the channel. Because very strong local magnetic fields can be applied (by on-chip or off-chip magnets), it is possible not only to move particles from one streamline to another, but even to immobilize particles at one wall of the channel until the field is turned off. The magnetic particles can be of any size that allows them to be at least temporarily suspended in the flow stream. These particles can now be purchased with surfaces that allow them to be modified with capture molecules such as antibodies. This has been used as an approach to microfluidic analyte capture [4] as it has been used commercially for some time in commercial macroscopic biochemical systems (see products offered by Dynal and Bangs Laboratories, for example).

7.2.4.3   Hydrodynamic Focusing (i.e., Fluid Velocity Momentum Effects on Particles that Are at least 1/100 Size of Channel)

At high flow rates, a set of hydrodynamic effects known collectively as hydrodynamic lift effects can also become very important for suspended particles (see, e.g., the recent Ph.D. Thesis by Huang, [5]). Because of the influence of the flow field on a particle of finite size, there are forces within the channel that push particles away from the center of the channel and stronger forces that push them away from the walls. As a consequence, in every channel there is a surface (parallel at all points to the flow direction) of zero force to which all particles will eventually migrate. The speed at which they migrate to this surface depends on the square of the flow rate as well as the square of the particle diameter, but the accessible rates are such that substantial migration is often seen for many particle types. A good example of practical application of these lift effects was to focus latex microspheres to a point on the zero-force surface near the bottom of a triangular flow channel to aid in aligning those particles with a laser beam [5].

7.2.4.4   Viscosity Preferential Positioning

Note that if two solutions of different viscosity are placed in a channel, a number of interesting effects can take place. The most important one occurs at any Reynolds number. When two solutions of different viscosities are pumped at equal volumetric flow rates into a common channel, a dramatic effect is observed when the contact area with the walls for each fluid is much larger than the areas of contact between the fluids. In this case when flow is fully developed, the fluids will occupy a fraction of channel that is proportional to the ratio of their viscosities — the more viscous fluid will occupy a proportionately larger fraction of the channel [6, 7]. As a consequence, the interface between the two fluids is displaced from the position expected if the viscosities were equal, which affects diffusional exchange between the fluids beyond the effect of viscosity on diffusivity alone. Because the two fluids in this case occupy different fractions of the channel, the velocity distribution will also differ, being faster overall in the stream that occupies a smaller portion of the channel.

Furthermore, the petroleum industry has long exploited the fact that under normal Poiseuille flow, most of the energy is dissipated at the walls where the shear rate is highest. If there are fluids of two viscosities in a common channel, there will be less net resistance to flow if the more viscous fluid is sheathed by the less viscous one. As a consequence, if two fluids are introduced side-by-side into a channel, there is a tendency for the more viscous fluid to separate from the wall(s) to which it is initially in contact and move to the center of the channel.

7.3   OPTICAL MONITORING OF MICRODEVICES

Of the various ways to collect information on the position-dependent concentration of molecules in microfluidic devices, the most convenient and versatile is optical microscopy. The primary view of interest for the study of transverse diffusion is of the interface between two adjacent fluids that contain components with different diffusion coefficients, as in the T-sensor (Figure 7.3). This region of interest generally includes the region of fully developed flow in a long interdiffusion channel, although it is often also of interest to view the region in which flow is in transition from the full-developed flow in an entry channel. Of course, in some cases, due to the geometry of the interdiffusion region, flow never becomes fully developed. To a lesser extent, there is occasionally also reason to observe the regions immediately before they are in contact, or, in the case of H-filter-like systems, the region just after the fluids have been separated on their way out of the interdiffusion channel.

Since the concentrations of analytes at any point in a continuously flowing systems are constant for at least a few seconds, depending on the size of the sample and flow rates used, this means that weak optical signals can be integrated using any of several types of cameras. The most commonly used such camera is a cooled CCD camera capable of integrating signals for as

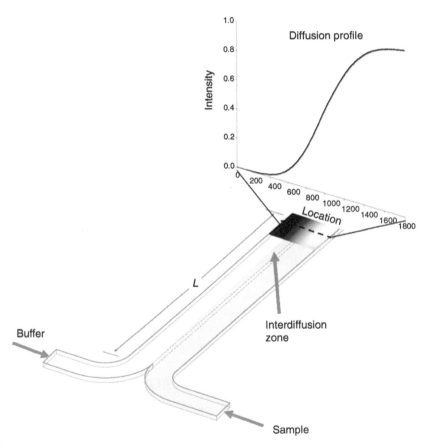

FIGURE 7.3 Schematic of the method of optical interrogation of T-sensor, with a typical interdiffusion experiment of a fluorescent analyte. At a particular point along the channel, a region is imaged onto a camera. A portion of that image is binned along the L axis, producing an array of intensities vs. position along d. This array of intensities is called an intensity profile.

short a time as a few milliseconds to as long as minutes. The area that can be imaged depends on the magnification of the microscope, the illuminated spot size, and the dimensions of the CCD chip in the camera. That area, in turn, determines the length of time that a molecule remains in the imaging region. Use of a cooled CCD camera is ideal for this application, although noncooled CCD cameras can work equally well when the concentration of the fluorescent analog is high enough that little integration is required.

Typically, a conventional fluorescence microscope is used as the imaging tool, as shown in Figure 7.4. This allows detection of very low concentrations of analytes. As a consequence, the three-dimensional information about the distribution of chemicals along the line of sight is lost as the data are compressed into a two-dimensional image. Note that there are examples of excellent use of full three-dimensional imaging of flow in microchannels that allow direct comparison with three-dimensional modeling data, generally employing confocal microscopy [8]. However, the expense of the technique is currently prohibitive for use in practical microfluidic systems for research or clinical applications. Therefore, with conventional microscopy, the curvature of the diffusion interface described in Section 7.4.2 is observed only as a broadening of the diffusion profile. This broadening is a source of error when image data are to be used to measure a property of the sample by comparison with a one-dimensional diffusion model. Furthermore, since the degree of curvature of the interface depends on both the diffusion coefficient of the analyte in question, as well as the geometry of the channel and flow rate, each experiment requires a separate model. In general, the use of a full three-dimensional model, followed by projection of the analyte concentration into the image plane is required to properly analyze such image data. Alternatively, if such model data are unavailable, it is always possible to calibrate a particular device–analyte pair with known samples and rely on the calibration curve.

While the geometry seen in Figure 7.3 and Figure 7.4 is ideal for analytical devices in which observing the interdiffusion of small fluid volumes is the aim, its throughput is generally very low. If the principles of microfluidics are to be used to prepare large samples of fluids, the two fluid streams must be so placed as to have the largest contacting surface area. In such a geometry is it impossible to image the interface with conventional microscopy. Hence, the positional information alone is of no use to determine the degree to which interdiffusion has occurred. This is a particular problem in the comparison of different geometries for mixers [9]. The challenge was to use interdiffusion of a dye like fluorescein from a dye-laden solution into a stream consisting of the same buffer with no dye. Initially it was proposed that the degree of mixing in a particular mixing geometry and operating conditions could be observed by the positional variation of the concentration of the dye; the assumption was that the greatest variation would be observed when part of the channel was at 100% dye concentration, and the other portion of the channel was at 0%. As

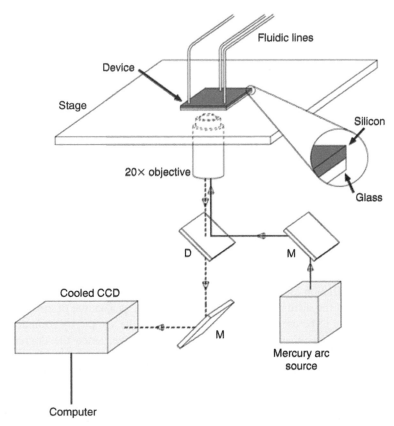

FIGURE 7.4 Schematic of a fluorescence microscope used for viewing a T-sensor. Illumination from a mercury arc source is introduced off a mirror (M) into a filter set containing a dichroic mirror (D). Emitted light passes through the dichroic mirror and is reflected into a cooled CCD camera. A digital signal from the camera is acquired by a PC, where frames of data are captured. The device is placed with its glass side down and fluids are delivered to the silicon side. In this case, the three fluid lines are for two inputs and one output. Not shown is the manifold used to align the fluidic lines and the device. The device shown is a silicon–glass sandwich, but many other materials have been used for microfluidic devices in the Yager laboratory, including PDMS and laminates of many layers of Mylar.

mixing occurred, more and more of the channel would be at dye concentra-tions closer to the geometric mean of the dye concentrations. This should be true if the device kept the two streams side-by-side in the same orientation. However, since the mixers in question rotated the orientation of the flow relative to the optical axis, it was found that the spatial variation of the fluorescence intensity was of no use in quantifying the degree of mixing (see Figure 7.5). The problem is that there is no way of differentiating the projected image of a fully mixed volume of dye and buffer from a set of completely unmixed laminae of the two starting solutions in which the laminae are oriented perpendicular to the axis of observation.[2]

It has proven to be very important to design experimental apparatuses in which phenomena that affect the concentration of analytes along the optical axis are suppressed. For example, at low ionic strength, it is possible to observe electroösmotic flow along planar surfaces in channels (see below). This flow causes randomization of the fluid in the channels by motion of fluid elements along surfaces. This can be suppressed using any of several methods that reduce surface charge or its influence on flow in the channel. Sedimentation of particles in a channel can also confound measurements of flow rates when the particles are no longer randomly positioned in the flow lines. A simple way to suppress the influence of gravity is to orient any long flow channels vertically. The curved diffusion interface effect described below can be minimized experimentally by using channels in which w is very small (so that equilibration along w occurs rapidly), or by observing the interdiffusion zone sufficiently far downstream that the curvature is no longer severe. When these methods are not applicable, one must either compare observed data to a

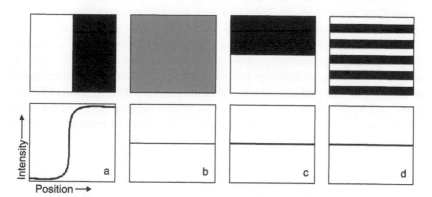

FIGURE 7.5 Schematic representation of the difficulty of using the projected image of layered fluids for determination of the degree of mixing. When the two streams are side-by-side and the plane between them is parallel to the axis of observation (a), the projected image (bottom row) can be used to determine the degree of mixing. However, the completely mixed case (b) is indistinguishable from the same case as in (a), but rotated by 90° (c), whether there are two or more laminae (d).

---

[2] This problem ultimately led to the discovery of the importance of junction potentials between solutions of different ionic composition in moving charged components across the channel by electrophoresis without the use of electrodes or application of an external field [10].

full three-dimensional model, or fall back on the empirical solution of preparing a calibration curve with multiple closely spaced points.

7.4 DEVICES THAT RELY ON TRANSVERSE DIFFUSIONAL TRANSPORT IN FLOW: FIELD-FLOW FRACTIONATION, THE H-FILTER, AND THE T-SENSOR

7.4.1 FIELD-FLOW FRACTIONATION

The first extensive use of microfluidic devices that use transport processes that are transverse to the flow direction to achieve a separation or measurement substantially predates the current microfluidics era. In its original form, field-flow fractionation (FFF) was never associated with the word "microfluidic," although since most of the devices had at least one dimension smaller than 1 mm, these applications should be considered microfluidic in retrospect. Years later, concepts of FFF were combined with new knowledge of microfabrication and microscale phenomena in novel classes of microfluidic devices designed to take advantage of the same types of effects.

FFF devices separate constituent molecules, particles, vesicles, or cells in a sample based on some differentiable physical property. The most common implementation, shown in Figure 7.6, introduces the sample by pressure-driven flow into a duct. Such devices are often microscale (10 μm to 1 mm) in the x-dimension and mesoscale (1 mm to 1 m) in the y-dimension. The cross-sectional shape imposes a parabolic velocity profile across the x-dimension. The fractionation process is ultimately achieved by distributing sample constituents at different locations across x, resulting in a distribution in elution time for different classes of molecules or particles. Thus, FFF is a chromatographic technique similar to CE, in that sample fractionation relies on temporal elution variation. In contrast, the majority of methods currently

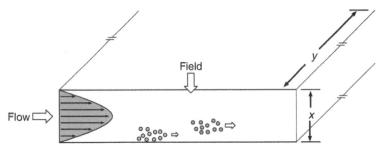

FIGURE 7.6 Typical arrangement of device for field-flow fractionation. The aspect ratio (y/x) is large (10 to infinity). The field is applied transverse to the direction of flow. Different types of analytes or particles have different equilibrium positions in the x-dimension and therefore have different elution times due to the parabolic velocity profile.

being developed using transverse separation rely on spatial elution variation and are therefore well suited for continuous applications.

A transversely applied field induces the spatial distribution across x by taking advantage of differences in a particular physical property of the sample constituents. A very thorough review article covers the major types of FFF [11]. The most frequently used field is that of gravity in what is termed gravitational, or more commonly, sedimentation FFF [12–22]. In this scenario, the entire flow chamber is spun to achieve centrifugation, resulting in a spatial distribution of particles based on density. In thermal FFF [23–27], a temperature gradient is applied across the x-dimension and constituent distribution is determined by thermal diffusivity. Electrical FFF [28–32] achieves separation by electrophoretic mobility, while dielectrophoretic FFF (DEP-FFF) [33–37] applies a retractive dielectrophoretic force on particles that, when balanced by gravitational force, results in a unique spatial location for each population of particle. In addition to one of these primary FFF methods, many devices also make use of steric FFF [38–43], where the spatial distribution of large particles is determined by the proximity to the wall that particles of different size can attain.

While various types of FFF have been applied mainly to perform cell separation and analysis [43–47] and particle sizing [48–54], there have been many studies on other types of analysis, including diffusion coefficient measurement [55–60], separation of magnetic particles [55, 61–64], and various types of sample purification [65–67].

7.4.2 THE T-SENSOR

The T-sensor, a recent microfluidic invention that builds on the success of FFF, has proven to be quite useful as an analytical tool [68, 69]. The basic format of all T-sensors is shown in Figure 7.7. Two continuous fluid inputs come together to flow side-by-side. The Re is very low (0.01 to 1), ensuring that flow is entirely laminar. Therefore, no convective mixing occurs between the two streams, and molecular transport occurs solely by diffusion transverse to the flow direction. In such devices, the aspect ratio (d/w) is greater than 1, usually higher than 3 and sometimes as high as 300. The width dimension (w) is typically 10 to 500 μm.

Of the two T-sensor input streams, one is the sample fluid and the other contains a probe molecule intended to interact with a target contained in the sample. Measurements are made with the T-sensor by monitoring a signal generated by the interdiffusion and reaction of the two streams. This signal is most often a fluorescence emission, resulting either from an increase in the quantum efficiency of a probe fluorophore upon binding of the target, or simply by the concentration of the fluorescent probe as it binds to the target. This latter mechanism occurs by a substantial decrease in the diffusivity of the probe upon binding.

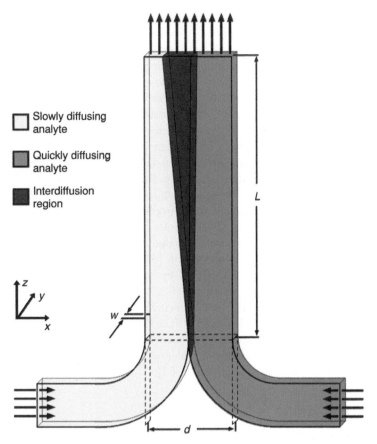

FIGURE 7.7 The basic geometry of the T-sensor. Two flow inputs are shown, although three or more are often used in practice. In the particular scenario illustrated here, each input contains one diffusing species. The faster diffusing analyte moves to the left more quickly than the slower diffusing analyte moves to the right. This induces an asymmetric development of the interdiffusion. Note that to be consistent with previous publications from the Yager laboratory, here and in other captions in this chapter, the variables d, w, and L are used to refer to the diffusion dimension, the width, and the interaction length within the channel, respectively.

The use of the T-sensor[3] has already been demonstrated for monitoring pH [70], measuring sample viscosity [6, 7], and measuring the concentration of various fluorescent species [71, 72]. The addition of electrodes along the

---

[3] In some uses of the T-sensor, the device is more appropriately called a H-filter. The H-filter is used specifically for sample fractionation and therefore device output is controlled. In the T-sensor, measurements are made on chip and all output fluids are disposed of.

sidewalls of the T-sensor has allowed for concentration measurements of electrochemically active molecules [73] and electrophoretic separation and concentration [74]. A review of T-sensor applications has recently been published [69]. T-sensor-type devices have been made in many materials, including silicon and glass [3, 6, 75] all glass [76–78], plastic laminate [79–84], and elastomeric polymer [8, 85].

One notable feature of T-sensor geometry is the shape of the inlet channels. The preferred embodiment employs curved inlets as shown in Figure 7.3, Figure 7.7, and Figure 7.8A. This geometry provides a well-defined starting point of the interface between the two input fluids. Also, the length of the d-dimension is constant along the entire length. The curvature of the inlets also allows more development of the inlet flow profiles upon merging than other geometries used in earlier T-sensors. These advantages are in comparison with straight Y and T input junctions (Figure 7.8B and C). Details of flow development and velocity profiles inside the curved-inlet geometry have been published elsewhere [75].

7.4.2.1 Diffusion Measurements in the T-Sensor

The typical use of the T-sensor, as shown in Figure 7.7, introduces two or more streams flowing side by side. Interaction of the streams occurs through lateral diffusion. The simplest measurement that can be made from such a scenario is the diffusion coefficient of a single species. The easiest experi-

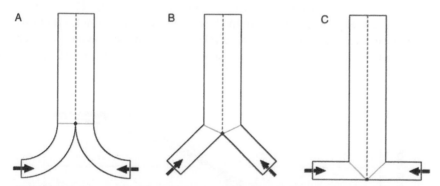

FIGURE 7.8 The preferred embodiment of T-sensor inlet geometry (A) has curved inlets. At the point where the two fluids begin to interdiffuse (bottom of dotted line), the channel cross section a perfectly perpendicular to the flow. In (B), the stagnation point (black circle) is recessed, skewing inlet geometry. This means that the input fluids will necessarily have some lateral velocity after interdiffusion has begun. In (C), the geometry is further skewed. Such skewing makes quantitative analysis very difficult because the flow near the stagnation point dominates diffusion across the interface. Details of flow development and velocity profiles inside the curved-inlet geometry have been published elsewhere [75].

ment from which to make such a deduction is a two-stream scenario where each stream is composed of the same high ionic strength buffer and one stream contains a fluorescent analog of the molecule to be studied. The buffers should have the same composition to avoid passive electrophoretic effects [10]; the high ion content mitigates a number of potential electrostatic effects. The analytes to be tracked must be observable by optical means, most conveniently by optical absorbance or fluorescence. If the molecular of interest is not inherently strongly absorbing or fluorescent, the use of a fluorescently labeled version of the molecule of interest is in order, if somewhat limiting.

If the field of view of the microscope is set to a known distance downstream in the T-sensor, an image can be acquired showing the spatial distribution of fluorescence across the d-dimension of the channel, as in Figure 7.3. Furthermore, if the concentration of the fluorophore is low enough such that its fluorescence is directly proportional to concentration, the captured digital image of the channel can be converted into a concentration map through multiplication by a single factor. In practice, the use of background and uniform images, as well as measuring the relationship between concentration and fluorescence, produces the best quantitative results [6].

Typical data generated from a diffusion experiment in a T-sensor are shown in Figure 7.9. In this case, the molecule of interest was a fluorescent analog of biotin. The data show the expected features based on mathematical solutions of this type of diffusion problem [86]: the concentration is uniform at its maximum value on one side of the channel, uniform at zero on the other side, and is a sigmoid with perfect symmetry in between. The symbols represent the experimental data from three separate trials completed at different flow rates. The broadest curve (circles) is the slowest flow rate — since all measurements were made at the same distance downstream, the slower

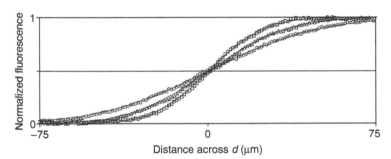

FIGURE 7.9 Example diffusion coefficient measurements in the T-sensor. The fluorescent analyte is initially on the right side of the channel. Shown are the fluorescent intensities observed from a diffusible analyte at different locations along a flow channel (or at the same location but at different flow rates).

moving flow allowed more time for interdiffusion and hence the curve is closer to uniform concentration. The steepest curve (squares) was at the fastest flow rate, and correspondingly is the most similar to the input condition. Diffusion coefficients can be determined from these data by fitting a simple model of the diffusion equation (solid lines). The nuances of this method have been discussed elsewhere [75].

7.4.2.2 Effects of Velocity Profile on Diffusion

It is convenient and intuitive to conceive of T-sensor behavior as a one-dimensional diffusion problem, with molecules showing net displacements across parallel streams where concentration gradients exist. As discussed earlier, however, the nonuniform velocity profile in the T-sensor induces a position-dependent spatial variation on the extent of interdiffusion (Figure 7.10).

Based on the method of fabrication and orientation of the device relative to the optical system, it is usually the case, the imaging is done through the w-dimension, meaning that the subtleties of interdiffusion (specifically the curved interface) cannot ordinarily be explicitly viewed. In one case, this limitation was overcome by using a confocal microscope to provide complete two-dimensional imaging of the channel cross section [8]. In addition to this study, theoretical studies have demonstrated that a one-dimensional solution

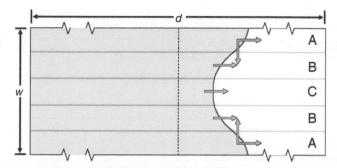

FIGURE 7.10 The interdiffusion interface in a microfluidic channel when $d > w$. With flow into the page, diffusible analyte (gray) starts on the left side of the channel with the original interface shown by the dotted vertical line. Net analyte flux by diffusion occurs in the directions of the gray arrows, caused by either the primary concentration gradient between the left and right side of the channel or by the secondary concentration gradients across the w-dimension. All three regions (A, B, and C) show net diffusion to the right caused by the primary concentration gradient. In region C, there is no net diffusion in the w-direction because there is no concentration gradient. In region A, there is a net flux of analyte into the interior of the channel because of the secondary concentration gradient. Region B is the recipient of this flux of material from region A.

to the diffusion equation across the d-dimension can have substantial error [75, 87]. However, if flow proceeds far enough down a channel, diffusion across the w-dimension tends to eliminate secondary concentration gradients induced by the velocity profile (Figure 7.11).

If measurements are made far downstream after interface curvature has been substantially eliminated, the two-dimensional effects caused by the parabolic velocity profile become less significant [3].

7.4.2.3   Chemical Detection: The T-Sensor

The conditions that can be created in a T-sensor allow some rather surprising chemical detections to be performed. One of the most promising uses of the T-sensor has proven to be the creation of a novel type of molecular binding assay: it requires the use of at least three types of molecules: a slowly diffusing capture molecule, a smaller ligand (or analyte), and a version of the ligand that has been made visible to the observer, typically by covalent modification with a fluorophore. The first published embodiment of the assay is called a diffusion immunoassay (DIA) [82], specifically a competition immunoassay. In it, one stream consists of an antibody stream. The other consists of a buffer containing the antigen (sample analyte) and a labeled form of the same analyte (labeled analyte). The label is typically a covalently

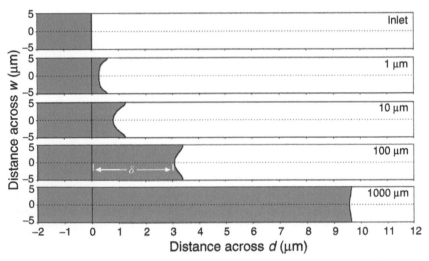

FIGURE 7.11 Two-dimensional simulation of diffusion in a microchannel. Two fluids begin with the original interface at $d = 0$. One contains a diffusing analyte (gray) and the other does not. Flow is into the page, and the distance downstream is indicated in the upper right of each frame. The interface between the two fluids starts as a vertical line. By 1 μm downstream, the interface is considerably curved. Eventually, this curvature lessens as flow proceeds downstream.

attached fluorophore. When the two streams are brought into contact, the labeled and unlabeled analytes, which must be at least two times more rapidly diffusing than the antibody, compete for antibody binding sites. If there is no sample analyte present, the labeled analyte is almost entirely captured by the slowly diffusing antibody, and there is a "pile-up" of the label near the interface between the two streams. On the other hand, if there is sample analyte present, it competes for and occupies the antibody binding sites, so the labeled analyte is free to diffuse past down its concentration gradient over time. The net result is that the degree to which the labeled analyte accumulates near the interface between the two streams is a measure of the concentration of the sample unlabeled analyte.

For the analyte phenytoin (known commercially as hydantoin) the DIA is sensitive down to less than 1 nM concentration of analyte, and has a dynamic range of over 3 orders of magnitude with one set of reagent concentrations. The sample can be very "dirty," in the sense that there is no need to remove cells and other scattering debris from the samples before the assay is run; it has been demonstrated, for example, on diluted whole blood. If one ignores the time required to mix the sample with buffer containing the labeled analyte, the assay is complete in less than 30 sec. This is very promising for point-of-care applications.

7.4.3 Extraction: The H-Filter

The earliest diffusion-based microfluidic device studied in this laboratory was the H-filter, which, as described above, allows extraction of components from one stream to another on the basis of their diffusivity [7, 88–93]. No intervening membranes are necessary; this maximizes the transport rate of the diffusible species and eliminates both preferential permeability of the membrane material for one species or another, as well as eliminating degradation of device function over time caused by clogging of the membrane. It is an alternative to the use of centrifugation for a wide range of sample preconditioning steps. There are, however, two primary modes of use for the H-filter. In the first mode, the aim is to extract as much as possible of a diffusible component present in a complex mixture that also contains more slowly diffusing species. In this case, the sample is brought into contact with a buffer stream, and during adjacent flow, the analyte of interest crosses by diffusion into the second stream. Of course, all components of similar or higher diffusivity will also cross the streamlines into the "extraction stream," so this method (as a single step) is best for extracting low molecular weight components. This is particularly useful for removing an analyte up to the size of a protein from a complex fluid containing particles, such as blood or other cell-containing suspensions. This device has recently been demonstrated for extraction of a drug from whole blood prior to HPLC analysis [93].

The second mode for use of the H-filter is in removing small molecules from a solution that contains slowly diffusing species that one wishes to retain for subsequent analysis (or use); desalting of DNA is one example of such a process that has been demonstrated successfully (Thomas Schulte, Micronics, Inc., personal communication).

In a recently published work, it was demonstrated that a H-filter could be run in tandem with a T-sensor on the same polymeric laminate device to perform a complex biological analysis [94, 95]. In the first of the two devices, the H-filter, a detergent solution diffused from a mild detergent stream into a stream holding an intact bacterial suspension. In this first channel, at least some of the bacterial cells were permeabilized, resulting in a release from the cells of high molecular weight components, including the protein β-galactosidase. At the output of the H-filter, the cell bodies were discarded and the detergent stream, carrying a finite fraction of the enzyme (as well as smaller components) was pumped directly into one inlet of a T-sensor. In that T-sensor the stream of enzyme-laden solution contacted a stream containing a fluorogenic substrate — one that is not fluorescent itself, but is cleaved by the enzyme into two products, one of which is intensely fluorescent. In this second device, it was possible to observe the progress of the enzymatic reaction. By comparing the position and intensity of the fluorescence that developed over time and space in this second microfluidic channel, it was possible to quantify the concentration of the enzyme released in the first device.

7.5 TRANSVERSE ELECTROKINETIC PROCESSES AND DEVICES

7.5.1 Fundamentals (Including Resolution Limitations)

7.5.1.1 Basic Physics of the System

One can differentiate between molecules or particles on the basis of many properties, such as density, and use such differences to separate, identify, and concentrate them. One such property, surface charge, forms the basis for many well-developed sample handling and analytical methods, referred to as "electrokinetic" methods. Examples of electrokinetic techniques include zone electrophoresis (ZE) and isoelectric focusing (IEF). ZE, the most widely known electrokinetic technique, differentiates on the basis of the velocity of the molecule or particle in an electric field. Capillary electrophoresis is perhaps the most well-known implementation of ZE in the world of micro-fluidics. In contrast, IEF selects on the basis of isoelectric point; that is, the pH at which a given molecule or particle has no net surface charge. A third technique, dielectrophoresis (DEP), does not differentiate based on surface charge but rather on differences in electrical conductivity between the particle of interest and the surrounding fluid.

7.5.1.2 Surface Charge and Electrophoretic Mobility

The effective charge of a particle, called the "zeta potential," is defined as the potential difference between the bulk solution and the zone of shear [96]. This zone of shear, located between the Stern and Gouy–Chapman layers, is on the order of several Ångstroms from the surface of the particle [97]. In addition to the charge of the particle surface, three fluid parameters influence the zeta potential: pH, dielectric constant, and ionic strength [98]. The electrophoretic mobility (EPM) of a particle is a function of its zeta potential, the dielectric constant, viscosity of the surrounding fluid, and the ratio of the particle radius to the electric double layer thickness [96]. Typically, the zeta potential is calculated from the experimentally measured EPM, rather than theoretically determined. The Helmholtz–Smoluchowski equation for EPM as a function of zeta potential, particle size, and fluid characteristics, initially developed in the context of electroosmosis, has also been shown to be applicable to electrophoretic movement of particles. The EPM can be thought of as the ratio between the electromotive forces and the viscous drag forces experienced by the particle. A generalized version of the equation for non-conducting spherical particles is as follows:

$$\text{EPM} = \frac{\zeta d}{6\pi\eta} f(\kappa r) \tag{7.4}$$

where ζ is the zeta potential, d the dielectric constant of surrounding fluid, η the kinematic viscosity of surrounding fluid, κ the inverse of the thickness of the electric double layer, r the radius of the particle, and $f(\kappa r)$ the empirically calculated function that varies from 1.0 for small κr to 1.5 for large κr. (Modified from Sherbert's *The Biophysical Characteristisation of the Cell Surface* [96].) The preceding discussion applies to surfaces at a solid–fluid interface as well as to particles.

For an excellent discussion of diffusion and electrophoresis, also known as "diffusion with drift," the reader is encouraged to read Berg's *Random Walks in Biology* [99].

7.5.1.3 Electroneutrality and Field Strength

The velocity of a charged particle in an electric field is simply the product of its EPM under local conditions and the local field strength. On a first pass, calculating field strength appears to be trivial: divide the applied voltage by the distance between the electrodes. Upon deeper analysis, however, several complications await. Variations in conductivity will affect both the field direction and magnitude. Perhaps more critically, the double layer proximal to each electrode will itself represent a large drop in voltage, such that the effective voltage applied to the bulk fluid between the two electrodes is significantly lower than that applied to the system as a whole.

For the typical electrode configuration used in transverse separations, the electrochemical cell can be thought of as two capacitors (the electrodes) bracketing a resistor (the bulk fluid). However, while the bulk fluid remains electrically neutral upon application of an electric field, the regions immediately adjacent to the electrodes do not.[4] According to modern theory, at each electrode surface there is a monolayer of counter-ions (the Helmholtz or Stern layer) [97]. Moving away from the electrode, the counter-ions are more diffuse (Gouy–Chapman layer) [97]. Within the Gouy–Chapman layer the fluid charge exhibits a Poisson distribution [102]. The development of these regions is essentially the result of capacitive charging, as reflected by an initial spike in current when a voltage is first applied. For a constant DC voltage, a steady-state current will ultimately be achieved, reflecting equilibrium between the rates of ion generation, due to electrolysis at the electrodes, and migration.

The nonelectrically neutral regions proximal to the electrodes are typically on the order of nanometers to Ångstroms in thickness [97]. The voltage drop across these regions typically ranges from 0.5 to 1.0 V and is attributed to both the electrolysis of water and the electric double layer [103]. The voltage drop across the bulk fluid is a function of the ohmic resistance of the bulk fluid and the current through the device; this relationship is a more reliable way of estimating the actual field strength in a channel during an experiment.

7.5.1.4 Relevant Equations

A mathematical description of the forces relevant to microfluidic transverse electrokinetic transport begins with the equation of continuity for incompressible flow, expressed in terms of mass conservation, as described earlier in Section 7.2.3. To adapt that derivation to account for electrokinetic transport, the right-hand side of the equation, which describes the flux of reactants in the system (J), must be modified to account for electrophoretic motion, in addition to diffusion, as shown below:

$$\frac{\partial c_N}{\partial t} + \left(v_x \frac{\partial c_N}{\partial x} + v_y \frac{\partial c_N}{\partial y} + v_z \frac{\partial c_N}{\partial z} \right) = -\nabla J \qquad (7.5)$$

where the flux is defined as

$$J = -u c_i \nabla \Phi - D \nabla c_i + R_i \qquad (7.6)$$

and u is the EPM of species "i", Φ the electric potential (the gradient is the electric field), D the diffusion coefficient of species "i," and R_i the rate of reaction (production or consumption) of species "i."

---

[4] Note that this assumption of electroneutrality may not hold at small size scales. A full discussion of electroneutrality is beyond the scope of this chapter; the interested reader is encouraged to reflect on Gauss' law and seek out other resources [10,101,102].

The addition of electric field to the flux component significantly complicates the solution of this problem, since the electric field is itself a function of local ion concentrations, both through the conductivity term and through the current-generation caused by local ion fluxes:

$$\frac{d\Phi}{dx} = -\frac{1}{\sigma}\left[I + F\sum_{i=1}^{n} z_i D_i \frac{\partial c_i}{\partial x}\right] \qquad (7.7)$$

where σ is the conductivity of the solution, a function of the local concentration of all charged species, and I the current density.

This problem does not easily yield an analytical solution. In addition, for IEF, buffering reactions in the system must also be considered, which adds a set of nonlinear conditions that must also be met to generate an accurate solution. A mathematical model of this system, for a constant DC voltage, has been developed for a two-dimensional system (y-axis is assumed to be homogenous) [79]. Similar equations apply to DEP, with the added complication that the electric field is not constant with time.

7.5.1.5 Basics of Electrophoresis

Electrophoresis is the movement of charged particle in an electric field. This velocity is the product of the strength of the local electric field and the EPM of the particle or molecule itself. ZE can achieve sample separations based on differences in both the magnitude and the direction of the electrokinetic velocity of the constituents. However, for two major classes of particles, DNA and proteins, ZE is typically performed under conditions such that there is no significant differences in the velocity of different particles of the same class. For uncoiled DNA, this invariance of electrophoretic velocity is due to the fact that both the zeta potential and the viscous drag scale linearly with size, so that the ratio of the two forces, the EPM, becomes size invariant. For proteins, which do have widely varied EPMs under nondenaturing conditions, one is often interested in separating on size rather than on EPM, so the protein solution is treated with a reducing agent, which confers essentially a uniform charge–mass ratio for all proteins and a detergent, typically sodium dodecyl sulfate (SDS), which denatures the proteins to greatly reduce variations in shape that would otherwise cause proteins of the same size to have variable viscous drag. With this treatment, the majority of proteins behave like DNA, in that both zeta potential and viscous drag scale linearly with size, so that the EPM is invariant with size. To perform a size-based chromatographic separation on these classes of particles, electrophoretic force is used to drive particles through a sieving matrix (such as polyacrylamide or agarose) that imposes size selection.

Implemented in transverse mode, ZE is performed without a sieving matrix, often referred to as "free flow electrophoresis (FFE)." The electric field is applied perpendicular to flow, and particles move towards the electrode of complementary sign, at a rate that depends on both the EPM of the particle and the field strength applied to the chamber. By the exit of the channel, the solution will have become fractionated by EPM, such that splitting the output results in spatial isolation of these different fractions.

7.5.1.6 Basics of Isoelectric Focusing

IEF is a fractionation and purification technique that differentiates particles based on isoelectric point (Figure 7.12). Neglecting the effect of dielectric constant, the pH at which a particle has no net surface charge is called its "isoelectric point" (pI). At its pI, the EPM of a particle is zero; it will no longer migrate in an imposed electric field. The majority of biological particles are amphoteric; that is, their zeta potential can vary from positive through neutral to negative, depending on the local pH. In IEF, one generates a pH gradient that spans the range in which a given particle or set of particles switches the polarity of its zeta potential. An electric field is then applied parallel to the pH gradient such that, the anode is proximal to the acidic extreme of the gradient and the cathode is proximal to the basic extreme. At the acidic extreme, there is a surfeit of protons and the particles of interest should have a net positive charge, which drives them away from the anode. As the particles travel through the increasingly basic gradient, propelled by

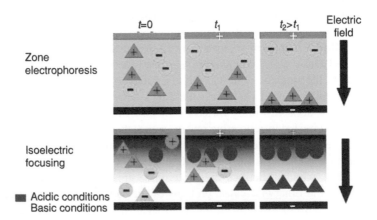

FIGURE 7.12 (See color insert) Comparison between zone electrophoresis (ZE) and isoelectric focusing (IEF). The two different species of particles are represented by triangles and circles. The pH gradient created in IEF is represented by a gradient of color between the electrodes in the lower images. (Figure adapted from Cabrera, C., *Microfluidic Electrochemical Flow Cells: Design, Fabrication, and Characterization.* Doctoral dissertation in Bioengineering, 2002, University of Washington: Seattle, WA.)

electrophoresis, they become gradually less charged, until they reach the point in the gradient that corresponds with their isoelectric point. At this location, the particle has no net charge, experiences no net electrophoresis, and is essentially in an electrokinetic trap. If the particle diffuses towards the more basic side, they lose protons, become negatively charged, are repelled by the cathode, and move back towards the point of neutrality. Similarly, if the particle diffuses toward the more acidic side, it gains protons, becomes positively charged, is repelled by the anode, and returns toward the point of neutrality. The width of the steady-state focused band is therefore a balance between diffusion and electrophoresis. The larger a given molecule, the more tightly it can be focused due to smaller spatial fluctuations induced by diffusion.

Unlike ZE and other rate-based chromatographic techniques, samples treated by IEF will eventually reach steady state, because the selection mechanism does not rely on velocity. Therefore, the ultimate resolution of IEF is theoretically independent of time and is instead determined by several factors: field strength, pH gradient slope, and rate at which the particle's zeta potential varies with pH. In practice, due to the lack of electrostatic repulsion, the neutral charge when the particles are at their pI, and the high local concentration, particles focused with IEF tend to clump together over time. Therefore, IEF fractionation should be monitored to determine the upper limit of experiment duration.

When implemented in a transverse fashion (Figure 7.13), the electric field and pH gradient are perpendicular to the direction of flow.

Assuming that the electric field and particle diffusion coefficient are constant within the focused band, the concentration of focused particles as a function of distance from the position in the pH gradient that corresponds to its pI can be calculated as follows:

$$C(x) = C_0 \exp\left[\frac{\dfrac{d\mu(x)}{d(pH)} \dfrac{d(pH)}{dx} E(x)x^2}{2D}\right] \tag{7.8}$$

where $C(x)$ is the protein concentration, $\mu(x)$ the electrophoretic mobility, $E(x)$ the electric field, and D the diffusion coefficient. Note that $x = 0$ at the pI of the protein of interest.

Equation (7.8) represents a Gaussian distribution. The inflection points of the curve can be calculated as follows:

$$\sigma = \pm \sqrt{\frac{D}{E\left(-\dfrac{d\mu}{d(pH)}\right)\left(\dfrac{d(pH)}{dx}\right)}} \tag{7.9}$$

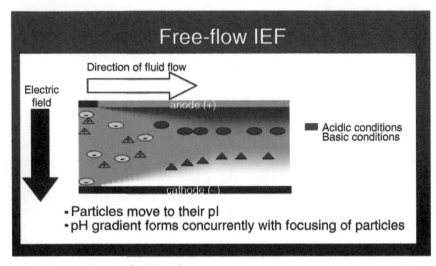

FIGURE 7.13 (See color insert) Schematic of transverse IEF implemented in a microfluidic device. Note the self-generation of the pH gradient as OH⁻ is produced at the cathode and H⁺ at the anode. (Figure adapted from Cabrera, C., *Microfluidic Electrochemical Flow Cells: Design, Fabrication, and Characterization.* Doctoral dissertation in Bioengineering, 2002, University of Washington: Seattle, WA.)

The advantages of IEF include increased selectivity, broad range of applicability, and relatively mild operating conditions, which allow samples to retain biological activity. Since the majority of biological particles are amphoteric, IEF has a broad range of potential applications, as demonstrated by the variety of analytes to which it has been successfully applied. IEF has become a common technique for high-resolution analysis of biological samples, particularly protein and peptide mixtures [104, 105].

7.5.1.7 Basics of Dielectrophoresis

Unlike ZE and IEF, DEP can operate on uncharged, as well as charged, particles, and can be implemented with either AC or DC fields. Biological particles and the surrounding fluid can become polarized when placed in an electric field. DEP exploits this phenomenon by applying a nonuniform electric field to a system in which the solution and suspended particles have different degrees of polarizability. For particles less polarizable than the surrounding media, the induced dipole is aligned in opposition to the applied field, which causes the particles to migrate to local field strength minima ("negative" DEP) [106]. Conversely, if the particles are more polarizable than the surrounding media, the induced dipole will be in alignment with the applied field and the particle will move towards local field maxima. See Gascoyne and Vykoukal for an excellent in-depth review [107]. A recent twist on DEP includes spatial and temporal variation of fluid conductivity, which varies the local field strength without changing the applied voltage [108].

7.5.1.8 Basics of Electric Field-Flow Fractionation

As discussed earlier, pressure-driven flow will exhibit a nonuniform flow profile in steady-state conditions, with the velocity increasingly pseudoparabolically from zero at the channel walls to a maximum velocity at the center of the channel. When a bolus of sample is introduced into a pressure-driven flow, particles in the center of the channel will exit the device before those proximal to the walls. By applying a field transverse to the direction of fluid flow, one can control the position of different particles in the nonuniform flow field and therefore control the order in which the particles exit the channel [109]. This technique is one embodiment of FFF (see Section 7.4.1 for further discussion).

Although gravity is the field most commonly associated with FFF, substantial work has been done with applying electrical fields. Giddings and colleagues, pioneers in the field of FFF, proposed the idea of applying an electric field transverse to the direction of fluid flow for the purposes of microfluidic sample fractionation [110]. In 1989, two groups published successful demonstrations of IEF in combination with FFF. For the reader interested in pursuing this history, the resulting acronym was IEF4. Chmelik and colleagues separated a binary mixture of proteins [111] and Thormann and colleagues concentrated a single protein [112]. The two groups soon began to collaborate and published a series of papers characterizing the behavior of these devices, using the focusing of cytochrome c as a model system [113–115]. The results of the multiple protein separation and cytochrome c work were compared to the classic model of IEF [116], which neglects fluid flow, and found to be in qualitative agreement. Chmelik and colleagues published the first study on pH gradient formation in an electric FFF device using phenol red as a representative ampholyte [117, 118].

DEP was first suggested as a new force for FFF in 1997 [33]. As with gravity and ZE, the basic idea is to exploit differential movement in an applied field to isolate particles into different lamina of a parabolic flow stream. This technique has been shown to separate breast cancer cells from normal human blood cells in ~5 min [119].

7.5.2 Device and Experimental Design

7.5.2.1 Design Constraints Unique to Electrokinetic Processes

In addition to the design concerns inherent in all microfluidic devices, such as fluidic interconnects, working with electrokinetic forces adds a new level of complexity to device design. Adding electrokinetic forces to microfluidic devices can be quite straightforward. Most strikingly, capillary electrophoresis at its simplest requires only that wires be inserted into holes at either end of the channel. However, to incorporate electrokinetic forces in a microfluidic device such that the field is *perpendicular* to fluid flow represents several significant design challenges. The electrodes must be electrically insulated from each other as well as the rest of the device, yet be easy to connect to a power supply. The gases generated at the electrodes as a result of electrolysis must be vented or otherwise isolated from the main body of the channel, or kept at a low enough level to avoid nucleation of bubbles, the bane of most microfluidics experiments. Similarly, any heat evolved at the electrodes or through resistive heating of the fluid itself must be removed, to prevent convective transport driven by a thermally induced density gradient that could lead to disruptive fluid flows. Macroscopic devices typically address these concerns by physically isolating the electrodes from the main separation chamber, typically through the use of membranes, and recirculating buffer through the two electrodes, independent of each other, to remove both heat and gas. Due to the complexity of constructing such a system using microfluidic fabrication techniques, combined with the reduced levels of heat and gas generation that correspond to the reduced size scale, microfluidic transverse electrokinetic separation units often eliminate the recirculating buffer streams and do not isolate the electrodes from the separation chamber.

The choice of electrode material presents another design challenge. Most importantly, the metal chosen should undergo minimal electrochemical reactions (either oxidation or reduction) upon application of a voltage, particularly if the device is to be operated at a relatively low voltage (<3 V). The use of copper, for example, is contraindicated, since it undergoes oxidation at low applied voltages [120], which would divert electron flux from generating H^+ while slowly eroding the electrode material. Better metals include gold, platinum, and palladium, and other noble metals.

As discussed earlier, EOF can be used to pump fluids in microfluidic devices. EOF can also be an undesirable side effect of transverse electrokinetic transport when only particle electrophoresis is desired. Because the field is applied perpendicular to flow, any resulting EOF will itself be perpendicular to flow, which can result in problematic recirculating flows, among other possible flow disturbances (Figure 7.14). EOF suppression methods include covalent modification of channel wall surfaces, often with polyethylene glycol, or more transient surface coating methods, such as incubation with a detergent prior to running an experiment [80]. The literature on capillary electrophoresis contains extensive information on EOF control and suppression. Higher conductivity buffers will also act to reduce EOF, but can cause other problems during the application of the electrokinetic technique.

For IEF, there are additional challenges with regard to creation of a pH gradient parallel to the electric field. The formation of pH gradients in commercial devices has predominantly been achieved either through the

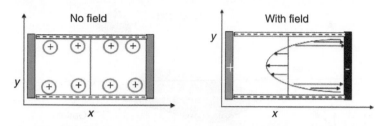

Note: Pressure-driven flow is into the screen for all cases

FIGURE 7.14 (Left) Schematic representation of the charges present on the walls of a microfluidic device in relation to the electrodes. (Right) Schematic representation of the bidirectional flow velocities (present at the vertical line shown in the device) that is induced by electroosmotic pumping at the walls. This recirculatory flow is in the plane perpendicular to the net flow through the channel. (Figure adapted from Cabrera, C., *Microfluidic Electrochemical Flow Cells: Design, Fabrication, and Characterization.* Doctoral dissertation in Bioengineering, 2002, University of Washington: Seattle, WA.)

migration of buffering compounds to their pI upon application of an electric field or through integration of large reservoirs of acid and base into the device design. These approaches are implemented in IEF devices that shield the bulk fluid from the products of electrolysis of water, typically via large electrolyte reservoirs and ion-selective membranes at the electrodes. Such isolation in a microfluidic device operating in transverse mode greatly increases the difficulty of device design and fabrication. An alternative takes advantage of these electrolysis products, essentially using them as self-generated reservoirs of acid and base, and can therefore be implemented in a much simpler device, in which the electrodes are in intimate contact with the bulk fluid [79] (Figure 7.13).

7.5.2.2   Selected Device Designs

7.5.2.2.1   Free-Flow Electrophoresis and Isoelectric Focusing

The device requirements for IEF and FFE are essentially the same: a pair of parallel electrodes integrated into a flow cell such that the direction of flow is perpendicular to the applied field. Functional designs are seen in Figure 7.15. DEP can require significantly more complicated electrode arrays but, when performed with AC fields, is not as prone to bubble generation as the other methods. Commercially available devices for FFE, also used for free-flow IEF [121], typically have separation chambers that consist of a shallow slot, on the order of centimeters in length (flow direction dimension) and width (distance between electrodes), but less than a millimeter in depth. The narrow gap results in laminar flow, which reduces the effects of convective disturbances caused by localized resistive heating [122]. Macroscopic versions of

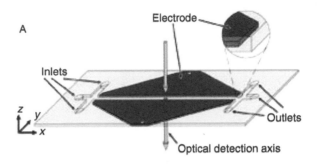

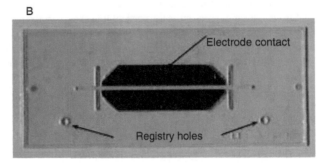

FIGURE 7.15 (A) Schematic of early electrokinetic polymeric laminate device in which the electrodes were made of gold evaporated onto folded Mylar. Note that the device was designed to have three outlets and three inlets. (B) Photograph of a more recent electrokinetic polymeric laminate device in which the electrodes consist of the cut edges of solid palladium foil pieces. (From Macounová, K., C.R. Cabrera, and P. Yager, *Analytical Chemistry*, 2001, 73(7): 1627–1633. With permission.)

FFE and free-flow IEF [123] separation devices have been used for many years and typically include a recycling component so that the particles make several passes through the chamber, with their position relative to the two electrodes shifted further with each pass. Such recirculation can be necessary to obtain sufficient sample fractionation to permit physical separation of the different lamina. Construction of a microfluidic device that permits such an operation, even without recycling of streams, is nontrivial.

An early review of the FFE field, by Hannig [124], is so often cited as the seminal paper in the field that similar FFE devices are still frequently referred to as "Hannig" devices. The original Hannig FFE device consisted of two sheets of glass, 55 cm × 10 cm, separated by a 0.5 mm gap, through which buffer was continually pumped, with the output collected into 90 fractions after 3 to 7 min of exposure to the electric field [124]. Later versions, larger in scale, had up to 197 fraction collection ports and a 3 mm gap [125]. In all such devices, the relatively large electrode gap (on the order of tens of centimeters) requires a large applied voltage to deliver a strong electric field, which in turn results in significant gas and heat production, both of which must be removed. Scaled-down devices with smaller electrode gaps do not necessarily require such additional engineering.

One of the first microfluidic FFE devices was described by Giddings and colleagues, who presented the idea of a continuous flow electrophoretic binary separator device, using a split-flow thin cell in which the electric field was imposed transverse to the direction of flow [126–128]. Although the device separated particles on the basis of isoelectric point, the separation was actually an electrophoretic separation, in which the particles migrated in a highly buffered system to the oppositely charged electrode and the outlet stream was split into two. No pH gradient was intentionally formed. As with the IEF devices discussed above, this device cell required *in situ* cooling of circulating electrolyte and degassing components.

Related work led to the development of a microfluidic electrochemical flow cell in the electrodes themselves formed the channel walls [129]. Like the split-flow cell, no pH gradient was deliberately formed. A recent publication presents the results of using such a device in conjunction with electric FFF to fractionate a heterogeneous mixture of polystyrene beads [130]. The authors use unbuffered water for their experiments and note an unexpected "wall repulsion" effect that caused experimental results to deviate from theory predictions. It is probable that this repulsive effect is actually a result of pH gradient formation leading to IEF.

Other microfluidic free-flow IEF and FFE devices have been built in which the electrodes form the walls of the channel, using pieces of palladium sandwiched between layers of laser-machined laminate and held together with pressure-sensitive adhesive [80]. The dimensions of the separation channel were $40 \, \text{mm} \times 0.3 \, \text{mm}$, with a $1.2 \, \text{mm}$ electrode gap.

A microfluidic transverse electrokinetic transport device can be modeled as an RC circuit, with each electrode acting as a capacitor and the channel itself acting as both a resistor and a capacitor (see Figure 7.16). Immediately upon

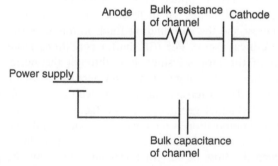

FIGURE 7.16 Circuit diagram of microfluidic device designed for transverse electrokinetic separations.

application of a potential, a large, nonfaradaic current is generated, caused by transient capacitive charging. This current does not result in any oxidation–reduction reactions and its duration depends on the capacitance of the circuit. Given the low conductance of the solutions typically used transverse eletrokinetic transport, this charging current is expected to be much higher than the faradaic current [120]. The faradaic current is accompanied by chemical changes in solution caused by electrolysis, typically of water, at the electrodes.

7.5.2.2.2 Dielectrophoresis

DEP devices typically have more complicated electrode designs, often with two interdigitated sets of electrodes [33, 106–108, 131, 132]. The increased complexity of electrode design is a result of the nature of DEP-mediated transport, which relies on variation of field strength, not field polarity. In a typically FFE or transverse IEF device, the electrode separation distance is constant and the field strength is therefore essentially uniform throughout the separation chamber (neglected small variations caused by conductivity differences), with sharp increases immediately proximal to the electrodes themselves. Such a geometry is not amenable to selection on the basis of field strength, whereas more elaborate geometries, in which the electrode separation distance varies significantly, produce useful field strength gradients.

7.5.2.3 Experimental Design Constraints: Importance of Buffer Selection, Suppression of EOF

As with all experiments, the selection of appropriate buffer conditions is important. However, once electricity is introduced into a system, buffer conditions become even more important. FFE is typically performed under uniform pH conditions, so the buffer must be chosen to maintain pH such that the target particles have the necessarily zeta potential to facilitate transport. That is, the pH should be higher or lower than the pI of the target particle(s). For FFE, as well as IEF, a low conductivity buffer is essential to maximizing field strength for a given applied voltage while minimizing resistive heat generation [125]. DEP relies on differences in conductivity between the particles of interest and the surrounding fluid; in this case the conductivity, not pH, is the critical factor. For IEF, buffer conditions represent the most important part of the process, since it is through the buffer that the pH gradient is generated. The range, slope, and persistence of the pH gradient are determined by the buffer composition, as well as other factors, such as field strength and channel geometry. Of the factors that affect pH gradient formation, buffer composition is the easiest to control and customize.

Methods of pH gradient formation can be divided into three broad classes: natural, in which the flow of current forms and maintains the pH gradient; artificial, which relies on the interdiffusion between solutions of different pHs; and immobilized, in which buffer constituents are covalently bound to a polymeric matrix. The third common method, immobilized pH gradients [105], while useful in many applications such as two-dimensional gel protein analysis, is not well suited to free-flow applications. Because natural pH gradients provide the greatest degree of flexibility and are by far the most common method used today, they merit further discussion.

As described by Svensson, the founder of modern IEF techniques, natural pH gradients form as a result of applying current to a solution containing one or more ampholytes and the simultaneous focusing of the amphoteric compounds in solution to their pIs [133]. These amphoteric compounds were either commercially available mixtures or simple amino acids. Until the early 1990s, the standard method of pH gradients involved the use of carrier ampholytes, a mixture of polyamino polycarboxylic acids, first produced by Vesterberg [134] and later commercialized, most commonly under the trade name Ampholine. These mixtures are well characterized with respect to the pH gradient formed upon application of electric current, but poorly characterized with respect to the actual composition of each particular mixture. An alternative to carrier ampholytes is simple buffer systems, comprised of a limited number of buffers, termed "buffer electrofocusing" (BEF). Because it uses buffers that are readily available and well characterized, BEF is less expensive than use of carrier ampholytes and reduces contamination of the target ampholyte [135]. Nguyen and Chrambrach first introduced BEF and demonstrated pH gradient formation with both amphoteric and nonamphoteric solutions [135] and showed that the resulting pH gradients could be reproducibly modified through addition of neutral, basic, or acidic compounds to the original amphoteric buffer system [136, 137]. The approach did not gain wide acceptance until the mid-1990s, when Bier and colleagues published a series of papers on the use of BEF, performed with one or two amino acids only, to create pH gradients of such shallow slope that they are referred to as "pH windows" [138].

7.5.3 Transverse Electrokinetic Separations of Molecules

7.5.3.1 Small Molecules

The first studies of microfluidic FFE of small molecules were performed in the Manz laboratory in the early 1990s. There it was demonstrated that fluorescently labeled amino acids could be separated and resolved in a 1 cm × 5 cm silicon device [139]. The amino acids were electrophoresed perpendicular to the flow, and position was read out by a scanning laser about a minute after they were introduced to the channel. Voltages as high as 50 V were applied through thin wires inserted manually into side-channels in the device. In later studies, IEF operating in transverse mode has been used to focus and transport small molecules, as seen in Figure 7.17. However, for small molecules, if the fields are limited by the application of less than 2.5 V, the forces applied are too low to make this a practical method.

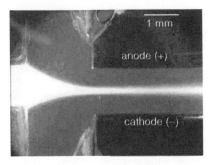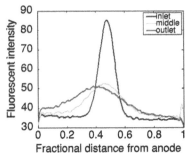

FIGURE 7.17 (See color insert) At left is an image of three solutions entering the channel between the two electrodes in a zone electrophoresis experiment. The central stream is entering from the left. At right is a set of fluorescence intensity profiles taken across the channel at a different point. Note the effects of both electrophoresis (to the left) and diffusion, which broadens the band with time. (Figure adapted from Cabrera, C., *Microfluidic Electrochemical Flow Cells: Design, Fabrication, and Characterization*. Doctoral dissertation in Bioengineering, 2002, University of Washington: Seattle, WA.)

7.5.3.2 Proteins

Given the strong variation in EPM with pH exhibited by proteins makes them quite amenable to fractionation by either FFE or IEF. The Manz group also performed microfluidic FFE of proteins and tryptic digests thereof in a silicon device similar to that used in their previous FFE work [140]. There it was demonstrated that proteins could be separated in a silicon device, and that the resolving power of the device (~8 bands/cm) was comparable to that of macroscopic devices. However, in this and all such microdevices, without going to extremely high voltages there is not enough room to resolve many bands.

Transverse (continuous) IEF was used over 30 years ago to fractionate a mixture of bovine serum albumin, gamma globulin, and horse myoglobin [141]. More recently, polymeric laminate electrokinetic devices were used to perform IEF separations of fluorescently labeled proteins by pI [84]. While these initial results were promising, several time-dependent effects on the intensity of fluorescence were observed that have yet to be resolved (Figure 7.18). Simple devices that allow application of higher voltages in similar polymeric laminate format would be of great utility.

7.5.3.3 Eukaryotic Cells

The suite of electrokinetic separation techniques described herein (FFE, IEF, and DEP) offers an important alternative to fluorescence-activated cell sorting (FACS) and other label-based methods of cell sorting, because the selection mechanism is already present. In applications for which antibodies

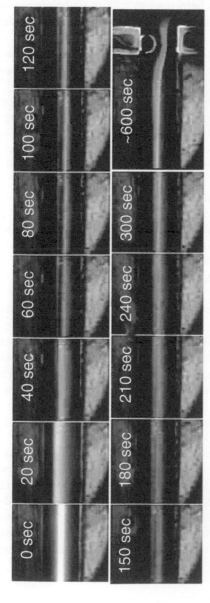

FIGURE 7.18 (See color insert) Images of the effect of time of application of voltage on the continuous separation of bovine serum albumin (BSA) (orange) and lectin (green) fluorescent conjugates over time in a flow channel. The proteins were pumped into the channel together (at left). Fluorescent images near the exit were made at the times shown after the application of 2.0 V. Note that separation is most pronounced about 60 sec after application of the voltage, followed by progressive loss of resolution and, perhaps, loss of fluorescence from the BSA. Note that reversing the voltage in this experiment immediately restores the "lost" fluorescence, implying that the BSA fluorophores were only reversibly quenched, not chemically altered. (From Macounová, K., C.R. Cabrera, and P. Yager, *Analytical Chemistry*, 2001, 73(7): 1627–1633. With permission.)

are not available or an undesirable perturbation to the system under investigation, such a native-characteristic fractionation tool can be useful [142].

Prior to the development of FACS, FFE was commonly used to isolate lymphocytes [142]. FFE has been used to fractionate heterogeneous populations of lymphocytes, from a variety of sources (e.g., rodents, chimpanzees, humans) into numerous distinct subpopulations [124, 143–145]. One review of electrophoretic fractionation of eukaryotic cells includes numerous examples of both low-volume FFE and scaled-up FFE for the analysis and preparation of heterogeneous eukaryotic cell populations [125]. However, the low ionic strength of FFE buffers tends to cause eukaryotic cells to explode after a short incubation. New techniques for ameliorating this effect through increasing the osmolarity of FFE buffers do offer a solution. In addition, the EPM of most eukaryotic cells is fairly close to that of human erythrocytes, which requires high-resolution FFE for effective fractionation [142]. DEP and IEF may be better suited for eukaryotic analysis.

Eukaryotic cells vary significantly enough in pI to allow successful fractionation and characterization with transverse IEF. For example, spermatozoa from boars either with or without intact seminal vesicles were effectively differentiated on the basis of pI, for reasons perhaps best left to the imagination [146]. Transverse IEF has been used to fractionate and analyze red blood cells from a mixed population of cells from humans, rabbits, and mice [147, 148].

A heterogeneous mixture of yeast was separated into viable and nonviable fractions using DEP in batch mode [131]. Also in batch mode, DEP was recently used to successfully isolate malaria-infected human erythrocytes from uninfected erythrocytes, using either an interdigitated electrode array or a more unusual spiral electrode pattern [149]. As mentioned above, DEP–FFF has been used to separate breast cancer cells from normal human blood cells [119]. This successful demonstration of DEP–FFF suggests that other combinations of particles that can be separated using DEP in batch mode could be fractionated with DEP–FFF.

Others have demonstrated the successful separation of human peripheral blood mononuclear cells from 6 μm latex beads, suspended in low conductivity sugar solution, during continuous flow through a microfluidic device constructed of SU8 and glass [132]. Matching pairs of electrodes deposited on the top and bottom glass plates were used to focus all particles into the center of the channel, relative to the electrodes. A second electrode, patterned only on the bottom glass plate, was used to selectively isolate the human blood cells from the flowing fluid.

7.5.3.4 Prokaryotes

Using DEP, *Escherichia coli* were collected onto electrodes from a flowing stream, followed by subsequent elution of the concentrated bacteria in a bolus after turning off the electric field [150]. In related experiments, DEP operated in static conditions was used to separate *Bacillus globigii* spores from *E. coli*, suggesting that the two could also be separated under flowing conditions. Using a DEP variant in which field strength is moderated by varying the fluid conductivity as well as the electrode separation distance, a heterogeneous mixture of *E.coli*, *Bacillus subtilis*, and *Micrococcus luteus* was fractionated into homogenous populations [108].

Transverse IEF and FFE have been used to concentrate Gram⁻ bacteria, *Erwinia herbicola*, under flowing conditions [80]. Varying the buffer composition resulted in changing the thickness of the focused band of bacteria and the position of the band relative to the two electrodes. Typical electrophoretic mobility values for bacteria are on the order of $2 \times 10^{-4}\,cm^2/sec$, obviously depending on the local buffer conditions (pH, conductivity, viscosity) [151]. Again depending on local buffer conditions, the pI of bacteria can range from <1.9 (for *A. aceti* [152]) to 6.4 (for *E. coli* [153]), with typical values between 3 and 5 [152–160].

7.5.3.5 Other Biological Molecules and Particles

Gel IEF has been used to fractionate a mixture of viruses, with pI values ranging from 4.0 to 6.8 [161], indicating that this population would also be amenable to fractionation via transverse IEF. A microfluidic FFE device was used for the separation of genomic and plasmid DNA from albumin [162]; preliminary data suggest that nucleic acid separations and extractions from complex samples should be readily performed, as well.

7.6 FUTURE DIRECTIONS

Work in the last decade has just scratched the surface of what can be done in this small region of microfluidics. More importantly, devices that incorporate transverse transport can be incorporated into systems that are based on other principles, such as chromatographic separation, sample capture, nucleic acid amplification, etc. The next frontier is integration of multiple types of devices into complete systems, as seen in other chapters in this volume.

One little-explored but very exciting potential application for transverse microfluidic separations is that of making rapid kinetic measurements. Such measurements are currently done with stop–flow apparatuses, which bring two fluids together under controlled conditions and a fixed mixing time. Such devices can be difficult to work with, particularly given the time pressure of having to take the needed measurement the instant the two fluids are mixed. Similar measurements could be made using a T-sensor type device, in which two (or more fluids) are again brought together under fixed conditions. By properly designing the channel, the z-axis can function as a pseudo-time dimension, allowing the user to monitor rapid kinetics under steady-state conditions. Since a fixed distance from the channel inlet represents a known amount of mixing time, that position can be monitored indefinitely, given sufficient reagents, thus allowing multiple measurements. In addition, by looking at different positions from the channel inlet, a range of mixing times can be surveyed.

ACKNOWLEDGMENTS

Portions of this chapter have been reprinted or derived from "Quantitative Analysis of Diffusion and Chemical Reaction in Pressure-Driven Microfluidic Channels," a doctoral dissertation by Andrew Evan Kamholz, and from "Microfluidic Electrochemical Flow Cells: Design, Fabrication, and Characterization," a doctoral dissertation by Catherine Regina Cabrera, with permission from the authors.

REFERENCES

1. Fox, R.W. and A.T. McDonald, *Introduction to Fluid Mechanics.* 1998, New York: John Wiley & Sons, p. 776.
2. Bird, R., W. Stewart, and E. Lightfoot, *Transport Phenomena.* 1960, New York: John Wiley & Sons, p. 780.
3. Kamholz, A.E., E.A. Schiling, and P. Yager, Optical measurement of transverse molecular diffusion in a microchannel. *Biophysical Journal,* 2001, 80(4): 1967–1972.
4. Jiang, G. and D.J. Harrison, mRNA isolation in a microfluidic device for eventual integration of cDNA library construction. *Analyst,* 2000, 125(12): 2176–2179.
5. Huang, M.-C., Silicon Microfabricated Device for Non-Sheath-Flow Cytometer-Based Chemical Analysis and Microchannel Flow Sensing. Doctoral dissertation in Electrical Engineering. 1998, University of Washington: Seattle, WA, 77 p.
6. Kamholz, A.E., B.H. Weigl, B.A. Finlayson, and P. Yager, Quantitative analysis of molecular interaction in a microfluidic channel: the T-sensor. *Analytical Chemistry,* 1999, 71(23): 5340–5347.
7. Galambos, P.C., Two-Phase Dispersion in Microchannels. Doctoral dissertation in Mechanical Engineering. 1998, University of Washington: Seattle, WA, 154 p.
8. Ismagilov, R.F., A.D. Stroock, P.J.A. Kenis, G. Whitesides, and H.A. Stone, Experimental and theoretical scaling laws for transverse diffusive broadening in two-phase laminar flows in microchannels. *Applied Physics Letters,* 2000, 76(17): 2376–2378.
9. Munson, M.S. and P. Yager, Simple quantitative optical method for monitoring the extent of mixing applied to a novel microfluidic mixer. *Analytica Chimica Acta,* 2004, 50(1): 63–71.
10. Munson, M.S., C.R. Cabrera, and P. Yager, Passive electrophoresis in microchannels using liquid junction potentials. *Electrophoresis,* 2002, 23(16): 2642–2652.
11. Giddings, J.C., Field-flow fractionation: analysis of macromolecular, colloidal, and particulate materials. *Science,* 1993, 260(5113): 1456–1465.

12. Assidjo, E. and P.J.P. Cardot, Sedimentation field-flow fractionation at gravitational field of red blood cells: systematic studies of injection conditions. *Journal of Liquid Chromatography & Related Technologies*, 1997, 20(16/17): 2579–2597.

13. Assidjo, E., T. Chianéa, M-F. Dreyfuss, and P.J.P. Cardot, Validation procedures of sedimentation field-flow fractionation techniques for biological applications. *Journal of Chromatography B*, 1998, 709(2): 197–207.

14. Metreau, J.M., S.G. Philippe, J.P. Cardot, V. LeMaire, F. Dumas, A. Hernvann, and S. Loric, Sedimentation field-flow fractionation of cellular species. *Analytical Biochemistry*, 1997, 251(2): 178–186.

15. Athanasopoulou, A., G. Karaiskakis, and A. Travlos, Colloidal interactions studied by sedimentation field-flow fractionation. *Journal of Liquid Chromatography & Related Technologies*, 1997, 20(16/17): 2525–2541.

16. Blau, P. and R.L. Zollars, Sedimentation field-flow fractionation of non-spherical particles. *Journal of Colloid and Interface Science*, 1996, 183(2): 476–483.

17. Moon, M.H. and S.H. Lee, Sedimentation steric field-flow fractionation: a powerful technique for obtaining particle size distribution. *Journal of Microcolumn Separations*, 1997, 9(7): 565–570.

18. Martin, M., Relative velocity profile and flow-rate in sedimentation field-flow fractionation. *HRC Journal of High Resolution Chromatography*, 1996, 19(9): 481–484.

19. Williams, P.S., M.H. Moon, and J.C. Giddings, Influence of accumulation wall and carrier solution composition on lift force in sedimentation/steric field-flow fractionation. *Colloids and Surfaces A — Physicochemical and Engineering Aspects*, 1996, 113(3): 215–228.

20. Stephen Williams, P., M.H. Moon, Y. Xu, and J.C. Giddings, Effect of viscosity on retention time and hydrodynamic lift forces in sedimentation/steric field-flow fractionation. *Chemical Engineering Science*, 1996, 51(19): 4477–4488.

21. Schure, M.R., K.D. Caldwell, and J.C. Giddings, Theory of sedimentation hyperlayer field-flow fractionation. *Analytical Chemistry*, 1986, 58(7): 1509–1516.

22. Levy, G.B., Sedimentation field flow fractionation — an overview. *American Laboratory*, 1987, 19(6): 84–87.

23. Myers, M.N., W.J. Cao, C.I. Chen, V. Kumar, and J.C. Giddings, Cold wall temperature effects on thermal field-flow fractionation. *Journal of Liquid Chromatography & Related Technologies*, 1997, 20(16/17): 2757–2775.

24. Ko, G.H., R. Richards, and M.E. Schimpf, Enhanced mass selectivity in thermal field-flow fractionation due to the temperature dependence of the transport coefficients. *Separation Science and Technology*, 1996, 31(8): 1035–1044.

25. Nguyen, M. and R. Beckett, Calibration methods for field-flow fractionation using broad standards. 1. Thermal field-flow fractionation. *Separation Science and Technology*, 1996, 31(3): 291–317.

26. Belgaied, J.E., M. Hoyos, and M. Martin, Velocity profiles in thermal field-flow fractionation. *Journal of Chromatography A*, 1994, 678(1): 85–96.

27. Gunderson, J.J., K.D. Caldwell, and J.C. Giddings, Influence of temperature-gradients on velocity profiles and separation parameters in thermal field-flow fractionation. *Separation Science and Technology*, 1984, 19(10): 667–683.

28. Gale, B.K., K.D. Caldwell, and A.B. Frazier, A micromachined electrical field-flow fractionation (μ-EFFF) system. *IEEE Transactions on Biomedical Engineering*, 1998, 45(12): 1459–1469.

29. Palkar, S.A. and M.R. Schure, Mechanistic study of electrical field flow fractionation.2. Effect of sample conductivity on retention. *Analytical Chemistry*, 1997, 69(16): 3230–3238.

30. Dunkel, M., N. Tri, R. Beckett, and K.D. Caldwell, Electrical field-flow fractionation: a tool for characterization of colloidal adsorption complexes. *Journal of Microcolumn Separations*, 1997, 9(3): 177–183.

31. Schimpf, M.E. and K.D. Caldwell, Electrical field-flow fractionation for colloid and particle analysis. *American Laboratory*, 1995, 27(6): 64–68.

32. Caldwell, K.D. and Y.S. Gao, Electrical field-flow fractionation in particle separation. 1. Monodisperse standards. *Analytical Chemistry*, 1993, 65(13): 1764–1772.

33. Huang, Y., X.B. Wang, F.F. Becker, and P.R. Gasoyne, Introducing dielectrophoresis as a new force field for field-flow fractionation. *Biophysical Journal*, 1997, 73(2): 1118–1129.

34. Huang, Y., J. Yang, X.B. Wang, F.F. Becker, and P.R.C. Gascoyne, The removal of human breast cancer cells from hematopoietic CD34(+) stem cells by dielectrophoretic field-flow-fractionation. *Journal of Hematotherapy & Stem Cell Research*, 1999, 8(5): 481–490.

35. Markx, G.H., J. Rousselet, and R. Pethig, DEP-FFF: field-flow fractionation using non-uniform electric fields. *Journal of Liquid Chromatography & Related Technologies*, 1997, 20(16/17): 2857–2872.

36. Muller, T., T. Schnelle, G. Gradl, S.G. Shirley, and G. Fuhr, Microdevice for cell and particle separation using dielectrophoretic field-flow fractionation. *Journal of Liquid Chromatography & Related Technologies*, 2000, 23(1): 47–59.

37. Wang, X.B., J. Yang, Y. Huang, J. Vykoukal, F.F. Becker, and P.R.C. Gascoyne, Cell separation by dielectrophoretic field-flow-fractionation. *Analytical Chemistry*, 2000, 72(4): 832–839.

38. Moon, M.H., M.N. Myers, and J.C. Giddings, Evaluation of pinched inlet channel for stopless flow-injection in steric field-flow fractionation. *Journal of Chromatography*, 1990, 517: 423–433.

39. Chen, X.R., K.G. Wahlund, and J.C. Giddings, Gravity-augmented high-speed flow steric field-flow fractionation — simultaneous use of 2 fields. *Analytical Chemistry*, 1988, 60(4): 362–365.

40. Tomida, T. and B.J. McCoy, Separations in conventional, hyperlayer, and steric field-flow fractionation. *AIChE Journal*, 1988, 34(2): 341–346.

41. Giddings, J.C., X. Chen, K.G. Wahlund, and M.N. Myers, Fast particle separation by flow steric field-flow fractionation. *Analytical Chemistry*, 1987, 59(15): 1957–1962.

42. Koch, T. and J.C. Giddings, High-speed separation of large (greater-than–1 mu-m) particles by steric field-flow fractionation. *Analytical Chemistry*, 1986, 58(6): 994–997.

43. Caldwell, K.D., Z.Q. Cheng, P. Hradecky, and J.C. Giddings, Separation of human and animal-cells by steric field-flow fractionation. *Cell Biophysics*, 1984, 6(4): 233–251.

44. Jiang, Y., M.N. Myers, and J.C. Giddings, Separation behavior of blood cells in sedimentation field-flow fractionation. *Journal of Liquid Chromatography & Related Technologies*, 1999, 22(8): 1213–1234.

45. Gao, Y.S., S.C. Lorbach, and R. Blake, Separation of bacteria by sedimentation field-flow fractionation. *Journal of Microcolumn Separations*, 1997, 9(6): 497–501.

46. Giddings, J.C., B.N. Barman, and M.K. Liu, Separation of cells by field-flow fractionation. *ACS Symposium Series*, 1991, 464: 128–144.

47. Bigelow, J.C., Y. Nabeshima, K. Kataoka, and J.C. Giddings, Separation of cells and measurement of surface-adhesion forces using a hybrid of field-flow fractionation and adhesion chromatography. *ACS Symposium Series*, 1991, 464: 146–158.

48. Reschiglian, P., D. Melucci, G. Torsi, and A. Zattoni, Standardless method for quantitative particle-size distribution studies by gravitational field-flow fractionation. Application to silica particles. *Chromatographia*, 2000, 51(1/2): 87–94.

49. Koliadima, A., E. Dalas, and G. Karaiskakis, Simultaneous determination of particle-size and density in polydisperse colloidal samples by sedimentation field-flow fractionation. *HRC Journal of High Resolution Chromatography*, 1990, 13(5): 338–342.

50. Koliadima, A. and G. Karaiskakis, Sedimentation field-flow fractionation — a new methodology for the concentration and particle-size analysis of dilute polydisperse colloidal samples. *Journal of Liquid Chromatography*, 1988, 11(14): 2863–2883.

51. Jones, H.K. and J.C. Giddings, Separation and characterization of colloidal materials of variable particle-size and composition by coupled column sedimentation field-flow fractionation. *Analytical Chemistry*, 1989, 61(7): 741–745.

52. Dalas, E., P. Koutsoukos, and G. Karaiskakis, The effect of carrier solution on the particle-size distribution of inorganic colloids measured by steric field-flow fractionation. *Colloid and Polymer Science*, 1990, 268(2): 155–162.

53. Giddings, J.C., Retention (steric) inversion in field-flow fractionation — practical implications in particle-size, density and shape analysis. *Analyst*, 1993, 118(12): 1487–1494.

54. Colfen, H. and M. Antonietti, Field-flow fractionation techniques for polymer and colloid analysis. *New Developments in Polymer Analytics I.*, 2000, 151: 67–187.

55. Fuh, C.B. and S.Y. Chen, Magnetic split-flow thin fractionation: new technique for separation of magnetically susceptible particles. *Journal of Chromatography A*, 1998, 813(2): 313–324.

56. Nguyen, M., R. Beckett, L. Pille, and D. H. Solomon, Determination of thermal diffusion coefficients for polydisperse polymers and microgels by ThFFF and SEC-MALLS. *Macromolecules*, 1998, 31(20): 7003–7009.

57. Jeon, S.J. and D.W. Lee, Thermal-diffusion and molecular-weight calibration of poly(ethylene-co-vinyl acetate)s by thermal field-flow fractionation. *Journal of Polymer Science Part B — Polymer Physics*, 1995, 33(3): 411–416.

58. Levin, S. and G. Tawil, Effect of surfactants on the diffusion-coefficients of proteins, measured by analytical SPLITT fractionation (asf) in the diffusion mode. *Journal of Pharmaceutical and Biomedical Analysis*, 1994, 12(4): 499–507.

59. Liu, M.K., P. Li, and J.C. Giddings, Rapid protein separation and diffusion coefficient measurement by frit inlet flow field-flow fractionation. *Protein Science*, 1993, 2(9): 1520–1531.
60. Schimpf, M.E., C. Rue, L.M. Wheeler, P.F. Romeo, and G. Mercer, Studies in the thermal-diffusion of copolymers using field-flow fractionation. *Journal of Coatings Technology*, 1993, 65(822): 51–56.
61. Tsukamoto, O., T. Ohizumi, T. Ohara, S. Mori, and Y. Wada, Feasibility study on separation of several tens nanometer-scale particles by magnetic field-flow-fractionation technique using superconducting magnet. *IEEE Transactions on Applied Superconductivity*, 1995, 5(2): 311–314.
62. Mori, S., Magnetic field-flow fractionation using capillary tubing. *Chromatographia*, 1986, 21(11): 642–644.
63. Gorse, J., T.C. Schunk, and M.F. Burke, The study of liquid suspensions of iron-oxide particles with a magnetic field-flow fractionation device. *Separation Science and Technology*, 1984, 19(13–1): 1073–1085.
64. Schunk, T.C., J. Gorse, and M.F. Burke, Parameters affecting magnetic field-flow fractionation of metal-oxide particles. *Separation Science and Technology*, 1984, 19(10): 653–666.
65. Bouamrane, F., N.E. Assidjo, B. Bouteille, M.F. Dreyfuss, M.L. Darde, and P.J.P. Cardot, Sedimentation field-flow fractionation application to toxoplasma gondii separation and purification. *Journal of Pharmaceutical and Biomedical Analysis*, 1999, 20(3): 503–512.
66. Bernard, A., C. Bories, P.M. Loiseau, and P.J.P. Cardot, Selective elution and purification of living trichomonas- vaginalis using gravitational field-flow fractionation. *Journal of Chromatography B — Biomedical Applications*, 1995, 664(2): 444–448.
67. Merinodugay, A., P.J.P. Cardot, M. Czok, M. Guernet, and J.P. Andreux, Monitoring of an experimental red-blood-cell pathology with gravitational field-flow fractionation. *Journal of Chromatography B — Biomedical Applications*, 1992, 579(1): 73–83.
68. Yager, P., M.R. Holl, B.H. Weigl, J.P. Brody, Microfabricated Diffusion-Based Chemical Sensor. U.S. patent No. 5,716,852 (1998).
69. Weigl, B.H. and P. Yager, Microfluidic diffusion-based separation and detection. *Science*, 1999, 283: 346–347.
70. Galambos, P., F.K. Forster, and B.H. Weigl. A method for determination of pH using a T-sensor. in *Transducers '97*. 1997, New York, NY: IEEE.
71. Weigl, B.H. and P. Yager, Silicon-microfabricated diffusion-based optical chemical sensor. *Sensors and Actuators B — Chemical*, 1997, 39(1–3): 452–457.
72. Weigl, B.H., J. Kriebel, K. Mayes, P. Yager, C.C. Wu, M. Holl, M. Kenny, and D. Zebert, Simultaneous self-referencing analyte determination in complex sample solutions using microfabricated flow structures (T-sensors). *Proceedings of MicroTAS 98*, 1998, pp. 81–84.
73. Darling, R.B., P. Yager, B. Weigl, J. Kriebel, and K. Mayes, Simultaneous self-referencing analyte determination in complex sample solutions using microfabricated flow structures (T-sensors). *Proceedings of MicroTAS 98*, 1998, pp. 105–108.
74. Yager, P., D. Bell, J.P. Brody, D. Qin, C. Cabrera, A. Kamholz, and B. Weigl, Simultaneous self-referencing analyte determination in complex sample

solutions using microfabricated flow structures (T-sensors). *Proceedings of MicroTAS 98*, 1998, pp. 207–212.

75. Kamholz, A.E. and P. Yager, Theoretical analysis of molecular diffusion in pressure-driven laminar flow in microfluidic channels. *Biophysical Journal*, 2001, 80: 155–160.

76. Oleschuk, R.D., L.L. Shultz-Lockyear, Y.B. Ning, and D.J. Harrison, Trapping of bead-based reagents within microfluidic systems: on-chip solid-phase extraction and electrochromatography. *Analytical Chemistry*, 2000, 72(3): 585–590.

77. Figeys, D. and R. Aebersold, Nanoflow solvent gradient delivery from a microfabricated device for protein identifications by electrospray ionization mass spectrometry. *Analytical Chemistry*, 1998, 70(18): 3721–3727.

78. Lagally, E.T., I. Medintz, and R.A. Mathies, Single-molecule DNA amplification and analysis in an integrated microfluidic device. *Analytical Chemistry*, 2001, 73(3): 565–570.

79. Cabrera, C.R., B. Finlayson, and P. Yager, Formation of natural pH gradients in a microfluidic device under flow conditions: model and experimental validation. *Analytical Chemistry*, 2001, 73(3): 658–666.

80. Cabrera, C.R. and P. Yager, Continuous concentration of bacteria in a microfluidic flow cell using electrokinetic techniques. *Electrophoresis*, 2001, 22(2): 355–362.

81. Cabrera, C.R. and P. Yager, Mapping of pH gradients in microfluidic electrokinetic devices, in *Micro Total Analysis Systems 2001*, 2001. Monterey, CA: Kluwer Academic.

82. Hatch, A., A.E. Kamholz, K.R. Hawkins, M.S. Munson, E.A., Schilling, B.H. Weigl, and P. Yager, A rapid diffusion immunoassay in a T-sensor. *Nature Biotechnology*, 2001, 19(5): 461–465.

83. Macounová, K., C.R. Cabrera, M.R. Holl, and P. Yager, Generation of natural pH gradients in microfluidic channels for use in isoelectric focusing. *Analytical Chemistry*, 2000, 72(16): 3745–3751.

84. Macounová, K., C.R. Cabrera, and P. Yager, Concentration and separation of proteins in microfluidic channels on the basis of transverse IEF. *Analytical Chemistry*, 2001, 73(7): 1627–1633.

85. Vahey, P.G., S.H. Park, B.J. Marquardt, Y.N. Xia, L.W. Burgess, and R.E. Synovec, Development of a positive pressure driven micro-fabricated liquid chromatographic analyzer through rapid-prototyping with poly(dimethylsiloxane): optimizing chromatographic efficiency with sub-nanoliter injections. *Talanta*, 2000, 51: 1205–1212.

86. Crank, J., *The Mathematics of Diffusion*. 1st ed. 1956, London: Oxford University Press, p. 347.

87. Kamholz, A.E. and P. Yager, Molecular diffusive scaling laws in pressure-driven microfluidic channels: deviation from one-dimensional Einstein approximations. *Sensors and Actuators B*, 2002, 82(1): 117–121.

88. Holl, M.R., P. Galambos, F.K. Forster, J.P. Brody, M.A. Afromowitz, and P. Yager, Optimal design of a microfabricated diffusion-based extraction device. *Proceedings of ASME Meeting*, 1996, pp. 189–195.

89. Brody, J.P., T.D. Osborn, F.K. Forster, and P. Yager, A planar microfabricated fluid filter. *Sensors and Actuators A — Physical*, 1996, 54(1–3): 704–708 (Proceedings of Transducers '95).

90. Brody, J.P., P. Yager, R.E. Goldstein, and R.H. Austin, Biotechnology at low Reynolds numbers. *Biophysical Journal*, 1996, 71(6): 3430–3441.
91. Brody, J.P. and P. Yager. Low Reynolds number micro-fluidic devices, in *Proceedings Hilton Head MEMS Conference, Solid-State Sensor and Actuator Workshop*, 1996, Hilton Head, SC.
92. Brody, J.P. and P. Yager, Diffusion-based extraction in a microfabricated device. *Sensors and Actuators A — Physical*, 1997, 58(1): 13–18.
93. Jandik P., B.H. Weigl, N. Kessler, J. Cheng, C.J. Morris, T. Schulte, and N. Avdalovic, Initial study of using laminar fluid diffusion interface for sample preparation in HPLC. *Journal of Chromatography A*, 2002, 954: 33–40.
94. Schilling, E.A., A.E. Kamholz, and P. Yager, Cell lysis and protein extraction in a microfluidic device with detection by a fluorogenic enzyme assay. *Analytical Chemistry*, 2002, 74(8): 1798–1804.
95. Schilling, E.A., E.A. Kamholz, and Y. P. Cell lysis and protein extraction in a microfluidic device with detection by a fluorogenic enzyme assay, in *Micro Total Analysis Systems 2001*. 2001, Monterey, CA: Kluwer Academic.
96. Sherbet, G.V., *The Biophysical Characterisation of the Cell Surface*. 1978, New York: Academic Press.
97. Atkins, *Physical Chemistry*. 4th ed. 1990, New York: Freeman, p. 995.
98. Boltz, R.C. and T.Y. Miller, A citrate buffer system for isoelectric focusing and electrophoresis of living mammalian cells, in *Electrophoresis '78*, N. Catsimpoolas, Editor. 1978, New York: Elsevier North-Holland, pp. 345–355.
99. Berg, H., *Random Walks in Biology*. 1983, Princeton, NJ: Princeton University Press.
100. Cabrera, C., Microfluidic Electrochemical Flow Cells: Design, Fabrication, and Characterization. Doctoral dissertation in Bioengineering. 2002, University of Washington: Seattle, WA.
101. Lindgren, E., R. Rao, and B. Finlayson, Numerical simulation of electrokinetic phenomena, in *Emerging Technologies in Hazardous Waste Management V*, D. Tedder and F. Pohland, Editors. 1995, Washington, DC: American Chemical Society.
102. James, A., J. Stillman, and D. Williams, Finite element solution of the equations governing the flow of electrolyte in charged microporous membranes. *International Journal for Numerical Methods in Fluids*, 1995, 20(10): 1163–1178.
103. Palkar, S.A. and M.R. Schure, Mechanistic study of electrical field flow fractionation. 1. Nature of the internal field. *Analytical Chemistry*, 1997, 69(16): 3223–3229.
104. Rodriguez-Diaz, R., T. Wehr, and M. Zhu, Capillary isoelectric focusing. *Electrophoresis*, 1997, 18: 2134–2144.
105. Righetti, P.G., M. Fazio, C. Tonani, E. Gianazza and F.C. Celentano, pH gradients generated by polyprotic buffers. II. Experimental validation. *Journal of Biochemical and Biophysical Methods*, 1988, 16: 129–140.
106. Hughes, M., AC electrokinetics applications for nanotechnology. *Nanotechnology*, 2000, 11: 124–132.
107. Gascoyne, P.R.C. and J. Vykoukal, Particle separation by dielectrophoresis. *Electrophoresis*, 2002, 23(13): 1973–1983.
108. Markx, G., P. Dyda, and R. Pethig, Dielectrophoretic separation of bacteria using a conductivity gradient. *Journal of Biotechnology*, 1996, 51: 175–180.

109. Schure, M., K. Caldwell, and J. Giddings, Theory of sedimentation hyperlayer field-flow fractionation. *Analytical Chemistry*, 1983, 58: 1509–1516.
110. Caldwell, K.D., L.F. Keener, M.N. Myers, and J.C. Giddings, Electrical field-flow fractionation of proteins. *Science*, 1972, 176: 296–298.
111. Chmelik, J., M. Deml, and J. Janca, Separation of two components of horse myoglobin by isoelectric focusing field-flow fractionation. *Analytical Chemistry*, 1989, 61: 912–914.
112. Thormann, W., M.A. Firestone, M.L. Dietz, T. Cecconie, and R.A. Mosher, Focusing counterparts of electrical field flow fractionation and capillary zone electrophoresis. Electrical hyperlayer field flow fractionation and capillary iso-electric focusing. *Journal of Chromatography*, 1989, 461: 95–101.
113. Chmelik, J. and W. Thormann, Isoelectric focusing field-flow fractionation and capillary isoelectric focusing with electroosmotic zone displacement. Two approaches to protein analysis in flowing streams. *Journal of Chromatography*, 1993, 632: 229–234.
114. Chmelik, J. and W. Thormann, Isoelectric focusing field-flow fractionation IV. Investigations on protein separations in the trapezoidal cross-section channel. *Journal of Chromatography*, 1992, 600: 306–311.
115. Chmelik, J. and W. Thormann, Isoelectric focusing field-flow fractionation III. Investigation of the influence of different experimental parameters on focusing of cytochrome c in the trapezoidal cross-section channel. *Journal of Chromatography*, 1992, 600: 297–304.
116. Bier, M., O.A. Palusinski, R.A. Mosher, and D.A. Saville, Electrophoresis: mathematical modeling and computer simulation. *Science*, 1983, 219(4590): 1281–1287.
117. Chmelik, J., Isoelectric focusing field-flow fractionation. Experimental study of the generation of pH gradient. *Journal of Chromatography*, 1991, 539: 111–121.
118. Chmelik, J., Isoelectric focusing field-flow fractionation. II. Experimental study of focusing of methyl red in the trapezoidal cross-section channel. *Journal of Chromatography*, 1991, 545: 349–358.
119. Yang, J., Y. Huang, X.B. Wang, F.F. Becker, and P.R.C. Gascoyne, Cell separation on microfabricated electrodes using dielectrophoretic/gravitational field flow fractionation. *Analytical Chemistry*, 1999, 71(5): 911–918.
120. Bard, A. and L. Faulkner, *Electrochemical Methods: Fundamentals and Applications*. 2001, New York: John Wiley & Sons.
121. Baygents, J.C., B.C.E. Schwarz, R.R. Deshmukh, and M. Bier, Recycling electrophoretic separations: modeling of isotachophoresis and isoelectric focusing. *Journal of Chromatography A*, 1997, 779: 165–183.
122. Kuhn, R., H. Wagner, R.A. Mosher, and W. Thormann, Experimental and theoretical investigation of the stability of stepwise pH gradients in continuous flow electrophoresis. *Electrophoresis*, 1987, 8: 503–508.
123. Bier, M., Recycling isoelectric focusing and isotachophoresis. *Electrophoresis*, 1998, 19: 1057–1063.
124. Hanning, K., Continuous free-flow electrophoresis as an analytical and preparative method in biology. *Journal of Chromatography*, 1978, 159(1): 183–191.
125. Bauer, J., Electrophoretic separation of cells. *Journal of Chromatography*, 1987, 418: 359–383.

126. Levin, S., M. Myers, and J. Giddings, Continuous separation of proteins in electrical split-flow thin (SPLITT) cell with equilibrium operation. *Separation Science and Technology*, 1989, 24(14): 1245–1259.

127. Levin, S., Field-flow fractionation (FFF) and related techniques for the separation of particles, colloids, and macromolecules. *Israel Journal of Chemistry*, 1990, 30: 257–262.

128. Fuh, C. and J. Giddings, Isoelectric split-flow thin (SPLITT) fractionation of proteins. *Separation Science and Technology*, 1997, 32(18): 2945–2967.

129. Liu, G. and J. Giddings, Separation of particles in nonaqueous suspensions by thermal-electrical field-flow fractionation. *Analytical Chemistry*, 1991, 63: 296–299.

130. Tri, N., K. Caldwell, and R. Beckett, Development of electrical field-flow fractionation. *Analytical Chemistry*, 2000, 72(8): 1823–1829.

131. Markx, G., M. Talary, and R. Pethig, Separation of viable and non-viable yeast using dielectrophoresis. *Journal of Biotechnology*, 1994, 32: 29–37.

132. Holmes, D. and H. Morgan. Particle focusing and separation using dielectrophoresis in a microfluidic device, in *Micro Total Analytical Systems*, 2001. Monterey, CA: Kluwer Academic.

133. Svensson, H., *Acta Chemica Scandinavica*, 1961, 15: 325.

134. Vesterberg, O., *Acta Chemica Scandinavica*, 1969, 23: 2653.

135. Nguyen, N. and A. Chrambach, Nonisoelectric focusing in buffers. *Analytical Biochemistry*, 1976, 74: 145–153.

136. Nguyen, N. and A. Chrambach, Stabilization of pH gradients in buffer electrofocusing on polyacrylamide gel. *Analytical Biochemistry*, 1977, 82: 54–62.

137. Nguyen, N.Y., A. Salokangas, and A. Chrambach, Electrofocusing in natural pH gradients formed by buffers: gradient modification. *Analytical Biochemistry*, 1977, 78: 287–294.

138. Bier, M., J. Ostrem, and R. Marquez, A new buffering system and its use in electrophoresis and isoelectric focusing. *Electrophoresis*, 1993, 14: 1011–1018.

139. Raymond, D.E., A. Manz, and H.M. Widmer, Continuous sample pretreatment using a free-flow electrophoresis device integrated onto a silicon chip. *Analytical Chemistry*, 1994, 66: 2858–2865.

140. Raymond, D.E., A. Manz, and H.M. Widmer, Continuous separation of high molecular weight compounds using a microliter volume free-flow electrophoresis microstructure. *Analytical Chemistry*, 1996, 68: 2515–2522.

141. Seiler, N., J. Thobe, and G. Werner, *Hoppe-Seyler's Zeitschrift für Physiologische Chemie*, 1970, 351: 865.

142. Bauer, J., Advances in cell separation: recent developments in counterflow centrifugal elutriation and continuous flow cell separation. *Journal of Chromatography B*, 1999, 722: 55–69.

143. Zeiller, K., H. Liebich, and K. Hannig, Free-flow electrophoretic separation of lymphocytes. Two thoracic duct lymphocyte subpopulations studied after prolonged cannulation and immunization. *European Journal of Immunology*, 1971, 1: 315–322.

144. Zeiller, K., G. Pascher, G. Wagner, H.G. Liebich, and E. Holzberg, Distinct subpopulations of thymus-dependent lymphocytes. *Immunology*, 1974, 26: 995–1012.

145. Zeiller, K. and G. Pascher, Differentiation antigens in rat lymphocyte subpopulations recognized by rabbit antisera against electrophoretically separated cells. *European Journal of Immunology*, 1978, 8: 469–477.
146. Moore, H. and K. Hibbitt, Isoelectric focusing of boar spermatozoa. *Journal of Reproduction and Fertility*, 1975, 44: 329–332.
147. Just, W., J. Leon-V, and G. Werner, Continuous-flow isoelectric focusing: studies on the separation of mixed red blood cells of different species and of the light mitochondrial fraction of rat liver, in *Progress in Isoelectric Focusing and Isotachophoresis*, P. Righetti, Editor. 1975, Amsterdam: North-Holland Publishing Company.
148. Just, W., J. Leon-V, and G. Werner, Isoelectric focusing in continuous-flow electrophoresis. *Analytical Biochemistry*, 1975, 67: 590–601.
149. Gascoyne, P., C. Mahidol, M. Ruchirawat, J. Satayavivad, P. Watcharasit, and F.F. Becker, Microsample preparation by dielectrophoresis: isolation of malaria. *Lab on a Chip*, 2002, 2: 70–75.
150. Huang, Y., K.L. Ewalt, M. Tirado, T.R. Haigis, A. Forster, D. Ackley, M.J. Heller, J. P. O'Connell, and M. Krihak, Electric manipulation of bioparticles and macromolecules on microfabricated electrodes. *Analytical Chemistry*, 2001, 73(7): 1549–1559.
151. Pfetsch, A. and T. Welsch, Determination of the electrophoretic mobility of bacteria and their separation by capillary zone electrophoresis. *Fresenius Journal of Analytical Chemistry*, 1997, 359: 198–201.
152. Mozes, N., A.J. L'Eonard, and P.G. Rouxhet, On the relations between the elemental surface composition of yeasts and bacteria and their charge and hydrophobicity. *Biochimica et Biophysica Acta*, 1988, 945(2): 324–334.
153. Longton, R.W., J.S. Cole, III, and P.F. Quinn, Isoelectric focusing of bacteria: species location within an isoelectric focusing column by surface charge. *Archives of Oral Biology*, 1975, 20: 103–106.
154. Collins, Y.E. and G. Stotzky, Heavy metals alter the electrokinetic properties of bacteria, yeasts, and clay minerals. *Applied and Environmental Microbiology*, 1992, 58(5): 1592–1600.
155. Gittens, G. and A. James, *Bichimica et Biophysica Acta*, 1963, 66: 237–249.
156. Lemp J., Jr., E. Asbury, and E. Ridenour, Electrophoresis of colloidal biological particles. *Biotechnology and Bioengineering*, 1971, 13: 17–47.
157. Schott, H. and C.Y. Young, Electrokinetic studies of bacteria I: effect of nature, ionic strength, and pH of buffer solutions on electrophoretic mobility of *Streptococcus faecalis* and *Escherichia coli*. *Journal of Pharmaceutical Sciences*, 1972, 61(2): 182–187.
158. Sonohara, R., N. Muramatsu, H. Ohshima, and T. Kondo, Difference in surface properties between *Escherichia coli* and *Staphylococcus aureus* as revealed by electrophoretic mobility measurements. *Biophysical Chemistry*, 1995, 55(3): 273–277.
159. Mangia, A., L. Teixeira, and F. Filho, The electrokinetic surface of five enteropathogenic *Escherichia coli* serogroups. *Cell Biophysics*, 1995, 26: 45–55.
160. Sherbet, G.V. and M.S. Lakshmi, Characterization of *Escherichia coli* cell surface by isoelectric equilibrium analysis. *Biochimica et Biophysica Acta*, 1973, 298: 50–58.

161. Rice, R. and J. Horst, Isoelectric focusing of viruses in polyacrylamide gels. *Virology*, 1972, 49: 602–604.
162. Shinohara, E., N. Tajima, J. Funazaki, K. Tashiro, S. Shoji, Y. Utsumi, and T. Hatttori, Simultaneous self-referencing analyte determination in complex sample solution using microfabricated flow structures (T-sensors). *Proceedings of MicroTAS* 98, 1998, pp. 125–126.

8 Electro-Driven Separation Methods on Chips

Rikke P. H. Nikolajsen, Gabriela S. Chirica,
Yolanda Fintschenko, and Jörg P. Kutter

CONTENTS

8.1 INTRODUCTION

The effectiveness of a chemical analysis method is generally measured by five figures of merit: sensitivity, limit of detection, accuracy, precision, and throughput or speed of analysis. While the name suggests that portability or size is the driving force, lab-on-a-chip, or micro total analysis systems (μ-TAS), science and technology is driven by the belief that miniaturization of (primarily) separation techniques will lead to an improvement in these figures of merit. Widmer et al. first outlined how different separation and detection methods scale down [1]. This led to the proposal of miniaturizing and integrating sample introduction, separation, and detection, elevating a separation technique to a quasicontinuous generic chemical sensor [1]. Manz et al. first reported capillary zone electrophoresis (CZE) in open channels etched in a glass chip in 1992 [2, 3]. However, it was the demonstration of the on-chip baseline separation of two fluorescent dyes within 150 msec by Jacobson et al. that best illustrates the impact of lab-on-a-chip technology on chemical analysis [4].

The chip format provides a platform for integrated functions because of the ability to use photolithography to define small volume channel networks that integrate all of the different functions required for μ-TAS. For example, in the case of CZE, the integration of the injection channel with the separation channel provides a platform for a rapid valveless injection method. This requires the electronic control of the voltages applied to each channel rather than the physical movement of the separation channel from sample to waste as in a typical capillary-based CZE system [4]. An illustration of a simple injection cross for electrokinetically driven separations is shown in Figure 8.1A. A diagram of the equivalent system in capillaries is shown in Figure 8.1B. It is seen that a capillary-based system similar to those etched in chips requires a four-way interconnect, which presents the problem of dead volume and serves as an additional point of stress, often resulting in capillary breakage.

The tools provided by microfabrication extend chip design capabilities beyond channel size definition and open doors to complex architectures for multiplex separations and integration that were impossible to achieve with traditional or capillary separation techniques. Cross-disciplinary collaborations between engineers, polymer chemists, and separation scientists have demonstrated proof of concept for on-chip micrometer size valves, membranes, mixers, and pumps [5].

In addition to miniaturization and potential simplified operation, on-chip integration also minimizes operator error, sample contamination, and loss due to transfer from one processing step to another. The small dead volumes, automatable flow control, and analysis are features inherent to microfabricated devices, which make them particularly suited to operation by non-skilled persons. Additionally, by targeting low-cost and specialized analyses, one would expect to see in the future these devices used much like disposable, easy to use glucose testing units. Therefore, once the efficient separation capability has been demonstrated on-chip in serial or parallel format, in one- or multidimensional separations, the sample collection, preparation, and labeling steps need to be addressed. Electrophoretic mobility inherent to constituents of biological samples, coupled with electroosmotic flow and the ability to selectively pattern surfaces afford sample preparation procedures to be performed in the same general format as the separation step. Presently, in most instances only one or two components of this integration potential list for microdevices have been demonstrated [6–10]. One example is the preconcentration technique which uses on-chip salt bridges to increase protein concentration prior to electrophoretic separations [11].

There are a number of challenges to using a chip-based system. Interconnects to the chip can pose the greatest problem. This has largely been solved by using pipette tips and more complicated fittings that allow for low- and high-pressure interfaces [2, 12, 13]. Evaporation is minimized in

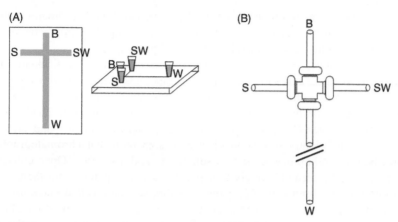

FIGURE 8.1 Injector integrated with separation channel. (A) Microfabricated in a chip. Chip dimensions: 8.5 × 2.0 cm. Typical channel dimensions are: 50 μm width and 25 μm depth. The separation channel is 7.5 cm long and the injection channel is 1 cm long. B = running buffer, W = waste, S = sample, SW = sample waste. Sample is introduced by applying voltage across S (100% total V applied) and SW (0%) with B at 40% and BW at 60%, and injected onto the separation column by switching voltage to B = 100% and W = 0% with S and SW at 60%. A switch box or computer is typically used to control voltages. (B) Formed using capillaries. The equivalent integrated injector and capillary column using capillaries and a commercially available cross fitting. Typical capillary sizes are 365 μm outer diameter and 50 to 100 μm inner diameter. (Reprinted from Fintschenko, Y.; Kirby, B. J.; Hasselbrink, E. F.; Singh, A. K.; Shepodd, T. J. In *Monolithic Materials*; Svec, F., Tennikova, T. B., Deyl, Z., Eds.; Elsevier: New York, 2003; Vol. 67, pp 659–683. With permission.)

these small volume reservoirs by covering the reservoirs, while problems like pH changes and bubbles can be addressed by using salt-bridges between the solution and channel [10]. Detection issues are similar to those in a fused silica capillary, with both the reduced path length and the detection volume resulting in a reduced limit of detection if using absorbance detection [1]. As a result, most systems employ laser-induced fluorescence (LIF) detection [3, 4, 6, 10, 14–21]. Alternatives, including electrochemical and chemiluminescence detection, have also been reported [22–30].

In order to realize the full potential of integrated microfluidic systems a number of components at dimensions from 10 to 100 μm have to be developed. For example, flow control has been valveless, an advantage in some cases, but largely a result of the absence of reliable valves for microfluidic systems. CZE and other low-pressure techniques have been pursued in the absence of a small high-pressure pump suitable for either integration with or connection to a chip-based system. High-pressure liquid chromatography (HPLC), an industry standard, has been largely untapped in chip separation systems due to the lack of reliable valves and pumps [13].

Electrophoretic techniques are limited to systems where a difference in charge to size ratio exists [31]. For many compounds of interest, for example, explosives, polycyclic aromatic hydrocarbons, and drugs, a difference in hydrophobicity or hydrophilicity is the best or only physical handle for separations. Other charged targets, such as proteins, peptides, and amino acids, are sufficiently similar in charge to size ratio that chromatography must be used to improve separation selectivity and resolution. Open column methods can be used effectively to separate these compounds, but there is a sample capacity limitation [32]. Consequently, separations that make use of a stationary phase, via a packed or coated channel, have been investigated for on-chip separations. The advantages of using porous polymer monoliths as a stationary phase have been described for capillary methods [33]. The challenges of the planar format of on-chip separations make full use of the unique advantages of porous polymer monoliths in a way that capillary systems do not.

There are several examples of chromatographic-type separations in open channels on a chip. Rapid open channel separations of explosives by micellar electrokinetic chromatography (MEKC) on-a-chip have recently been reported [10]. While MEKC has the advantage that the conditions, including the stationary phase, can be rapidly adjusted, it also has two inherent limitations: a dependence on solubility of compounds and a limited elution window [34]. In MEKC, there is an optimum elution time outside which maximum resolution cannot be achieved because strongly retained compounds are pushed to the end of the chromatogram and elute after or close to the micellar zone. Organic solvents added to the running buffer, typically used to extend the elution range of MEKC, can disrupt the micellar pseudostationary phase [35].

Gradient elution MEKC, demonstrated on a chip by Kutter et al., shows promising results with coumarin dyes; however, relatively low organic (30%) buffers were used, limiting the sample range to less hydrophobic compounds [36].

Coated open channels in chips have given good results both with gradient and isocratic conditions [4, 37]. No limitations in terms of organic content in buffer composition exist and more hydrophobic samples can be analyzed. Reduction of the channel depth to dimensions on the order of 3 to 5 μm afforded higher efficiencies characterized by low plate heights, but channels of these dimensions proved difficult to make in a controlled and reproducible fashion [4, 37]. Additionally, the issue of capacity remains for any open channel technique, particularly as dimensions decrease.

The technical difficulties associated with the fabrication of packed columns and the low sample capacity limitation of the open channel design prompted the development of polymeric stationary phases. Fabrication of cast-in-shape chromatographic media entails a simple *in situ* procedure: the channel is filled with a mixture of monomers and solvent, followed by a UV, redox, or thermal initiated polymerization. Furthermore, entirely new polymeric formulations are developed, tailor-made for specific separation methods. To date, most electrokinetic driven separations have been demonstrated on glass or quartz (fused silica) substrates. They are optically transparent, can readily generate electroosmotic flow and provide a clean surface after etching. However, the fabrication process is rather laborious, slow, requires high temperature for bonding and produces limited aspect ratios. In particular, for separations of biological relevance, nonspecific interactions of macromolecules with the negatively charged walls can pose serious problems. Consequently, a topic of great interest and continuous innovation is the development of new materials and fabrication procedures for microdevices. Laser ablation was used to produce channels in polystyrene, polycarbonate, cellulose acetate and poly(ethylene terephthalate) [38]. Polymethymethacrylate (PMMA) devices are already employed in electrophoretic separations [39]. Using cationic surfactant solution the generation of reversed electroosmotic flow was demonstrated [40]. Procedures for channel definition in polydimethoxysiline (PDMS) as well as subsequent surface treatment have been investigated. Collocated monolith support structures (COMOSS) made in PDMS were used for generation of packed beds as well as modified surfaces for electrochromatographic separations [41]. Hobo et al. studied the potential of polyester chips for electrochromatographic applications [42, 43]. They reported the separation of two dyes using MEKC and CZE on polyester chips.

8.2 ELECTROOSMOSIS, ELECTROOSMOTIC FLOW, AND ELECTROKINETIC TRANSPORT

As described in Chapter 2 (and in Chapter 3), the interfacial phenomenon of electroosmosis can be exploited to supply a pumping action, moving the entire bulk of an electrolyte solution (often, a buffer solution) through a channel system. In the classic case of a glass or silicon dioxide surface in contact with an aqueous borate buffer (pH > 7), the charge distribution at the interface will be such that there is an excess of mobile positive charge carriers at the shear plane of the double layer (diffuse or Gouy–Chapman layer; see also Figure 2.3), counter-ions to the immobilized negative charge carriers close to the surface (adsorbed or Stern layer), thereby maintaining electrical neutrality for the entire bulk solution. Upon applying an external electric field through the channel (i.e., in the axial direction), the movement of these positive charge carriers along with their hydration shells (or, alternatively, the impossibility of the excess negative charge carriers in the Stern layer to move) will result in a bulk movement from the anode towards the cathode.

Since the movement is initiated at the walls and propagated by viscous drag it results in a flat flow profile over basically the entire cross section of the channel. The velocity associated with this movement is called electroosmotic velocity, u_{eo}, and can be expressed as

$$u_{eo} = \mu_{eo}E = \frac{\zeta \varepsilon}{4\pi\eta}E \tag{8.1}$$

where μ_{eo} is the electroosmotic mobility, ζ the zeta potential (potential at the shear plane), ε the dielectric constant, η the viscosity, and E the electric field strength.

As (negative) surface charges are naturally occurring, it is readily understood that it is fairly easy to generate this type of flow. However, it is an entirely different game to control, modify, or even suppress the electroosmotic flow. There are a large number of physical and chemical parameters, which affect the electroosmotic flow (EOF). Apart from the electric field strength, temperature will influence the electroosmotic mobility (mainly via the viscosity), but in particular the ionic strength and the pH value have a strong effect on the zeta potential, and hence on the magnitude of the electroosmotic flow. Any process affecting the parameters in Equation (8.1) will change the electroosmotic flow. This could be dilution, a chemical reaction, or an adsorption process to the wall. While this seems to offer many possibilities to influence the EOF, it also shows how difficult it is to control and maintain a constant EOF during, for example, a separation. Often, electrolysis at the terminal electrodes leads to a pH change in the buffer reservoirs, which can cause a pH gradient along the channel altering the EOF over time. This process, called buffer depletion, can be observed in capillaries over long periods of time, but due to the small volumes of the reservoirs associated with microfluidic chips, becomes a much more pronounced issue for on-chip electrically driven separations. In essence, controlling the EOF entails having the freedom to choose certain conditions (direction, magnitude) and maintaining these conditions over an extended period of time, to ensure reproducibility of all processes making use of the EOF.[1]

In the simplest case (glass surface with negative surface charges and a borate or phosphate buffer with pH > 3) an EOF in the direction of the cathode is given. But, depending on the application (see further below and

---

[1] Since much of the work on EOF control and manipulation has been developed in capillaries and ported to microchips later detailed information on the various procedures can be found in the CE literature. The theory behind electroosmosis is also discussed in Chapter 2, while more on electroosmotic transport can be found in Chapter 3.

in Chapter 12), it might be more favorable to have a reduced EOF, a reversed EOF, or even no EOF at all. The pH value of the buffer solution is a tool, reducing the EOF when going to lower pH values. In the case of a Pyrex glass surface, the EOF is basically negligible below a pH value of about 2.5 [44]. Increased ionic strengths will also reduce the magnitude of the EOF, probably because of an increased shielding of the electrical charges in the double layer. However, low pH values and high ionic strength buffers might not be the optimum conditions to solve a certain separation problem, and these conditions might entail other disadvantages (such as, e.g., increased Joule heating). Methods to reverse the EOF without reversing the electric field have been proposed and tested, making use of, for example, tertiary alkylammonium salt solutions [45, 46]. The positively charged alkylammonium ions physisorb to the negatively charged surface by electrostatic interaction. Their alkyl moieties then "capture" the alkyl chains of a next layer of alkylammonium ions by hydrophobic interaction. Thus, a new positively charged immobilized layer is created leaving an excess of freely moving negatively charged counter-ions, which will produce a bulk flow towards the anode upon application of an electric field. This is conceptually different from switching the polarity at the electrodes, as it leaves the electrophoretic movement unchanged (see also further below). A number of differently substituted alkylammonium compounds have since been suggested and tested, mainly in the capillary format [47].

There are some applications where only electrophoresis is required, and the presence of EOF reduces separation efficiency (cf. Equation 2.41). In such cases, the complete suppression of the EOF is desired. Based on a closer examination of Equation (8.1), Hjertén suggested that a drastic increase of the viscosity just within the double layer should result in a strong reduction of the electroosmotic mobility. In the first experiments Hjertén used a hydrophilic methylcellulose solution cross-linked with formaldehyde [48]. This completely suppressed the EOF in the glass tubes he was using then. Many variations of this principle have since then been devised and also used successfully on microchips for DNA analysis [49, 50]

While the EOF is very dependent on the chemical environment, this fact can be exploited by carefully engineering chemically modified surfaces that allow more control over surface charge density and physical and chemical adsorption than is possible on native substrate surfaces. Particularly in microsystems, where the surface-to-volume ratio is steadily increasing with decreasing channel dimensions, more and more attention must be paid to surfaces. Controlling the properties of surfaces can be accomplished by chemical means (dynamic or static coating with chemicals) or by physical means, typically by manipulating the surface charge density on the surface electrically. Because in microsystems many more substrate materials than just fused silica and glass are used (namely, silicon and various plastics), the chemical modification methods had to be extended from the tetraalkyl ammonium ion approach mentioned above.[2] Typically, these modifications serve two purposes: (a) a more controllable (hence, more reproducible) EOF, and (b) a reduced analyte–wall interaction to minimize band broadening effects due to adsorption. Two recent examples demonstrated in capillaries, but not yet in microchips, are the UV-induced polymerization of a zwitterionic monomer $(N, N$-dimethyl-N-methacryloxyethyl-N-(3-sulfopropyl) ammonium betaine) onto the capillary inner wall [52] and the wall modification with tentacle-like oligourethanes [53]. In the first case, a drastic decrease in the EOF was found after the coating, together with the possibility to further influence the EOF by addition of divalent cations or chaotropic anions through their interaction with the polymeric zwitterionic layer. In the case of the oligourethanes the pH dependence of the EOF was altered as compared to native fused silica, thus allowing reduction and reversal of the EOF at pH values more suitable for the separation of certain analytes.

Soper and coworkers described various surface modifications for channels on polymer chips [made from poly(methylmethacrylate) (PMMA) or polycarbonate (PC)] and show how these treatments led to changes in the magnitude of the EOF and the shape of the curve describing the pH dependence of the EOF on these modified materials [54, 55]. In a different approach, Liu et al. used polyelectrolyte multilayers to arrive at PDMS-based microchips with very stable electroosmotic flows, virtually independent of the pH value of the used buffer solutions [56]. They described the use of a polymeric bilayer consisting of a cationic layer of polybrene (hexadimethrine bromide) and an anionic layer of dextran sulfate. As a final example, Kirby and coworkers reported on the use of self-assembled monolayers to facilitate surface modifications on silica microchips [57]. Combined with localized photoinduced polymerization the channel surfaces could be patterned as desired giving precise control over surface charge and polarity as well as adhesion behavior at arbitrary locations within the fluidic network.

Two excellent reviews detailing a broad range of surface modification techniques for microfabricated devices and in particular microchips for electrophoretic separation of biomolecules were published in 2003 and are recommended for further information on this subject [58, 59].

Among the more physical methods to control the EOF is the use of an additional external electric field, applied radially to influence the surface charge density in the double layer [60, 61]. While such an approach was already described in the early 1990s for capillary systems, where, for example, the capillaries were externally coated with a metallic paint or a conductive polymeric film (e.g., Ref. [62]), an implementation of this means of EOF control on chips was demonstrated successfully only in 1999 by Schasfoort et al. [63]. One

---

[2] Often two (or more) different materials are used for the fabrication of microchannel systems. This leads to channels, where the zeta potentials are not the same on all channel walls, which in turn has implications for the EOF profile and dispersion in such a system (see, e.g., [51]).

disadvantage of the capillary-based implementation was that relatively large voltages had to be applied to produce sufficiently high electric fields, due to the relatively thick capillary wall. At higher pH values the influence of the external field was no longer "felt" strongly enough, as there was already a high surface charge under these conditions. By the skillful application of microfabrication technologies Schasfoort et al. succeeded in producing free-standing channels with extremely thin insulating silicon nitride walls (390 nm thick). With such thin channel walls the application of small external voltages (-37.5 to 25 V) was enough to manipulate the magnitude and direction of the EOF inside the channel. Combinations of several of these flow-control elements (also dubbed flow field effect transistor) allowed more complex fluidic manipulation, such as the definition of plugs for an injection. These systems suffer from the restriction of working best at relatively low pH values without the use of special modifiers or surface treatments [64].

A different physical way to influence the surface charge density, and thereby the zeta potential and the EOF, was demonstrated by Johnson et al. [65]. They manufactured polymer chips by hot embossing (see also Chapter 5) using a silicon master tool, which resulted in sloped sidewalls. Afterwards they employed an excimer laser, as it is also used for laser ablation, and selectively illuminated only one of the two sloped sidewalls of their polymer microchannel. The power of the laser was set below the ablation threshold. However, the light intensity was strong enough to change the surface charge density through ionization processes. In a channel treated in such a way, the zeta potential and hence, the EOF, were different on the two opposite channel walls, with a stronger EOF on the treated sidewall. Such spatially localized surface treatment can be used to reduce the band broadening effects of U-turns (see also Chapter 2) by increasing the EOF only on the outside "lanes" of these turns.

One of the great advantages of the presence of EOF is that its magnitude is often sufficient as to force all species present in a sample plug to move in the same direction. In order to determine the migration direction of a particular species (charged or neutral molecule), we can just linearly superimpose the contributions from electroosmosis (Equation 8.1) and electrophoresis (Equation 8.2, further below). In other words, whenever the absolute value of the electroosmotic velocity (or, alternatively, the electroosmotic mobility) is larger than the absolute value of the electrophoretic velocity (mobility) the molecule will move in the same direction as the EOF. Separation is still possible as the differences in the electrophoretic mobilities are not cancelled out, but merely offset.[3] This has the practical advantage that all species can be injected on the same end of the channel and detected near the opposite

---

[3] Of course, this statement is only true within reasonable limits. Clearly, the difference between 10 and 50 μm/sec is more pronounced than between 1010 and 1050 μm/sec (see also Equation 2.41).

end. There is a caveat; however, which is that all injection methods making use of this transport mechanism (i.e., the combination of electroosmotic and electrophoretic transport, often dubbed electrokinetic transport) will lead to biased injections, where ions with an electrophoretic movement in the same direction as the EOF will be enriched in the injected sample compared to neutrals and ions with an electrophoretic movement opposed to the EOF. Chapter 2 provides a more detailed discussion of electroosmotic flow and related theory.

8.3 ELECTROPHORETIC TECHNIQUES

In this section, examples of miniaturized electrophoretic devices will be presented and discussed. A range of different electrophoretic modes has been demonstrated in more traditional formats, and many of those have been transferred to the microchip format (free zone CE, isotachophoresis, isoelectric focusing, gel-CE, affinity-CE). While it is beyond the scope of this book to give a comprehensive overview over all those techniques, we would like to focus on some selected examples, either to highlight particular problems and challenges encountered when working on planar microdevices, or to demonstrate electrophoretic microchips which offer new, hitherto unavailable functionalities and performance features. For more detailed information and discussion on the electrophoretic techniques mentioned here, the reader is referred to recent reviews for lab-on-a-chip applications [66–71] or to one of the many monographs on capillary electrophoresis [44, 72–75].

8.3.1 FREE ZONE CAPILLARY ELECTROPHORESIS

Free zone capillary electrophoresis (FZCE), also sometimes referred to as capillary zone electrophoresis (CZE) or free solution capillary electrophoresis (FSCE), is the simplest form of CE, where separation of analytes is based on the mobility differences of ions in free solution.

A photograph of a typical chip as used for electrophoretic separations is shown in Figure 2.1 (Chapter 2). In CZE, the separation channel is filled with separation buffer, sample is injected and separation of analytes occurs according to their charge to size ratios (i.e., their mobilities), when applying an electrical field across the channel. The separated analytes can then be detected by a number of different techniques, for example, fluorescence (cf. Chapter 6).

The electrophoretic velocity, u_{ep} (cm/sec), of a charged molecule is given by Equation (8.2) (analogous to Equation 8.1)

$$u_{ep} = \mu_{ep}E = \frac{q}{6\pi\eta r_i}E \tag{8.2}$$

where μ_{ep} (cm^2 V/sec) is the electrophoretic mobility[4] and E (V/cm) is the electrical field strength, which is easily calculated by dividing the applied separation voltage by the length of the channel or capillary. The electrophoretic mobility is a physical constant characteristic for a given molecule/ion in a given buffer or medium. When calculating μ_{ep}, the degree of ionisation, α, which determines the effective charge of the species, should also be taken into account, hence, $\mu_{ep} = \alpha q/6\pi\eta r_i$. In the presence of electroosmotic flow, the measured mobility is called *apparent* and is designated μ_a so $\mu_a = \mu_{ep} + \mu_{eo}$, where μ_{eo} is the electroosmotic mobility (cf. Section 8.2). The apparent mobility can be calculated directly from experiments by dividing the velocity of the analyte, u, by the applied electric field, E; μ_{eo} is typically determined by detecting a neutral compound, such as acetone (i.e., $\mu_{ep} = 0$) or by employing the so-called "current monitoring method" [76].

CE has the potential for highly efficient separations, since the separation mechanism only relies on differences in migration behavior in electric fields. No other secondary equilibria (as in chromatography) are required, and hence no band broadening due to mass transfer issues occurs. A simple model describing the efficiency in electrophoretic separations (expressed as number of theoretical plates, N) is given in the following equation:

$$N = \frac{\mu_a E L}{2D} = \frac{\mu_a U L}{2DL_{tot}} \tag{8.3}$$

where L_{tot} is the total channel length, L the length to the detector, and U the applied voltage. The inverse dependence on the diffusion coefficient makes CE a preferred method for the separation of large biomolecules. It is also seen that N is directly proportional to E, which implies that higher applied field strengths should result in higher efficiencies. In fact, very high field strengths can be achieved on microfabricated devices. However, at higher field strengths, other band-broadening mechanisms become more pronounced and often lead to a roll-over of the N vs. E plot. Amongst these other band-broadening mechanisms, the most important ones are contributions from the injection process, from the detector, and from Joule heating (see Chapter 2). When comparing capillary formats and planar microfabricated devices it becomes evident that the latter are less susceptible to these band-broadening sources. Due to the potential direct integration of injectors and detectors and a careful design and operation much higher field strengths are possible on microchips before detrimental effects become apparent. Effects due to Joule heating are also less pronounced due to the higher thermal mass of many

---

[4] Other symbols in the equation are q: ion charge, η: viscosity of fluid, and r_i: hydrodynamic radius of the ion. In this equation the migrating molecule is modelled as a "solid sphere." Other molecular models (long rod, random coil, etc.) may be more appropriate.

microdevices, a better thermal conductivity of some of the used substrate materials and more favorable surface to volume ratios, which improve the dissipation of heat [77].

The simplest, and to date most frequently used chip design is a cross design depicted in Figure 2.1, where two channels overlap in a cross, with the longest part of the channel network constituting the separation channel. However, more sophisticated designs are typically utilized when integrating other functionalities, such as on-chip derivatization or sample preparation [66, 67, 71]. Additional channels and reservoirs enable changing of buffer composition (e.g., solvent programming) [36, 37], entirely switching the buffer or adding new reagents or analytes to the system [78]. Serpentine channels increase separation length [79] and parallel channel designs allow multiplexing [80]. Furthermore, specialized designs exist, for example, for high-speed separations [81] or cyclic CE [82].

The first microfluidic CE devices were made in glass or quartz mainly for three reasons. First, using glass surfaces meant direct transfer of hitherto known surface chemistry and therefore to some extent CE methods from bench top set-ups. Second, glass has desirable material properties (optical transparency, chemical inertness, good electrical insulation). And third, standard clean room processes (such as etching and bonding) for glass materials were known. Recent years, however, have seen polymer materials emerge, such as poly(methylmethacrylate) (PMMA), polycarbonate (PC), polystyrene (PS), cycloolefincopolymer (COC) and poly(dimethoxysilane) (PDMS), showing great potential, which is also reflected in the number of publications where polymer chips are employed in CE applications (e.g., Refs. [83–86]). A strong motivation for investigating polymer substrate materials is the cost of the final devices and the processing time required to come from a design to a working device. Not surprisingly, therefore, polymers are the material of choice for one-time use devices and for rapid prototyping, where designs are iterated to find the best solution. On the other hand, it should also be clear that different EOF and adsorption characteristics have to be expected from new materials, and new chemistries have to be developed to realize surface modifications (see, e.g., Refs. [54, 59, 87, 88]).

In the following, a few recent examples of CZE separations on chips are presented. It is noteworthy that in many recently published papers involving a CE separation, the emphasis is often on "additional features," such as a new on-chip detection method, a combination of different CE techniques, or CZE combined with other analytical methods on chip, rather than on the development of the actual CE method. This is due to the fact that many of the intrinsic advantages (improved efficiency with respect to sample size, response time, cost, analytical throughput, automation, possibility for parallel analysis) that come from transferring conventional CE methods to chips have already been shown and published in the early years of μ-TAS and lab-on-a-chip.

Examples of simple CZE separations are described in a paper focusing on a fully disposable, low-cost microanalytical device based on a combination of a polymer (PMMA) CE microchip[5] and a thick-film amperometric detector [89]. Separation of dopamine (DA) and catechol (CA) is accomplished using an MES[6] buffer (25 mM, pH 6.5). At the chosen buffer pH catechol ($M = 110$ g/mol) is neutral and dopamine ($M = 153$ g/mol) is predominantly positively charged. The different charges explain why dopamine migrates faster towards the cathode than catechol,[7] which actually can act as EOF marker, since it is a neutral molecule. These analytes were used to examine the performance characteristics of the system. Figure 8.2 shows the influence of field strength on the electropherogram (E was increased from 132 to 439 V/cm).

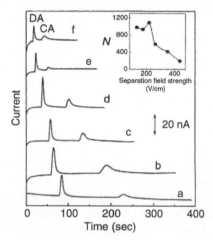

FIGURE 8.2 Influence of the separation field strength on the response from a mixture containing 100 μM dopamine (DA) and 200 μM catechol (CA). Separations performed using the following field strengths: (a) 132 V/cm; (b) 175 V/cm; (c) 219 V/cm; (d) 263 V/cm; (e) 351 V/cm; and (f) 439 V/cm. Sample injection preformed at 175 V/cm. Electrochemical detection at 0.7 V using carbon screen-printed electrode; MES buffer (25 mM, pH 6.5) was used as electrophoretic buffer. Also shown (inset) is the effect of the separation potential on the number of theoretical plates (N) of DA. (Reprinted from Wang, J.; Pumera, M.; Chatrathi, M. P.; Escarpa, A.; Konrad, R.; Griebel, A.; Dorner, W.; Lowe, H. *Electrophoresis* **2002**, *23*, 596–601. With permission.)

---

[5] Standard cross design. Channels: 50×50 μm^2 cross section; lengths: 57 and 18 mm, respectively.

[6] MES: 2-(4-morpholino)ethanesulfonic acid hydrate.

[7] It is safe to assume that the hydrodynamic radii of DA and CA do not differ much. Hydrodynamic radii can be estimated from the Stokes–Einstein equation using tabulated values of diffusion coefficients.

As expected from theory (cf. Equation 8.3) the speed of the analysis increases and peak widths decrease with higher field strengths. Migration times are reduced by roughly 80%, resulting in migration times of 18.4 sec for catechol and 43.0 sec for dopamine at $E = 439$ V/cm. The separation efficiency, N, was plotted against the separation field strength, E, in the same figure. The separation efficiency decreases with higher field strengths, which is the opposite of what theory predicts (N should be proportional to E). The authors attribute this drop in efficiency to Joule heating, end-column broadening and incomplete isolation of the detection system. The peak-to-peak background noise level also increased, when field strengths higher than 219 V/cm were applied, which is why $E = 219$ V/cm was finally chosen as the optimum for further experiments, where linearity and detection limits were investigated.

In another recent publication, FZCE is utilized for the indirect determination of nitric oxide production by monitoring nitrate and nitrite [78]. Nitrite can directly be determined electrochemically, while nitrate is reduced to nitrite first (using copper-coated cadmium granules) and then detected. The amount of nitrate is determined by calculating the difference in the amount of nitrite in the sample before and after the reduction of nitrate. The PDMS microchip design[8] includes the possibility for on-chip sample preparation (i.e., reduction) and electrochemical detection. As nitrite is an anion, the analysis is run with reversed polarity (i.e., negative voltage is applied). Additionally, the EOF is reversed by adding a positively charged surfactant, DTAB,[9] to the boric acid running buffer, pH 9.0. This resulted in nitrite being separated from the background and detected in about 45 sec (at $E = 200$ V/cm). Even in these very controlled experiments a large interfering peak, which was suspected to be a copper-glycine complex, appeared in the electropherograms (see Figure 8.3).[10] Application to a test compound (SIN-1, which releases NO, which is converted to nitrate and nitrite) was demonstrated, and the system has potential for biological applications, such as *in vivo* monitoring of NO.

Alterations of solute–wall interactions and EOF using different buffer additives for different chip materials receive increased interest (see Section 8.2). Additives and modifiers are also necessary to facilitate separations, where secondary equilibria need to be exploited, as for example in the case of chiral electrophoretic separations on microchips through the addition of cyclodextrins to the background buffer (e.g., Ref. [90]), or in the case of the separation of neutrals using electrokinetic forces, as demonstrated in micellar electrokinetic chromatography (MEKC, see Section 8.4). Surfactants

---

[8] Channels: 30 μm deep and 30 μm wide, from cross to detection point is 5.9 cm.

[9] DTAB: dodecyl trimethylammonium bromide.

[10] This peak could, however, be removed by adding EDTA to the running buffer.

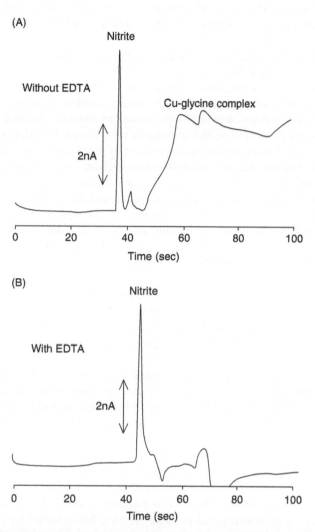

FIGURE 8.3 Electrochemical detection of nitrite produced by on-chip reduction of nitrate using chip B. Sample: nitrite produced by the reduction of 100 μM nitrate diluted with 5 mM boric acid including 7.5 mM DTAB. Separation conditions: 15 mM boric acid, pH 9.0, with 7.5 mM DTAB and (A) not including or (B) including 1 mM EDTA. Separation voltages: −1200 V (buffer to waste, 200 V/cm) and −960 V (sample to sample waste). Injection: 1 sec. Detection conditions: E = +0.9 V vs. Ag/AgCl. (Reprinted from Kikura-Hanajiri, R.; Martin, R. S.; Lunte, S. M. *Analytical Chemistry* **2002**, *74*, 6370–6377. With permission.)

(anionic, cationic, zwitterionic, nonionic) are very often used additives, and below the critical micelle concentration (i.e., not performing MEKC separations) they can act as solubilizing agents for hydrophobic solutes, as ion-pairing reagents, or as wall modifiers.

One example is the ultrafast analysis of oligosaccharides with fluorescence detection on a PMMA microchip [84]. In preliminary experiments, the analytes (oligosaccharides labelled with the fluorophore APTS[11]) were found to strongly adsorb to the channel surface, making analysis impossible. To circumvent this problem different classes of dynamic coating additives (amines, surfactants, and neutral polymers) were tested. A variety of amines and surfactants[12] were investigated, but none of these suppressed the adsorption of the labeled analytes. Neutral polymers with hydroxyl groups (PEG, HEC, HPMC, and MC)[13] were employed and addition of 0.1 wt% MC gave satisfactory results in suppressing adsorption and the electroosmotic flow. Under optimal conditions, 15 oligosaccharides in dextrin hydrolysate can be separated within 45 sec with high efficiency and high reproducibility (400,000 theoretical plates and relative standard deviation [RSD] of migration times of 0.50% between six different channels).

Another microchip CE issue that has received increased attention is electrokinetic sample handling (improved injections, utilizing stacking, etc.). Electrokinetic injections (i.e., using electroosmotic and electrophoretic transport) have always been very reproducible, as the timing is usually done via computer-controlled high-voltage power supplies, and the injected amount is governed by both the timing and the geometry of the chip (i.e., its design). It is thus possible on chips to implement injection techniques that heavily rely on accurate and reproducible timing, such as cross-correlation electrophoresis [91]. Here, the gated injection is used, but instead of performing just one injection, a pseudorandom binary sequence of injections is executed. In essence, a "1" in the sequence means "inject" while a "0" corresponds to "no inject." The multiple injections of analyte within the same run lead to a detector output, which looks more like random noise. However, the cross-correlation of the readout with the known injection sequence results in an electropherogram with peaks corresponding to the expected migration times. The benefit of this technique is an increased signal-to-noise ratio and hence useful for very dilute samples. This only works, however, if the timing dictated by the binary sequence is exactly reproduced by the injection technique, otherwise the subsequent mathematical data treatment only results in "data garbage."

---

[11] APTS: 8-aminopyrene-1,3,6-trisulfonic acid, trisodium salt.

[12] Amines: ethylamine, diethylamine, triethylamine, and triethanolamine. Surfactants: sodium dodecyl sulfate (SDS), dodecyltrimethylammonium chloride (DTAC), and cetyltrimethylammonium bromide (CTAB).

[13] PEG, polyethylene glycol; HEC, hydroxyethyl cellulose; HPMC, hydroxypropylmethyl cellulose; MC, methylcellulose.

Despite the precision already achieved using gated, pinched, and double-tee injection schemes (see also Chapter 2 for more details), these techniques are nonetheless continuously reevaluated and improved upon. Lee and co-workers describe a combination of an upstream electrokinetic focusing with a subsequent cross injector for pinched-type injections [92, 93]. Their designs allow for an improved geometrical definition of the sample plug, and it also makes it possible to inject different amounts (volumes), within certain limits. However, while allowing better injection plug definition and control, more fluidic and electrical connections are required, making operation more complicated and potentially increasing the chances of chip failure.

Earlier, Zhang and Manz proposed a modified injector design, where the channels carrying the sample solution are much narrower than the main separation and buffer channels, typically by a factor of 10 [94]. With their design the requirements for elaborate injection protocols to avoid leakage during injection are greatly relieved, rendering chip operation simpler and more user-friendly. The advantage of the narrow side channels is that uncontrolled flow (either dilution of the sample through the buffer, or "bleeding" of sample into the buffer) is reduced to an extent that makes active leakage control unnecessary. An additional feature of narrow sample channels is that when using the simplest possible design, namely a tee injector with a narrow sample channel, it is possible to design parallel systems where the total number of reservoirs (and hence voltage control points) only equals $N + 2$, where N is the number of parallel systems.

Similar protocols as for injection purposes can also be exploited to perform sample preparation, in particular enrichment or concentration procedures. This is commonly referred to as "sample stacking." Here, concentration boundaries are created, for example, by injecting a sample dissolved in a diluted buffer into a more concentrated buffer, and thereby generating an electric field discontinuity. As this discontinuity charged analyte molecules experience a velocity change, which, when set up appropriately, stacks all charged analytes from within the sample plug at one of the two concentration boundaries, depending on the effective mobility of the analytes [95]. Clever design of the channel layout and the use of computer-controlled voltage protocols allow different stacking variants to be realized, making it possible to stack from large sample volumes and achieving enrichment rates of 100- or 1000-fold [9, 95, 96]. A combination of a stacking technique and a narrow sample channel injector (see above) for the analysis of biogenic amines on microchips was reported recently by Beard et al. [97]. By using fluorescence detection the labelled amino acids could be detected down to a (original sample) concentration of $20\,pM$ thanks to the fact that the sample was concentrated by stacking prior to an electrophoretic separation and the detection. More details on how to employ sample preparation techniques together with separations on microchips are given in Chapter 9.

8.3.2 ISOTACHOPHORESIS

Isotachophoresis (ITP), also called *displacement electrophoresis*, is a so-called moving boundary technique, where the separation and zone focusing occurs because of a conductivity gradient. ITP is an analytical technique in its own right, but is very often used as a preconcentration step prior to CE and also in column-coupling separations on chip. In a review from 2000 [98], ITP is mentioned as one of several online preconcentration methods for CE, while a review from 2002 describes recent progress in capillary ITP, and concludes that ITP principles are becoming an interesting possibility for microchip devices [99].

In ITP, a combination of two buffer systems is employed, the so-called *leading* and *terminating* electrolytes (LE and TE, respectively), and the sample remains sandwiched between these two buffers. LE and TE have, respectively, high and low effective mobilities with regard to the analytes. The analytes migrate in order of decreasing mobility, and are focused in discrete zones between LE and TE. Usually, an ITP experiment is run in a "constant current mode." It is possible to fill 30 to 50% of the capillary or channel with sample and still get a good separation. However, it should be noted that in a single ITP experiment either cations or anions can be analyzed.

Two unique aspects of ITP are that all solute zones migrate at the same velocity (isotachophoresis means migration at the same speed) and that the concentration of the leading electrolyte determines the concentrations in the other zones. Hence, the leading ion defines both the velocity and the concentration in each zone. Since the analytes have different mobilities (i.e., also different conductivities), the field varies across each analyte-zone; that is, the lowest field is across the zone with highest mobility. If an ion from one zone diffuses into another zone, its velocity will change because of the different field strength, and it will immediately return to the previous zone. This means that ITP zones are sharp and self-adjusting. Because of the "constant current mode," a constant ratio between the concentration and the mobility in each zone must exist, and the concentration of analyte in the individual bands will adjust according to the concentration of the leading electrolyte.

A couple of practical issues have to be taken into account when setting up an ITP experiment: one issue is that it can be difficult to find suitable buffer systems that contain both leading and terminating ions and also have the desired pH. Another issue is that it is often advantageous to suppress the EOF which can interfere with the ITP separation by moving the analytes out of the ITP channel before complete zone sharpening has occurred.

A Slovakian group in collaboration with Merck has published several papers concerning PMMA chips with coupled columns (for ITP and FZCE, respectively) and integrated conductivity detection (e.g., Refs. [100–103]). A recent example is the online coupling of ITP and free zone electrophoresis for direct determination of drugs in serum, as explored in a feasibility study

[103]. The antiepileptic drug, valproic acid (an anion), was used as the test analyte. The authors state that ITP provided multitask sample pretreatment, accomplishing separation of the analyte(s) from the matrix, removal of the migrating matrix constituents form the separation compartment, and well-defined concentrations of the analytes [103]. Figure 8.4 depicts the chip layout.

A run consists of pumping the leading electrolyte (containing an anion with an effective mobility higher than the mobilities of all analytes) into the channel, followed by the sample, and finally, the terminating electrolyte. Then voltage is applied and ITP occurs. Part of the ITP-zone is then injected into the FZCE channel, and separation of the analyte from other sample constituents in the injected segment is achieved. In the mentioned paper, EOF was suppressed using methylhydroxyethylcellulose (MHEC) added to the electrolyte solutions. Food and cosmetics analysis was the topic of an

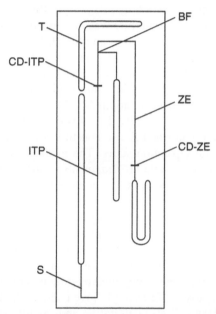

FIGURE 8.4 Arrangement of channels on a poly(methyl methacrylate) column-coupling chip. T = terminating electrolyte channel; S = sample injection channel (0.95 μl volume; 24 mm × 0.2 mm × 0.2 mm [length, width, depth]); ITP = ITP separation channel (3.04 μl volume; 76 mm × 0.2 mm × 0.2 mm) with a platinum conductivity sensor (CD-ITP); ZE = ZE separation channel (1.68 μl volume; 42 mm × 0.2 mm × 0.2 mm [length, width, depth]) with a platinum conductivity sensor (CD-ZE); BF = bifurcation section. (Reprinted from Ölvecka, E.; Konikova, M.; Grobuschek, N.; Kaniansky, D.; Stanislawski, B. *Journal of Separation Science* **2003**, *26*, 693–700. With permission.)

earlier paper, by the same group [101]. Both papers show that real samples can be analyzed with a minimum of preparation.

A similar approach was taken by Grass et al., who presented a PMMA microfluidic device isotachophoresis with integrated electrodes for conductivity detection [104]. The chip design includes two separation columns hereby allowing for coupling of ITP and CE. The authors tested two different chip designs, and pointed out that the analytical performance could be improved, for example, by optimizing the injection procedure. The choice of material, design, and detection scheme shows promise for mass fabrication of disposable polymer chips.

Another example is a paper on transient isotachophoresis capillary gel electrophoresis[14] (Tr-ITP-CGE), where the ITP step improves the sensitivity of the analysis by preconcentrating the sample [105]. Standard quartz chips[15] and a MCE–2010 chip electrophoretic system with a UV PDA[16] detector (all from Shimadzu) were used for the experiments. A single channel microchip (i.e., no injection cross) was preferred for the analysis, where the sieving matrix, that is, the "gel," was 2% hydroxyethylcellulose (HEC). Because of the ITP step[17] a special buffer mixture was developed, which included both a leading ion (50 mM HCl–Tris, pH 8.1) and the HEC sieving matrix. Briefly, the procedure works as follows: first, the channel is filled with leading electrolyte (LE). Then DNA sample and the terminating electrolyte (TE; 20 mM glycine–Tris, pH 8.1) are introduced electrokinetically. While the TE is introduced (i.e., voltage is applied), ITP occurs thereby sharpening the bands of DNA fragments. Finally, LE (this time without HEC) is reintroduced to the inlet reservoir (the LE also constitutes the CGE run buffer), and the gel electrophoretic separation starts. The method was tested on a commercial DNA stepladder sample (16 fragments in the range from 50 to 800 base pairs), and baseline separation was achieved in 120 sec (signal-to-noise ratio 2–5). An advantage of this particular method is that the syringes in the automated system can be used for loading the electrolytes.

A final example of on-chip isotachophoresis, where ITP is actually used as an analysis technique and not as a pretreatment method, is the determination of four metal cations (Li^+, La^{3+}, Dy^{3+}, and Yb^{3+}) [107]. A silicone elastomer (PDMS) device with integrated on-column conductivity detector is employed, and, similarly to the example given above, the EOF is suppressed (HEC, MW = 250,000). Both quantitative and qualitative information can potentially be extracted from such an analysis. However, a manual injection

---

[14] Capillary gel electrophoresis is described in Section 8.3.5.
[15] Channel length, 40.5 mm; width, 110 μm; depth, 50 μm; effective length, 25 mm.
[16] UV PDA: ultraviolet photodiode array.
[17] The authors use the term *electrokinetic supercharging*, which is essentially electrokinetic injection with a transient ITP process; see [106].

method used in this experiment prevented the possibility of obtaining quantitative information.

In general, transferring ITP to microchips results in the well-known "μ-TAS advantages," such as faster separations, less consumption of sample and reagents, potential of portability, and low-cost single-use polymer devices. But a unique feature of on-chip ITP is that it allows easy coupling of ITP with other analytical techniques, as exemplified in some of the publications mentioned earlier. This means that analytes in real samples can be separated from other matrix constituents and preconcentrated on-chip before further analysis.

8.3.3 ISOELECTRIC FOCUSING

Isoelectric focusing (IEF) is the electrophoretic separation of ampholytes in a pH gradient. The basis for IEF separations is the isoelectric point[18] and the technique is often used for "high resolution" separations of peptides and proteins, but it is applicable to all ampholytes, hence also to viruses and bacteria. Zones with different pH values are formed when voltage is applied across a capillary or channel filled with one or more ampholyte (zwitterionic) buffer. There is an acidic solution at the anode (anolyte) and a basic solution at the cathode (catholyte). The analytes, which have to be zwitterionic themselves, migrate towards zones where the analytes have no net charge, and are hence distributed according to their isoelectric point, pI. No net migration occurs at the isoelectric point, that is, a steady state is reached and no current flows. EOF should be eliminated in IEF, since this driving force could flush the ampholytes from the capillary before focusing is complete. Reduction of EOF can be accomplished by the use of dynamic or covalent coatings. Since the analyte is loaded together with the ampholytes it is possible to load larger amounts of sample [73].

A recent review by Kilár [108] describes applications of IEF from a general point of view, and also has a short section on IEF on chips, where whole-column imaging is mentioned as a promising approach when using the chip format. However, in the paper referred to, the authors image a 1.2 cm portion of a capillary, thereby only mimicking a microchip IEF system [109]. IEF, like ITP, is often used as a first dimension in a multidimensional separation approach. Below, two examples of this are mentioned.

In a paper published by Herr et al., IEF is coupled to free zone capillary electrophoresis [110]. The authors state that IEF is attractive as the first dimension of a microdevice-based separation system for several reasons: it has great preconcentration potential, high speeds can be achieved due to the short length scales involved, once focused sample bands stay focused, and the possibility to defocus and refocus those bands. The device employed was a

---

[18] Isoelectric point: the pH where a molecule has no net charge (e.g., the pH where there are equal amounts of the acidic and basic form of an ampholyte).

cross-chip fabricated in PMMA by hot embossing. The separation medium in both dimensions was a mixture of ampholytes (Bio-Lyte pH 3–10). Focused ampholytes were used for IEF and unfocused ampholytes were used for CE (without an anolyte and a catholyte solution the ampholyte mixture is reported to behave as a pH 8.5 buffer). For the IEF step, the catholyte was $40\,\text{m}M$ NaOH, pH 10, and the anolyte $20\,\text{m}M$ phosphoric acid, pH 3. The analytes were three fluorescing molecules, which were detected by full-field fluorescence imaging. Both IEF and ampholyte based CE on chip were studied separately, after which the two-dimensional analysis was characterized. Obviously, the sampling step from IEF to CE is crucial: one fluid volume at a time was sampled from the IEF channel and injected and separated by CE (see Figure 8.5). A two-dimensional analysis of a fluid volume spanning 15% of

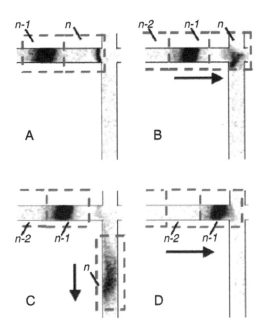

FIGURE 8.5 CCD images during species sampling. (A) Species are focused by IEF in the first dimension (dark bands in horizontal channel). Simultaneously, the bands are mobilized toward the catholyte reservoir by low-dispersion EOF. (B) Once a fluid volume of interest, n, reaches the microchannel intersection, all electrodes are switched to electrically float for 3 sec. (C) High voltage is then applied at reservoir B and reservoir W is grounded, initiating sample separation in the second dimension. (D) Upon completion of the CE separation, IEF/EOF is reinitiated causing sample species to refocus and the next fluidic volume $(n-1)$ to migrate to the intersection. This sequence is repeated until all fluidic volumes are sampled from the first dimension into the second. (Reprinted from Herr, A. E.; Molho, J. I.; Drouvalakis, K. A.; Mikkelsen, J. C.; Utz, P. J.; Santiago, J. G.; Kenny, T. W. *Analytical Chemistry* **2003**, *75*, 1180–1187. With permission.)

the total IEF channel was completed in less than 5 min (preconcentration factor >70). The authors mention that a manifold of CE channels (i.e., parallelization) would be beneficial and reduce analysis time.

In the next example such a parallelization has been realized (see Figure 8.6). An integrated protein concentration–separation system, combining non-native isoelectric focusing with sodium dodecyl sulfate (SDS) gel electrophoresis[19] on a polycarbonate microfluidic chip was described by Li et al. [111]. Focused proteins were electrokinetically transferred into an array of orthogonal microchannels, where parallel, size-dependent, separations of each fraction of proteins were achieved by SDS gel electrophoresis. Here, only the IEF-part will be highlighted. The polycarbonate device was flushed with bovine serum albumin (BSA) prior to use, in order to reduce protein (i.e., analyte) adsorption, and at the same time the electroosmotic flow was eliminated, which is also desirable. IEF occurs between reservoirs A and B, gel electrophoresis in the parallel channels between reservoirs D and C.

A commercial carrier ampholyte mixture (Pharmalyte 3-10) was employed for the IEF. The catholyte solution was 30 mM ammonium hydroxide at pH 10.5, while 10 mM phosphoric acid at pH 2.8 was used as the anolyte. IEF occurred in 90 sec in a 1 cm long channel at an electric field strength of 500 V/cm over a pH gradient from 3 to 10. This resulted in protein bandwidths of about 130 μm. The authors tested two different approaches to

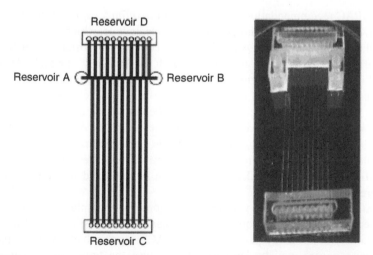

Reservoir D

Reservoir A

Reservoir B

Reservoir C

FIGURE 8.6 Schematic and image of a two-dimensional protein separation platform using plastic microfluidics. (Reprinted from Li, Y.; Buch, J. S.; Rosenberger, F.; DeVoe, D. L.; Lee, C. S. *Analytical Chemistry* **2004**, *76*, 742–748. With permission.)

[19] Capillary gel electrophoresis is described in Section 8.3.5.

this two-dimensional separation scheme. One used a single gel-based separation medium (ampholytes and gel in a single medium) and another one used two different separation media: an aqueous solution with the Pharmalyte 3–10 for IEF and a gel solution (1.5% poly(ethylene oxide)) for the SDS gel electrophoresis. The latter requires pressure filling of the aqueous solution (between reservoirs A and B, Figure 8.6) after filling the whole chip with the gel separation medium. Using two separation media enables analyte stacking at the gel–solution interface, which means that the bandwidths of the focused proteins were further reduced by about a factor of 2. This resulted in better resolving power in the gel electrophoresis step, and the separation channel could be reduced from 4 to 2.5 cm. The authors state that IEF has a typical preconcentration factor of 50 to 100, and that a two-dimensional separation (including fluorescent labeling) can be completed in about 10 min. An obvious way of improving the system is to employ additional parallel channels for gel separation, that is, more individual IEF segments could be analyzed.

While there are relatively few papers describing IEF-on-a-chip, it is a powerful first-dimension technique in multidimensional separations on microdevices, and also showcases what can be achieved by electrokinetic fluid control.

8.3.4 AFFINITY CAPILLARY ELECTROPHORESIS

An IUPAC recommendation from 2002 defines *affinity capillary electrophoresis* (ACE) as "an electrophoretic separation technique, which takes place in a capillary, with the *background electrolyte* containing substances capable of specific, often biospecific, interactions with the analytes." Loosely defined, ACE is a broad term referring to the electrophoretic separation of substances that participate in specific or nonspecific affinity interactions during separation. The interacting molecules can either be in solution or immobilized to a solid support [112]. According to Heegaard et al. [112], ACE may be classified into three modes: (1) nonequilibrium electrophoresis of equilibrated sample mixtures,[20] (2) dynamic equilibrium affinity electrophoresis,[21] and (3) affinity-based CE or CEC separation on immobilized selectors (also see Figure 8.7).

ACE has been used for various applications, for example, purification and concentration of analytes, quantitation of analytes, and estimation of binding constants, but not many papers regarding ACE-on-chip have been published (or these papers may be hard to find because the term affinity capillary

---

[20] Receptor and ligand in sample, electrophoresis separation buffer void of both receptor and ligand.
[21] Sample contains receptor protein, electrophoresis separation buffer contains ligand. Binding and separation of the affinity components occur within the channel/capillary.

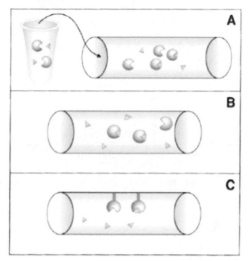

FIGURE 8.7 Schematic diagram of the three main modes of affinity capillary electrophoresis. (A) Nonequilibrium electrophoresis of equilibrated sample mixtures mixed in a test tube. Receptor and ligand in sample, electrophoresis separation buffer empty. An aliquot of the sample is introduced into the capillary for separation. (B) Dynamic equilibrium affinity electrophoresis. Sample contains receptor protein, separation buffer contains ligand. Binding and separation of the affinity components occur within the capillary. (C) Receptor protein is immobilized on, for example, the walls of a portion of the capillary, or on microbeads, membranes, or microchannels. Ligand present in a diluted solution or in a complex mixture is affinity-captured on receptor, nonrelated matrix components are washed away, specific ligand is released from adsorbed state, and separation is performed by capillary electrophoresis. (Reprinted from Heegaard, N. H. H.; Nilsson, S.; Guzman, N. A. *Journal of Chromatography B* **1998**, *715*, 29–54. With permission.)

electrophoresis is not often used). For reviews on ACE, see, for example, the aforementioned paper [112] or other reviews regarding affinity interactions [113–115]. Furthermore, there is a paper describing bioaffinity interactions in electrokinetically controlled assays on microfabricated devices [116].

Harrison's group explored the development of a six-channel device to perform reagent and sample mixing, an immunological reaction and an ACE separation [117]. The so-called "SPIDY" device was fabricated in glass, and consisted of six independent mixer, reactor, injector, and separation manifolds, and LIF was used for detection (layout shown in Figure 8.8).

The paper addresses several design issues, which will not be dealt with here. However, ensuring uniform EOF proved to be an important issue in obtaining good results. An antiestradiol assay was used as a model system, for demonstrating simultaneous quantitative immunoassays. The fluorescein-labeled antigen (Ag), estradiol (E2), was present in excess, while its antibody

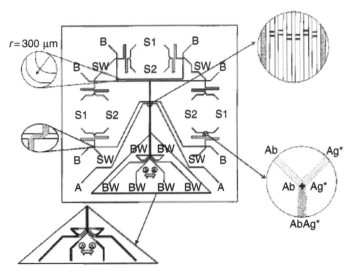

FIGURE 8.8 Illustration of the overall layout of the flow channel manifolds in the SPIDY device. Each reaction cell has reservoirs for sample (S1), antibody (S2), sample injection waste (SW), and running buffer (B). Six reaction cells are replicated around the outside of the wafer. The expansion on the left illustrates the double-T injector design for loading sample from the reactor into the separation column. In the bottom right expansion, the antibody (Ab) and antigen (Ag) mixer in the reaction cell is illustrated. An expansion in the lower left illustrates the pattern of channels at the buffer waste (BW) reservoirs used during separations. The thicker lines identify channel segments that are 300 μm wide, while the thinner lines represent 50 μm wide segments. The expansion in the upper right of the detection zone, across which a laser beam is swept, illustrates a separation of two components in the six separation channels. Two optical alignment channels filled with fluorescent dye sandwich the separation channels. (Reprinted from Cheng, S. B.; Skinner, C. D.; Taylor, J.; Attiya, S.; Lee, W. E.; Picelli, G.; Harrison, D. J. *Analytical Chemistry* **2001**, *73*, 1472–1479. With permission.)

(Ab) was added in varying concentration (20 to 50 n*M*). E2 and Ab were mixed on chip, the fluorescing species E2 and the E2–Ab complex were separated[22] according to their different mobilities, and finally, fluorescence was measured for all six channels in the detection zone by sweeping the laser beam using a galvano-scanner. With this chip it is possible to perform measurements, where standards and samples are present at different concentrations in each channel. Therefore, the entire analysis can be performed within 30 sec in a single run on the chip, including mixing, reaction and separation. This so-called cross-channel calibration was demonstrated by

---

[22] Buffer: 10 mM, pH 8.2 borate with 0.01% v/v Tween 20.

loading three different concentrations of antiestradiol into two channels each (i.e., the measurements were performed in duplicate). The quantification was based on measuring the peak height of the E2–Ab complex. Other examples of homogenous ACE methods on chip are the determination of cortisol in saliva [118] and the determination of serum theophylline [15].

Guijt et al. [116] identified a number of advantages in the implementation of immuno- (and enzyme-) assays on chip:

- Fast analysis compared to conventional immunosorbent techniques (favorable reaction kinetics on the microscale).
- Reduced reagent consumption, which is very important since immunoassay reagents are often costly, and sample volume is often limited.
- Enables automated sample handling and pretreatment (labeling, mixing, etc.).
- Possibility for parallel analyses (as shown in the example above [117]).
- Has the potential for point-of-care analysis.

A different way of performing ACE is to use molecular imprints [119]. Molecular imprinting is achieved by forming a synthetic cavity around the analyte of interest such that the cavity is complementary to the analyte in terms of shape and chemical functionality. In other words, a synthetic memory with selectivity towards the analyte molecule is formed ("plastic antibodies") [112]. However, this particular ACE mode is most often referred to as a capillary electrochromatography (CEC) mode [120, 121].

8.3.5 Capillary Gel Electrophoresis

Capillary gel electrophoresis (CGE) is directly comparable to traditional slab gel electrophoresis, which has been used for size-based separation of macromolecules (e.g., proteins and peptides, oligonucleotides, DNA fragments). As the name implies, in CGE the capillary or channel is filled with a "gel," which most often means a "polymer network structure." The polymers employed can be covalently cross-linked polymers, linear polymers, or hydrogen bonded polymers. Separation is obtained by electrophoresis of the analytes through a suitable "gel," which acts as a molecular sieve, that is, larger molecules are more hindered than smaller ones, and have lower migration velocities. The separation mechanism is mainly based on size, since the typical macromolecules analyzed have charge–size ratios that do not vary with increasing size [73].

Zhao et al. described the fast and relatively easy production of a PDMS or glass chip for CGE [86]. The authors demonstrated the chip functionality by separation of a DNA restriction ladder (26 to 700 base pairs) and DNA fragments from a PCR experiment using LIF as the detection technique. The chips had the standard cross design, and 2% agarose solution was used as sieving matrix in the 3.5 cm separation channel. The separation efficiency was better than 250,000 theoretical plates per meter. The authors stated that they could design and fabricate the system in less than 8 h with common instrumentation from machine shops, meaning clean-room facilities were not necessary and mass production is a real option.

Photo-defined polyacrylamide gels and electrode-defined sample injection were employed for double-stranded DNA separations [122]. The device consisted of an etched glass channel network bonded to a silicon or quartz substrate containing the microelectrodes. Photopolymerization allows precise definition of position and shape of the gel within the channel network, and furthermore, requires a shorter curing time than conventional chemical polymerization. The use of microelectrodes to define the size and shape of the injection plug allows extremely small but highly concentrated sample plugs to be generated in a simple way.

Briefly, an electric field is initially applied between electrodes E1 and E2 (see Figure 8.9). DNA collects at the positively charged electrode E2, and the

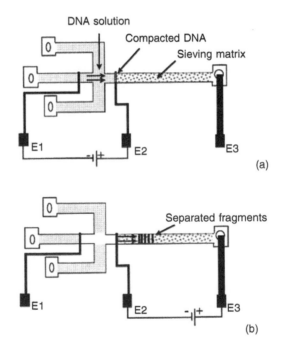

FIGURE 8.9 Sample compaction, definition, and release using on-chip electrodes. (a) Negatively charged DNA molecules are compacted by applying a low positive potential to a thin electrode (Pt) of definite width. (b) Sample is released by switching the polarity of the electrode and injected onto a separation matrix. (Reprinted from Brahmasandra, S. N.; Ugaz, V. M.; Burke, D. T.; Mastrangelo, C. H.; Burns, M. A. *Electrophoresis* **2001**, *22*, 300–311. With permission.)

field is applied until sufficient sample has accumulated at the liquid–gel interface. The field is then switched between electrodes E2 and E3 spanning the gel portion of the channel, and hence, gel electrophoretic separation occurs. The authors further refined this preconcentration[23] and injection system by redesigning the electrode arrangement. Their final, improved, injection scheme allowed the separation distance to be reduced by nearly 30% without any loss of resolution. They demonstrated the resolving power of the system by a baseline separation of a 20 base pair double-stranded DNA ladder and showed complete resolution from 20 to 500 base pairs in about 12 min using only a few millimeters of separation length.

Griebel et al. described a PMMA chip for two-dimensional CGE protein separation [123]. The first dimension used isoelectric focusing, which was realized using immobilized pH gradients (i.e., a gel). The second dimension was sodium dodecylsulfate-CGE: the focused proteins were transferred to 300 parallel channels (50 μm × 50 μm, 64 mm length) for gel electrophoresis using 15% pullulan (natural polysaccharide) as sieving matrix. The two techniques were demonstrated separately, and the first experimental results of protein transfer from the first to the second dimension were described. This work illustrates the power of parallelization afforded by the chip format.

As is obvious from the examples above, CGE is often employed in conjunction with other electrophoretic techniques, and recently, a three-dimensional gel electrophoresis[24] was published [124] (although not adapted to a microchip format). The most prominent advantage when transferring CGE to the chip format is the high analysis speed, and the possibility to include preconcentration on-chip. It should also be noted that CGE was the first of the aforementioned techniques to be applied successfully in commercial lab-on-a-chip instrumentation (Agilent 2100 bioanalyzer). In Chapter 12, which focuses on applications, more examples of CGE on chip are given.

8.4 MICELLAR ELECTROKINETIC CHROMATOGRAPHY

Micellar electrokinetic chromatography (MEKC), first described by Terabe et al. in 1984, is a simple way to transform the electrophoretic analytical method of capillary electrophoresis (CE) to a chromatographic analytical method for the separation of neutral (mostly) and charged compounds by adding a stationary phase to the buffer [125, 126]. This technique is also referred to in the literature as MECC (micellar electrokinetic capillary chromatography). For the purposes of this book, it will be referred to only as MEKC.

---

[23] The authors refer to it as "compaction."

[24] First dimension, separation by IEF; second and third dimensions, separation by molecular weight, focusing on separation of low- and high-molecular weight proteins, respectively.

MEKC, like CE, requires a channel made in an insulating material, an electrolyte, an applied electric field to drive the solution flow, and a detection unit. However, the key difference between the two methods is the addition of a highly charged pseudostationary phase comprised of micelles to the running buffer. These micelles migrate in the electric field at a velocity proportional to their charge-to-size ratio while uncharged analytes with different micelle-running buffer partition coefficients associate for different times with the lipophilic interior of the micelles. The difference between the neutral molecules, partition coefficients into the micelles constitutes the basis for the separation mechanism, and therefore MEKC is a chromatographic method. Ionic compounds have also been separated by MEKC in which case a combination of electrophoretic migration and interaction with the micellar phase determines the separation selectivity. MEKC combines the selectivity advantage of chromatographic techniques with the high efficiencies, rapid analysis times, small solvent and sample consumption, and no moving parts characteristic to electrophoresis. Because MEKC offers the possibility of chromatography without the presence of packing material, it is uniquely suited to the planar format of chip-based μ-TAS systems.

The selectivity, and therefore innovation, in MEKC lies in the selection and synthesis of the pseudostationary phase. Such innovations are independent of the separation platform — chip or capillary — and the microfabricated devices can capitalize on the considerable progress attained by MEKC in capillary format. For example, alternatives to the anionic surfactant used in the first MEKC separations include nonionic, zwitterionic, and cationic surfactants, macroemulsions, vesicles, dendrimers, and molecular micelles. Each variation extends the elution window of the separation, the stability of the pseudostationary phase, the selectivity or solvent range in effect widening the application range of electrokinetic chromatography.

Several recent reviews give an excellent overview of the state-of-the-art of MEKC in general, which is outside the scope of this work [34, 68, 127]. However, what is of interest here is a discussion of how MEKC has been implemented for chip-based separations, issues that have been reported, and performance. Additionally, the planar architecture and use of photolithography to define the channel geometry has resulted in some interesting innovations in separations.

The first application of MEKC to microchip separations was by Moore et al. to the separation of three coumarin dyes [128]. Baseline resolution of the neutral dyes was achieved on serpentine and straight channels. The short straight channel provided efficiencies of 3200 to 8800 theoretical plates in 3 min. In more recent experiments, a significant improvement in efficiency was demonstrated on a spiral-shaped separation channel as 1,000,000 plates were obtained for dichlorofluorescein [129]. An important environmental and forensic application that illustrates the opportunity of portable devices is the detection of explosives. Wallenborg and Bailey. employed chip-based MEKC to resolve 10 explosives within 60 sec [10]. Near infrared (near-IR) indirect LIF was used to detect concentrations on the order of 1 ppm [10]. The work by Wallenborg and Bailey [10, 130] provides excellent data for comparison between the performance of a MEKC method for the chip and the capillary because most experimental conditions were held constant for both methods. Figure 8.10 shows the IDLIF explosives separation on-chip and in a capillary. Table 8.1 summarizes the results from the chip and the capillary MEKC experiments.

The change in the fluorophore and diode laser for detection did not seem to affect the limit of detection (LOD) or linear range. The difference between the running buffers also should not dominate. Following are the important considerations in selecting a chip over a capillary method. For example, the resolution is slightly affected with two less compounds resolved on chip compared to the capillary. However, in comparison, the enormous advantage in throughput, with the shortened channel and higher field strengths achieved by using a chip resulting in an improvement of run time from 35 min in the capillary to 1 min on chip, the sacrifice in resolution has an extremely high payoff. Moreover, this result shows that optimization, for example, by lengthening the channel, could improve the chip results and still allow for faster times. This is due to the fourfold increase in field strength that the heat transfer abilities of the thick glass chip, superior to that of the thin-walled fused silica capillary, afford. Other potential advantages, such as the effect of the integration of the injection arm with the separation channel, are more difficult to evaluate because the relevant data for comparison were not given in the description of the capillary-based experiment. A detailed study of the effect of the injection plug size and

---

TABLE 8.1
MEKC/IDLIF Analysis of EPA 8330 Explosives Standard (14 Compound Mixture) [10, 130]

Parameters	Chip [10]	Capillary [130]
Theoretical plates (max)	60,000	56,000 to 129,000
Applied electric field (V/cm)	538 V/cm	100 V/cm
Analysis time (min)	1	35
Buffer composition	50 mM borate, pH 8.5, 50 mM SDS, 5 µM Cy-7	10 mM borate, 50 mM SDS, 1 µM Cy-5, 3% isopropanol
Column length (cm)	6.5	50
Number of peaks resolved	10/14 IDLIF N/A	12/14 (IDLIF) 14/14/(ID UV ABS)
Fluorophore	5 µM Cy-7	1 µM Cy-5
LOD	1 ppm	1 ppm 1,3,5-trinitrobenzene
Linear range	1 to 5 ppm	1 to 5 ppm

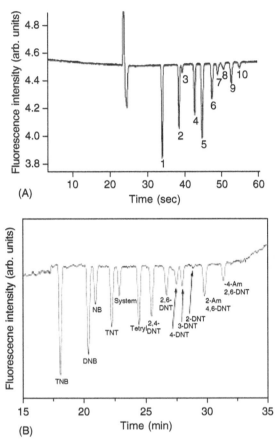

(A)

(B)

FIGURE 8.10 (A) Chip-based MEKC-IDLIF electropherogram of the EPA 8330 mixture of nitroaromatics and nitramines. Analytes: 20 ppm each of TNB (1), DNB (2), NB (3), TNT (4), tetryl (5), 2,4-DNT (6), 2,6-DNT (7), 2-, 3-, and 4-NT (8), 2-Am-4,6-DNT (9), and 4-Am-2,6-DNT (10). Conditions: MEKC buffer, 50 mM borate, pH 8.5, 50 mM SDS, 5 μM Cy7, separation voltage 4 kV, separation distance 65 mm. (B) MEKC separation on a 50 cm capillary. Running buffer was 10 mM sodium borate, 50 mM SDS, 3% isopropanol, 1 mM Cy-5. Injections of 10 mg/l (each component) were performed electrokinetically for 5 sec at 10 kV. (Reprinted from Wallenborg, S. R.; Bailey, C. G. *Analytical Chemistry* **2000**, *72*, 1872–1878 and Wallenborg, S. R.; Lurie, I. S.; Arnold, D. W.; Bailey, C. G. *Electrophoresis* **2000**, *21*, 3257–3263. With permission.)

the effect of the counter voltages applied during injection and run are useful in planning an experiment in a similar chip. The results presented support the conclusion that the offset in the structure of the injection arm itself (Figure 8.11) has no effect on the amount of sample injected, while the authors postulate that control of the counter voltages applied immediately after injection could maximize the sample plug injected.

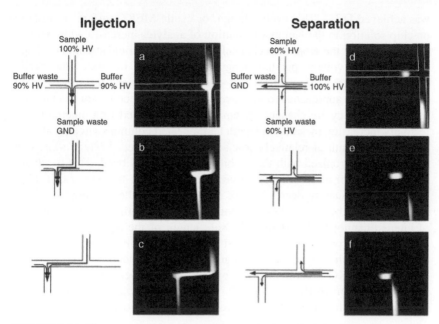

FIGURE 8.11 Fluorescence microscope images of microchip injections, using different injector configurations. Marker: 0.4 mM fluorescein. Injections were done at 1 kV (a–c) and separations at 1 kV (d–f). Injector configurations: straight cross (a, d), 100 μm offset double-T (b, e), and 250 μm offset double-T (c, f). The illustrations to the left of the photographs show the injector configurations and applied voltages during injection and separation modes. Arrows indicate flow directions. The microfabricated channels on the chip have been outlined in (a) and (d) for clarification. (Reprinted from Wallenborg, S. R.; Bailey, C. G. *Analytical Chemistry* **2000**, *72*, 1872–1878. With permission.)

Of interest to pharmaceutical and biotechnology applications is the ability to differentially alter the mobilities of enantiomers by incorporating chiral additives in the running buffer. Rodriguez et al. demonstrated that using microchips with cyclodextrin-modified micellar electrokinetic chromatography (CD-MEKC) for the separation of fluorescein isothiocyanate (FITC)-labeled amino acid enatiomers it was possible to achieve lower dispersion, higher resolution with analysis times 10 times lower than the comparable capillary-based system [131, 132]. Chiral and achiral separations of amphetamine and related compounds were studied on straight, U-folded, and S-folded channels etched on glass chips. Optimized separations were achieved on the S-folded channel as the presence of small amounts of SDS dramatically improved chiral resolution [131].

By using the planar format, von Heeren et al. demonstrated that an extremely short column could be used effectively by making it into a square path that the analytes would pass through again and again until the separation was achieved [133]. This cyclic design, or cyclic MEKC, offers an original on-chip solution to the fact that resolution of analytes increases with increasing the length of the separation channel and the applied field strength (defined as the applied voltage per unit length). Figure 8.12 illustrates the cyclic MEKC experiment. As practical and safety considerations limit the applied voltage the two approaches to improve performance become somewhat contradictory. The cyclic channel system permits unlimited time operation effectively increasing the column length and repeated column switching affords high field strength at relatively low separation voltages. Voltage switching has to be synchronized with the mobility of a selected band. This is advantageous if one or a few closely spaced analytes are to be detected. It does, however, significantly decrease peak capacity. Also, the running time is in practice limited by the band broadening, that is, the width of the band or peak of interest cannot be larger than the side length of the square column. The presence of sharp turns further exacerbates band spreading. Nonetheless they prove that much higher field strengths can be applied (up to 2 kV/cm), shortening MEKC separation times by 50 to 100 times while maintaining (in the shortest microfluidic channel), or increasing separation efficiency by greater than twofold in the 10 cm microfluidic channels compared to the same

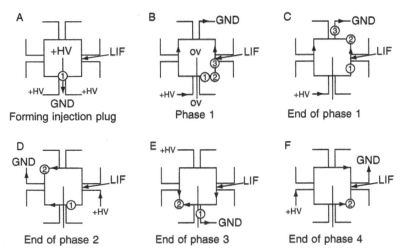

FIGURE 8.12 The principle of synchronized cyclic CE. The separation of three sample components, 1–3, over one cycle is illustrated. The direction of fluid flow and sample transport are indicated by the arrows. LIF refers to the location where samples are detected. The voltage switching protocol is synchronized to cycle component 2: (A) injection phase, (B) time point during phase 1, (C) end of phase 1, (D) end of phase 2, (E) end of phase 3, and (F) end of phase 4 and end of first cycle. (Reprinted from von Heeren, F.; Verpoorte, E.; Manz, A.; Thormann, W. *Analytical Chemistry* **1996**, *68*, 2044–2053. With permission.)

separation in a 50 cm fused silica capillary [133]. Using LIF, nanomolar limits of detection at a signal-to-noise ratio of 3 were obtained for fluorescein isothiocyanate-labeled arginine. By using the cyclic system, the researchers were able to optimize channel length by selecting the number of cycles rather than by having to have chips with channels of different lengths. This is a distinct advantage over capillary and most chip-based methods.

As mentioned previously, detection sensitivity in open-channel separation techniques is rather poor. Online preconcentration techniques such as ITP, sample stacking and sweeping, extensively studied for capillary systems, can also be applied to chip based separations. ITP can be performed in MEKC with nonionic surfactants as the leading and terminating background electrolytes are properly selected. Another concentration method, sample stacking, discovered and largely employed in CZE, is achieved when the sample solution has a considerably lower conductivity than the surrounding background electrolyte. Ten- to 1000-fold increase in concentration have been reported for capillary-based systems. The preconcentration effect of sweeping was demonstrated on chip by Sera et al. [134]. This phenomenon, first characterized in capillaries by Quirino et al., uses the mobility difference between SDS micelles and neutral analytes to focus the sample that does not have micelles in the sample buffer. The principle works as follows, and is illustrated in Figure 8.13. Under neutral or alkaline conditions SDS micelles migrate slower than neutral samples when samples are injected from the anode and migrate to the detector close to the cathode. The injected sample zone, containing no SDS is sandwiched between the preceding and the following lower velocity running buffer zones containing SDS. The slower SDS micelles intrude into the sample zone. If the sample is

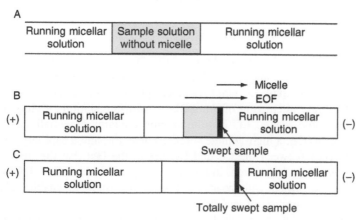

FIGURE 8.13 Schematic diagram of concentration by sweeping. (Reprinted from Sera, Y.; Matsubara, N.; Otsuka, K.; Terabe, S. *Electrophoresis* **2001**, *22*, 3509–3513. With permission.)

hydrophobic, the sample remains associated with the micelles. Eventually, the entire sample is focused at the boundary between the sample zone and running micellar zone. The width of the neutral analyte band is predicted to be narrowed by $1/(1 + k)$, where k is the retention factor. If the sample is hydrophilic, the sample remains in the aqueous phase rather than in the micellar phase; therefore the sweeping concentration phenomenon is not significant. Rhodamine B, sulforhodamine, and cresyl fast violet were concentrated by sweeping and separated by chip-based MEKC. Due to sweeping the observed enhancement in sensitivity on-chip was 170-, 90-, and 1500-fold, respectively. The advantage described here in moving to the chip was the flexibility of the detection point. This was critical to the easy determination of the concentration profiles during sweeping. They also used this to their advantage in determining the best detection point for their separation. Unlike the cyclic MEKC system, however, the longest dimension possible is physically limited by the absolute length of the channel.

In chromatography, the method of choice for tailoring selectivity and solving the general elution problem is the use of solvent programming. In the case of capillary systems, the limited availability of gradient elution is considered a major reason that precludes their use in routine applications. Consequently, the demonstration by Kutter et al. [36] that gradient elution can be readily achieved on microfluidic devices is of great significance. Figure 8.14 shows the chip design allowing control of the solvent composition at the mixing tee. Precise gradients of various shapes (linear, concave, convex, sinusoidal) were demonstrated by simply adjusting the voltage settings applied to the two buffer reservoirs (Figure 8.15). Given that organic modifiers have such complex effects in MEKC, the use of systems equipped with gradient capabilities is greatly anticipated for fast and efficient method development.

8.5 CAPILLARY ELECTROCHROMATOGRAPHY

Electrochromatography in capillaries (CEC) or chips (ChEC) is an electrokinetic separation technique, a hybrid between capillary electrophoresis and liquid chromatography. EC employs electrophoretic and electroosmotic motion to facilitate interaction of the analytes with the stationary phase of the chromatographic media. EC instrumentation contains essentially the same components as CE: a high-voltage source, a detector, and a capillary column. The difference between the CE and EC lies in the fact that the fused silica capillary is either derivatized, in the case of open tubular EC (OT CEC and OT ChEC), or filled with beads or porous polymers to enable selective partition and separation of analytes.

The use of EOF in packed beds has demonstrated three notable advantages over pressure-driven flow. The first advantage stems from the fact that the radial uniformity of the mobile phase velocity profile, that is, its independence of the distance to the walls of the column, has a minimal effect on

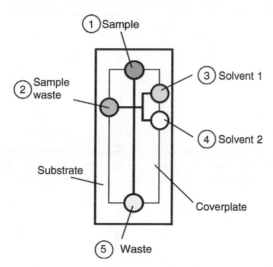

FIGURE 8.14 Schematic of the chip with channels and reservoirs shown. The effective length of the main channel from the cross to the detection point close to the waste
reservoir is approximately 25 mm. The channel depth is 9 μm, and its width at half-
depth is 50 μm. (Adapted from Kutter, J. P.; Jacobson, S. C.; Ramsey, J. M. *Analytical
Chemistry* **1997**, *69*, 5165–5171.)

the chromatographic peak dispersion. The second advantage of EOF is that, in
the absence of double layer overlap, its velocity is independent of the radius
of the channel, that is, the geometry and size of the packing (Figure 8.16). The
third advantage of EOF is that no significant pressure drop is associated with
this form of bulk transport. Therefore, very small particles (particles smaller
than 1 μm have been employed) can be used to significantly improve the
separation efficiency. Consequently, narrower peaks, higher resolution, and
larger sample capacity can be observed as EOF replaces the pressure-driven
flow. These capabilities are important particularly when undertaking complex
samples in which minimization of matrix effects and baseline separation of
multiple analytes is desired.

The increased separation power in electrochromatography relies mostly
on the ability to use a wide variety of surface chemistries to support separations based on reversed-phase, ion exchange, affinity, and other type of
interactions. Such sorbents, extensively used in liquid chromatography, are
readily available in bead or particulate format in a variety of diameters, pore
sizes, and base materials (i.e., silica and organic polymer). More specialized
continuous porous columns made of both silica and organic-based polymers
are being studied in many laboratories [135–137]. A comprehensive analysis
of the state-of-the-art in this field is presented in a recent book dedicated to
monolithic materials [138].

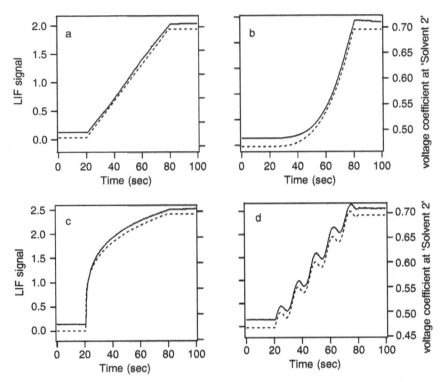

FIGURE 8.15 Verification of gradient shapes with rhodamine B-doped buffer; dashed traces, input voltage coefficient at "Solvent 2," solid traces, output LIF response measured between mixing tee and injection cross (see Figure 8.14). (a) Linear, (b) concave, (c) convex, and (d) sinusoidal gradients. (Adapted from Kutter, J. P.; Jacobson, S. C.; Ramsey, J. M. *Analytical Chemistry* **1997**, *69*, 5165–5171.)

Electroosmotic flow
Velocity profile is flat and independent of channel diameter

Pressure-driven flow
Velocity profile is parabolic and dependent on channel diameter

FIGURE 8.16 Flow profile and its dependence on channel diameter for EOF and pressure-driven flow.

The physical stability of the stationary phase to organic solvents gives EC an advantage over MEKC for the analysis of neutral and charged molecules. From the instrumentation perspective, the fact that EC, like CE, requires only voltage potential, no pumps or other moving parts for operation, makes EC particularly suited to on-chip operation.

Based on the structure of the chromatographic media there are three major categories of EC columns: open channel, packed structures, polymeric rods, or monoliths. The open-channel columns are readily obtained in chip, as in capillary format, by chemically modifying column walls with reagents endowed with functional groups that act as stationary phase. In the first EC separation on a chip, Jacobson et al. [139] followed a standard derivatization procedure which uses chlorodimethyloctadecylsilane to functionalize the surface of a glass chip. Longer capillaries are typically employed in OT-CEC to increase capacity factors and improve resolution. In order to retain the small dimensions of the chip, longer columns can be obtained by etching channels in a spiral, serpentine, or racetrack-type geometry. For example, Jacobson et al. used a 17 cm serpentine column to achieve baseline resolution of three coumarin dyes in less than 3 min [139].

Resolution of complex samples necessitates the ability to optimize retention factors. This is generally achieved by varying the composition of the mobile phase during elution (gradient elution or solvent programming). The first open tubular chip electrokinetic chromatography (OT-ChEC) separations using gradient elution were reported by Kutter et al. [37]. The system consists of a microchip and two computer programmable power supplies. The chip design is based on a typical channel cross which contains a separation column, the customary channels and via holes for sample injection, as well as a modified side channel which incorporates mixing region and two via holes which connect to the reservoirs with the two solvents. The hydrophobic surface of the column was obtained by reacting the silica channel with octadecyltrimethoxysilane. Electrosmotic fluid pumping is controlled by the two programmable power supplies which deliver time-variable voltages. The summation of the EOFs generated from the two mobile phase reservoirs defines the shape of the gradient. Linear gradients with various duration times, slopes were obtained reproducibly. The ability to rapidly adjust solvent strength is a highly desirable feature for method development of any chromatographic separation. In this case, optimization of channel geometry and fine-tuning the elution conditions significantly improved separation efficiency and reduced analysis time. As a result, four coumarin dyes were baseline resolved in less than 20 sec (Figure 8.17).

Packed ChEC is an area that received a lot of interest in the past years. Columns packed with sorbent particles are the most effectively used media for electrochromatographic separations in capillary format. Their most attractive feature is the significantly higher surface area which provides increased sample capacity over open channels. However, packing

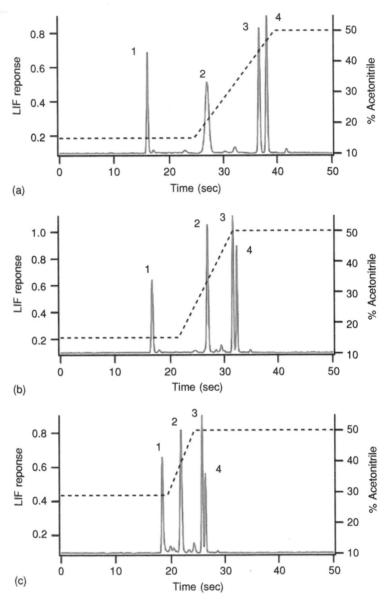

FIGURE 8.17 Gradient elution in open-channel chromatography on microchips. Channel depth, 5.2 μm; stationary phase, octadecylsilane; mobile phase, 10 m*M* borate buffer (pH 8.4) with different amounts of acetonitrile, (a) linear from 15 to 50% within 15 sec, starting 8 sec after injection, (b) linear from 15 to 50% within 10 sec, starting 5 sec after injection, and (c) linear from 29 to 50% within 5 sec, starting 3 sec after injection; analytes (1) C 440, (2) C450, (3) C 460, and (4) C480; field strength, 500 V/cm; dotted line, gradient trace. (Adapted from Kutter, J. P.; Jacobson, S. C.; Matsubara, N.; Ramsey, J. M. *Analytical Chemistry* **1998**, *70*, 3291.)

micrometer size beads is not trivial and packing channels etched in chips is especially difficult because frits — structures able to retain the beads while allowing for liquid flow — must be incorporated in the structure of the separation column [140–142]. To start with, slurry packing requires interconnects from a pump to the chip and the packing pressures are limited by the chip material. For example, glass or quartz chips can take up to 1500 psi while most plastic chip becomes defective at about 500 psi. In CEC, typical packing pressures for high efficiency tightly packed beds are on the order of 3000 to 5000 psi. Consequently, packed beds are difficult to obtain in the chip channels while packing small diameter particles is simply not practical. These might be the main reasons why packed ChEC has not yet attained the performance of CEC. Other packing procedures, such as electrokinetic packing, have also been tested without success on columns longer than 1 mm [142].

The difficulties encountered in packing the chips are progressively addressed. One solution for stabilizing the packed beds is adapted from capillary format [143] and considers entrapment of the packed beads in a highly porous polymer matrix. This procedure eliminates void formation, increases column lifetime and separation performance [142]. An innovative exploit of the characteristics of chip fabrication materials other than fused silica (the capillary of choice for CEC) employed thermal treatment of a bed packed in PDMS to stabilize the column by eliminating the need for a second retaining frit [140]. Lastly, the microfabrication advantage, the most ingeniously used resource in chip development, offers the ability to design and fabricate micrometer-size features that could not be realized in capillary format. In the case of packed beds this translates into structures such as tapered columns, shelves, or weirs, which retain the beads with minimal backpressure, effectively acting as frits. A simple design commonly used for fabrication of packed columns is comprised of a channel that has shelves with much smaller depths at inlet and outlet and a third port in the middle of the column to be connected to the bead slurry (Figure 8.18). Liquid pressure is applied to pack the particles against the shelves; the third port is closed when the entire channel is packed.

Another demonstration of the microfabrication advantage in chip-based devices is the introduction by the group of Regnier of micrometer size, particle like structures which constitute separation columns [144–146]. From the chromatographic perspective, the so-called collocated monolith support structures (COMOSS) are a hybrid between open channel and packed columns (Figure 8.19); they require postfabrication derivatization, have minimal back pressures and higher sample capacity than OT-ChEC without the need to pack the channels. Initial band broadening studies on COMOSS columns on quartz substrates showed promising results for electrophoretic and electrochromatographic separations. However, the design and manufacturing of these columns is quite complex. COMOSS separation columns were

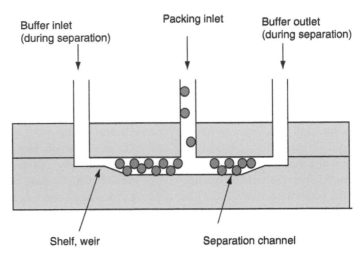

FIGURE 8.18 Chip layout for separation channel containing packed beads. A slurry of packing material is pumped through the middle inlet into the deep etched separation channel. The shallow etched shelves or weirs effectively act as frits; they retain the particles while allowing the packing fluid to pass through.

also made from less expensive PDMS substrate. Typical derivatization procedures generate a hydrophobic surface enabling the separation of peptides in complex samples such as tryptic digests [146]. In a demonstration of the ease with which integration of multiple functions can be achieved on microdevices, a combination of two particle-packed columns and a COMOSS unit on a PDMS substrate performed protein digestion followed by affinity and electrochromatographic separations of the peptide fragments [41].

Another option for on-chip electrochromatography is to polymerize acrylamide gel supports and incorporate various additives to impart the desired chromatographic interaction [147, 148]. The acrylamide gels show promise as on-chip separation media due to the ease of fabrication and very high separation efficiencies. The main drawback is that the gel-like composition reduces their structural stability and limits their applicability. Relevant examples of ChEC with acrylamide gels include the work of Ericson et al., which used polyacrylamide gels modified with sulfonic groups (for the generation of EOF) and isopropyl groups (to support hydrophobic interactions) for the separation of neutral compounds with high efficiencies (up to 350,000 plates/m) [147]. If chiral ChEC separations are targeted, β-cyclodextrans (β-CD) can be covalently attached to acrylamide gels. For example, dansyl D,L aminoacids were separated on this media with efficiencies up to 150,000 plates/m [148].

A more robust alternative to achieve higher capacities and efficient separations lies in the porous polymer monoliths (PPMs) introduced in the

FIGURE 8.19 Scanning electron micrographs of microfabricated collocated monolith support structures: (A) a section of the COMOSS column, (B) higher resolution micrograph showing high degree of homogeneity of the COMOSS monolith structures and the channels, (C) the column–wall interface, and (D) high-magnification scanning electron micrograph showing the surface roughness of the COMOSS monolith structures. (Reprinted from He, B.; Tait, N.; Regnier, F. E. *Analytical Chemistry* **1998**, *70*, 3790–3797. With permission.)

1990s by Svec and Frechet originally for HPLC analysis. In recent years, these polymers proved to be ideal for capillary and chip electrochromatography due to the combined advantages of ease of introduction, patternability, as well as tunability of porosity and surface chemistry [149]. The manufacturing of PPM is very simple and requires no special equipment. Channels with virtually any geometry are filled with a homogeneous mixture of monomers, crosslinkers, polymerization initiators, and solvents. Polymerization is initiated as the chip is exposed to UV radiation or increase in temperature. The

PPM form as the growing polymer phase-separates from the solvent to create a continuous monolith. Immediate flushing with high organic solvent removes unreacted reagents. In cases where the monolith must be covalently bonded to glass or silica walls, prior to the introduction of the polymerization mixture, the chip is pretreated with silanes. To this effect, reagents with double functionality, such as 3-(trimethoxysilyl)propyl acrylate, are commonly used. They have groups, like trimethoxysilyl, that react with the surface silanols as well as groups with double bonds, like acrylate, which during the polymerization step react with the monomers or crosslinker of the PPM solution and anchor the monolith to the walls of the channel.

Before employing PPM in etched channels one should consider the few shortcomings of porous organic polymers. Swelling and shrinking effects are observed as the polymers are exposed to different solvents. Extensive cross-linking significantly reduces these effects; polymerization mixtures containing up to 30% crosslinker are commonly used. Another downside is the fact that PPMs should be kept wetted at all times. Drying, even if superficial, makes removal of air bubbles very difficult. In most cases, upon polymer rewetting initial polymer performance is not recovered entirely. In the case of silica substrates, prolonged exposure to basic solutions (pH > 9.5) must be avoided. The wall-anchoring bonds at the silica surface are slowly hydrolyzed causing the movement or extrusion of the monolith from the channel. Nonetheless, the PPM qualities derived from *in-situ* polymerization, such as ease of fabrication, photopatternability, simultaneous control of surface chemistry and porosity, make PPM one of the most promising media for on-chip electrochromatography.

The surface tunability and increased sample capacity advantages are evident as 16 polyaromatic hydrocarbons are baseline resolved on PPM patterned glass chip (Figure 8.20). Reports on the reproducibility of separation performance of PPMs are very encouraging. For example, the polymer formulations proposed by Ngola et al. delivered capacity factors that varied by 0.6% for a single column and 10% from column-to-column [150].

Two methods are commonly used to initiate the phase separation and polymerization. Thermal polymerization is the most often found method in the literature. Complications were reported when moving to the chip, where evaporation of the porogenic solvent from the reservoirs mounted on the channel was observed during the polymerization process [151]. Photoinitiation is more desirable for monoliths in chips because it is much faster and avoids polymerization at elevated temperatures or any change in phase behavior caused by the variations in temperature. More importantly, photolithography can be used to define the regions on the chip where the monolith forms, as reported by Fintschenko et al. [151]. Briefly, butyl acrylate-based solution in a porogenic solvent was introduced into the channels of a chip using capillary action after overnight pretreatment of the walls with 3-(trimethoxysilyl)propyl acrylate. The chip was then exposed to 365 nm

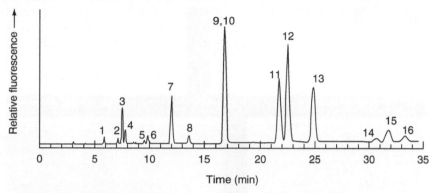

FIGURE 8.20 Electrochromatographic separation of 16 polyaromatic hydrocarbons (PAHs) on the butyl stationary phase in 75:25 v/v acetonitrile/5 mM Tris, pH 8, at field strength of 833 V/cm. Numbering of the PAHs: (1) naphthalene, (2) acenaphthalene, (3) acenaphthene, (4) fluorene, (5) phenanthrene, (6) anthracene, (7) fluoranthene, (8) pyrene, (9) benz[a]anthracene, (10) chrysene, (11) benz[b]fluoranthene, (12) benzo[k]fluoranthene, (13) benzo[a]pyrene, (14) dibenz[a,h]anthracene, (15) benzo[ghi]-perylene, and (16) indeno[1,2,3-cd]pyrene. (Reprinted from Ngola, S. M.; Fintschenko, Y.; Choi, W.-Y.; Shepodd, T. J.; *Analytical Chemistry* **2001**, *73*, 849–856. With permission.)

light from a UV lamp after the injection arms were masked with black tape [151]. The photopatterned monolith is shown in Figure 8.21.

Extensive research studies demonstrate that the monolith mean pore size and distribution can be changed by adjusting the monomer–solvent mixture composition [149–152]. In addition, varying the amount and nature of the monomers readily predetermines the surface chemistry [152, 153]. Such versatility gives the scientist the ideal handle for tuning the separation media and running conditions to their specific application.

One critical element in all stationary phases to be used in electrochromatographic applications is the requirement for ionizable groups that are able to generate EOF. Silanol groups intrinsic to silica or glass-based materials have customarily provided the support for EOF. In PPM, charged monomers need to be incorporated to generate the double layer which upon application of voltage generates the EOF. The most commonly used monomers with ionizable groups include sulfonic acid containing reagents. For example, augmenting the surface charge of the monolith significantly increases EOF, thereby reducing the analysis time. Apart from solving the issue of generation of EOF in organic polymers, these polymer-integrated charged groups provide the advantage of higher stability and EOF support at low pH.

Shediac et al. demonstrated that in the synthesis of polymer monoliths charge and hydrophobicity can be simultaneously tuned for a given application [152]. A demonstration to this effect is the separation of complex peptide

FIGURE 8.21 Pictures of monoliths cast in channels on an off-set-T fused silica chip. (A) Injection and separation channels completely filled. (B) Monolith cast only in separation region. (Reprinted from Fintschenko, Y.; Choi, W.-Y.; Ngola, S. M.; Shepodd, T. J. *Fresenius Journal of Analytical Chemistry* **2001**, *371*, 174–181. With permission.)

containing samples (Figure 8.22). Most notably, scanning electron microscope (SEM) images indicate that the geometrical structure of the polymer can be maintained essentially the same for polymers with various surface properties.

To take full advantage of the separation power of chromatographic inter-actions, a chip-based CEC instrument must offer the option of using gradient elution. Complex liquid mixtures — such as biological samples, a crucial application for microfluidic devices — are typically resolved using gradient HPLC or CEC. Adaptation to microchips is particularly challenging due to the stringent requirements for minimized mixing volume. Moreover, on-chip gradients must be fast and be generated in a reproducible manner. Recently, Singh et al. demonstrated gradient elution electrochromatography in chips containing photopatterned acrylate polymer [153]. Solvent programming was achieved using an eight-channel computer-controlled high-voltage power supply and a mixing manifold consisting of sharp right-angle turns. The system takes full advantage of the small dead volumes attainable in chip format. Consequently, linear and nonlinear solvent gradients with minimized

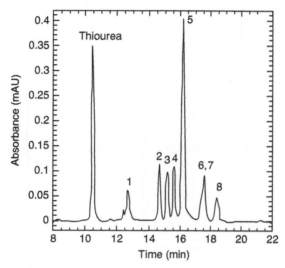

FIGURE 8.22 Electrochromatographic separation of eight bioactive proteins and peptides on positively charged butyl stationary phase incorporated with 0.5% charge in acetonitrile–12.5 mM phosphate (30:70, v/v), pH 2.8. The pH of the mobile phase should be chosen so as to minimize the electrostatic attraction between most of the peptides and the stationary phase. UV detection at 214 nm. Field strength, 93 V/cm. Capillary: total length 32.5 cm; length to detector distance 18.5 cm; inner diameter 100 µm. The numbers labeling the peaks correspond to (1) Leu-Enkephalin-Lys, (2) Met-Enkephalin-Arg-Phe, (3) α-Casein, fragment 90–95, (4) Met-Enkephalin-Lys, (5) β-Lipotropin, fragment 39–45, (6) Thymopentin, (7) Splenopentin, and (8) Kyotorphin. (Reprinted from Shediac, R.; Ngola, S. M.; Throckmorton, D. J.; Anex, D. S.; Shepodd, T. J.; Singh, A. K. *Journal of Chromatography A* **2001**, *925*, 251–263. With permission.)

dwell volume were obtained rapidly and reproducibly. Finally, the advantage of using a lauryl acrylate-based porous polymer monolith with high peak capacity and high retention capabilities was exploited in the separation of peptides from a tryptic digest of cytochrome c (Figure 8.23). Fine-tuning of selectivity by controlling both the stationary phase and mobile phase composition combined with the multiplexing potential of chip devices uniquely empower ChEC to address challenging applications such as complex proteomic analyses.

8.6 MORE DIMENSIONAL SEPARATIONS

Most of the techniques described so far have been one-dimensional, like simple CE. However, a couple of examples of two-dimensional separations, where two different techniques are employed sequentially, have already been mentioned in the sections on isoelectric focusing [110, 111], isotachophoresis [103, 104], and capillary gel electrophoresis [123].

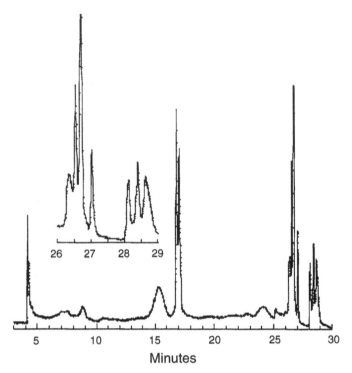

FIGURE 8.23 Gradient elution electrochromatography of tryptic digest of cytochrome *c*. The digested peptides were labeled with NDA (naphthalene 2,3-dicarboxaldehyde). The gradient was from 20:80 ACN:5.5 m*M* borate pH 8.5 to 60:40 ACN:10 m*M* borate (ACN = acetonitrile). Detection was performed using the 413 nm line of a Kr-ion laser. (Reprinted from Singh, A. K., Throckmorton, D. J., Brennan, J. S., Shepodd, T. J., Squaw Valley, Gradient-elution reversed-phase electrochromatography in microchips, in *Proceedings of μ-TAS 2003 Seventh International Conference on Micro Total Analysis Systems*, Volume 2; Northrup, M.A.; Jensen, K.F.; Harrison, D. J., Eds.; Transducers Research Foundation, Inc., Cleveland Heights, OH, 2003, pp. 1163-1166. USA, 2004. With permission.)

The number of peaks that can be resolved by a separation system (the peak capacity) can be increased by combining two orthogonal separation techniques, that is, by creating a two-dimensional separation system [154]. In this context orthogonal means that the techniques chosen are (ideally) totally independent (uncorrelated), that is, separations are based on different physiochemical properties. Hence, theoretically, the peak capacity of the two-dimensional method is the product of the peak capacities of the individual one-dimensional methods, although in reality this is never achieved [155] (see also Chapter 2).

---

[25] See Section 8.5 on capillary electrochromatography.

The Ramsey group published a couple of papers on two-dimensional separation devices. They combined MEKC and CE for peptide mixtures and tryptic digests [156, 157], and they further presented a chip, where open-channel electrochromatography (OCEC)[25] was the first dimension and capillary electrophoresis the second dimension [7]. OCEC (performed in a 25 cm spiral channel with C18-coating) and CE (in a 1.2 cm straight channel) were serially coupled on a glass chip to analyze a fluorescently labeled tryptic digest of β-casein. At the beginning of an analysis a single gated injection (see Chapter 2) of sample was made into the OCEC channel. Aliquots of the effluent from the OCEC were injected into the CE channel by a series of gated injections at regular intervals; about 9% of the total elution volume of the OCEC was analyzed by CE. When coupling two techniques in a *serial* manner, it is important to make a representative injection from the first to the second dimension. Hence, it is advantageous if the analysis in the second dimension is relatively fast, so a large portion (i.e., many sequential injections) of the effluent from the first dimension can be sampled. An example of a two-dimensional contour plot resulting from a separation is seen in Figure 8.24.

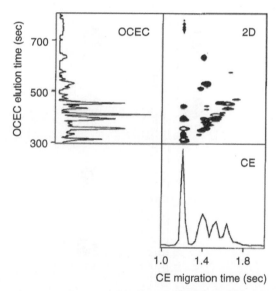

FIGURE 8.24 Two-dimensional separation of TRITC-labeled tryptic peptides of β-casein. The projections of the two-dimensional separation into the first dimension (OCEC) and second dimension (CE) are shown to the left and below the two-dimensional contour plot, respectively. The field strengths were 220 V/cm in the OCEC channel and 1890 V/cm in the CE channel. The buffer was 10 mM sodium borate with 30% (v/v) acetonitrile. An injection of the effluent from the OCEC channel into the CE channel was made every 3.2 sec, and the sample was detected 0.8 cm into the CE channel. (Reprinted from Gottschlich, N.; Jacobson, S. C.; Culbertson, C. T.; Ramsey, J. M. *Analytical Chemistry* **2001**, *73*, 2669–2674. With permission.)

The authors [7] discuss that OCEC and CE are not truly orthogonal, which is also not expected since CEC is actually a hybrid between CE and HPLC. This is confirmed as some of the peaks in the two-dimensional plot line up along a diagonal indicating correlation between the two separation mechanisms. The authors conclude that the information content increased by about 50% when going from one-dimensional OCEC to the two-dimensional OCEC or CE separation, and that increasing the sampling rate from the first to the second dimension as well as improving the phase ratio (e.g., by employing monolithic stationary phases) would improve the method.

In the group's most recent paper on MEKC and CE [157], a serpentine channel of 19.6 cm for MEKC and a 1.3 cm long straight CE separation channel were produced on a glass substrate (see Figure 8.25). The authors further investigated the so-called asymmetric turns in order to minimize geometric dispersion caused by turns in microchannels (racetrack effect). This approach proves advantageous in the serpentine MEKC channel.

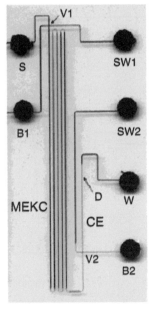

FIGURE 8.25 Image of a microchip with a serpentine channel for two-dimensional separations. Injections were made at valve 1 (V1) for the first-dimension MEKC separation and at valve 2 (V2) for the second dimension CE separation. The sample was detected 1 cm downstream from V2 at point D using LIF. The reservoirs are labeled sample (S), buffer 1 (B1), sample waste 1 (SW1), buffer 2 (B2), sample waste (SW2), and waste (W). The channels and reservoirs were filled with black ink for contrast. (Reprinted from Ramsey, J. D.; Jacobson, S. C.; Culbertson, C. T.; Ramsey, J. M. *Analytical Chemistry* **2003**, *75*, 3758–3764. With permission.)

Efforts were made to decrease the CE analysis time, including shortening the length of the injection plug, increasing the electric field strength in the CE channel, and overlapping analysis times to maximize sample throughput. Still, the authors follow again the principle of serial coupling. An example of the other alternative, namely parallel coupling, was already mentioned in the section on isoelectric focusing [111], where the focused bands from the first dimension (IEF) are simultaneously injected into the second dimension and analyzed in several parallel channels. This approach is obviously easier when the fist dimension is a focusing technique like IEF or ITP; since otherwise timing the moment of injection becomes quite crucial.

A couple of recent papers [158, 159] on two-dimensional separations still employed standard bench-top equipment for either the first or both dimensions. The on-chip approach using electrokinetic microliquid handling facilitates the combination of different techniques with minimal dead and sample volumes and no "parking" or collection of fractions. Furthermore, more-dimensional ($n > 2$) separations also become a real option, for example, when using mass spectrometry as the third discriminating dimension.

REFERENCES

1. Manz, A.; Graber, N.; Widmer, H. *Sensors and Actuators B — Chemical* **1990**, *1*, 244–248.
2. Harrison, D. J.; Manz, A.; Fan, Z. H.; Ludi, H.; Widmer, H. M. *Analytical Chemistry* **1992**, *64*, 1926–1932.
3. Manz, A.; Harrison, D. J.; Verpoorte, E. M. J.; Fettinger, J. C.; Paulus, A.; Ludi, H.; Widmer, H. M. *Journal of Chromatography* **1992**, *593*, 253–258.
4. Jacobson, S. C.; Hergenroder, R.; Koutny, L. B.; Warmack, R. J.; Ramsey, J. M. *Analytical Chemistry* **1994**, *66*, 1107.
5. Fintschenko, Y.; Kirby, B. J.; Hasselbrink, E. F.; Singh, A. K.; Shepodd, T. J. In *Monolithic Materials*; Svec, F., Tennikova, T. B., Deyl, Z., Eds.; Elsevier: New York, 2003; Vol. 67, pp 659–683.
6. Gottschlich, N.; Culbertson, C. T.; McKnight, T. E.; Jacobson, S. C.; Ramsey, J. M. *Journal of Chromatography B* **2000**, *745*, 243–249.
7. Gottschlich, N.; Jacobson, S. C.; Culbertson, C. T.; Ramsey, J. M. *Analytical Chemistry* **2001**, *73*, 2669–2674.
8. Liang, Z.; Chiem, N.; Ocvirk, G.; Tang, T.; Fluri, K.; Harrison, D. *Analytical Chemistry* **1996**, *68*, 1040–1046.
9. Lichtenberg, J.; Verpoorte, E.; de Rooij, N. F. *Electrophoresis* **2001**, *22*, 258.
10. Wallenborg, S. R.; Bailey, C. G. *Analytical Chemistry* **2000**, *72*, 1872–1878.
11. Khandurina, J.; Jacobson, S. C.; Waters, L. C.; Foote, R. S.; Ramsey, J. M. *Analytical Chemistry* **1999**, *71*, 1815.
12. Manz, A.; Miyahara, Y.; Miura, J.; Watanabe, Y.; Miyagi, H.; Sato, K. *Sensors and Actuators B — Chemical* **1990**, *1*, 249–255.
13. Ocvirk, G.; Verpoorte, E.; Manz, A.; Grasserbauer, M.; Widmer, H. M. *Analytical Methods and Instrumentation* **1993**, *2*, 74–82.

14. Seiler, K.; Harrison, D. J.; Manz, A. *Analytical Chemistry* **1993**, *65*, 1481–1488.
15. Chiem, N.; Harrison, D. J. *Analytical Chemistry* **1997**, *69*, 373–378.
16. Hadd, A. G.; Raymond, D. E.; Halliwell, J. W.; Jacobson, S. C.; Ramsey, J. M. *Analytical Chemistry* **1997**, *69*, 3407–3412.
17. Effenhauser, C. S.; Bruin, G. J. M.; Paulus, A.; Ehrat, M. *Analytical Chemistry* **1997**, *69*, 3451–3457.
18. Colyer, C. L.; Tang, T.; Chiem, N.; Harrison, D. J. *Electrophoresis* **1997**, *18*, 1733–1741.
19. Crabtree, H. J.; Kopp, M. U.; Manz, A. *Analytical Chemistry* **1999**, *71*, 2130–2138.
20. Jiang, G. F.; Attiya, S.; Ocvirk, G.; Lee, W. E.; Harrison, D. J. *Biosensors & Bioelectronics* **2000**, *14*, 861–869.
21. Schrum, K. F.; Lancaster, J. M.; Johnston, S. E.; Gilman, S. D. *Analytical Chemistry* **2000**, *72*, 4317–4321.
22. Hilmi, A.; Luong, J. H. T. *Environmental Science & Technology* **2000**, *34*, 3046–3050.
23. Tantra, R.; Manz, A. *Analytical Chemistry* **2000**, *72*, 2875–2878.
24. Gong, Q. L.; Zhou, Z. Y.; Yang, Y. H.; Wang, X. H. *Sensors and Actuators A — Physical* **2000**, *83*, 200–207.
25. Wang, J.; Tian, B. M.; Sahlin, E. *Analytical Chemistry* **1999**, *71*, 5436–5440.
26. Rossier, J. S.; Roberts, M. A.; Ferrigno, R.; Girault, H. H. *Analytical Chemistry* **1999**, *71*, 4294–4299.
27. Wang, J.; Jiang, M.; Mukherjee, B. *Analytical Chemistry* **1999**, *71*, 4095–4099.
28. Wang, J.; Tian, B. M.; Sahlin, E. *Analytical Chemistry* **1999**, *71*, 3901–3904.
29. Hashimoto, M.; Tsukagoshi, K.; Nakajima, R.; Kondo, K.; Arai, A. *Journal of Chromatography A* **2000**, *867*, 271–279.
30. Fiaccabrino, G. C.; deRooij, N. F.; KoudelkaHep, M. *Analytica Chimica Acta* **1998**, *359*, 263–267.
31. Weinberger, R. *Practical Capillary Electrophoresis*; Academic Press: New York, 1993.
32. Crego, A. L.; Gonzalez, A.; Marina, M. L. *Critical Reviews in Analytical Chemistry* **1996**, *26*, 261–304.
33. Jin, S.; Venkataraman, D.; DiSalvo, F. J.; Peters, E. C.; F. Svec; Frechet, J. M. J. *Abstracts of Papers of the American Chemical Society* **2000**, *219*, 479-POLY.
34. Khaledi, M. G. *Journal of Chromatography A* **1997**, *780*, 3–40.
35. Balchunas, A. T.; Sepaniak, M. J. *Analytical Chemistry* **1987**, *59*, 1466–1470.
36. Kutter, J. P.; Jacobson, S. C.; Ramsey, J. M. *Analytical Chemistry* **1997**, *69*, 5165–5171.
37. Kutter, J. P.; Jacobson, S. C.; Matsubara, N.; Ramsey, J. M. *Analytical Chemistry* **1998**, *70*, 3291.
38. Roberts, M.; Rossier, J.; Bercier, P.; Girault, H. *Analytical Chemistry* **1997**, *69*, 2035–2042.
39. Martynova, L.; Locascio, L.; Gaitan, M.; Kramer, G.; Christensen, R.; MacCrehan, W. *Analytical Chemistry* **1997**, *69*, 4783–4789.
40. Wang, S.; Perso, C.; Morris, M. *Analytical Chemistry* **2000**, *72*, 1704–1706.
41. Slentz, B.; Penner, N.; Regnier, F. *Journal of Chromatography A* **2003**, *984*, 97–107.

42. Uchiyama, K.; Xu, W.; Qiu, J.; Hobo, T. *Fresenius Journal of Analytical Chemistry* **2001**, *371*, 209–211.
43. Xu, W.; Uchiyama, K.; Shimosaka, T.; Hobo, T. *Journal of Chromatography A* **2001**, *907*, 279–289.
44. Li, S. F. Y. *Capillary Electrophoresis — Principles, Practice and Applications*; Elsevier: New York, 1993.
45. Galceran, M. T.; Puignou, L.; Diez, M. *Journal of Chromatography A* **1996**, *732*, 167.
46. Quang, C. Y.; Khaledi, M. G. *Analytical Chemistry* **1993**, *65*, 3354.
47. Dworschak, A.; Pyell, U. *Journal of Chromatography A* **1999**, *855*, 669.
48. Hjertén, S. PhD Thesis, Uppsala, 1967.
49. Waters, L. C.; Jacobson, S. C.; Kroutchinina, N.; Khandurina, J.; Foote, R. S.; Ramsey, J. D. *Analytical Chemistry* **1998**, *70*, 5172.
50. Waters, L. C.; Jacobson, S. C.; Kroutchinina, N.; Khandurina, J.; Foote, R. S.; Ramsey, J. D. *Analytical Chemistry* **1998**, *70*, 158.
51. Bianchi, F.; Wagner, F.; Hoffinann, P.; Girault, H. H. *Analytical Chemistry* **2001**, *73*, 829–836.
52. Jiang, W.; Awasum, J. N.; Irgum, K. *Analytical Chemistry* **2003**, *75*, 2768.
53. König, S.; Welsch, T. *Journal of Chromatography A* **2000**, *894*, 79.
54. Soper, S. A.; Henry, A. C.; Vaidya, M.; Galloway, M.; Wabuyele, M.; McCarley, R. L. *Analytica Chimica Acta* **2002**, *470*, 87–99.
55. Vaidya, B.; Soper, S. A.; McCarley, R. L. *Analyst* **2002**, *127*, 1289.
56. Liu, Y.; Fanguy, J. C.; Bledsoe, J. M.; Henry, C. S. *Analytical Chemistry* **2000**, *72*, 5939.
57. Kirby, B. J.; Wheeler, A. R.; Zare, R. N.; Fruetel, J. A.; Shepodd, T. J. *Lab on a Chip* **2003**, *3*, 5.
58. Doherty, E. A. S.; Meagher, R. J.; Albarghouthi, M. N.; Barron, A. E. *Electrophoresis* **2003**, *24*, 34.
59. Belder, D.; Ludwig, M. *Electrophoresis* **2003**, *24*, 3595–3606.
60. Culbertson, C. T.; Jorgenson, J. W. *Journal of Microcolumn Separations* **1999**, *11*, 167.
61. Chen, Y.; Zhu, Y. *Electrophoresis* **1999**, *20*, 1817.
62. Hayes, M. A.; Ewing, A. G. *Analytical Chemistry* **1992**, *64*, 512.
63. Schasfoort, R. B. M.; Schlautmann, S.; Hendrikse, J.; van den Berg, A. *Science* **1999**, *286*, 942.
64. Hayes, M. A. *Analytical Chemistry* **1999**, *71*, 3793.
65. Johnson, T. J.; Ross, D.; Gaitan, M.; Locascio, L. E. *Analytical Chemistry* **2001**, *73*, 3656.
66. Bruin, G. J. M. *Electrophoresis* **2000**, *21*, 3931–3951.
67. Dolnik, V.; Liu, S.; Jovanovich, S. *Electrophoresis* **2000**, *21*, 41–54.
68. Kutter, J. P. *Trends in Analytical Chemistry* **2000**, *19*, 352–363.
69. Reyes, D. R.; Iossifidis, D.; Auroux, P. A.; Manz, A. *Analytical Chemistry* **2002**, *74*, 2623–2636.
70. Auroux, P.-A.; Iossifidis, D.; Reyes, D. R.; Manz, A. *Analytical Chemistry* **2002**, *74*, 2637–2652.
71. Erickson, D.; Li, D. *Analytica Chimica Acta* **2004**, *507*, 11–26.
72. Guzman, N. A. *Capillary Electrophoresis Technology*; Marcel Dekker: New York, 1993.

73. Heiger, D. *High Performance Capillary Electrophoresis — An Introduction*; Agilent Technologies, Palo Alto 2000.
74. Rathore, A. S. *Electrokinetic Phenomena: Principles and Applications in Analytical Chemistry and Microchip Technology*; Marcel Dekker: New York, 2003.
75. Westermeier, R. *Electrophoresis in Practice*; Wiley-VCH: New York, 2001.
76. Huang, X.; Gordon, M. J.; Zare, R. N. *Analytical Chemistry* **1988**, *60*, 1837–1838.
77. Petersen, N. J.; Nikolajsen, R. P. H.; Mogensen, K. B.; Kutter, J. P. *Electrophoresis* **2004**, *25*, 253.
78. Kikura-Hanajiri, R.; Martin, R. S.; Lunte, S. M. *Analytical Chemistry* **2002**, *74*, 6370–6377.
79. Culbertson, C. T.; Jacobson, S. C.; Ramsey, J. M. *Analytical Chemistry* **1998**, *70*, 3781.
80. Emrich, C. A.; Tian, H.; Medintz, I. L.; Mathies, R. A. *Analytical Chemistry* **2002**, *74*, 5076.
81. Jacobson, S. C.; Culbertson, C. T.; Daler, J. E.; Ramsey, J. M. *Analytical Chemistry* **1998**, *70*, 3476–3480.
82. Burggraf, N.; Manz, A.; Verpoorte, E.; Effenhauser, C. S.; Widmer, H. M.; de Rooij, N. F. *Sensors and Actuators B* **1994**, *20*, 103–110.
83. Bilitewski, U.; Genrich, M.; Kadow, S.; Mersal, G. *Analytical and Bioanalytical Chemistry* **2003**, *377*, 556.
84. Dang, F.; Zhang, L.; Hagiware, H.; Mishina, Y.; Baba, Y. *Electrophoresis* **2003**, *24*, 714–721.
85. Fiorini, G. S.; Jeffries, G. D. M.; Lim, D. S. W.; Kuyper, C. L.; Chiu, D. T. *Lab on a Chip* **2003**, *3*, 158.
86. Zhao, D. S.; Roy, B.; McCormick, M. T.; Kuhr, W. G.; Brazil, S. A. *Lab on a Chip* **2003**, *3*, 93–99.
87. Abbasi, F.; Mirzadeh, H.; Katbab, A. A. *Polymer International* **2001**, *50*, 1279–1287.
88. Makamba, H.; Kim, J. H.; Lim, K.; Park, N.; Hahn, J. H. *Electrophoresis* **2003**, *24*, 3607–3619.
89. Wang, J.; Pumera, M.; Chatrathi, M. P.; Escarpa, A.; Konrad, R.; Griebel, A.; Dorner, W.; Lowe, H. *Electrophoresis* **2002**, *23*, 596–601.
90. Skelley, A. M.; Mathies, R. A. *Journal of Chromatography A* **2003**, *1021*, 191–199.
91. Fister, J. C.; Jacobson, S. C.; Ramsey, J. M. *Analytical Chemistry* **1999**, *71*, 4460.
92. Yang, R.-J.; Fu, L.-M.; Lee, G.-B. *Journal of Separation Science* **2002**, *25*, 996–1010.
93. Fu, L. M.; Yang, R. J.; Lee, G. B. *Analytical Chemistry* **2003**, *75*, 1905.
94. Zhang, C.-X.; Manz, A. *Analytical Chemistry* **2001**, *73*, 2656.
95. Yang, H.; Chien, R. L. *Journal of Chromatography A* **2001**, *924*, 155.
96. Quirino, J. P.; Shigeru, T. *Electrophoresis* **2000**, *21*, 355.
97. Beard, N. P.; Zhang, C.-X.; de Mello, A. J. *Electrophoresis* **2003**, *24*, 732.
98. Osbourn, D. M.; Weiss, D. J.; Lunte, C. E. *Electrophoresis* **2000**, *21*, 2768–2779.
99. Gebauer, P.; Bocek, P. *Electrophoresis* **2002**, *23*, 3858–3864.
100. Kaniansky, D.; Masar, M.; Bielcikova, J.; Ivanyi, F.; Eisenbeiss, F.; Stanislawski, B.; Grass, B.; Neyer, A.; Johnck, M. *Analytical Chemistry* **2000**, *72*, 3596–3604.

101. Bodor, R.; Zuborova, M.; Olvecka, E.; Madajova, V.; Masar, M.; Kaniansky, D.; Stanislawski, B. *Journal of Separation Science* **2001**, *24*, 802–809.
102. Masar, M.; Zuborova, M.; Bielcikova, J.; Kaniansky, D.; Johnck, M.; Stanislawski, B. *Journal of Chromatography A* **2001**, *916*, 101–111.
103. Ölvecka, E.; Konikova, M.; Grobuschek, N.; Kaniansky, D.; Stanislawski, B. *Journal of Separation Science* **2003**, *26*, 693–700.
104. Grass, B.; Neyer, A.; Johnck, M.; Siepe, D.; Eisenbeiss, F.; Weber, G.; Hergenroder, R. *Sensors and Actuators B* **2001**, *72*, 249–258.
105. Xu, Z. Q.; Hirokawa, T.; Nishine, T.; Arai, A. *Journal of Chromatography A* **2003**, *990*, 53–61.
106. Hirokawa, T.; Okamoto, H.; Gas, B. *Electrophoresis* **2003**, *24*, 498–504.
107. Prest, J. E.; Baldock, S. J.; Fielden, P. R.; Brown, B. J. T. *Analyst* **2001**, *126*, 433–437.
108. Kilár, F. *Electrophoresis* **2003**, *24*, 3908–3916.
109. Wu, X.-Z.; Sze, N. S.-K.; Pawliszyn, J. *Electrophoresis* **2001**, *22*, 3968–3971.
110. Herr, A. E.; Molho, J. I.; Drouvalakis, K. A.; Mikkelsen, J. C.; Utz, P. J.; Santiago, J. G.; Kenny, T. W. *Analytical Chemistry* **2003**, *75*, 1180–1187.
111. Li, Y.; Buch, J. S.; Rosenberger, F.; DeVoe, D. L.; Lee, C. S. *Analytical Chemistry* **2004**, *76*, 742–748.
112. Heegaard, N. H. H.; Nilsson, S.; Guzman, N. A. *Journal of Chromatography B* **1998**, *715*, 29–54.
113. Heegaard, N. H. H.; Kennedy, R. T. *Electrophoresis* **1999**, *20*, 3122–3133.
114. Heegaard, N. H. H. *Electrophoresis* **2003**, *24*, 3879–3891.
115. Guijt, R. M.; Frank, J.; van Dedem, G. W. K.; Baltussen, E. *Electrophoresis* **2000**, *21*, 3905–3918.
116. Guijt, R. M.; Baltussen, E.; van Dedem, G. W. K. *Electrophoresis* **2002**, *23*, 823–835.
117. Cheng, S. B.; Skinner, C. D.; Taylor, J.; Attiya, S.; Lee, W. E.; Picelli, G.; Harrison, D. J. *Analytical Chemistry* **2001**, *73*, 1472–1479.
118. Koutny, L. B.; Schmalzing, D.; Taylor, T. A.; Fuchs, M. *Analytical Chemistry* **1996**, *68*, 18–22.
119. Lavignac, N.; Allender, C. J.; Brain, K. R. *Analytica Chimica Acta* **2004**, *510*, 139.
120. Schweitz, L.; Spegel, P.; Nilsson, S. *Electrophoresis* **2001**, *22*, 4053–4063.
121. Schweitz, L.; Andersson, L. I.; Nilsson, S. *Journal of Chromatography A* **1998**, *817*, 5–13.
122. Brahmasandra, S. N.; Ugaz, V. M.; Burke, D. T.; Mastrangelo, C. H.; Burns, M. A. *Electrophoresis* **2001**, *22*, 300–311.
123. Griebel, A.; Rund, S.; Schonfeld, F.; Dorner, W.; Konrad, R.; Hardt, S. *Lab on a Chip* **2004**, *4*, 18–23.
124. Lee, B.-S.; Gupta, S.; Morozova, I. *Analytical Biochemistry* **2003**, *317*, 271–275.
125. Terabe, S.; Otsuka, K.; Ichikawa, K.; Tsuchiya, A.; Ando, T. *Analytical chemistry* **1984**, *56*, 111.
126. Terabe, S.; Otsuka, K.; Ando, T. *Analytical Chemistry* **1985**, *57*, 834–841.
127. Pyell, U. *Fresenius Journal of Analytical Chemistry* **2001**, *371*, 691–703.
128. Moore, A. W.; Jacobson, S. C.; Ramsey, J. M. *Analytical Chemistry* **1995**, *67*, 4184–4189.

129. Culbertson, C. T.; Jacobson, S. C.; Ramsey, J. M. *Analytical Chemistry* **2000**, *72*, 5814–5819.
130. Bailey, C.; Wallenborg, S. *Electrophoresis* **2000**, *21*, 3081–3087.
131. Wallenborg, S. R.; Lurie, I. S.; Arnold, D. W.; Bailey, C. G. *Electrophoresis* **2000**, *21*, 3257–3263.
132. Rodriguez, I.; Jin, L. J.; Li, S. F. Y. *Electrophoresis* **2000**, *21*, 211–219.
133. von Heeren, F.; Verpoorte, E.; Manz, A.; Thormann, W. *Analytical Chemistry* **1996**, *68*, 2044–2053.
134. Sera, Y.; Matsubara, N.; Otsuka, K.; Terabe, S. *Electrophoresis* **2001**, *22*, 3509–3513.
135. Svec, F.; Peters, E. C.; Sykora, D.; Frechet, J. M. J. *Journal of Chromatography A* **2000**, *887*, 3.
136. Svec, F.; Peters, E. C.; Sykora, D.; Yu, G.; Frechet, J. M. J. *HRC Journal of High Resolution Chromatography* **2000**, *23*, 3–18.
137. Tanaka, N.; Kobayashi, H.; Ishizuka, N.; Minakuchi, H.; Nakanishi, K.; Hosoya, K.; Ikegami, T. *Journal of Chromatography A* **2002**, *965*, 35–49.
138. Svec, F. T., Tennikova, T.B.; Deyl, Z. *Monolithic Materials*; Elsevier: New York, 2003.
139. Jacobson, S. C.; Hergenroder, R.; Koutny, L. B.; Ramsey, J. M. *Analytical Chemistry* **1994**, *66*, 2369–2373.
140. Ceriotti, L.; de Rooij, N. F.; Verpoorte, E. *Analytical Chemistry* **2002**, *74*, 639–647.
141. Jemere, A. B., Oleschuk, R. D., Harrison, D. J. *Electrophoresis* **2003**, *24*, 3018–3025.
142. Oleschuk, R.; Shultz-Lockyear, L.; Ning, Y.; Harrison, D. *Analytical Chemistry* **2000**, *72*, 585–590.
143. Chirica, G.; Remcho, V. T. *Electrophoresis* **1999**, *20*, 50–56.
144. He, B.; Tait, N.; Regnier, F. E. *Analytical Chemistry* **1998**, *70*, 3790–3797.
145. He, B.; Regnier, F. *Journal of Pharmaceutical and Biomedical Analysis* **1998**, *17*, 925–932.
146. Slentz, B. E.; Penner, N. A.; Lugowska, E.; Regnier, F. *Electrophoresis* **2001**, *22*, 3736–3743.
147. Ericson, C.; Holm, J.; Ericson, T.; Hjerten, S. *Analytical Chemistry* **2000**, *72*, 81.
148. Koide, T.; Ueno, K. *HRC Journal of High Resolution Chromatography* **2000**, *23*, 59–66.
149. Peters, E. C.; Petro, M.; Svec, F.; Frechet, J. M. J. *Analytical Chemistry* **1997**, *69*, 3646–3649.
150. Ngola, S. M.; Fintschenko, Y.; Choi, W.-Y.; Shepodd T. J. *Analytical Chemistry* **2001**, *73*, 849–856.
151. Fintschenko, Y.; Choi, W.-Y.; Ngola, S. M.; Shepodd, T. J. *Fresenius Journal of Analytical Chemistry* **2001**, *371*, 174–181.
152. Shediac, R.; Ngola, S. M.; Throckmorton, D. J.; Anex, D. S.; Shepodd, T. J.; Singh, A. K. *Journal of Chromatography A* **2001**, *925*, 251–263.
153. Singh, A. K.; Throckmorton, D.J.; Brennan, J.S.; Shepodd, T.J.; Squaw Valley, Gradient-elution reversed-phase electrochromatography in microchips, in *Proceedings of μ-TAS 2003 Seventh International Conference on Micro Total Analysis Systems*, Volume 2; Northrup, M. A.; Jensen, K. F.; Harrison, D. J., Eds.; Transducers Research Foundation, Inc., Cleveland Heights, OH, 2003, pp. 1163-1166, USA, 2004.

154. Giddings, J. C. *Unified Separation Science*; John Wiley and Sons: New York, 1991.
155. Meyer, V. R.; Welsch, T. *LC–GC International* **1996**, *9*, 670.
156. Rocklin, R. D.; Ramsey, R. S.; Ramsey, J. M. *Analytical Chemistry* **2000**, *72*, 5244–5249.
157. Ramsey, J. D.; Jacobson, S. C.; Culbertson, C. T.; Ramsey, J. M. *Analytical Chemistry* **2003**, *75*, 3758–3764.
158. Yang, C.; Zhang, L.; Liu, H.; Zhang, W.; Zhang, Y. *Journal of Chromatography A* **2003**, *1018*, 97–103.
159. Yang, X.; Zhang, X.; Li, A.; Zhu, S.; Huang, Y. *Electrophoresis* **2003**, *24*, 1451–1457.

9 Gas Chromatography on Microchips

Richard J. Kottenstette, Chung-Nin Channy Wong, and Curtis D. Mowry

CONTENTS

9.1 INTRODUCTION

The analysis of gases and vapors is of major importance in manufacturing, environmental monitoring, and public safety. The introduction of integrated circuit technology and the more recent micromachining technology has pushed gas-phase chemical sensing to smaller and more elaborate analysis tools. This chapter provides a review of developments in microanalytical separation systems that operate in the gas phase. Specifically, we will review the modern history of gas sensing and examine the evolution from discrete sensors to sensor array systems to microanalytical systems with miniaturized sampling, separation, and detection components.

9.1.1 GAS SENSING

A number of technologies have been used to detect gases and vapors. These consist of infrared (IR) analyzers, mass spectrometers, and gas chromatographs (GCs), which can use a wide variety of detectors such as flame ionization detectors (FIDs) and thermal conductivity detectors (TCDs). Most analytical systems in use today are part of bench-top instruments that have gradually been reduced in size with great improvements in reliability and quality. Portable miniaturized total analysis systems are recent and have been preceded by gas analysis using dedicated microsensors. Several review articles survey progress in microsensors.[1-3] There have also been several books describing the theory and operation of sensors for chemical detection.[4-7] A significant subset of these sensors is used for gas-phase detection. Pertinent reviews of microhotplates, metal oxide sensors, microcantilevers, surface acoustic wave (SAW) arrays and flexural plate wave (FPW) devices and, quartz crystal microbalances (QCMs) have also been published.[8-11] Many gas sensors are currently in commercial production and their fabrication has led the way to further miniaturization.[12] Current commercial sensors are used for specific industrial hygiene and safety applications. These sensors are used to detect volatile organic compounds (photoionization, flame ionization, and IR sensors), toxic chemicals (metal oxide and electrochemical sensors), oxygen (IR and electrochemical sensors), and, for safety purposes, the LEL or lower explosive limit (catalytic bead sensors). In general, these portable and miniaturized gas sensors can be categorized as shown in Table 9.1.

TABLE 9.1
Portable Gas Sensor Technologies

Sensor	Types	Mechanism
Electrochemical	Amperometric or potentiometric	Reduction or oxidation on electrode
Chemiresistors	Conductive polymer or emulsion	Impedance change
Metal oxide solid-state sensors (MOSS)[a]	Semiconductor bead, microhotplate	Impedance change
Calorimetric sensors	Catalytic bead	Catalytic combustion
Mass sensors	QCM, SAW, FPW, cantilever	Mass sorption, resonance change
Infra-red	Waveguide, fiber	Light absorption
Photoionization	Ultraviolet lamp	Ionization
CMOS sensors[a]	CHEMFET	Modified work function

[a]Technically CMOS sensors, MOSS sensors, and even chemiresistors can be classified as "electrochemical" sensors since they use the chemical interaction to modulate the charge flow or the potential change. Janata lumps these together.[5]

9.1.2 SENSOR ARRAYS AND ELECTRONIC NOSES

Although discrete sensors are currently the most important commercial gas detectors in use today, they suffer from cross-sensitivity and aging effects that affect the identification and quantitation of gaseous chemicals. Arrays of sensors operating together can sometimes address the nonspecificity of discrete sensors. "Electronic noses" are arrays of gas sensors with different selectivity patterns, a signal-collecting unit, and pattern recognition software. Electronic noses were first introduced in the 1980s at the University of Warwick, Coventry, England.[13] Electronic noses can be based on mass sensors such as SAWs, FPWs, and microcantilevers or electrochemical sensors like chemiresistors or metal oxide devices.

All electronic noses (and indeed most array-based detectors) rely on pattern recognition algorithms such as principle component analysis, partial least squares, or artificial neural networks to detect different smells. A great deal of research has been undertaken to improve the quality of the data produced by array-based systems. In order for the chemometric technique to work, the pattern recognition system must be "trained" using known quantities of individual chemicals or chemical mixtures that comprise odors.

Applications for which electronic noses have been used commercially are diverse and include detection of solvents (pure compounds and mixtures); detection of gases in confined spaces; purity and quality of foods such as coffee, fruit, and fish; grading and typing of beverages such as beer, wine, and whiskey; perfume analysis; and medical diagnostic analyses. In most cases of aroma analysis, specific chemicals are not identified, but specific odors (possibly consisting of a number of compounds) are identified. One example of this typing is the classification of grain into musty, acid, and rancid categories.[14] Other products are evaluated in the same aggregate way in what is referred to as an organoleptic classification scheme.

Janata et al.[1] have used the term "higher-order chemical sensing arrays" to refer to arrays of chemical sensors in which more than one transduction principle is employed; for example, work function modulation and impedance and mass modulation. Such arrays have been seen as sensor analogs of hyphenated analytical techniques, and are direct predecessors and competitors with microanalytical systems depending on the viewpoint.

9.1.3 Gas Sensing with Microanalytical Systems

Chemical vapors and gas samples are always mixtures. Air itself is a mixture and trace gases and pollutants in an air matrix always constitute a multicomponent sample. For this reason analysts use gas separation schemes such as selective preconcentration using adsorbent media and gas chromatography to separate components in a sample. Quantitative or qualitative analysis of a wide variety of important mixtures involves three basic operations: sampling, separation, and detection. Sensor arrays can be integrated with various other analytical devices like preconcentrators and microseparation columns to develop a true microanalytical system that uses different components in series or parallel to achieve specificity and sensitivity.

9.2 GAS CHROMATOGRAPHY — HISTORY AND BACKGROUND

Separation of gaseous and vapor mixtures using gas chromatography is the central feature of a gas-phase microanalytical system. Gas chromatography is a physical method of separation in which components of a mixture are distributed between a stationary phase and a mobile gas phase. The mobile phase is referred to as a carrier gas and sweeps through a column containing the stationary phase. Figure 9.1 shows a schematic representation of a gas chromatographic system. It consists of a carrier gas delivery system, a sample injector, a gas chromatographic column, a column oven or heater, a detector, and a data acquisition system. The sample is introduced into the instrument as either a gas or liquid "finite slug" through the injection port or by means of a sampling valve. The sample is carried into the column by the carrier gas (mobile phase) and is separated into bands of molecules by interacting with the stationary phase in the column. The separated gas stream then enters a detector for measurement. Gas chromatography can roughly be divided into two basic types: gas solid chromatography (GSC) and gas liquid chromatography (GLC).

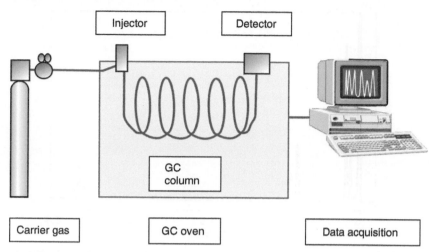

FIGURE 9.1 Schematic diagram of bench-top gas chromatograph (not to scale).

In GSC or adsorption chromatography, the stationary phase is a solid and the sample adsorbs differentially on the surface of the stationary phase to separate analyte molecules. Cremer first conceived adsorption chromatography in the gas phase in 1944, but events during the Second World War prevented timely publication of the work. Typically, GSC separates low-molecular-weight high-volatility compounds such as natural gas components and air. Although most GSC separations have been performed on columns packed with solid media, they can also be performed on long capillary columns referred to as porous-layer open-tubular (PLOT) columns that have a thin (typically 10 to 30 μm) layer of solid adsorbent on the inside of a fused silica capillary column. Figure 9.2 shows a chromatogram from a PLOT column illustrating a GSC separation of light hydrocarbons.

GLC (referred to as partition chromatography) separates sample molecules using a high-molecular-weight liquid stationary phase that is coated onto a solid support or bonded directly to a fused silica capillary tube. Sample molecules absorb into the polymer and are thus partitioned as they traverse the length of the column. Partition chromatography like early adsorption chromatography was performed on packed bed columns that were typically 0.125″ to 0.25″ in diameter. GLC was first proposed in a paper by James and Martin in 1952[15] and the implementation of this new technique led to the 1952 Nobel Prize in Chemistry for "the invention of partition chromatography." These early papers led to a groundbreaking paper by Golay[16] in 1958 describing separation of gas-phase analytes using capillary columns made from metals. Desty et al. introduced a method of extruding glass capillary columns in 1960[17] and subsequent research led to the more inert and sturdy fused silica columns that became commercially available in 1979.

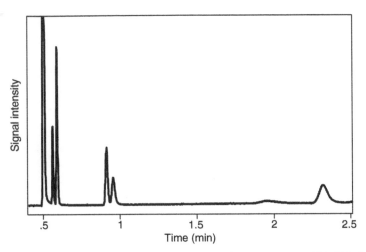

FIGURE 9.2 Separation of light hydrocarbons methane to pentane (C1–C5) demonstrating adsorption chromatography (with PLOT column).

These columns are often referred to as wall-coated open-tubular (WCOT) columns. Today, over 90% of gas chromatographic separations are performed on fused silica columns with a broad range of important applications. Petroleum refining, pharmaceutical and food production, as well as forensic and environmental analyses are heavily dependent on gas-phase chromatographic separations. This powerful technique has led to countless improvements in process control and manufacturing as well as improved methods for forensic and environmental science. GLC separations today are usually performed on fused silica capillary columns that are coated with a high-molecular-weight polymer phase (typical thickness 0.1 to 3 μm). Figure 9.3 shows a complex chromatogram that results from a GLC separation of gasoline vapor and Figure 9.4 shows cross sections of a packed column, a WCOT column, and a PLOT column.

9.2.1 History of Microanalytical Gas Chromatographic Systems

A recent review[18] describes the advent of microanalytical system design and fabrication. These first attempts produced a gas chromatographic column etched in a silicon wafer.[19] This column and associated developments led to liquid chromatographic and electrophoretic microfabricated systems that comprise the bulk of modern day microanalytical systems. The primary difference between gas-phase microanalytical systems and liquid-phase microanalytical systems is that the gas-phase miniaturization is aimed at producing portable instrumentation, while a major thrust of the liquid analysis systems is to produce high-throughput biochemical testing based on parallel processing

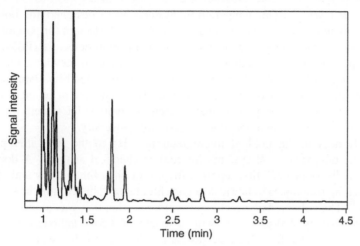

FIGURE 9.3 Separation of gasoline components propane to decane (C3–C10) demonstrating partition chromatography with a wall-coated capillary column.

and analysis. For this reason, the micro-GC system addresses issues such as temperature programming, speed of analysis, and miniature power supplies including battery capacity. These issues were reviewed by Overton et al. in 1995 and 1996.[20,21] Portable gas chromatographic systems were first based on packed columns and often performed at room temperature. Later portable systems were based on capillary megabore columns and could be performed

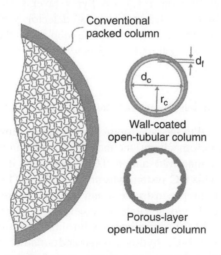

FIGURE 9.4 Scale drawings of a packed column (with packing), a wall-coated column showing film thickness, d_f (~0.1 to 3 μm), and column radius and diameter, r_c and d_c, and a porous-layer open-tubular column (10 to 30 μm phase).

at elevated yet constant temperature. Systems were designed for portable environmental research and even space missions.[22] Detectors for portable gas chromatographs have included TCDs, photoionization detectors (PIDs), and mass spectrometers, although a few of these systems are somewhat large to be considered easily portable. Prototype systems have included FID, flame photometric, and SAW sensors as the detector. The first system to incorporate microfabrication techniques to manufacture a commercial system was the MTI GC,[23–25] based upon the work at Stanford University.

Advances in the field of microelectronics led to the first attempts to fabricate microanalytical systems for routine chemical analysis. A doctoral thesis by Terry in 1975 first reported the use of photolithography and chemical etching techniques to fabricate a gas chromatographic column in a silicon wafer.[19] The column had a rectangular cross section that was approximately 30-μm deep by 200-μm wide with a length of 1.5 m, although 3-m-long columns were also fabricated. A gas sample loop and valve seats were also etched into the wafer to facilitate gas handling. A channel is formed with the column by sealing a Pyrex cover plate to the surface of the silicon substrate. This is achieved by applying heat and DC voltage (approximately 600 V at 400°C) to the silicon and Pyrex to form an anodic bond.[26] The column was first sealed, and then coated since the anodic bonding requires contamination-free conditions to produce a hermetic seal between adjacent channels. The column was then lined with an organosilane adhesion layer and a polydimethylsiloxane (PDMS: OV-101) or polyethylene glycol (Carbowax) stationary phase. A separate micromachined TCD was fabricated and subsequently bonded to the wafer at the exit of the separation column. The TCD consisted of a 100-nm nickel film on a 1.5-μm Pyrex layer suspended in a cavity between the silicon GC substrate and the detector substrate. Figure 9.5 shows the column and the associated sample loop and etched interconnecting channels. This work was featured in a *Scientific American* article in 1983[27] and led to a patent in 1984.[28]

9.2.2 Commercial Instrumentation Based on Micromachining

The work by Terry led to the development of the first commercial GC system based, in part, on micromachined technology. This technology uses a microfabricated injection manifold and a TCD; however, it uses conventional capillary columns. This GC system was produced by MTI and several articles were written describing the instrument and applications for its use.[23–25,29–31] The system provided two separate channels using narrow-bore capillary columns (WCOT or PLOT). Applications for this system include separations of permanent gases, C1–C5 hydrocarbons, and volatile organic compounds such as chlorinated solvents. A separate manufacturer (Chrompak International) produced a nearly identical system based on the same principles of design and fabrication. Eventually, Agilent would produce the MTI GC

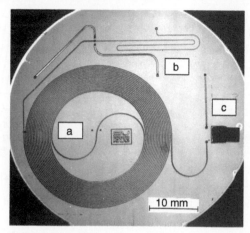

FIGURE 9.5 Photograph showing the very first microfabricated gas chromatographic system on a silicon wafer; includes column, sample loop, and detector well: (a) column; (b) sample loop; (c) detector well. (From De Mello, A. *Lab Chip*, 2002, *2*, 48N–54N. With permission.)

system (marketed as the 3000 MicroGC) and the Chrompak system would be produced by Varian (marketed as the CP2003 or CP4900 Micro-GC).

9.3   PROTOTYPE MICROANALYTICAL GC SYSTEMS

Subsequent to the development of commercial micro-GC systems, ongoing research on prototype micro-GCs has continued by using newly developed tools for micromachining such as high aspect ratio silicon etching and improved materials for fabrication of true microsystems. As mentioned, the three basic operations necessary to perform gas chromatography are sampling, separation, and detection. These unit operations are discussed in the following sections.

9.4   SAMPLE HANDLING AND INJECTION FOR MICRO-GC

9.4.1   Sample Valves and Loops

Sample preparation for gas-phase microanalytical systems usually involves drawing a vapor or gas sample into a sampling loop or across a solid-phase adsorbent. The first micro-GC system used a micromachined sample loop and a valve seat. The miniature valve seat, shown in Figure 9.6, was etched in silicon and uses a nickel diaphragm that is actuated by an external solenoid plunger to open and close the valve.[32] This injector, using only one valve, works by drawing in a sample through a check valve using a servo-driven

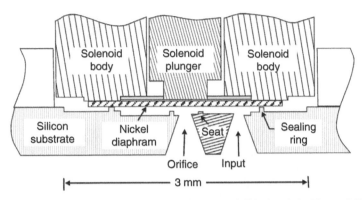

FIGURE 9.6 Valve seat diagram for microfabricatred GC shown in Figure 9.5.

piston pump, closing the valve, reversing the pump direction to compress the sample, and then opening the valve to inject into the column. The injection volume was a few nanoliters (at carrier gas pressure) in size and the injection time is quite rapid (approximately 30 msec). When MTI and Chrompack produced the micromachined system commercially, the etched silicon column was replaced with more traditional capillary columns. The sample is brought into the microfabricated sample loop with a sampling pump and then pressurized with carrier gas just before sample injection. In this system, the method of injection involved two silicon seat valves with 4 and 11 nl dead volumes, respectively (the sample injection volume range was 25 to 250 nl).[25] The updated valve used similar concentric ridges etched in silicon and a nickel–Teflon diaphragm. This basic valve scheme has been used and updated over the years to produce an injector system with two selectable sample loops (Figure 9.7). A great advantage is realized by using microfabrication combined with high-speed computer control of solenoid actuation. The reduction of dead volume and rapid injection allow for high-speed gas chromatography.

9.4.2 SOLID-PHASE MICROEXTRACTION

Another means of introducing samples into a microanalytical gas chromatographic system uses adsorbent beds to collect samples from an air sample and then introduce the sorbed chemicals into the GC by thermally desorbing them. Solid-phase microextraction (SPME) has been popularized by Pawliszyn et al.[33,34] and consists of coating a fused silica fiber (used as an injection needle) with an adsorbent material to "extract" and concentrate organic compounds from air and water samples. The needle with adsorbed sample is inserted into the hot injection port of a bench-top GC to effectively inject a "solventless" peak onto the column. This SPME principle can be used in microsystems by fabricating very low mass adsorbent beds and incorporating them in the sampling system. PLOT capillary columns have been used to

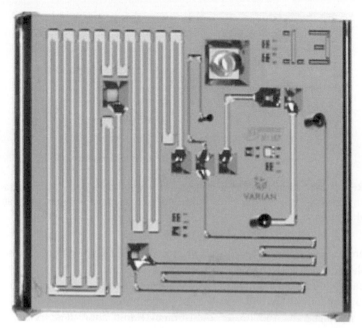

FIGURE 9.7 Micromachined sampling manifold including two sample loops for micro-GC application. (Courtesy of Varian Inc.)

extract sample from a flowing stream and thermally desorb it onto a SAW device for sensing.[35] This technique can be used on microfabricated substrates to achieve the same purpose. Frye-Mason et al.[36,37] built a membrane-based hotplate that can be coated with a variety of selective adsorbent films (sol–gel or carbon). An air sample is drawn across the membrane at room temperature, and organic vapors sorb into the film. Rapid heating of the membrane with metallic heaters injects the sample into a microfabricated column for analysis. Figure 9.8 shows an example of the hotplate setup.

9.4.3 OTHER TECHNOLOGIES

Another approach to preconcentration has been described by Bonne et al.[38,39] This technology, named PHASE, consists of an array of heatable gas adsorption–desorption microelements that are positioned so that sample gas components of interest can first adsorb and later be desorbed into the sample gas stream in a controlled or phased manner. Gas flows through a 100-μm wide channel over a continuous membrane heater. The sample is first equilibrated with the adsorbent film lining the entire concentrator and contiguous heaters are timed or phased to "pump" the preconcentrated pulse down the length of the channel. The preconcentrator has been designed with large numbers of heaters (20 to 100) to produce a concentration "wave" of the components

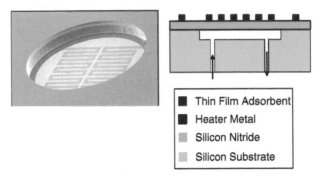

FIGURE 9.8 Membrane hotplate for vapor sampling by solid-phase microextraction; diagram shows adsorbent film, silicon nitride membrane, and meandering heater for thermal desorption.

of interest as the flow carries the gas volume to the end of the concentrator. As this plug exits the concentrator it enters a separator (GC column) that is manufactured in the same fashion as the concentrator (meandering membrane heater). Separation is realized as this concentrated plug interacts with the stationary phase on the heater membrane. Rapid short desorption pulses and high flow velocities relative to typical gas diffusion "velocities" enable separation in a column length between 20 and 50 cm. Separated components are detected with an on-chip differential TCD.

9.5 MICROFABRICATED GC COLUMNS

The primary separation component of typical gas-phase microanalytical systems is the gas chromatographic column. The GC separation column is to gas-phase microanalytical systems as electrophoretic separation columns are to liquid-phase microanalytical systems. Prototype systems that use microfabricated columns have been developed over the years and are summarized in Table 9.2.

9.6 THEORY OF MICROFABRICATED GC COLUMNS

The theory of separation for capillary columns with round cross sections has been well developed and is documented in most introductory textbooks on the subject.[40,41] Several papers address the separation properties of rectangular cross section GC columns.[42–44] We will present theory that models the transport and separation process in GC columns for rectangular columns as well.

9.6.1 TRANSPORT MODEL

The governing equations modeling the transport in the GC column are the three laws of conservation: conservation of mass, momentum, and species

TABLE 9.2
Summary of Microfabricated GC Column Research

System	Geometry	Length	Phase-Deposition Technique	Application
Terry 1975–1989	Rectangular 30-μm deep, 200-μm wide	1.5–3 m	PDMS — solvent PEG — solvent	VOCs
Kolesar 1994–1998	Rectangular 10-μm deep, 100-μm wide	0.9 m	Cu phthalocyanine plasma	Ammonia, nitrogen dioxide
Lehmann 1999	Rectangular 27-μm deep, 70-μm wide	2 m	PDMS plasma	C1–C5
Yu 1998	Round 100 μm	5 m	PDMS — solvent	VOCs
Hannoe 1998	Rectangular 10-μm deep, 100-μm wide	2 m	Amino acid plasma	Methane
Frye-Mason 1998	Rectangular 400-μm deep, 100-μm wide	0.9 m	PDMS — sol–gel PEG — sol–gel	CWA, VOCs
Frye-Mason 1999	Rectangular 300 μm × 300 μm	20-cm packed	Hayesep A Carboxen	Natural gas, C3
Zellers 2002	Rectangular 150 μm × 260 μm	3 m	PDMS — solvent	VOCs

concentration. Since our interest is a very dilute gas mixture, we can approximate that the properties of gas mixtures behave similarly to the carrier gas.

The conservation of mass equation is

$$\frac{\partial \rho}{\partial t} + \frac{\partial}{\partial x}(\rho v_x) + \frac{\partial}{\partial y}(\rho v_y) + \frac{\partial}{\partial z}(\rho v_z) = 0$$

where ρ is the density of the gas mixture and v_x, v_y, and v_z are the three components of velocity.

The conservation of momentum equation in the x_i direction (where x_i can be x, y, or z direction) is

$$\rho \frac{\partial v_i}{\partial t} + (\rho v_x)\frac{\partial v_i}{\partial x} + (\rho v_y)\frac{\partial v_i}{\partial y} + (\rho v_z)\frac{\partial v_i}{\partial z} = -\frac{\partial P}{\partial x_i} + \mu\left(\frac{\partial^2 v_i}{\partial x^2} + \frac{\partial^2 v_i}{\partial y^2} + \frac{\partial^2 v_i}{\partial z^2}\right)$$

where μ is the viscosity of gas mixture and P is pressure.

The conservation of species concentration equation is

$$\frac{\partial n_g}{\partial t} + v_x \frac{\partial n_g}{\partial x} + v_y \frac{\partial n_g}{\partial y} + v_z \frac{\partial n_g}{\partial z} = D_g\left(\frac{\partial^2 n_g}{\partial x^2} + \frac{\partial^2 n_g}{\partial y^2} + \frac{\partial^2 n_g}{\partial z^2}\right)$$
$$- n_g\left(\frac{\partial v_x}{\partial x} + \frac{\partial v_y}{\partial y} + \frac{\partial v_z}{\partial z}\right)$$

where n_g is the number density for the analyte in the gas mixture (molecules/ml) and D_g is the coefficient of diffusion.

In the mass and momentum equations given above, the variation of density of gas in the axial direction of the GC column is included in this analysis. The density variation is being considered because most GC columns are long and narrow channels — for example, the GC column in the μChem-Lab® is 100-μm wide, 400-μm deep, and 86-cm long, thus the length-to-diameter ratio is approximately 9000.[45] Such a high length-to-diameter ratio leads to gas density and pressure values at the inlet that are significantly higher than the values at the outlet.

Even though the flow velocity is much smaller than the speed of sound, the gas compressibility effects can be very important. This phenomenon is also known as the "compressible creep flow." Since the channel is very small, the viscous effect becomes dominant. The pressure required to pump gas along the channel increases as the channel size decreases. For a rectangular GC column, which has the following dimension: width $= 2a$, height $= 2b$ (such that $a \geq b$), and length $= L$, and for a given pressure gradient in the flow stream, the mean flow velocity at any axial location in the micro-GC column can be expressed as

$$\bar{v}_z \cong \frac{a^2}{3}\left(-\frac{1}{\mu}\frac{dP}{dz}\right)\left[1 - \frac{192}{\pi^5}\left(\frac{a}{b}\right)\tanh\left(\frac{\pi b}{2a}\right)\right]$$

Because of the gas compressibility effect, the pressure gradient is not constant along the GC column:

$$-\frac{dP}{dz} = \frac{P_i^2 - P_o^2}{2L\sqrt{P_i^2 - (z/L)(P_i^2 - P_o^2)}}$$

where P_i and P_o are the inlet and outlet pressures, respectively.

This leads to a nonlinear axial pressure distribution in the column, as follows:

$$P(z) = \sqrt{P_i^2 - \frac{z}{L}\cdot(P_i^2 - P_o^2)}$$

In addition to the enormous changes of pressure and density between the inlet and outlet, the gaseous flow will accelerate near the outlet of a micro-GC column. This results in a higher flow velocity at the outlet ($v_{z,L}$)

$$\text{mass flow rate} = \rho(z)\cdot v_z(z)\cdot\text{area} = \rho_0\cdot\bar{v}_{z,L}\cdot\text{area} = \text{constant}$$

For a given pressure difference between the inlet and outlet, the mean axial velocity at the outlet can be found as

$$\bar{v}_{z,L} \cong \frac{a^2}{3}\frac{1}{\mu}\left(\frac{P_i^2 - P_o^2}{2LP_o}\right)\left[1 - \frac{192}{\pi^5}\left(\frac{a}{b}\right)\tanh\left(\frac{\pi b}{2a}\right)\right]$$

In gas chromatography, an important parameter to measure in the separation process is the holdup time. This parameter represents the time taken by the carrier gas to travel through the GC column. Furthermore, we will define a linear, overall lengthwise average axial velocity for separation analysis such that this linear velocity is equal to the GC column length (L) divided by the holdup time (t_{holdup}):

$$\langle v_z\rangle = L/t_{holdup} = L/\left(\int\frac{1}{\bar{v}_z}dz\right)$$

This linear velocity can be related to the mean axial velocity at the outlet as follows:

$$\langle v_z\rangle = f_2\bar{v}_{z,L}$$

where

$$f_2 = \frac{3P_o(P_i^2 - P_o^2)}{2(P_i^3 - P_o^3)}$$

If the column size decreases even further into the sub-micrometer scale, the effect of slip flow at the boundary needs to be considered. From the kinetic theory of gases, if the characteristic length of the physical domain (d) is small and compatible with the mean free path of the molecules (λ), the continuum hypothesis and no-slip boundary condition become invalid. At the standard atmospheric conditions, the mean free path of air is about 0.065 μm. The deviation of the state of the gas from continuum is measured by the Knudsen number (Kn), which is defined as $Kn = \lambda/d$. As the value of Knudsen number increases, rarefaction effects become more important and the slip velocity at the boundary is greater than zero. For those cases where slip flow exists at the wall, the average flow velocity will be higher than the predicted velocity with a no-slip boundary condition.

For example, if the column is 40-μm wide and 1-m long, the maximum Knudsen number will be about 0.004, which indicates the flow is a continuum flow. The difference in the flow rates calculated with slip and no-slip boundary conditions is very small (less than 2%). However, slip flow does become important if the column size is reduced to below 10 μm. In this situation, the difference in volumetric flow rate, with and without slip, can be as high as 12%.

9.6.2 Separation Model

The separation model involves predicting the analyte transport and physico-chemical interaction with the coated surface or stationary phase in the micro-GC column to determine the performance of chromatography. As the analyte travels through the micro-GC column, the analyte band will be broadened and diluted. Hence, the extent of this band spreading is an important measure of the separation efficiency. The governing equation to model the analyte transport is the conservation of species concentration equation presented in Section 9.6.1.

To simplify the analysis, a popular approach is to perform a coordinate transformation such that the origin of the new coordinate system will move with the center of mass of the analyte. Because there exist physicochemical interactions between the analyte and the stationary phase, the average velocity of the analyte traveling downstream will be the fraction $1/(1 + k)$ of the mean axial velocity of the carrier gas mixture. Hence, the equation for the coordinate transformation is as follows:

$$z_1 = z - \frac{\bar{v}_z}{1 + k}t$$

where k is the retention factor or partition coefficient that characterizes the retentive property of the stationary phase.

The new analyte transport equation, which predicts the band spreading in a rectangular column, is as follows:

$$\frac{\partial n_g}{\partial t} + \bar{v}_z \left[\frac{2a \cdot 2b \cdot K_z(x,y)}{K_v} - \frac{1}{1+k} \right] \frac{\partial n_g}{\partial z} = D_g \left(\frac{\partial^2 n_g}{\partial x^2} + \frac{\partial^2 n_g}{\partial y^2} + \frac{\partial^2 n_g}{\partial z^2} \right)$$

where D_g is the diffusion coefficient of the analyte in the carrier gas, $K_z(x, y)$ is the local permeability of the column, and K_v is the overall permeability of the column for volumetric flow.

The required boundary condition is a time-dependent adsorption–desorption model to capture the physicochemical interaction between the analyte and the stationary phase. This boundary condition is

$$\frac{D_g}{b} \left(\frac{\partial n_g}{\partial y} \right) \bigg|_{\pm b} = -k \left(\frac{\partial n_g}{\partial t} \right) \bigg|_{\pm b} + \frac{k}{1+k} \cdot \bar{v}_z \cdot \left(\frac{\partial n_g}{\partial z_1} \right) \bigg|_{\pm b}$$

where y is normal to the surfaces.

A similar expression can be defined for the wall boundary conditions in the x-direction. This expression implies that the total mass flux of the analyte is adsorbed into the adsorbent coated onto the wall, and then desorbed back to the carrier gas stream with a time delay controlled by the retention factor (k). For simplicity, this surface interaction model has neglected any in-depth penetration and diffusion of analyte into the stationary phase.

In the present form, it is relatively difficult to solve analytically this analyte transport equation with this specified boundary condition because of the complicated mathematical expression involved. Computational techniques are good approaches to study this problem. However, it is important to impose very fine computational meshes to model the analyte distribution so that numerical diffusion can be minimized.

Since the primary focus in this subsection is to study the dispersion of analyte or band broadening, we will simplify our analysis by making a few assumptions such that an analytical solution can be derived and the performance of a micro-GC column can be evaluated. To simplify the problem, our assumption is that most microfabricated GC columns have a high aspect-ratio-flow geometry; that is, its width is much greater than its height ($2a \gg 2b$). This assumption neglects the end effect of the column with a finite width of $2a$ and gives the following approximation for the axial velocity:

$$v_z(x,y) = \frac{3}{2} \bar{v}_z \left(1 - \frac{y^2}{b^2} \right)$$

Applying this velocity approximation, we can solve for the incremental second moment and obtain the separation efficiency, which is expressed in terms of theoretical plate height. This expression is also known as the modified Golay equation:

$$h = f_1 \frac{B}{\bar{v}_{z,L}} + f_1 \cdot C_g \cdot \bar{v}_{z,L} + f_2 \cdot C_s \cdot \bar{v}_{z,L} + (f_2 \cdot \bar{v}_{z,L})^2 \cdot E$$

where h is defined as the height of an equivalent theoretical plate or the HETP value.

This modified Golay equation is for an open column of uniform cross section and for an isothermal operation. It includes a pressure correction (f_1 and f_2; same f_2 as in Section 9.6.1) and an additional term for the instrumental contribution to band broadening (E). These four terms in the Golay equation represent the static and dynamic diffusions of the analyte in the carrier gas, diffusion of the analyte in the stationary layer, and the band broadening owing to the sum of the instrumental dead times, respectively. The description of the coefficients in these terms can be found in Table 9.3.

The coefficient in the third term of the Golay equation, C_s, which describes diffusion in the stationary retentive layer, is sometimes neglected. Neglecting this term implies that the adsorption–desorption mode of retention takes place at the surface of the stationary phase and instantaneous (versus an actual slow) diffusion into the bulk of the material. In the earlier micro-GC column work, Reston and Kolesar attribute four orders of magnitude difference between their theoretical and experimental separation factors to this term.[46] In addition, Golay[47] indicates that this term can dominate in simple capillary columns.

Separation experiments using a conventional capillary column (100 μm diameter) with a length of 1 m and nitrogen as a carrier gas have been performed to assess this analytical model. A comparison of retention between decane and dodecane in the 0.1-μm thick, SPB-5-coated column are shown in Figure 9.9 as a function of pressure drop (proportional to flow velocity). The model compares reasonably well within the uncertainty on the data, and it predicts the proper trends.

TABLE 9.3
Analytical Expressions for the Coefficients in the Modified Golay Equation

B	C_g	C_s	E	f_1	f_2
$2D_g$	$\dfrac{(1 + 9k + \frac{51}{2}k^2)(2b)^2}{105(k+1)^2 D_g}$	$\dfrac{2kw^2}{3(k+1)^2 F^2 D_s}$	$\dfrac{(\Delta t)^2}{L(k+1)^2}$	$\dfrac{9(P_i^4 - P_o^4)(P_i^2 - P_o^2)}{8(P_i^3 - P_o^3)^2}$	$\dfrac{3P_o(P_i^2 - P_o^2)}{2(P_i^3 - P_o^3)}$

Note: w is the thickness of the stationary layer.

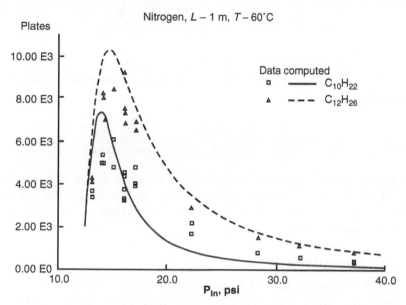

FIGURE 9.9 Comparison of predicted GC performance and data: number of theoretical plates versus inlet pressure for decane and dodecane (GC column is a capillary tube of 100 μm diameter and 1 m length; fluid is nitrogen at 60°C). (From FryeMason, G.C.; Kottenstette, R.; Heller, E.J.; Matzke, C.; Casalnuovo, S.A.; Lewis, P.; Manginell, R.P.; Schubert, W.K.; Hietala, V.M.; Shul, R. Proceedings at MicroTAS 2000, Enschede, the Netherlands; van den Berg, A., Olthins, W., Bergreld, P. Eds., Kluwer Academic Publishers, Amsterdam, 2000, pp. 229–232. With permission.)

9.6.3 IS SMALLER BETTER?

It can be shown from the Golay equation that there is an optimal flow velocity that will produce the best performance of the gas chromatography. To simplify our analysis, we will neglect both contributions of band broadening due to diffusion in the stationary phase and due to instruments, and only work with the simplified Golay equation

$$h_o = \frac{B}{\langle v_z \rangle} + C_g \cdot \langle v_z \rangle$$

where $\langle v_z \rangle$ is the average linear velocity. By taking the derivative with respect to the velocity, one can easily find the optimal velocity as

$$\langle v_z \rangle|_{\text{optimal}} = \sqrt{\frac{B}{C_g}}$$

At this optimal velocity, the separation efficiency in the GC column, in terms of the HETP, will be

$$h_{\min} = 2\sqrt{B \cdot C_g} = (2b)\frac{2}{\sqrt{105}}\frac{\sqrt{2 + 18k + 51k^2}}{(1 + k)}$$

The design criterion we are most interested in is the performance of the micro-GC column, which is expressed in terms of the number of theoretical plates:

$$N = \frac{L}{h_{\min}} = \frac{L}{2b} \cdot \frac{\sqrt{105}}{2} \cdot \frac{(1 + k)}{\sqrt{2 + 18k + 51k^2}}$$

As the size of the column decreases, one can achieve a similar GC performance (same number of theoretical plates) with a shorter column. This implies that a miniaturized GC column, in theory, can perform equally well as a conventional GC column.

9.6.4 THEORY DISCUSSION

The models presented here are based on the assumption that the flow in GC columns is a fully developed laminar flow with a parabolic velocity profile (a two-dimensional flow). This implies that the analyte transport in the transverse direction is by diffusion only. It neglects any convective transport in the cross-sectional plane, which may occur in some special cases such as a rectangular spiral column. For that geometry there may be a secondary flow at the corners and a recirculation flow generated from the curvature effect. To evaluate the effect of the geometry and complex flow dynamics, one may be required to apply computational fluid dynamics analysis to simulate the transport dynamics of species and fluid.

9.7 GC COLUMN FABRICATION AND COATING

After the groundbreaking work in microfabricating a GC column at Stanford, several different groups have fabricated planar GC columns using various wet and dry etching techniques followed by column coating. The term planar implies a column produced by etching channels into a substrate rather than conventional GC technology that uses capillaries that are produced by drawing out glass or fused silica tubes.

Kolesar and Reston made a planar gas chromatographic system in a silicon wafer.[48] The spiral column was photolithographically defined onto a silicon wafer using a conventional integrated circuit negative photoresist. The column was etched using a chemical isotropic etch (HF:HNO$_3$, CH$_3$COOH) to a depth of 9 μm, while the Pyrex cover plate was etched to a depth of 1 μm using HF:NH$_4$F. The silicon wafer was subsequently anisotropically etched with KOH to make the injection port interface and a dual detector cavity. A copper phthalocyanine stationary phase was sublimed onto the Pyrex cover plate and the silicon column. The stationary phase deposited on top of the sealing surfaces was removed with a polishing compound, cleaned, and then anodically bonded using a relatively low temperature (150°C at 1800 V for 24 h). A commercial TCD and a custom chemiresistor were used as the detectors in this system. The final GC column had a rectangular-shaped cross section that was 300-μm wide by 10-μm high and was 0.9 m in length.

Yu et al. fabricated a planar chromatographic column in a silicon wafer using isotropic etching to create a semicircular cross section in silicon.[49–51] The process uses a short plasma etch to remove an exposed nitride layer followed by an isotropic etch to etch a rounded profile into the silicon. The overhanging nitride that persisted is removed with another wet etch. Two wafers (mirror images) are aligned and fusion bonded to create channels in a silicon wafer that have a circular cross section.[49] This column is then coated using traditional coating means and siloxane stationary phases. Deviations from semicircular profile or slight misalignment of the wafer halves lead to less than optimal coating thickness near these defects. Columns with rectangular cross section that are coated with polymer solutions or sol–gel reactions are also prone to coating irregularities since capillary forces gather the liquid phase into the corners of the rectangular channel. Yu's column was incorporated into a portable GC that used a microfabricated TCD and a more sensitive glow discharge detector.

Lehmann et al. introduced a microfabricated column that used a plasma deposition process to deposit the stationary phase.[52,53] The column was reactive ion etched to 27-μm depth by 70-μm width and was 2-m long. Rather than being spiral the column consisted of a meandering pattern etched into a 20-mm by 25-mm chip. A PDMS stationary phase was deposited with plasma enhanced chemical vapor deposition and then a Pyrex cover plate was attached to the silicon chip using anodic bonding. This column was incorporated into an instrument that used four monolithically integrated microball valves to deliver carrier gas. A 10-nl internal volume TCD was also fabricated by using freestanding platinum heater strips. A prototype of this system is being produced by SLS Microtechnology and was described by Lehmann et al.[54]

Frye-Mason et al. fabricated another chip-based GC column.[36] This column, fabricated with a deep reactive ion etch (DRIE) process, has a relatively large aspect ratio (100-μm wide by 300-μm deep). This process is also referred to as high aspect ratio silicon etching (HARSE).[55] HARSE uses reactive gases to alternately etch the silicon in the vertical direction and then deposit passivation on the sidewalls of the evolving channel to prevent lateral (horizontal) etching during subsequent etch steps. This technique essentially turns the rectangular profile on edge so that the long dimension is into the silicon and the short dimension is side to side (shown in Figure 9.10). This method allows similar pressure drops to the previous columns, yet provides a much smaller footprint. The column itself weighs less than 350 mg since the sidewalls have

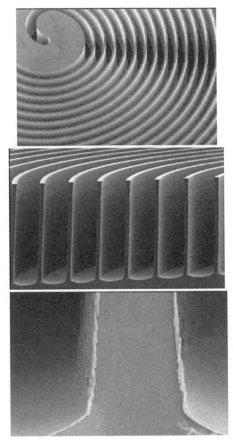

FIGURE 9.10 Scanning electron microscope images of high aspect ratio column; cross section of column is 100-μm wide by 400-μm deep; far right panel shows column coating.

been etched to be only 25-μm wide. This is important since the smaller footprint (1.2 cm by 1.2 cm) is more efficiently heated during temperature programming cycles. The column is coated with a modified sol–gel polymer deposition that provides for passivation and coating in one step by using the sol–gel chemistry to bond the phase to the column wall.[56] Several types of coating have been used ranging in polarity and including PDMS, phenyl PDMS, Carbowax, and a free fatty acid phase (OV-351). Samples are injected into the GC column with a microfabricated preconcentrator that consists of a silicon nitride membrane hotplate (described in Section 9.4). An array of SAW devices is used as the detector for this system and a typical separation for chemical warfare agents using this microanalytical system (μChemLab) is shown in Figure 9.11. Another version of a high aspect ratio column has been described and patented by Overton.[57]

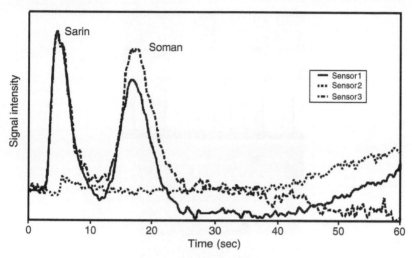

FIGURE 9.11 Separation of chemical warfare agents with high aspect ratio column and detection with a SAW array.

Zellers et al. subsequently used DRIE to etch a 3-m long column in silicon for use with a microfabricated preconcentrator and chemiresistor array.[58] The column was etched as a square cross section box-like spiral pattern, a Pyrex cover was anodically bonded to the silicon, and the resulting channel was then coated with a PDMS phase.

Micropacked columns have also been fabricated and tested.[59] In general, packed columns are shorter in length than wall-coated columns since the column packing causes significant flow restriction. The column shown in Figure 9.12 has a 300-μm by 300-μm cross section in silicon and is packed with a porous polymer stationary phase after a Pyrex lid is anodically bonded on. Silicon posts in the exit end of the column are used to retain the stationary phase. The stationary phase particles are significantly smaller than similar phases used in larger columns to maintain suitable separation efficiency. Adsorption chromatography has been demonstrated with these columns by inserting them into a commercial GC oven using an FID. Figure 9.13 shows a very high quality separation of C2 hydrocarbons using two different types of stationary phases.

9.8 MICRO-GC DETECTORS

Although fabricating and coating micro-gas chromatographic columns is an essential part of a microanalytical system, true analytical performance hinges on incorporating highly sensitive detectors into the analyzer. Many of the sensors used in microanalytical systems were developed in an evolutionary way as stand-alone detectors that were incorporated into sensor arrays (as in

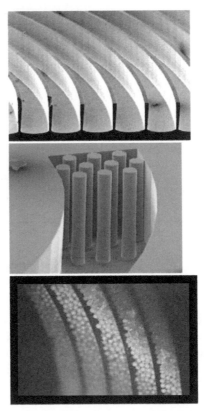

FIGURE 9.12 (See color insert following page 208) Microfabricated packed column; cross section is 300 μm by 300 μm; posts are used as "gate" material to retain packing; packing shown is a small diameter porous polymer (70 μm, Hayesep A).

electronic nose technology) and finally into microanalytical systems. Many of the detectors described in the following section have developed in such a fashion and consist of components described in Table 9.1 as well as new designs based on miniaturization of traditional gas chromatography detectors.

9.8.1 Thermal Conductivity

TCDs measure the change in thermal conductivity of gases using heated wires configured in a bridge circuit. The MTI TCD consisted of four precisely matched nickel filaments (two sample filaments and two reference filaments) deposited on a silicon chip and housed in a ceramic package. The cell dead volume of the detector was 1.5 nl with a thermal time constant of 200 μsec.[23–25,60] The very low volume detector preserves the chromatographic separation by minimizing band broadening. Subsequent improvements on the detector were evolutionary, yet the original detector comprises

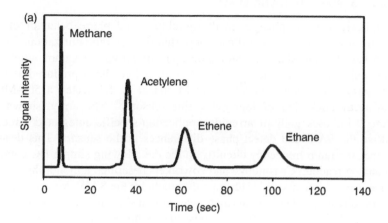

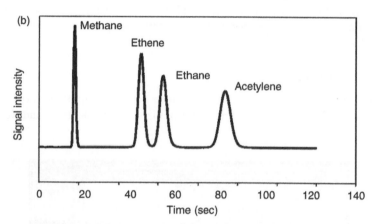

FIGURE 9.13 Microfabricated packed column separation of C1–C2 hydrocarbons: (a) using carbon molecular sieve adsorbent packing; (b) using porous polymer adsorbent packing.

the main commercial technology in microanalytical gas systems. One method of lowering detection levels and increasing the number of VOCs that could be analyzed with the TCD-based system was to attach a preconcentration unit onto the MTI system.[29] A miniature planar TCD was constructed in silicon and extensive modeling was performed to optimize the performance.[61] Sorge and Pechstein described a fully integrated TCD that minimizes dead volume using microfabrication and integration.[62] Hotplate membranes have been constructed to measure the thermal conductivity of gases as well.[63] Yu has also patented a microfabricated TCD design.[64]

9.8.2 SURFACE ACOUSTIC WAVE

SAW devices detect changes in the amplitude and phase behavior of an acoustic wave and offer significant opportunities for miniaturization. SAW arrays have been used as gas chromatographic detectors in vacuum outlet gas chromatography.[65] Complete analytical systems have been produced and in some cases commercialized.[66,67] Frye-Mason et al. fabricated a 500-MHz SAW array consisting of four delay line sensors.[36] The array, shown in Figure 9.14, uses gallium arsenide application-specific integrated circuits to drive the SAW and detect phase differences in the sensors. This design confines the radio frequency circuitry to the die allowing simple DC control and analog readout. Casalnuovo and coworkers described a monolithic SAW device in a gallium arsenide substrate that includes the SAW drive and phase detector circuitry.[68] Gallium arsenide was chosen since microelectronics fabrication and piezoelectric effect are possible in this substrate. SAW devices, which employ polymer films, contribute a degree of band broadening due to the time for mass transport into and out of the film; therefore, larger-molecular-weight (lower vapor pressure) compounds produce wider peaks on GC-SAW systems.

9.8.3 CHEMIRESISTORS

Chemiresistors detect changes in impedance between microfabricated electrodes and they are prime candidates for miniaturization. Kolesar and Reston

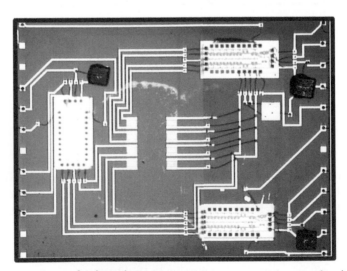

FIGURE 9.14 (See color insert) Surface acoustic wave array detector showing three prominent application-specific integrated circuits for radio frequency excitation and phase detection. Four delay line SAW devices shown in center.

used a copper phthalocyanine coated chemiresistor as one detector in their wafer-based GC.[46] Hughes et al. have combined a miniaturized preconcentrator with a chemiresistor based on conductive carbon particles suspended in a polymeric film between interdigitated electrodes.[69] Novel coatings based on colloidal gold nanoparticles seem to have high selectivity and sensitivity for chemiresistor-GC applications.[70,71] Polymer coatings on chemiresistors contribute to band broadening in GC applications for the same reasons they do with SAW devices.

9.8.4 Flame Ionization and Flame Atomic Emission Detector

Ionization detectors function by adding or removing electrons from organic molecules in an electric field. Zimmerman et al. fabricated a micromachined FID in 1998.[72,73] This detector was based on a microburner unit fabricated with a Pyrex–silicon–Pyrex sandwich that used an annular electrode design. The Pyrex was etched with hydrofluoric acid and the silicon was plasma etched. Stacking the different elements forms a planar, concentric annular jet. An upper annular electrode (anode) and a substrate electrode (cathode) are used to measure ionization current. The concentration of organic substances in the flame results in a proportional ionization current intensity. Figure 9.15 shows a cross section of the burner unit. The prototype FID had a nozzle diameter of approximately 60 μm and used an oxyhydrogen flow of 35 ml/min. Figure 9.16 shows a diagram of the device that includes schematic representation of an electrometer circuit. This same microburner has been used as the basis for an atomic emission detector.[72,74] This device is primarily used to optically measure atomic emission from the flame using a microspectrometer. Although this device has not been demonstrated for gas detec-

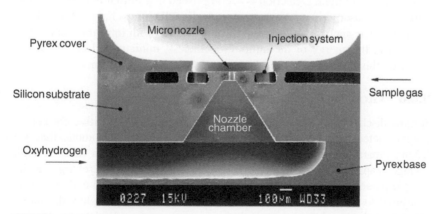

FIGURE 9.15 Cross section of micromachined flame ionization detector. (From Zimmermann, S.; Krippner, P.; Vogel, A.; Muller, J. *Sens. Actuators B—Chemi.* 2002, *83*, 285–289. With permission.)

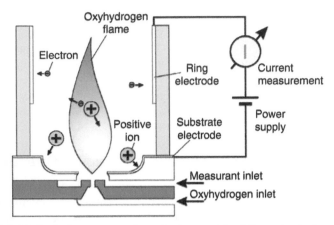

FIGURE 9.16 Stylistic representation of micromachined flame ionization detector. Detector gases are ignited at the nozzle defined by planar layers. (From Zimmermann, S.; Krippner, P.; Vogel, A.; Muller, J. *Sens. Actuators B—Chemi.* 2002, *83*, 285–289. With permission.)

tion, it may hold promise for element-specific detection of organic compounds that elute from a GC column. Zimmerman et al. also describe a microfabricated electrolysis cell for production of hydrogen for the flame analyzer.[72]

9.8.5 PHOTOIONIZATION

PIDs have become increasingly miniaturized and are marketed as stand-alone detectors for environmental work.[75] PIDs operate by ionizing gas molecules using ultraviolet light. Detection is accomplished in a similar manner to FIDs; current is measured in an electrometer circuit. PID lamps have a limited life span and their application is limited to those gases with ionization potentials lower than the energy of the lamp. Nonetheless, miniaturized GC detectors based on this principle can be quite useful for handheld systems.

9.8.6 GLOW DISCHARGE AND MOLECULAR EMISSION

A glow discharge detector has been designed by Yu as a sensitive chromatographic detector.[76] This detector consists of a cathode and anode that are 0.5 mm apart. The electrodes are contained in a small glass tube. A high voltage is applied across the electrodes resulting in a corona discharge; as the sample passes into the detector chamber it is partially ionized and current is measured to produce a signal. An atmospheric pressure DC glow discharge plasma is also used for an optical emission spectrometric detector for gas chromatography.[77–80] Optical emission offers the hope of element-specific detection for chromatography depending on the emission spectra of analyte fragments. The device has a 190-nl plasma chamber and uses helium carrier gas and the very small design allows the detector to operate with only 9 mW of power. Helium is used to generate a plasma and optical emission is measured via optical fibers (carbon is monitored as CO at 519 nm). Initial tests with this device have led to more robust designs that can operate continuously even though electrode sputtering and erosion have been noted. As in all microfabricated detectors, the connecting volumes and cell dead volume issues play an important role in peak shape and amplitude.

9.8.7 PHOTOACOUSTIC DETECTOR

A novel microfabricated photoacoustic detector has been fabricated as a detector for a microreactor.[81,82] In photoacoustic spectroscopy, incident light is modulated at an acoustic frequency. When light is absorbed by gaseous species passing through the cell a resultant gas expansion is detected using a microfabricated microphone. The sensitivity of photoacoustic spectroscopy scales with the inverse of cell dimensions, unlike absorption spectroscopy that becomes less sensitive when the absorption path length is reduced. Firebaugh et al. note that the increased sensitivity may not be realized in a resonant photoacoustic cell, but most certainly will be achieved in a nonresonant photoacoustic cell.[81] This sensor offers the potential for spectroscopy by using several different diodes in an array. A patent has been issued on micromachined optoacoustic sensors as well.[83]

9.8.8 MINIATURE MASS ANALYSIS

In addition to the detectors discussed so far, detection on the basis of molecular mass may prove to be the ultimate technology for microanalytical gas sensing, providing a true second dimension. Mass analysis techniques include mass spectrometry and ion mobility spectrometry (IMS). A complete spectrometer can be considered to consist of four subsystems: sample introduction, mass analysis, some level of reduced pressure (except IMS, which can operate at atmospheric pressure), and power. Sample introduction is provided by the gas chromatograph. The mass analysis portion depends on three fundamental physical processes: ionization, mass (or mobility) separation, and ion detection. Constructing a fully microfabricated spectrometer therefore requires microfabricated components that perform these fundamental physical processes while also performing the functions of the other subsystems. The current lack of miniature or microfabricated vacuum pumps has prevented the demonstration of a fully miniature mass spectrometer. This discussion presents progress only for the components for mass analysis since fully operational microfabricated systems have not yet been constructed.

While there are many "portable" or "miniature" mass spectrometers that are commercially available and reported in the literature, few have been constructed using truly microfabricated components, and none have fully integrated components that are all microfabricated. A full discussion of spectrometer types is beyond the scope of this chapter, but are covered in books on the subject,[84,85] and miniature versions are reviewed.[86]

The ionization process has not been the focus of much effort toward microfabrication, as it is typically achieved using the same methods used for larger systems. These include electron beam or radioactive source ionization. However, micromachined sources have been produced.[87,88]

Mass separation, in the case of a mass spectrometer, has been reported using the methods of quadrupole filter, ion trap, time-of-flight, and novel designs. While a fully microfabricated mass analyzer (which includes ionization, separation, and detection) has not been demonstrated at this time, significant progress is being made. Table 9.4 indicates components that have been microfabricated and demonstrated (not necessarily in combination, "macro" system components not listed).

While there are not many fully microfabricated mass separators at this time, the field of miniature devices is very active. These devices may be suitable for portable applications as associated electronics and pumping components become smaller. Quadrupole array designs range from the commercial array (rods 1 mm diameter $\times$ 10 mm) of Ferran and Boumsellek[89] to an array of 2×25 mm rods by Chututjian and coworkers[90]; the trade-offs of this design have been discussed.[91] Ion trap designs range from a single hyperbolic trap with $r_0 = 10$ mm,[92] to dual cylindrical traps with r_0 of 4 and 5 mm,[86] to an array of four (or ten) cylindrical traps with $r_0 = 2.5$ mm (or 1.5 mm),[86] to a single cylindrical trap with $r_0 = 0.5$ mm.[93] A unique torroidal trap design was demonstrated[94] as well as a portable double-focusing magnetic sector.[95] Portable time-of-flight analyses, despite a general loss of resolution at smaller sizes, have also been successfully demonstrated for air monitoring.[96,97]

Ion mobility spectrometers have also been reduced significantly in size from conventional models using some microfabricated parts. Eiceman et al. used Pyrex and silicon wafers to create a spectrometer with approximately 3-cm^3 volume (without electronics) and used it as a GC detector.[98,99] Also, a 1.7-mm diameter by 35-mm drift channel IMS was demonstrated by Ramsey et al.[100]

Much progress has been made in the design and fabrication of mass separator components for microanalytical systems including both mass spectrometer and ion mobility spectrometry hardware. Future near-term work will undoubtedly improve on the current designs. Although mass analysis may prove to be the most powerful detection technology for microanalytical gas sensing systems, significant challenges to miniaturization remain. Among those are the production of a small reliable ionization source and especially a powerful miniaturized vacuum system.

TABLE 9.4
Microfabricated Mass Separation Component Descriptions

Ionization	Separation	Detection	References and Notes
Novel microwave-stimulated plasma source 300 μm × 300 μm × 150 μm	100 μm × 100 μm separation channel, modified time-of-flight		Siebert et al.[114]; separation device design, no separation data
	200 μm × 1 cm channels, ExB Wien filter	Microfab wire	Baptist sillon and[115]; single mass detection, no separation demonstrated
Microfab tungsten tip	10 mm × 10 mm × 1 mm flight tube, time-of-flight		Yoon et al.[116]; separation demonstrated with multiphoton ionization
	ExB up to m/z 103		Diaz et al.[117]; 2 W (not incl. pumps) full working system demonstrated
	500 μm × 3 cm rods, quadrupole		Taylor et al.[118]; separation of He–air mixture

9.9 IMPROVEMENTS AND FUTURE DIRECTIONS

Although the pioneering work in microanalytical systems began in the late 1970s, gas-phase systems have not achieved widespread production and commercial development. Many exciting new developments have permitted new systems to be made smaller and potentially integrated into true "chip-based" instruments. The following discussion is by no means an exhaustive list of the potential opportunities and pitfalls of future gas-phase microanalytical systems, but it is hoped that it will serve as a starting point for new development and the goal of building and embedding microanalytical systems for the benefit of society.

9.9.1 COLUMN COATING

Several issues have been encountered in microfabricated GC work. Terry[19] first mentioned the difficulty of coating a microfabricated column due to poor stationary-phase wetting and uneven coating due to surface tension of the coating solution. Conventional column coating by dynamic or static means,[101] and a modified sol–gel phase deposition[56] can produce fillets in the corners of rectangular or square cross-sectional profiles. Producing a circular cross section can alleviate the coating nonuniformity, but this is especially difficult to achieve since imperfect symmetry and misalignment of semicircular channels leads to coating nonuniformity as well.[102] One solution in avoiding these coating anomalies has been to coat columns with polymer that has been plasma deposited.[48,103] To date, two of the plasma-deposited phases explored have been rather specialized (copper phthalocyanine and amino acid) and have rather unique applications. A plasma-deposited PDMS phase has been coated onto a shallow column.[53] Plasma deposition, polymer grafting, and synthetic pathways to column coating may improve the separation power of microfabricated columns in the future.

9.9.2 MICROFLUIDICS, PUMPS AND VALVES, TRANSFER TUBES

Another significant challenge for micro-GC systems is appropriate sizing and interfacing with connecting plumbing to interact with the outside world. One GC system uses air as a carrier gas using a small diaphragm pump to pull a sample into the column.[36] Another lab prototype uses vacuum to pull the sample into the column and indeed establishes that vacuum-outlet GC provides a distinct advantage for separations.[65] A downside to air as carrier gas, however, is the potential degradation of the stationary phase due to oxidation as the column is operated at higher temperatures. Traditionally, micro-GC systems have used carrier gas reservoirs to supply carrier gas using small and even microfabricated valves. Although valves have been microfabricated, they usually have large solenoid actuators associated with them and are thus not truly integrated into a microsystem. Still many other challenges exist in the area of sample loops and interconnections between sampler, GC columns, and detectors. Many of the microfabricated systems have been "hybridized," in that separate components are fluidically connected using customized connections. Significant challenges remain for integration including true microfabricated valves, selective coating deposition, and passivation of integrated vias and feed-throughs.

9.9.3 DETECTORS

The most popular detector for microfabricated GC has been the TCD. TCDs are very simple to microfabricate and have proved valuable in first-generation systems, but a significant number of applications (environmental and industrial hygiene to name two) require more sensitive detection. A good deal of research has been focused on microfabricated detectors that can provide more sensitivity and selectivity. Without question a second dimension of selectivity such as a mass spectral identification combined with GC retention time will provide more analytical power. For certain applications improved selectivity can be realized through element-selective devices like the flame photometric or thermionic detectors.

9.9.4 POWER

One significant engineering parameter that provides a challenge to portable microsystems is the consumption of power. Continuous use of a handheld GC requires several Watts of power per analysis. Power is primarily consumed by temperature ramping of the micro-GC column. Commercial valves and pumps (meso-scale) also consume a relatively large amount of power. Innovative solutions to this problem will involve diligent reduction of the heat capacity of different components while integration and dead volume reduction in the fluidic pathway will lead to further energy savings. Certainly the development of microfuel cells may increase energy storage capacity for continuous portable or autonomous use. Plant power and infrastructure could be used in industrial settings where a large number of embedded microsystems may be installed.

9.9.5 MATERIALS DEPOSITION AND FABRICATION

Gas detectors and detection systems have been microfabricated for several decades. During this time, many new microfabrication procedures have been invented. Various wet and dry etching techniques, surface micromachining, LIGA (lithography, electroplating, molding), HARSE, and other processes have led to advances in the types and quality of microstructural elements. Many of these processes have led to unique devices such as membrane hotplates, cantilevers, and high aspect ratio structures. These new platforms offer many exciting possibilities for analytical microsystems but provide challenges as well. These small devices require nontraditional materials (from an integrated circuit perspective) such as polymers, inorganic–organic hybrids, or selective membrane structures to make them fully functional. To this end several new materials and deposition techniques will be necessary in the future. Polymer application techniques will be very important. Polymer grafting, self-assembly, and molecular design will define these new capabilities. Microcontact printing has been described and involves using stamps to place materials in tiny zones on a substrate.[104] Chemical vapor deposition comprises another rich area of research along with laser machining. McGill describes a process known as matrix-assisted polymer laser ablation to form very high quality polymer films.[105] Plasma deposition has also been used to make functional chemical films. Photopolymerization has been described as a way of writing polymer onto surfaces.[106] Still other application techniques such as inkjet direct writing may also play a part. Wafer bonding processes will need to perform at lower temperatures to preserve chemical functionality of films deposited on the "open face" of a substrate before channels are formed and sealed. Passivation techniques will need to be used in transfer vias and connections between functional fluidic components.

9.9.6 INTEGRATION

A significant challenge to miniaturization and optimization of performance for gas-phase microsystems is monolithic integration. Certainly miniaturization would be optimized in a completely integrated system such that the entire analysis system would be on one chip. Indeed, the very first microsystem contained fluid handling, valve seats, separation columns, and detectors, all on a silicon substrate. This was a wafer-based approach rather than chip-based and subsequently the separation column was removed in favor of using traditional capillaries for separation. Integration provides other challenges in that contiguous components may need to operate in different temperature regimes. GC columns sometimes need to be temperature ramped; this is usually undesirable treatment for detector stability. Figure 9.17 shows a prototype chip[107] fabricated in silicon with a silicon nitride membrane preconcentrator and a GC column in a single substrate. Although it represents an example of a true chip-based system, difficulties in coating and incompatibility of operational temperature have made this particular device impractical. Another approach to integration is to assemble the microsystem in a hybridized fashion. Figure 9.18 shows a fluidic manifold made from a low-temperature co-fired ceramic (LTCC) process.[108,109] This substrate can use "pick and place" technology such as flip chip bonding to assemble a fluidic board that functions also as a printed circuit assembly. Separation columns, sample introduction, microactuators, and detectors can be attached to complete the system. The LTCC has vias to fluidically connect different components and the material itself functions as an electrical circuit board. Still other

FIGURE 9.17 (See color insert) Prototype chip integration of micromachined high aspect ratio column and membrane device (two devices shown). Membrane is used as solid-phase microextraction device and serves as GC injector.

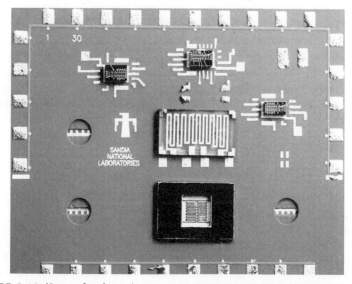

FIGURE 9.18 (See color insert) Prototype fluidic/circuit board fabricated in low-temperature co-fired ceramic. Membrane preconcentrator and SAW array attached in pick and place scheme. Fluidic vias are circular features.

fabrication techniques may be useful in defining tubes and complex shapes on substrates. Plastic columns made by embossing processes may be used for low-cost parts; columns can also be formed in SU-8, a light-sensitive epoxy, or Foturan,[110] a photoetchable glass. Microfuel cells may provide a higher-density power supply for autonomous or handheld systems and on-chip hydrolysis may provide a reliable source of hydrogen as a carrier, reactant, or detector gas. Low-power microactuators for valves and pumping will also facilitate true chip-based operation. Miniature gas valves must be able to hold off and regulate flow of pressurized gases.

9.9.7 New Technologies

Future applications for microanalytical systems will most certainly include microreactors. Several papers have described the use of microfabricated devices to perform chemical reactions in gas samples. A gas analysis system on-chip can be used to evaluate the efficiency and yield of such reactors. Becker et al. report using a tin oxide microhotplate as a platform to analyze gas mixtures,[111] and Jensen and co-workers have reported using a microreactor for catalyst testing of the oxidation of carbon monoxide.[112] A specialized form of a microreactor has been reported that describes the use of a membrane hot-plate to derivitize nonvolatile chemicals by methylating them.[113] Fatty acids in edible oils and bacterial samples have been processed this way to make volatile fatty acid methyl esters. Other derivitizing agents can be used to produce functionalized compounds including fluorinated species. Still further advancements in microactuators may allow multidimensional high-speed chromatography that will use a second column and a "heart-cutting" valve to provide more separation in a microfluidic system. Recent advances in materials science will undoubtedly contribute to miniaturization. Nanoscale technology using molecules and molecular architecture, as components of advanced sensing systems, will most certainly play a part in future intelligent gas-sensing systems.

REFERENCES

1. Janata, J.; Josowicz, M.; Vanysek, P.; DeVaney, D. M. *Anal. Chem.* 1998, *70*, R179–R208.
2. Wilson, D. M. *IEEE Sens. J.* 2001, *1*, 256–274.
3. Kohl, D. *J. Phys. D — Appl. Phys.* 2001, *34*, R125–R149.
4. Janata, J.; Huber, R. J. *Solid State Chemical Sensors*; Academic Press, New York, 1985.
5. Janata, J. *Principles of Chemical Sensors*; Plenum Press, New York, 1990.
6. Korvink, P. J. In *The MEMS-Handbook*; Kovink, J G., Ed., Noyes Publications, Park Ridge, NJ, 2003.
7. Mosely, P. T.; Norris, J.; Williams, D. *Techniques and Mechanisms in Gas Sensing*; IOP Publishing, Bristol, U.K., 1991.
8. Semancik, S.; Cavicchi, R. E.; Wheeler, M. C.; Tiffany, J. E.; Poirier, G. E.; Walton, R. M.; Suehle, J. S.; Panchapakesan, B.; DeVoe, D. L. *Sens. Actuators B — Chem.* 2001, *77*, 579–591.
9. Grate, J. W. *Chem. Rev.* 2000, *100*, 2627–2647.

10. Datskos, P. G.; Sepaniak, M. J.; Tipple, C. *Anal. Chem. A*, 2002, 568A–575A.
11. Isolde, S.; Barsan, N.; Bauer, M.; Weimar, U. *Sens. Actuators B — Chem.* 2001, *73*, 1–26.
12. Chou, J. *Hazardous Gas Monitors*; McGraw-Hill, New York, 2000.
13. Gardner, J. W., University of Warwick, Coventry, U.K., 1987.
14. Gardner, J. W., Bartlett, P. N. *Sens. Actuators B — Chem.* 1996, *33*, 60–67.
15. James, A. T.; Martin, A. J. P. *Analyst* 1952, *77*, 915–932.
16. Golay, M. J. E. *Gas Chromatogr.* 1958, 1–13.
17. Desty, D. H.; Haresnip, J. N.; Whyman, B. H. F. *Anal. Chem.* 1960, *32*, 302–304.
18. de Mello, A. *Lab Chip* 2002, *2*, 48N–54N.
19. Terry, S. C. PhD thesis, Stanford University, Palo Alto, CA, 1975.
20. Overton, E. B.; Stewart, M.; Carney, K. R. *J. Hazard. Mater.* 1995, *43*, 77–89.
21. Overton, E. B.; Dharmasena, H. P.; Ehrmann, U.; Carney, K. R. *Field Anal. Chem. Technol.* 1996, *1*, 87–92.
22. Raulin, F.; Sternberg, R.; Coscia, D.; VidalMadjar, C.; Millot, M. C.; Sebille, B.; Israel, G. *Adv. Space Res.* 1999, *23*, 361–366.
23. Siemers, R.; Heigel, D.; Spilkin, A. *Am. Lab.* 1991, *23*, 44L–44R.
24. Saadat, S.; Terry, S. C. *Am. Lab.* 1984, *16*, 90–101.
25. Lee, G.; Ray, C.; Siemers, R.; Moore, R. *Am. Lab.* 1989, *21*, 110–119.
26. Wallis, G.; Pomerantz, D. *J. Appl. Phys.* 1969, *40*, 3946–3949.
27. Angell, J. B.; Terry, S. C.; Barth, P. W. *Sci. Am.* 1983, *248*, 44–55.
28. Terry, S. C.; Jerman, J. H. U.S. Patent 4,474,889; Microsensor Technology Inc., 1984.
29. Bruns, M. *Am. Environ.* 1995, 29–34.
30. Etipoe, F. *J. Chromatogr. A* 1997, *775*, 243–249.
31. Lambert, R. H.; Owens, J. A. *Field Anal. Chem. Technol.* 1997, *1*, 367–374.
32. Terry, S. C.; Jerman, J. H.; Angell, J. B. *IEEE Trans. Electron Dev.* 1979, *ED–26*, 1880–1886.
33. Pawliszyn, J. *Solid Phase Microextraction — Theory and Practice*; Wiley, HCH, New York, 1997.
34. Koziel, J.; Jia, M. Y.; Pawliszyn, J. *Anal. Chem.* 2000, *72*, 5178–5186.
35. Zellers, E. T.; Morishita, M.; Cai, Q.-Y. *Sens. Actuators B — Chem.* 2000, *67*, 244–253.
36. Frye-Mason, G. C.; Kottenstette, R.; Heller, E. J.; Matzke, C.; Casalnuovo, S. A.; Lewis, P.; Manginell, R. P.; Schubert, W. K.; Hietala, V. M.; Shul, R. Proceedings of microTAS 1998, Bun Canada; Harrison, D. J., van der Berg., A., Eds., Kluwer Academic Publishers, Amsterdam, 1998.
37. Manginell, R. P.; Frye-Mason, G. C. U.S. Patent 6,171,378 B1; Sandia Corporation, Sandio, Livermore, CA, 2001.
38. Bonne, U.; Detry, J.; Higashi, R. E.; Rezachek, T.; Swanson, S., Orlando, FL, September 30, 2002.
39. Bonne, U.; Goetz, J.; Dasgupta, P. K. U.S. Patent 6,393,894; Honeywell International Inc., Honeywell, Morris Township, NJ, 2002.
40. Ettre, L. S.; Hinshaw, J. V. *Basic Relationships of Gas Chromatography*; Advanstar Publishing, Cleveland, OH, 1993.
41. Grob, R. L. *Modern Practice of Gas Chromatography*; John Wiley and Sons, New York, 1995.

42. Spangler, G. E. *J. Microcolumn Sep.* 2001, *13*, 285–292.
43. Spangler, G. E. *Anal. Chem.* 1998, *70*, 4805–4816.
44. Wong, C. C.; Adkins, D. R.; Frye-Mason, G. C.; Hudson, M. L.; Kottenstette, R.; Matzke, C. M.; Shadid, J. N.; Salinger, A. G. Proceedings of SPIE Conference on Microfluidic Devices and Systems II, Santa Clara, CA, 1999, pp. 120–129.
45. Frye-Mason, G. C.; Kottenstette, R.; Heller, E. J.; Matzke, C.; Casalnuovo, S. A.; Lewis, P.; Manginell, R. P.; Schubert, W. K.; Hietala, V. M.; Shul, R. Proceedings of microTAS 2000, Enschede, The Netherlands; van den Berg, A., Olthins, W., Bergveld, P. Eds., Kluwer Academic Publishers, Amsterdam, 2000, pp. 229–232.
46. Kolesar, E. S.; Reston, R. R. *IEEE Trans. Components Packaging Manufact. Technol.* 1998, *21*, 324–328.
47. Golay, M. J. E. *J. Chromatogr.* 1981, *216*, 1–8.
48. Kolesar, E. S.; Reston, R. R. *Surf. Coatings Technol.* 1994, *68–69*, 679–685.
49. Yu, C. M.; Hui, W. C. U.S. Patent 5,575,929; The Regents of the University of California, Oakland, CA, 1996.
50. Yu, C. M. U.S. Patent 6,068,780; The Regents of the University of California, Oakland, CA, 2000.
51. Felton, M. J. In *Today's Chemist at Work*, James F. Ryan, Ed., 2002, pp. 26–31.
52. Lehmann, U.; Plehn, H.; Krusmark, O.; Muller, J. Proceedings of Sensors '99, 1999, 155–158.
53. Lehmann, U.; Krusmark, O.; Muller, J. Vogel, A., Binz, D. Proceedings of microTAS 2000, Enschede, The Netherlands; van den Berg, A., Olthius, W., Bergveld, P. Eds., Kluwer Academic Publishers, Amsterdam, 2000, pp. 167–170.
54. Lehmann, U.; Mahnke, M.; Glampe, O.; Behrends, H.; Schiliffke, Schmidt, T.; Bollaparagada, B.; Nuremberg, Germany, May 13, 2003.
55. Ayon, A. A.; Branff, R.; Lin, C. C.; Sawin, H. H.; Schmidt, M. A. *J. Electrochem. Soc.* 1999, *146*, 339–349.
56. Wang, D. X.; Chong, S. L.; Malik, A. *Anal. Chem.* 1997, *69*, 4566–4576.
57. Overton, E. B. U.S. Patent 6,068,684; Board of Supervisors of Louisiana State University and Agricultural & Mechanical College, Baton Rogue, LA, 2000.
58. Zellers, E. T.; Wise, K. D.; Oborny, M.; Sacks, R., Waltham, M. A. *Technical Digest of the IEEE Conference on Technologies for Homeland Security*, November 13-14, 2002, Waltham, MA, IEEE, Boston, 2002, pp. 92–95.
59. Frye-Mason, G. C.; Kottenstette, R.; Mowry, C.; Morgan, C.; Manginell, R. P.; Matzke, C.; Lewis, P.; Dulleck, G.; Anderson, L.; Adkins, D. Proceedings of microTAS 2001, Monterry, CA; Michael Ramsey, J., van den Berg, A., Eds., Kluwer Academic Publishers, Amsterdam, 2001, pp. 658–660.
60. Shinar, R.; Liu, G. J.; Porter, M. D. *Anal. Chem.* 2000, *72*, 5981–5987.
61. Chen, K.; Wu, Y. E. *Sens. Actuators A — Phys.* 2000, *79*, 211–218.
62. Sorge, S.; Pechstein, T. *Sens. Actuators A — Phys.* 1997, *63*, 191–195.
63. Gajda, M. A.; Ahmed, H. *Sens. Actuators A — Phys.* 1995, *49*, 1–9.
64. Yu, C. M. U.S. Patent 6,502,983; The Regents of the University of California, Oakland, CA, 2003.
65. Whiting, J. J.; Lu, C. J.; Zellers, E. T.; Sacks, R. D. *Anal. Chem.* 2001, *73*, 4668–4675.
66. McGill, R. A.; Nguyen, V. K.; Chung, R.; Shaffer, R. E.; DiLella, D.; Stepnowski, J. L.; Mlsna, T. E.; Venezky, D. L.; Dominguez, D. *Sens. Actuators B — Chem.* 2000, *65*, 10–13.

67. Staples, E. J.; Watson, G. U.S. Patent 6,354,160; Electronic Sensor Technology, LLP, Thousand Oaks, CA, 2002.
68. Baca, A. G.; Heller, E. J.; Hietala, V. M.; Casalnuovo, S. A.; Frye-Mason, G. C.; Klem, J. F.; Drummond, T. J. *IEEE J. Solid-State Circuits* 1999, *34*, 1254–1258.
69. Hughes, R. C.; Manginell, R. P.; Kottenstette, R. 200th Electrochemical Society Meeting — Sensor Symposium, September 2001, San Francisco CA, 348.).
70. Wohltjen, H.; Snow, A. W. *Anal. Chem.* 1998, *70*, 2856–2859.
71. Cai, Q. Y.; Zellers, E. T. *Anal. Chem.* 2002, *74*, 3533–3539.
72. Zimmermann, S.; Wischhusen, S.; Muller, J. *Sens. Actuators* 2000, *63*, 259–166.
73. Zimmermann, S.; Riepenheusen, B.; Muller, J.; Kluwer Academic Publishers, Amsterdam, 1998, pp. 473–476.
74. Zimmermann, S.; Wischhusen, S.; Muller, J. The proceedings of microTAS 2000, Enschede, The Netherlands; van den Berg, A., Olthuis, W., Bergreld, P., Eds., Kluwer Academic Publishers, Amsterdam, 2000, pp. 135–138.
75. Wrenn, C.; www.raesystems.com, 2002.
76. Yu, C. M. U.S. Patent 6,457,347; The Regents of the University of California, Oakland, CA, 2002.
77. Eijkel, J. C. T.; Stoeri, H.; Manz, A. *Anal. Chem.* 1999, *71*, 2600–2606.
78. Eijkel, J. C. T.; Stoeri, H.; Manz, A. *Anal. Chem.* 2000, *72*, 2547–2552.
79. Eijkel, J. C. T.; Stoeri, H.; Manz, A. The proceedings of microTAS 2000, Enschede, The Netherlands; van den Berg, A., Olthuis, W., Bergreld, P., Eds., Kluwer Academic Publishers, Amsterdam, 2000, pp. 591–594.
80. Eijkel, J. C. T., Stoeri, H., Manz, A. *J. Anal. At. Spectrom.* 2000, *15*, 297–300.
81. Firebaugh, S. L.; Jensen, K. F.; Schmidt, M. A. *J. Microelectromech. Syst.* 2001, *10*, 232–237.
82. Firebaugh, S. L.; Jensen, K. F.; Schmidt. In *Micro Total Analysis Systems 2000*; Berg, A. v. d., Ed.; Kluwer Academic Publishers, Amsterdam, 2000, pp. 49–52.
83. Bonne, U.; Cole, B. E.; Higashi, R. E. U.S. Patent 5,886,249, U.S.A., 1999.
84. Herbert, C. G.; Johnstone, R. A. *Mass Spectrometry Basics*; CRC Press, Boca Raton, FL, 2002.
85. Eiceman, G. A.; Karpas, Z. *Ion Mobility Spectrometry*; CRC Press, Boca Raton, 1994.
86. Badman, E. R.; Cooks, R. G. *Anal. Chem.* 2000, *72*, 5079–5086.
87. Petzold, G.; Siebert, P.; Muller, J. *Sens. Actuators B — Chem.* 2000, *67*, 101–111.
88. Ghodsian, B.; Parameswaran, M.; Syrzycki, M. *IEEE Electron Dev. Lett.* 1998, *19*, 241–243.
89. Ferran, R. J.; Boumsellek, S. *J. Vac. Sci. Technol. A — Vac. Surf. Films* 1996, *14*, 1258–1265.
90. Orient, O. J.; Chutjian, A.; Garkanian, V. *Rev. Sci. Instrum.* 1997, *68*, 1393–1397.
91. Boumsellek, S.; Ferran, R. J. *J. Am. Soc. Mass Spectrom.* 2001, *12*, 633–640.
92. Orient, O. J.; Chutjian, A. *Rev. Sci. Instrum.* 2002, *73*, 2157–2160.
93. Kornienko, O.; Reilly, P. T. A.; Whitten, W. B.; Ramsey, J. M. *Rapid Commun. Mass Spectrom.* 1999, *13*, 50–53.
94. Lammert, S. A.; Plass, W. R.; Thompson, C. V.; Wise, M. B. *Int. J. Mass Spectrom.* 2001, *212*, 25–40.
95. Kogan, V. T.; Pavlov, A. K.; Chichagov, Y. V.; Tuboltsev, Y. V.; Gladkov, G. Y.; Kazanskii, A. D.; Nikolaev, V. A.; Pavlichkova, R. *Field Anal. Chem. Technol.* 1997, *1*, 331–342.

96. Berkout, V. D.; Cotter, R. J.; Segers, D. P. *J. Am. Soc. Mass Spectrom.* 2001, *12*, 641–647.

97. Syage, J. A.; Nies, B. J.; Evans, M. D.; Hanold, K. A. *J. Am. Soc. Mass Spectrom.* 2001, *12*, 648–655.

98. Miller, R. A.; Nazarov, E. G.; Eiceman, G. A.; King, A. T. *Sens. Actuators A — Phys.* 2001, *91*, 301–312.

99. Eiceman, G. A.; Nazarov, E. G.; Miller, R. A.; Krylov, E. V.; Zapata, A. M. *Analyst* 2002, *127*, 466–471.

100. Xu, J.; Whitten, W. B.; Ramsey, J. M. *Anal. Chem.* 2000, *72*, 5787–5791.

101. Grob, K. *Making and Manipulating Capillary Columns for Gas Chromatography*, Huethig Publishers, Heidelberg, 1986.

102. Madou, M. *Fundamentals of Microfabrication*; CRC Press, Boca Raton, FL, 1997.

103. Hannoe, S.; Sugimoto, I.; Katoh, T. Proceedings of microTAS 1998, Banff, Canada; Harrison, D. J., van den Berg, A., Eds., Kluwer Academic, Amsterdam, 1998, pp. 145–448.

104. Whitesides, G. M.; Grzybowski, B. A.; Haag, R.; Bowden, N. *Anal. Chem.* 1998, *70*, 4645–4652.

105. Pique, A.; McGill, R. A.; Chrisey, D. B.; Leonhardt, D.; Mslna, T. E.; Spargo, B. J.; Callahan, J. H.; Vachet, R. W.; Chung, R.; Bucaro, M. A. *Thin Solid Films* 1999, *356*, 536–541.

106. Hsieh, M. D.; Zellers, E. T. *Sens. Actuators B — Chem.* 2002, *82*, 287–296.

107. Manginell, R. P.; Okandan, M.; Kottenstette, R. J.; Lewis, P. R.; Adkins, D. R.; Bauer, J. M.; Manley, R. G.; Sokolowski, S.; Shul, R. J., Proceedings of microTAS 2003, Squaw Valley, CA; Allen Northup, M., Ed., Kluwer Academic Publishers, Squaw Valley, CA, 2003, pp. 1247–1250.

108. Chanchani, R., Scottsdale, AZ, Proceedings of IEEE — Components, Packaging and Manufacturing Technology Conference. May 8–10, 2001.

109. Gongora-Rubio, M. R.; Espinoza-Vallejos, P.; Sola-Laguna, L.; Santiago-Aviles, J. J. *Sens. Actuators A — Phys.* 2001, *89*, 222–241.

110. Hansen, W.; Fuqua, P.; Livingston, F.; Huang, A.; Abraham, M.; Taylor, D.; Janson, S.; Helvajian, H. *Industrial Physicist, 2002 8*, 18–21..

111. Becker, T.; Muhlberger, S.; Bosch-vonBraunmuhl, C.; Muller, G.; Meckes, A.; Benecke, W. *Sens. Actuators B — Chem.* 2001, *77*, 48–54.

112. Ajmera, S.; Delattre, C.; Schmidt, M. A.; Jensen, K. F. *Sens. Actuators B — Chem.* 2002, *82*, 297–306.

113. Mowry, C. D.; Morgan, C. H.; Manginell, R. P.; Kottenstette, R. J.; Lewis, P. R.; Frye-Mason, G. C. *Chemical and Biological Early Warning Monitoring for Water, Food, and Ground*, Boston, MA, October 28, 2001, pp. 83–90.

114. Siebert, P.; Petzold, G.; Hellenbart, Á.; Müller, J. *Appl. Phys. A* 1998, *67*, 155–160.

115. Sillon, N.; Baptist, R. *Sens. Actuators B — Chem.* 2002, *83*, 129–137.

116. Yoon, H. J.; Kim, J. H.; Choi, E. S.; Yang, S. S.; Jung, K. W. *Sens. Actuators A — Phys.* 2002, *97–98*, 441–447.

117. Diaz, J. A.; Giese, C. F.; Gentry, W. R. *J. Am. Soc. Mass Spectrom.* 2001, *12*, 619–632.

118. Taylor, S.; Tunstall, J. J.; Leck, J. H.; Tindall, R. F.; Jullien, J. P.; Batey, J.; Syms, R. R. A.; Tate, T.; Ahmad, M. M. *Vacuum* 1999, *53*, 203–206.

10 Sample Preparation on Microchips

Jan Lichtenberg*, Sander Koster*,
Laura Ceriotti, Nico F. de Rooij,
and Elisabeth Verpoorte

CONTENTS

---

* Both authors contributed equally.

10.1 INTRODUCTION

Both qualitative and quantitative chemical information about a sample can be obtained using separation techniques such as high-performance liquid chromatography or capillary electrophoresis (CE). As in all branches of analysis, however, the quality of that information will very much depend on the matrix in which the analyte of interest is determined. In laboratory standards, the composition of the matrix is very well defined and generally kept very simple, allowing the measured response of the detector to be directly related back to the original concentration of the analyte. With real samples, of course, the situation very rapidly becomes more complicated. Components in the matrix can often mask the presence of an analyte through complexation or other chemical interactions. Alternatively, there may be components present that induce an additional detector response, making it difficult to attribute results solely to the analyte(s) of interest. Matrix components may also affect the analysis in nonspecific ways. For instance, particulate matter may clog a column, or matrix components may foul surfaces in the separation system. Generally, the analyst is aware of a sample's origin, and thus has a general idea of matrix composition. A defined sample preparation procedure is implemented before separation to remove the different types of matrix components known to adversely affect the quality of the separation, and hence of the resulting data. This procedure is often relatively laborious and time consuming, requiring various manual manipulations.

    A different analytical issue that may arise when performing analysis using separations is that the analyte of interest is present at concentrations below the detection limit of the available detector. Biochemical analytes such as nucleic acids or low-abundance proteins present a particular challenge, since these analytes are often present at very low concentrations in sample volumes of not more than a couple of microlitres. Again, sample preparation procedures whose aim it is to preconcentrate the analyte of interest can be applied to ensure that the analyte is detected. As is the case with the removal of matrix components, these procedures are generally not automated.

While the sophistication of analytical separation instrumentation has increased tremendously in recent years, instrument developers have generally not focused on the incorporation of automated sample preparation. Clearly, a high demand would be placed on analytical instrumentation developed to perform completely automated chemical analysis of complex samples using a separation approach. The concept of micrototal chemical analysis systems (μ-TAS), which deals with the miniaturization of analytical chemistry instrumentation, has been proposed as a route to the development of flow systems for fast, automated chemical analysis [1]. Interconnected microchannel networks formed in the surfaces of planar glass, polymer, or silicon substrates are the means by which these systems are realized. The integration of separation techniques such as CE and capillary electrochromatography onto microchip formats have been particularly successful [2, 3]. An integral part of the μ-TAS concept is the integration of small-volume sample preparation for the removal of matrix compounds and/or the increase of the analyte concentration in the sample. Ideally, sample preparation and separation could be combined into a single microfluidic device. The necessity for sample preparation for the removal of matrix compounds is particularly great when downscaling of analytical instrumentation is explored, to avoid, amongst other things, channel clogging. Determining, for example, the concentration of aflatoxin in a shipload of pistachio nuts requires a proper pretreatment of the sample before chemical analysis can take place. This chapter does not deal with how a sample representative for a lot should be collected, e.g., the number of pistachio nuts that should be collected and how from the shipload. It rather deals with how to treat a representative bulk sample from this shipload for further analysis using microfabricated devices.

For our purposes, sample pretreatment can be roughly divided into two categories, as shown in Figure 10.1. The first is off-chip sample pretreatment. Though many types of sample preparation have been demonstrated on chips (as will become clear below), there are a few procedures that will be difficult to integrate onto chips, and which may be better done off-chip in any case, for practical reasons. One such operation is the grinding of solid samples into smaller particles, a task that requires considerable force and a substantial amount of sample. To perform such a procedure on a chip may never be as practical or advantageous as by the conventional means it is now done, simply because of the size of the sample. If we consider the pistachio nut example, it is not difficult to imagine that the analysis of pistachio nuts requires some kind of sample pretreatment such as destruction (e.g., grinding) and dissolv-

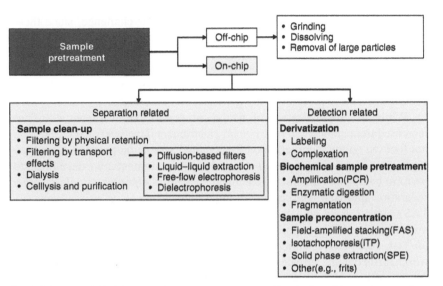

FIGURE 10.1 Off-chip sample pretreatment typically deals with crude samples, which require a robust, particle-tolerant handling system. Also, a considerable physical force is often required to perform the treatment, e.g., for pulverization or grinding. On-chip techniques, on the other hand, benefit from low sample volumes, fast pretreatment times, and straightforward integration with subsequent analysis steps. There are two main applications for on-chip sample pretreatment. First, sample clean-up is used to extract the analyte of interest from its matrix to make it available to the analysis process and to avoid analytical problems caused by interfering molecules. Second, the sample might be treated for improved detection by derivatization to make it detectable, by amplification or transformation to yield enough material in a suitable form, or by preconcentration to lower the detection limits of the instrument.

ing or extraction, which should be performed off-chip since the dimension of a pistachio nut is much larger than the microchannels on a chip.

The second category of sample pretreatment comprises methods of on-chip sample pretreatment [4, 5], which is the subject of this chapter. An analytical chemical experiment involves the measurement of a physical or chemical property or process with a detection system, frequently in combination with a chromatographic separation. Proper sample conditions are required in order not to influence the performance of the microchip device in a negative way. For example, the separation of a mixture of aflatoxins detected with mass spectrometry (MS) requires that the sample be free of particles to avoid channel clogging. In addition, the sample may contain only a low concentration of salts to improve the mass-spectrometric signal-to-noise ratio. On-chip sample pretreatment can be further subdivided into two categories of sample pretreatment, as shown in Figure 10.1. One category involves sample pretreatment to avoid problems in the separation process, whereas the other category focuses on the sample pretreatment necessary to facilitate and improve detection. Sample clean-up, or the separation of sample from the sample matrix, is mainly used to improve the performance of the separation before detection takes place. However, certain techniques listed here may also be used in the context of improving detection efficiency. Dialysis is one example of this, as it is also used to reduce the salt concentration in samples for improved mass-spectrometric analysis. Derivatization, biochemical sample pretreatment, and sample preconcentration are typical steps that are performed to improve detection, though again, many of these methods may also be used upon occasion to prevent problems in the separation itself. Labeling, complexation, and solid-phase extraction can also be used to improve the detection and separation efficiency:

- *Sample clean-up:* Sample clean-up is an important procedure in analytical chemistry in order to prepare the sample for the separation. Filtering of particles is especially important in microchip systems to avoid channel clogging. When the contents of cells are of interest, cell lysis and removal of outer cell membrane components are important aspects of the sample clean-up.
- *Sample preconcentration:* Except in quality control of large batches of a material or chemical process, there is seldom a kilogram of the analyte of interest available for analysis. Particularly in bioanalytical applications, sample size is more frequently in the order of a milligram or microliter with analyte concentrations of nanomoles or even picomoles. To be able to detect such small amounts of analyte, a preconcentration procedure whereby the few available analyte molecules are squeezed into a smaller volume is often a necessity.
- *Derivatization:* Having a sufficient amount of sample is no guarantee that a chemical analysis will work. The analyte of interest must, of course, be detectable with the detection technique used. If this is not the case, the chemical structure of the analyte can be modified to enable its detection. Alternatively, a complex can be formed with a compound that is sensitive for the detection technique used. Derivatization is one of the topics in Chapter 11 and will therefore not be discussed here.
- *Biochemical sample pretreatment:* The molecules studied in biochemistry are often large but present in low concentrations. On-chip amplification, using, for example, the polymerase chain reaction (PCR), can be performed to increase the concentration of DNA. Nucleic acids, in general, may be cleaved into smaller fragments that are more suitable for detection using enzymatic reactions or physical means. Enzymatic digestion is also used to break proteins up into peptides of shorter lengths for protein analysis.

A sample frequently requires several pretreatment steps. This is typically accomplished by carrying out several techniques sequentially. A main advantage of μ-TAS over conventional analytical instrumentation is that μ-TAS allows the monolithic integration of various analytical steps with in-line detection, with negligible dead volume connections, on a single chip. However, most studies performed so far on microfluidic devices demonstrate single sample pretreatment steps only. Few completely integrated monolithic microfluidic devices exist. In this chapter, the sample preparation processes that have been successfully transferred to microchips are described. Each process is discussed with some state-of-the-art examples from the literature. Note that this chapter is not an extensive review and more literature on the subject can be found in peer-reviewed journals, for example, in Refs. [4–7].

10.2 BASIC SAMPLE CLEAN-UP

Sample clean-up comprises a range of physical and chemical techniques that remove sample constituents that might interfere with the subsequent analysis. These constituents can have sizes ranging from the molecular level, such as dissolved salts, to physical particles, including dust or spores. Depending on the clean-up required, different techniques such as physical particle filtering, diffusion-based filters, dialysis, and cell lysis have been implemented on-chip. Sample clean-up techniques are typically used toward the front-end of an analysis system just after the sampling process itself, but intermediate sample clean-up steps further down the line might be required in some situations.

A recent addition to the microfluidic technology base are centrifugally pumped microfluidic systems, which utilize microchannel networks laid out on compact disc-like platforms [8]. These can be spun at high rotation rates to drive solutions through the microchannels. This technology should allow the on-chip implementation of analytical steps typically performed in a centrifuge.

10.2.1 PHYSICAL PARTICLE FILTERING

Clogging of channels due to particle contamination is a common problem when working with miniaturized analysis systems. The small cross sections of microfluidic channel networks require filtering of most liquid samples prior to analysis. Although this can be achieved off-chip using syringe filters and the like, integrated filtering can be desirable for automatic analysis systems. An overview of the size ranges relevant to filtering systems is given in Figure 10.2.

A number of microstructures has been developed to prevent particles from entering into a microfluidic system. Before going into detail, it should be pointed out that most of these structures can also be deployed for the opposite purpose, namely to retain particles like chemically coated polymer beads or biological cells in a reaction volume. The examples given in this section are taken from both filtering and retention applications. What these devices all have in common is the use of advanced fabrication technologies to realize

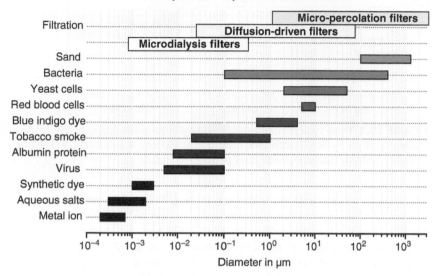

FIGURE 10.2 Typical dimensions of particles or analyte entities encountered in microfluidic systems. (Reprinted from J. Lichtenberg, N.F. de Rooij, E. Verpoorte, *Talanta* **2002**, *56*, 233. With permission. Copyright 2002 Elsevier Science. Based on GE Osmonics' "Filtration and Separation Spectrum", GE Osmonics Inc., Minnetonka, MN.)

arrays of structures with well-defined, reproducible micrometer and even submicrometer gaps.

10.2.1.1 Filters Based on Lithographically Defined Constrictions

A straightforward approach to the filtering problem is the implementation of flow restrictions in the flow path of the channel network. Thus, particles having a cross section larger than the gap between the barriers are retained, while solution can freely pass through the restriction. For efficient filtering, a large number of gaps should be fabricated in parallel so that blocked filter gaps do not adversely affect the fluid flow. These so-called *axial percolation filters* can be fabricated by simple one-mask photolithography processes. For example, formation of filter structures in a silicon wafer, consisting of arrays of narrow channels, can be accomplished using one photopatterning step followed by transfer of the channel array into the silicon surface by either dry or wet etching. Dry etching using a process known as deep reactive-ion etching (DRIE) can yield channels that are narrow and deep [9], as shown in Figure 10.3 (leftmost diagram). The depth-to-width (or height-to-width) ratio of microfabricated structures is known as the aspect ratio. If features are narrow and tall, they are referred to as high-aspect-ratio structures, whereas

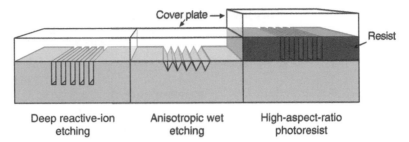

FIGURE 10.3 Three typical fabrication technologies for microfilters: bulk micromachining of silicon by deep reactive-ion etching (DRIE) (left) or anisotropic wet etching (middle); additive structures made from photoresist (right).

low-aspect-ratio features are wide and short. DRIE makes possible the formation of high-aspect-ratio gap structures in silicon, capable of retaining particles but allowing for sufficient fluid flows. Conventional wet chemical etching by immersion of the silicon wafer in an alkaline solution yields channels or channel arrays as shown in the middle diagram of Figure 10.3 [10]. In both cases, the etching process is described as anisotropic, since the etch rate is clearly not the same in all directions into the wafer bulk. The profile of the channel cross section obtained by wet etching will be determined by the crystal orientation of the silicon wafer, with the more-difficult-to-etch crystal planes defining the walls of the microchannel [11]. While the cross-sectional profile of a channel obtained by anisotropic wet etching of a regular (1 0 0)-oriented silicon wafer is typically triangular or trapezoidal as depicted in the middle diagram, specially cut (1 1 0) wafers may also yield vertical side walls as shown in the left sketch [11]. High-aspect-ratio structures may also be formed in thick (>5 μm) layers of photoresist [12], as depicted in the rightmost diagram of Figure 10.3. Polymeric replication techniques like casting and hot embossing have also been used to fabricate structures similar to those obtained by silicon substrate etching [13].

In an early example of liquid chromatography integrated onto a silicon-glass device, arrays of V-shaped grooves 3 μm wide and 2 μm deep, formed by anisotropic wet etching, defined the frit at the end the microchannel serving as the separation column. In this way, it was possible to retain 5-μm C18 particles in the column to form a packed bed for separation [14]. Anisotropic wet etching was also the first technique employed for on-chip filters [10]. White blood cells could be efficiently isolated by filtering them through a network of silicon pillars with a gap of 5 μm. Due to the elasticity of the cells, the pillars had to be rather long in the flow direction (175 μm) to retain the cells, since the cells deformed and easily slipped through the channels between shorter pillars. A disadvantage of anisotropic wet etching is the low-aspect ratio of the etched features, which is always less than 1. For optimum filtering, one dimension should be small (to retain the particles), while the other should be larger in order to allow sufficient fluid flow at low pressures. Numerical mathematical methods have been employed to predict flow characteristics in microfabricated filters [15].

As mentioned above, DRIE of silicon has been used to work around the problem of low-aspect-ratio features by alternating between anisotropic etching and polymeric side wall passivation, which leads to extraordinary aspect ratios of up to 30:1 [16]. Arrays of 50-μm-high posts (3 μm × 10 μm) with spacings as close as 2 μm have been used to capture polystyrene beads with a diameter of 5.5 μm [9, 17] (Figure 10.4). The aim here was not to remove particles from the analyte solution, but rather to pack the complete filter volume with surface-coated particles to perform a chemical reaction. It should be noted that alternative methods for packing microbeads into microfluidic devices without a requirement for filter structures have been developed recently as well [18, 19].

An alternative approach to fabricating high-aspect-ratio structures is the use of special photoresists like the epoxy-based EPON SU-8. These resists can be deposited in thick layers (up to a millimeter), show excellent aspect ratios (up to 20:1) and, once cross-linked by exposure to light, exhibit high mechanical strength and chemical resistance. A grid of SU-8 pillars was used to retain functionalized glass beads 135 to 180 μm in diameter in a micro reactor, the pillars being 300 μm high, 120 μm wide, and spaced at 80 μm [12]. Replication techniques are an alternative route to photopatterning to obtain polymer structures. Russo et al. [13] fabricated an axial filter sieve with a cutoff size of 5 μm by hot embossing polystyrene.

Although the majority of microfluidic devices today have a single layer of fluidic conduits on the top side of the substrate, vertical channels through the

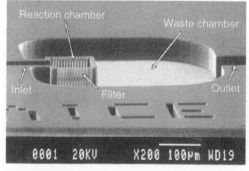

FIGURE 10.4 A flow-through reactor featuring a filter structure composed of a row of pillars fabricated by deep reactive-ion etching. The pillar dimensions are $3 \times 10 \times 50 \, \mu m^3$ with a spacing of 2 μm. (Reprinted from H. Andersson, W. van der Wijngaart, P. Enoksson, G. Stemme, *Sens. Actuators B* **2000**, *67*, 203. With permission. Copyright 2000 Elsevier Science.)

substrate are used in more complex devices [20, 21]. Along the same lines, solutions flowing perpendicular to the substrate plane can be filtered through perforated membranes [22–24]. Again, the minimum particle cutoff size is determined by the resolution of the photopatterning process. For instance, Kuiper et al. [24] were able to fabricate an array of small holes with diameters down to 65 nm and a pitch of 200 nm in thin silicon nitride membranes about 100 nm thick. These researchers used laser-interference photolithography to create a pattern of parallel fringes on the wafer, rather than illuminating a photoresist layer directly through a mask containing the two-dimensional pattern of the desired structure. In this way, it was possible to achieve sub-micrometer resolution of the filter pores. A thin layer of photosensitive resist deposited on the wafer served as mask for the subsequent etching of filter pores into the membrane. By adjusting the exposure dose to half of the necessary dose and superimposing two fringe patterns at a 90° angle, all areas of the resist where two fringes crossed received a double dose of light and were thereby polymerized. After development, a regular array of square posts remained. However, a general limitation of all membrane-type filters is the maximum pressure applicable before the fragile membrane breaks. The fracture of filter membranes has been studied for 0.5-μm-thick silicon nitride membranes perforated by circular holes with 5 μm diameter and 25% perforated area [23]. Under airflow, these membranes broke at 300 mbar for a 2-mm-wide, rectangular membrane and at 1 bar for a 500-μm-wide one.

A different approach to the filtering problem was adopted by He et al., who designed a *lateral percolation filter* [25], which is based on the retention of the particles in a bed of posts, through which the fluid flows laterally. In the given example, an array of posts 1.5 μm wide and 10 μm high was machined into a quartz wafer by reactive-ion etching at the bottom of the reservoirs leading into the microchannels. Liquid entering the system from above passes the network of channels between the posts and then exits laterally into the channel. Particles larger than the 1.5-μm distance between the posts are held back without clogging the filter, as many other surrounding channels still allow liquid flow.

10.2.1.2 Filters Based on Sacrificial Layer Technology

Most of the design strategies mentioned above are limited to filter cutoff sizes of about 1 μm, which corresponds to the resolution limit of typical photolithography processes used for microfluidic device fabrication. However, to achieve pore sizes comparable to standard syringe filters, the feature size of the elements forming the flow restriction network has to be less than 0.5 μm. Microelectronic foundries have long been working in the sub-micrometer feature size range for planar devices (etching depth <0.5 μm). However, the fabrication of microfluidic structures of that size with micrometer depths is extremely difficult, because of the poor aspect ratios of most etching processes at these dimensions.

To overcome this problem, several groups have proposed sacrificial layer (SL) techniques to create tortuous flow paths with restrictions in the range of tens of nanometers for efficient filtering [26–29]. Here, the smallest feature size is not directly defined by lithography, but rather by deposition of a thin, intermediate material layer. This is subsequently selectively removed, and the gap that remains acts as a filter "pore." Deposition techniques like chemical or physical vapor deposition (CVD and PVD) can be controlled very precisely to yield surface layers in the nanometer range, so that filters with pore sizes of down to 10 nm can be fabricated [27].

Figure 10.5 shows the schematic fabrication procedure for an SL filter as proposed by Kittilsland et al. [26]. The filter principle is based on two planes perforated by a regular hole pattern, which are offset by half of the hole diameter. A thin gap between the two planes allows liquid flow from one side

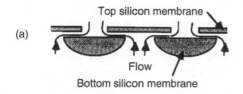

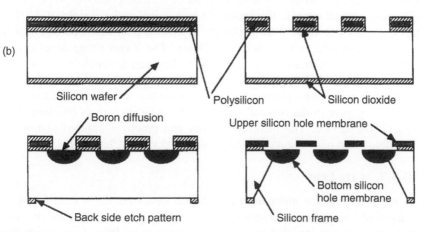

FIGURE 10.5 (a) Schematic principle of a sub-micrometer sacrificial layer filter. (b) A thin layer of silicon dioxide acts as spacer for the flow restriction opening. It is removed (or sacrificed) at the end of the fabrication procedure. (Reprinted from G. Kittilsland, G. Stemme, B. Norden, *Sens. Actuators A* **1990**, *23*, 904. With permission. Copyright 1990 Elsevier Science.)

of this membrane structure to the other. To fabricate this device, a thin layer of thermal silicon oxide, covered by polysilicon, serves as the SL. Access holes are etched into the polysilicon layer by dry etching, followed by removal of the SL to open the filtration pores. Subsequently, holes are etched into the backside of the filter by dopant-selective etching in potassium hydroxide solution [30]. The final structure has a 1.5-μm-thick filter membrane with inlet and outlet holes in the range of 10 μm. The cutoff size of the filter can be tailored by changing the SL thickness, which is 50 nm in the example given above.

Table 10.1 compares different fabrication approaches for physical particle filters.

10.2.2   FILTERING BY DIFFUSION

A unique feature of microfluidic systems is the dominance of laminar flow conditions at the flow rates normally used (nanoliters per second to microliters per second), due to the small dimensions of the channels. An interesting quality of laminar flow is that material transport between two adjacent streamlines can only take place by diffusion. Turbulent mixing, as it is done in a test tube or beaker by stirring or active agitation of the liquid vessel, is not possible in microchannels. Diffusional filtering can be achieved if laminarly flowing sample and buffer streams are brought in contact for a given time. This contact time can be tuned such that lighter molecules having a high mobility can cross over into the buffer stream, while heavier particles remain in the source flow and are finally washed out of the system. In fact, rates of diffusion for small molecules differ substantially from those for larger molecules and particles. To illustrate, a small organic dye molecule with a molecular weight of 330 Da takes roughly 0.2 sec to diffuse 10 μm, whereas a polymer bead of 0.5-μm diameter travels the same distance in 200 sec. The Yager group developed a number of filter systems based on diffusion between a sample and a filtrate stream flowing side by side in a laminar fashion [31]. These devices do not require sub-micrometer features for particle retention and show a reduced risk of clogging as both the original sample and the filtrate are in constant flow through the device. The theoretical basis and practical implementation of diffusion-based microfluidics are discussed in detail in Chapter 7.

10.2.3   CELL LYSIS

An ultimate goal for microanalytical device development is the integrated analysis of intracellular components like nucleic acids or proteins on-chip. However, this requires that the analyte of interest be freed from inside the surrounding cell membrane and related structural material by lysis or disruption. Four approaches for cell lysis have been implemented on-chip, namely chemical lysis using a detergent, thermal lysis, ultrasonication to physically break open spores, and electroporation.

TABLE 10.1

Comparison of Different Fabrication Techniques Used to Form Microstructures for Physically Retaining Particles for Filtering or Bead-Capture Applications

Fabrication of the Filter Structure	Material of the Filter Structure	Filter Type	Smallest Cutoff Size	References
Anisotropic wet etching	Silicon	Axial	5 μm	[9]
Deep RIE	Silicon	Axial	2 μm	[10]
High-aspect ratio photo resist structures	SU-8, epoxy-based photo resist	Axial	80 μm	[12]
Hot embossing	Polystyrene	Axial	5 μm	[13]
Membrane perforated by RIE etching	Silicon nitride	Axial	0.5 μm [22], 4 μm [23]	[22, 23]
Perforated membrane using RIE and interference lithography	Silicon nitride	Axial	0.065 μm	[24]
RIE	Quartz	Lateral	1.5 μm	[25]
Sacrificial layer	Silicon	Axial	0.05 μm [26], 0.01 μm [27]	[26–29]

Li and Harrison demonstrated the transport and manipulation of biological cells on planar glass substrates [32]. Electric fields were applied to generate electroosmotic flow (EOF) of the solution containing the cells, and thus move the cells themselves. Cell lysis was demonstrated by mixing a sample stream containing canine erythocytes in an isotonic buffer with a 3-mM sodium dodecyl sulfate (SDS) detergent solution at a T-intersection. The use of other lysis reagents proved to be less suitable for cell lysis on chip. For instance, the addition of copper(II) ions resulted in a lysis process that was too slow to be monitored over a short channel distance. The use of acidic solutions also was not practical, because of the EOF reversal associated with low pH. A different fluidic approach for partial cell lysis and protein extraction, based on diffusion transfer, has been presented as a T-shaped flow-through system [33]. This device relies on two adjacent, laminar streams, one of which carries the sample cell suspension, while the other contains a lysis agent. The latter diffuses into the sample stream, partially lysing the cell membrane to extract proteins contained inside the cells. These diffuse back into the lysis stream, which is guided into the downstream analysis module of the microchip.

Aside from chemical treatment, cell lysis can also be induced by thermally disrupting the cell membrane at temperatures above 90°C for short times (typically ~1 min). This technique can be conveniently combined with the PCR (see Section 10.4.1), which also requires heating of the sample solution. Devices integrating cell lysis and PCR have been fabricated by attaching a Peltier-type heater [34] or by placing the chip in a conventional PCR thermocycler [35]. Cell lysis induced by short YAG laser pulses has been used in conventional microcolumn analysis [36].

Lysis of targets like bacterial spores is more challenging, as the spore outer cortex is very resistant to both chemical and physical treatments. Belgrader et al. [37], therefore, selected a harsh physical force to break the spores open. Ultrasonic waves (power 40 W, frequency 47 kHz) were coupled into the microfluidic system containing the sample in a suspension with 106-μm-diameter glass beads. An external ultrasonic transducer based on a stack of piezoelectric transducer disks was mechanically interfaced via a flexible membrane to the polymer chip. A titanium horn transferred the ultrasonic energy into the lysis chamber. It was proposed that gaseous cavitation by collapsing bubbles from dissolved gas creates localized high-pressure waves that disrupt the spores. The efficiency of the disruption was verified by subsequent on-chip PCR of the lysis product of a sample containing 10^5 *Bacillus subtilis* spores per milliliter. TaqMan analysis was used to monitor the formation of PCR product in real time, using fluorescence to determine the threshold cycle number, C_T, the point at which a significant increase in fluorescence signal is suddenly observed. Purified spores ultrasonically disrupted for 30 sec yielded a C_T of 28 (14 min), while the C_T for the amplification of untreated, physically intact spores was 40 (21 min). The efficiency of the miniaturized sonicator was nearly identical to a conventional 1-h germination procedure used in laboratories today. An improved, self-contained version of the cartridge developed for portable applications has also been presented [38]. Further optimization of the ultrasonic energy transfer via a flexible membrane into the chamber by applying a static pressure to the sample-bead suspension in the lysis chamber reduced the C_T further down to 24 [39].

Applied electric fields can be used to render cell membranes permeable for transfection, in a process known as electroporation [40]. In fact, if the electric field is high enough, complete disruption of the cell membrane can result [41]. A microfluidic device integrating electrokinetic focusing and lysis of cells together with fluorescent labeling and electrophoretic separation of cell contents was presented recently [42]. A field strength of 900 V/cm was required to induce lysis of Jurkat cells derived from human T-cell leukemia. To reduce Joule heating effects caused by the high field strength and the high-conductivity extracellular buffer typically used in cell analysis, the lysis field was applied in AC mode at 75 Hz. Electrically induced cell lysis at field strengths between 1 and 10 kV/cm was achieved using metal electrodes integrated into a microfluidic device [43]. The close spacing between the electrodes of a few micrometers required only low absolute voltages (10 V) to obtain the necessary fields.

For automated analysis, transport and handling of cells inside a microfluidic device is required before the actual cell lysis. A number of approaches for cell transport have been presented, including electroosmotic pumping [32], elektrokinetic focusing of cells [44], and hydrodynamic cell handling by integrated peristaltic pumps fabricated in an elastomeric substrate [45]. Fluorescence-activated cell sorting on-chip allows a cell population to be sorted according to whether a fluorescent marker molecule has been incorporated into the cell or not [46]. A device capable of capturing single cells from a suspension for perfusion with a sample reagent was also reported recently [47].

10.2.4 DIALYSIS

Dialysis is a process that allows the separation of molecules in solution by means of their differential rates of diffusion through a semipermeable membrane. As the diffusion coefficient depends directly on the molecular weight of a molecule, dialysis is often used to separate molecules of different size. The first dialysis procedure reported in the literature dates from almost one century ago. This was a hemodialysis experiment in which "diffusible substances were removed from the circulating blood of living animals" [48]. Since then, dialysis has served as a vital treatment for patients with poor kidney function. In the field of chemical analysis, the introduction of microdialysis probes in 1974 [49] started a new trend toward miniaturized sampling. These devices were capable of transferring the analyte with higher efficiency into the analysis systems for both *in vitro* and *in vivo* experiments. The tailor-made design of microdialysis probes also facilitated coupling of this sampling technique with modern instrumentation and newly developed sensors. After introducing the dialysis process in the next section, the discussion will focus on the continuing efforts to improve the performance of microdialysis probes and their integration with μ-TAS devices.

10.2.4.1 The Dialysis Process

Dialysis is typically performed in a continuous-flow system as schematically shown in Figure 10.6. The sample, a mixture of different compounds, is pumped through the donor channel past a semipermeable membrane that permits the transfer of hydrophilic molecules with a molecular weight below the molecular-weight cutoff (MWCO) of the dialysis membrane. A second liquid stream, the acceptor solvent, is guided along the other side of the membrane in the recipient channel. Other terms often used for the acceptor solvent are perfusion fluid or perfusate (before the dialysis membrane) and dialysate (after the dialysis membrane). Under steady-state conditions, the liquids do not mix as there is no convective flow over the membrane. However, due to the semipermeable nature of the membrane, mass transport of small molecules is possible by diffusion across it from the sample stream to the acceptor solvent or the other way around. Only molecules with a molecular weight above the membrane MWCO are retained in the sample stream.

Diffusion processes can be described by the differential transport governed by Fick's second law of molecular diffusion, relating temporal and spatial derivatives of the concentration:

$$\frac{\partial C}{\partial t} = D \frac{\partial^2 C}{\partial x^2} \tag{10.1}$$

where C is the concentration of the analyte, t is the diffusion time, D is the diffusion coefficient of the analyte in the medium, and x is the diffusion distance.

Fick's law can be employed for dialysis setups where the pore diameter of the membrane is much larger than the size of the diffusing solute. As can be seen from Equation (10.1), diffusion is driven by a spatial concentration gradient, i.e., efficient dialysis requires that the sample and acceptor fluids have different analyte concentrations. A stagnant acceptor solution stream would result in a steadily decreasing concentration gradient as analyte concentrations in the two streams become increasingly similar with time. Therefore, the acceptor solvent is continuously flushed past the membrane to maintain a concentration gradient between sample and acceptor solutions, which is sufficient to drive diffusion of analyte from sample to acceptor. Note

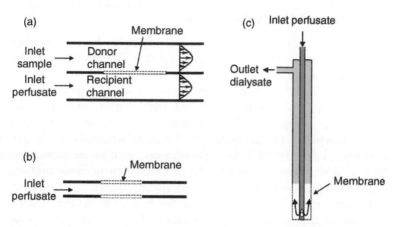

FIGURE 10.6 (a) Schematic of a continuous-flow dialysis experiment. A sample is flushed through a donor channel along a semipermeable membrane that permits the transfer of hydrophilic molecules with a molecular weight smaller than the MWCO into the acceptor solvent on the other side of the membrane. (b) Linear probe consisting of a hollow fiber permitting small molecules to pass the membrane. (c) Microdialysis probe with concentric tube design.

that molecules may also travel in the other direction, from perfusate into the sample solution. This latter situation arises when it becomes desirable to deliver compounds to the system under study.

Dialysis is employed in two ways, either for transfer of a low-molecular-weight analyte of interest into the acceptor or removal of unwanted, low-molecular-weight species from the sample stream. In the first case, the sample stream is discarded after dialysis, since the low-molecular-weight analyte of interest has been dialyzed into the acceptor fluid. The acceptor fluid containing the analyte, known as dialysate, can be fed into the subsequent analysis step. This approach is often used to perform *in vivo* sampling of biologically important molecules from living systems using low-volume microdialysis techniques. The second approach is used when the analyte of interest has a high molecular weight. Now it is the sample stream that is guided to the analyzer, while the acceptor liquid loaded with the low-molecular-weight contaminant is discarded. As an example, desalting by dialysis is a typical pretreatment step to reduce ion concentrations before performing a mass-spectrometric analysis of oligonucleotides. This is required because a high salt content masks the analytes and reduces the detection limits of the spectrometer.

Dialysis is not always the best choice for sample pretreatment, because diffusion is a relatively slow process and dialysis is not highly specific. The speed of sample pretreatment can be very important, especially for experiments where molecules are continuously monitored in a sample stream as will be discussed in the next section.

10.2.4.2   Microdialysis Probe Designs

For the discussion in this chapter, several examples have been selected from the large number of microdialysis probe designs reported in the literature. A more complete overview can be found in Refs. [50, 51].

The planar flow-through probe design presented in Figure 10.6(a) can be used for samples of both biological and nonbiological origin [52]. While planar structures can generally be easily integrated into microfluidic devices, their main drawback is that the sample fluid must be removed from the system under study. In the case of *in vivo* experiments, such as the monitoring of analytes in subcutaneous tissues or the determination of concentrations of clinically important molecules in a blood stream, the removal of fluids is generally undesirable as it might disturb the biological system. Therefore, different probe designs have been introduced for *in vivo* applications, of which two important ones are presented in Figure 10.6(b) and (c). The design shown in Figure 10.6(b) is a linear probe that consists of a hollow fiber permitting small molecules to pass from the outside to the inside. The fiber is connected on both ends with regular tubing for interfacing with the instrument and the perfusate is pumped through the inner space of the fiber. The probe can be positioned in the sample medium or implanted in tissue, allowing samples to be extracted from a spatially well-defined position.

Linear probes can also be constructed in concentric designs, where the hollow fiber is placed inside a second tube, the so-called shunt, carrying the sample solution. Such a setup has been used for the investigation of phenol and its metabolites in a bile duct, which was connected to the shunt [51].

To improve localized sampling further, concentric, needle-shaped microdialysis probes that can be directly inserted into the sample are used. Usually, the probe is formed by two concentric tubes with the perfusate entering through the inner, impermeable tube as shown in Figure 10.6(c). The perfusate is guided into the tip of the probe, where the hydrophilic semipermeable membrane is mounted for analyte collection. Finally, the resulting dialysate exits the probe through the gap between the inner tubing and the outer tubing. The needle can be inserted into a liquid sample or positioned in subcutaneous samples, which allows the analysis of small molecules from well-defined regions [50, 51, 53].

10.2.4.3   Factors Influencing the Performance of Microdialysis

The performance of a microdialysis probe for a given analyte is expressed as the relative recovery (RR) or extraction efficiency, which is a measure of the ratio of analyte concentration in the dialysate coming from the probe and that in the actual sample:

$$\mathrm{RR} = \frac{C_\mathrm{d} - C_0}{C_\mathrm{s} - C_0} \tag{10.2}$$

where C_d is the analyte concentration in the dialysate, C_S is the actual concentration of analyte in the sample, and C_0 is the concentration of analyte in the perfusate before microdialysis.

Note that the difference between C_S and C_0 is the driving force for diffusional mass transport across the membrane, and that C_S has to be larger than C_0 to assure diffusion toward the perfusate.

Microdialysis is a process that can be used in two directions, i.e., molecules diffuse from the sample through the membrane into the perfusate or, inversely, molecules diffuse into the sample (C_S being lower than C_0). Thus, endogenous molecules can be sampled at the same time that exogenous molecules, for example, drugs, are delivered into the subcutaneous tissue [51].

One of the main factors that influences the recovery is the flow rate, which is inversely proportional to the recovery. Lower flow rates lead to longer contact times between the analyte and the membrane surface, and a higher concentration of the analyte of interest in the dialysate as a result. Typical flow rates used for optimized recovery are in the order of microliters per minute or even nanoliters per minute for both the donor and acceptor channels. However, some important considerations must be made when decreasing the flow rate:

- The development of analytical instrumentation with low dead and connection volumes is required. Dispersion will be significant if the dead volume is large in comparison to the flow rate.
- The final dialysate volume available for further analysis decreases with the flow rate. The sample volumes required for the analytical instrumentation used further down the line will therefore dictate the minimum flow rates which may be used.
- Decreasing the flow rate will increase the delay from sample site to detector, which compromises real-time monitoring applications. To quantify this delay, the lag time, τ, can be calculated as [54]:

$$\tau = \frac{V_{\text{active}} + V_{\text{connector}}}{q} \tag{10.3}$$

where q is the flow rate, V_{active} is the swept volume of the dialysis site, and $V_{\text{connector}}$ is the swept volume of interconnection to the downstream analyzer.

Microfabricated systems can help alleviate problems associated with connection volumes, dead volumes, and flow rates, since it is possible to fabricate microflow systems with extremely low dead volumes, and to handle minute amounts of liquids precisely and at low flow rates.

Besides flow rate, other factors influencing the recovery are:

- Membrane type and characteristics: surface area, surface chemistry, porosity, and thickness are important parameters here.
- Diffusion rate in the sample medium and membrane: the distance x traveled on average by an analyte molecule in time t is $x = (2Dt)^{1/2}$. Since diffusion coefficients of small molecules are in the order of $10^{-5}\,cm^2/sec$, such a molecule travels typically ~50 μm in 1 sec.
- Medium viscosity, η: this is because the diffusion coefficient, D, is inversely proportional to the viscosity of the medium as described by the Stokes–Einstein equation:

$$D = \frac{kT}{6\pi\eta a} \tag{10.4}$$

where k is the Boltzmann constant, T is the temperature, and a is the effective hydrodynamic radius of the molecule.
- Temperature: temperature also has a direct effect on the diffusion coefficient as can be seen from Equation (10.4). Note that the viscosity of a liquid decreases roughly exponentially with temperature.
- Geometry of the microdialysis unit: concentric tube designs as shown in Figure 10.6(c) generally lead to higher recoveries and lower lag times.
- Direction of the flow: counter-current flows have been observed to lead to higher recoveries than concurrent flows [52, 55, 56].
- Adsorption processes: membranes with ionic or polar groups may give rise to adsorption of analytes, leading to fouling that will affect diffusion over the membrane, and hence recovery.
- Nature of the sample matrix: this influences the diffusion coefficient. It is possible that the analyte is bound to the matrix so that the analyte remains unavailable for dialysis.
- Physiological processes: related to the environment under study.

Membranes are commonly made of cellophane, cellulose acetate, polycarbonate (PC), polyvinyl chloride, polyamides, polysulfone, polyethersulfone, polyvinylidene fluoride, copolymers of acrylonitrile and vinyl chloride, polyacetal, polyacrylate, polyelectrolyte complexes, cross-linked polyvinyl alcohols, and acrylic copolymers such as Nafion [57]. Membranes that are used for aqueous sample media and hydrophilic analytes are hydrophilic in nature. The pores of the membrane are filled with water from the sample or dialysate. Due to the incompatibility with hydrophobic compounds, their recovery is, in general, lower than the recovery of more polar and ionic compounds.

10.2.4.4 Examples of Microdialysis on a Chip

All microfluidic implementations of dialysis processes rely on conventional, semipermeable membranes, which are typically integrated in a sandwich structure between two substrates structured with channel networks. Smith and co-workers used two membranes on a single chip, one with a high and the other with a low MWCO, to obtain a sample free of both cellular residues and low-molecular-weight salts for mass-spectrometric analysis. The chip consisted of three structured PC layers with a cellulose ester membrane in between each pair (Figure 10.7). Channel networks were ablated into the PC by an excimer laser micromachining system. The sample is introduced through the top PC layer into the serpentine channel 1 with a flow rate between 0.5 μl/min and 200 nl/min. The high MWCO membrane ensures that only molecules with a molecular weight of below 50,000 Da can diffuse into channel 2, while the remainder goes to a waste reservoir. A hole is drilled through the middle PC plate at the end of channel 2 connecting it with channel 3. A counter-current flow of buffer solution (10 mM ammonium acetate, 10 μl/min) runs in channel 4, which is in contact with channel 3 via a low MWCO membrane (<8,000 Da). On their way to the on-line electrospray tip, the low-molecular-weight compounds diffuse into the acceptor solvent in channel 4 and are thus removed from the sample. As a result, all compounds finally reaching the mass spectrometer have a molecular weight between

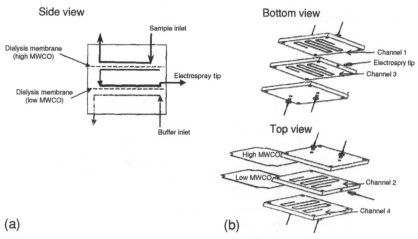

FIGURE 10.7 (a) Schematic of the dual-microdialysis device showing the flow directions inside the channels. (b) Three-dimensional view illustrating the design and assembly of the device. Two microdialysis membranes are sandwiched between three polymer layers containing serpentine flow channels. (Reprinted from F. Xiang, Y. Lin, J. Wen, D.W. Matson, R.D. Smith, *Anal. Chem.* **1999**, *71*, 1485. With permission. Copyright 1999 ACS.)

8,000 and 50,000 Da. This low dead-volume dual microdialysis approach was applied to remove undesired compounds from a cellular lysate of *Escherichia coli*. Additionally, the high concentration of salts originally present in the sample was reduced, leading to a 20-fold increase in the signal-to-noise ratio of the mass spectrum. The detection sensitivity was sufficient to perform additional MS/MS analysis for the identification of biomarkers [58–61].

A combination of dialysis and preconcentration has been realized as a three-layer chip made by silicon-template imprinting with integrated high-molecular-weight ($>80,000$ Da) cutoff membranes. The sample, a reaction mixture containing an aflatoxin B_1 antibody and the aflatoxins B_1, B_2, G_1, G_2, and G_{2a}, is introduced into channel 2 as depicted in Figure 10.8. In this channel, the antibody binds to the aflatoxins to form complexes with molecular weights above 80,000 Da, which remain in channel 2 as they cannot pass through the dialysis membrane. Unbound aflatoxins, on the other hand, are carried away through channel 1.

Sample flow rates of less than 100 nl/min corresponding to a 2-min interaction time with the membrane were required in order to remove all nonbinding low-molecular-weight molecules. For subsequent processing, the outlet of channel 2 is connected to the preconcentration channel 3 by a through-hole. The channels 3 and 4 are again separated by a high MWCO membrane, which allows evaporation of water from the sample stream into a dry-air counterflow flushed through channel 4. Before the sample is ionized in an electrospray ionization source for mass-spectrometric analysis, the antibody–aflatoxin complex is dissociated at a microdialysis junction. Figure 10.9 illustrates the mass spectra of a 4:1 (a) and 1:4 (b) molar binding

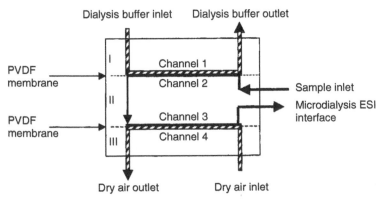

FIGURE 10.8 Side-view schematic of a miniaturized affinity dialysis and concentration system. I–III indicate top, middle, and bottom imprinted copolyester pieces (piece II is imprinted on both sides). Two PVDF membranes separate the copolyester channels. (Reprinted from Y. Jiang, P.-C. Wang, L.E. Locascio, C.S. Lee, *Anal. Chem.* **2001**, *73*, 2048. With permission. Copyright 2001 ACS.)

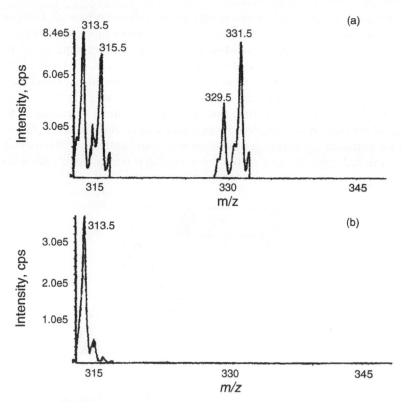

FIGURE 10.9 Positive ESI mass spectra of an aflatoxin B1 antibody–aflatoxins reaction mixture eluted from the miniaturized affinity dialysis and concentration system in Figure 10.8: (a) 4:1 and (b) 1:4 molar binding ratios of antibody to each aflatoxin. It is easily visible that the aflatoxin antibody has a preference for aflatoxin B_1 (*m/z* value of 313.5). (Reprinted from Y. Jiang, P.-C. Wang, L.E. Locascio, C.S. Lee, *Anal. Chem.* **2001**, *73*, 2048. With permission. Copyright 2001 ACS.)

ratio of antibody to each aflatoxin. A 2-week repeatability experiment for the analysis of aflatoxins proved that the robustness of the system was very good with day-to-day variations of 5 to 10%, most likely due to the electrospray ionization conditions [62].

While the planar designs described so far are well adapted to continuous processing of an already available sample, microfluidic devices coupled to microdialysis probes for direct sampling have also been developed. Freaney et al. presented a system consisting of a mini-shunt microdialysis unit and a biosensor for the continuous *in vivo* monitoring of glucose and lactate in dogs [63, 64]. The microdialysis probe assured that a protein-free sample was obtained. A delay time of 5 min at a flow rate of 3 μl/min led to a relative recovery of almost 100%.

Böhm et al. connected a conventional microdialysis probe with a silicon-based biosensor system [54]. A concentric microdialysis probe was integrated with a flow cell containing solid-state sensors for the monitoring of the chloride concentration in the dialysate flow. The flow cell consisted of two fusion-bonded silicon wafers with fluidic features created by a multistep, anisotropic wet etching process (Figure 10.10). The outer capillary of the microdialysis probe was inserted into the deeper channel extending to the edge of the chip. A through-hole in the lower silicon chip served as an inlet for the perfusate liquid between the outer and inner capillaries. The dialysate was directed from the exit of the inner capillary into a small connection

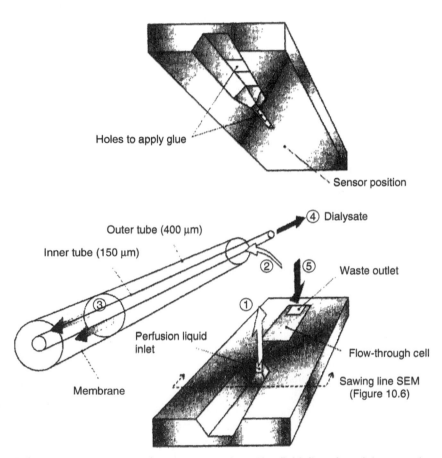

FIGURE 10.10 Connector/sensor construction. The fluid flow from inlet to outlet is marked with numbers 1–5 (1–3: perfusion liquid, 4: dialysate entering the flow-through cell, 5: dialysate leaving the system). (Reprinted from S. Böhm, W. Olthuis, P. Bergveld, *Sens. Actuators B* **2000**, *63*, 201. With permission. Copyright 2000 Elsevier Science.)

channel leading into the flow cell. The latter featured an Ag/AgCl electrode serving as chloride sensor. The microdialysis probe was held in place by glue that was dispensed into two through-holes etched into the upper silicon chip. While conventional probes externally coupled with the chloride sensors had lag times of 60 sec, the lag time could be reduced to only 10 sec with the same relative recovery by using the integrated design.

10.3 ANALYTE EXTRACTION/PRECONCENTRATION

10.3.1 Liquid–Liquid Extraction

In liquid–liquid extraction one or more compounds are transferred over a liquid–liquid interface from one solvent to the other. Extraction is carried out either to isolate or concentrate the analyte of interest, or to separate it from interfering species. It is essential for this type of separation that the analyte of interest in the sample stream has a preference for the accepting liquid phase. Conventional liquid–liquid extraction is performed by mixing the sample solution with an immiscible solvent. Usually, one phase is aqueous, while the other is an organic solvent; it is the aqueous phase that is extracted with the organic solvent in the majority of cases. The analytes partition between the two phases until equilibrium is reached. Partitioning is enhanced by stirring or shaking to accelerate the transfer process. The equilibrium at the end of the extraction process can be described using the partition constant, K_p:

$$K_p = \frac{[Z]_{org}}{[Z]_{aq}} \tag{10.5}$$

where $[Z]_{org}$ and $[Z]_{aq}$ are the concentrations of analyte in the organic and aqueous phase, respectively. K_p is also variously known as the distribution constant or partition coefficient. A better way to describe the extraction process is through introduction of the distribution ratio, D, which takes into account that the analyte of interest can appear in several forms in either or both of the phases:

$$D = \frac{[\text{total concentration of all forms of Z}]_{org}}{[\text{total concentration of all forms of Z}]_{aq}} \tag{10.6}$$

D is also known as the distribution coefficient. The use of D in conjunction with ionic compounds is appropriate, for instance, since such compounds may be present in both their neutral or ionic forms. When dealing with mixtures of similar compounds of which only one is to be extracted, selective extraction can be achieved by exploiting differences in the chemical properties of the species involved. The properties of the mixture to be extracted or the extracting solvent can be adjusted so that the analyte of interest is converted to a form that is preferentially transferred. Multiple extractions can also be performed, either with aliquots of the same extracting solvent or by varying the composition of this solvent to ultimately achieve separation of the analyte of interest.

Due to the low surface-to-volume ratio in conventional liquid–liquid extraction setups, the specific surface area of the liquid–liquid interface is comparatively small leading to slow extraction speeds. Although this issue can be alleviated by turbulent mixing, this may lead to foam and emulsion formation, which requires additional time to let the mixture settle. A possibly faster and more efficient way to perform this sample pretreatment step is through miniaturization of the extraction vessel, leading to a higher contact surface per processed liquid volume under well-defined, laminar flow conditions, as will be discussed in the examples section. A complete treatment of the theory underlying liquid–liquid extraction can be found in Refs. [65–67].

10.3.1.1 Factors Influencing Extraction Efficiency and Selectivity

The nature of the analyte and type of organic solvent are the most important parameters that influence K_p and D. Ionic compounds, for example, generally do not dissolve well in organic solvents. However, an ionic analyte can be extracted into the organic phase through interaction or reaction with a species that converts it to an uncharged species. Formation of an ion pair or of a complex with a chelating agent are examples of such interactions. In this way, the solubility of the analyte increases in the organic phase while it decreases in the aqueous phase. This is one way of increasing the extraction selectivity, since use of a selective complexing agent will ensure transfer of only one species. Changing the pH of the aqueous phase can also have an effect on the K_p of acidic and basic species, since charge will be a strong function of pH. To take the effect of temperature into account, the thermodynamic partition coefficient (κ_p) can be introduced, which is the ratio of the activities of analyte in the organic phase, $(a_i)_{org}$, and in the aqueous phase, $(a_i)_{aq}$. It can be shown that this ratio is equal to $\kappa_p = \{(a_i)_{org}/(a_i)_{aq}\} = \exp(-\Delta G^0/RT)$, where ΔG^0 is the free energy change of the process, T is the temperature, and R is the ideal gas constant [65–67]. Other factors that influence the extraction efficiency are the concentration of the analyte and the effect of masking agents present in the sample.

10.3.1.2 Examples on Chip

Tokeshi et al. demonstrated that miniaturizing a liquid–liquid extraction set-up leads to a reduction of the extraction time by one order of magnitude compared to conventional techniques. They used an ion-pairing agent added to the organic phase (chloroform) to extract an iron complex from the aqueous phase using a parallel two-phase laminar flow. However, the increased surface-to-volume ratio of their microfluidic channels also resulted in an increased adsorption of the ion-pairing agent on the channel walls. The resulting extraction efficiency is therefore smaller than that reported for conventional liquid–liquid instrumentation [68].

Burns and Ramshaw used an immiscible slug flow produced in a T-junction structure with channels 380 μm deep and 380 μm wide to perform liquid–liquid extraction. Slug formation was accomplished by pumping an organic phase (a mixture of silicon oil and kerosene) through one of the inlets (lower vertical channel with dark solution, Figure 10.11) and an aqueous phase through the other inlet (upper vertical channel with light solution, Figure 10.11). This resulted in the formation of alternating oil and water plugs in the channel as shown in Figure 10.11. To quantify the extraction efficiency of the device, KOH or NaOH and the pH indicator phenol red were added to the water phase. Acetic acid was added to the organic phase together with a dye to visualize the flow. After slug formation, extraction of acetic acid into the liquid phase begins, where it reacts with hydroxyl ions to change the pH. Internal circulation, as shown in Figure 10.11, enhances the speed of the extraction and should provide a reaction performance similar to or better than that possible by simple diffusion between parallel flow streams [69].

10.3.2 SOLID-PHASE EXTRACTION

Solid-phase extraction (SPE) is a widely used technique for the preconcentration of analytes and/or the separation of analytes from the matrix compounds in a sample [70, 71]. Again, the technique relies on the partitioning of the analyte of interest between the sample and the extraction material. If the partitioning process is well designed, the analyte of interest prefers to be adsorbed to or absorbed into the extraction material rather than remaining in the sample matrix. After the analyte is trapped on the extraction material, it can be desorbed by washing with a solvent that reverses the partitioning process, such that the analyte preferably dissolves into the elution solvent. Alternatively, an increase

(a) 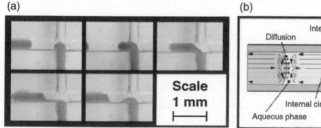 (b)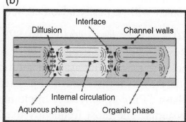

FIGURE 10.11 (a) Slug formation by alternating injection of two immiscible liquids into a microchannel. (b) Illustration of internal circulation generated within immiscible slug flow. (Adapted from J.R. Burns, C. Ramshaw, *Lab Chip* **2001**, *1*, 10. Reprinted with permission. Copyright 2001 RSC.)

in temperature can be used to desorb compound, as is done when SPE is used to extract and preconcentrate analytes in GC.

SPE can be performed in static and dynamic modes [70]. In static SPE, the sample and extracting material are mixed in a confined space for a given period of time, during which neither the sample nor the extraction material are renewed. As a consequence, static SPE would typically be used as an off-chip sampling technique, which is beyond the focus of this chapter. SPE is mostly used in the dynamic mode, where the sample is flushed continuously over the extraction material. Often, a dynamic SPE system will consist of a column with a packed bed of extraction material for preconcentration and sample clean-up.

10.3.2.1 Adsorption Mechanisms

Several adsorption mechanisms are available for the transfer of the analyte in the sample to the extraction material: dipole–dipole interactions and hydrogen bonding (normal-phase SPE), van der Waals interactions or nonpolar–hydrophobic interactions (reversed-phase SPE), and ionic interactions (ion-exchange SPE). While these techniques are rather selective toward bulk properties of the analyte, a more specific extraction technique combining molecular recognition and noncovalent reversible adsorption mechanisms is heterogeneous (immuno)affinity chromatography.

10.3.2.1.1 Normal-Phase SPE
Normal-phase SPE is performed with a stationary phase that is more polar than the solvent or sample matrix. Analytes with a relatively high polarity will be adsorbed and preconcentrated on the extraction material based on dipole–dipole interactions, hydrogen bonding, π–π interactions, and induced dipole–dipole interactions. Typical stationary phases are silica, alumina magnesium silicate, and bonded phases including aminopropyl, cyanopropyl, and propyldiol. Normal-phase SPE is not very common in combination with microfluidic devices as it is difficult to implement with EOF, which is often used for liquid handling on chips. The aqueous buffers required for EOF pumping are not compatible with the polar, stationary phase as water molecules will be easily adsorbed, leaving no place for the analyte of interest.

10.3.2.1.2 Reversed-Phase SPE
Reversed-phase SPE is performed with an apolar extraction material and a polar (e.g., aqueous) sample solution, which makes the partitioning of apolar analytes with the extractant possible. As aqueous buffers are used, reversed-phase SPE can be readily implemented in miniaturized, EOF-driven analysis systems. The most frequently used packing materials for this technique are octylsilane (C-8) or octadecylsilane (C-18) immobilized on a silica support. Other adsorption materials contain alkyl chains with one, two, or four carbon atoms, cyclohexyl, and phenyl groups. Graphite and polymeric sorbents, such as styrene–divinylbenzene (SDB), are also used. Structures for retaining

beads or particles in microchannels were mentioned in Section 10.2.1. Further discussion of this issue will be presented in Section 10.3.2.2. To harvest the adsorbed analyte, solvents like methanol, acetonitrile, or ethylacetate are typically used. It might also be necessary to apply two different solvents to elute all analytes from the surface. Solvent molecules can form hydrogen bonds with the free silanols of the silica surface and lead to a disruption of the van der Waals or nonpolar–hydrophobic interactions between the analyte and SPE material [71].

Although reversed-phase SPE is generally used for apolar analytes, a method exists that allows polar analytes, which are often ionic, to bind to reversed phases as well. In this method, termed ion-pair (IP) SPE, an additional IP-marker molecule is added to the sample [72]. This marker molecule comprises both nonpolar and polar moieties. The polar part of the marker interacts with the polar analyte, and in doing so, also imparts nonpolar properties to the analyte–marker complex through virtue of the nonpolar moiety. This ensures increased interaction of the analyte–marker complex with the extracting phase. Some examples of IP markers are propanesulfonic acid, trifluoroacetic acid, triethylamine, and tetraoctylammonium bromide.

10.3.2.1.3 Ion-Exchange SPE

Ion-exchange SPE is performed with an extraction phase that is capable of exchanging an analyte ion from the sample with an ion of the same charge type from the hydrophilic ion-exchange material. Often copolymer resins (SDB) or silica particles with diameters in the nano- to micrometer range are used as support material for the ion-exchange groups.

Ion-exchange resins are called strong exchangers if they have a permanent positive or negative charge, independent of the pH. A strong cation exchanger is, for example, a resin functionalized with sulfonic acid groups and protons as counter ions. A competition for the available negative sulfonic acid groups arises when a sample is introduced containing cations, such as potassium ions, and an exchange of protons for cations will take place until equilibrium is established. A typical support material of strong anion exchangers is silica functionalized with quaternary ammonium ions.

Weak ion exchangers are typically materials that become uncharged when the pH of the solution is adjusted. If the pH is adjusted to about 2 pH units below the pK_a of the weak cation-exchange resin, its ion-exchange capabilities will be lost. The same accounts for weak anion resins when the pH is adjusted to 2 units above the pK_a value.

The capacity of an ion-exchange resin is typically given in milli-equivalent (meq) per gram of material and is of the order of 0.5 to 1 meq/g. The meq is 1/1000 of the gram-equivalent weight of the material (e.g., 0.039 g for potassium). The ion-exchange capabilities of an exchanger for a given ion depend on the charge of the analyte ion and its competing ions. Multiply charged ions have a stronger interaction with the charge on the resin than singly charged ions, and large ions are retained better than small ions. Some ions have a stronger interaction with the resin than others and are therefore more selectively exchanged. Protons interact weakly with the negative groups of the resin and are therefore easily exchanged for cations like copper, which have a stronger interaction with the resin.

10.3.2.1.4   Heterogeneous (Immuno)affinity

In extraction processes based on heterogeneous (immuno)affinity, probe molecules (ligands or antibodies) are immobilized on the surface of the support material, and can be regarded as the primary active sites in the extraction [73–75]. Contact of a sample containing the analyte of interest (target) with the support material leads to capture of the analyte at these sites. Extraction by affinity interactions involves the formation of biomolecular ligand–receptor complexes such as lectin–sugar, drug–protein, and enzyme–substrate complexes. Extraction making use of the selectivity of immunological reactions, such as antigen–antibody interactions, is referred to as immunoaffinity extraction. The name has its root in the mammalian immune system, which produces antibodies in the presence of a potentially harmful foreign substance (antigen). These antibodies can be used as highly selective active sites for sample preconcentration and clean-up of compounds, which led to antibody production in the first place.

The ligands or antibodies are chosen for their specificity of interaction with the analyte to be extracted. The intermolecular interactions involved are based on hydrogen bonding, and noncovalent ionic, hydrophobic, and van der Waals interactions. The spatial arrangement of the molecular interactions is unique for each pair of reactants considered. The binding energy between ligand or antibody and analyte can be large even though individual interactions are often weak, since a large number of these interactions are involved in the formation of the active site–analyte complex.

The ligands or antibodies are bound to the surface of a support material by covalent or noncovalent bonding. These support materials can be organic in nature, such as cellulose and dextran. Inorganic supports, such as silica and synthetic polyacrylamide, are also frequently used. The choice of the packing material depends on several parameters. The most important of these are listed below:

- Possibility to derivatize the surface of the capillary or packing material to immobilize the probe molecules necessary for (immuno)affinity interactions.
- Chemical and biological resistance to nonspecific adsorption (NSB) of both the analyte of interest and other undesired sample components. NSB can be avoided to some extent if hydrophilic support materials are used. Agents which passivate binding sites not due to the probe species, so-called blocking agents, may also be used to improve performance.
- Large surface areas and pore sizes are desirable.

- High mechanical stability to prevent degradation of the material, which could result in blockages by small particles originating from the material and a degradation of sample flow through the packed bed. Increased pressure drops over the packed bed is one example of the latter.
- Possibility to immobilize probe molecules in such a way that the interaction with analyte is affected as little as possible by steric hindrance. Spacer molecules are often inserted between the support surface and the probe molecule to improve accessibility of the immobilized species to the species in solution.
- Uniform size of the packing material for good flow characteristics.

A dissociation of the active site–analyte complex can be induced by washing the extraction material with a solvent containing a displacer or chaotropic agent, or through variation of the buffer pH or use of a water–organic modifier mixture (e.g., water–acetonitrile). An increase of the temperature can also be used to dissociate the complex.

An important aspect of (immuno)affinity extraction is the density of active sites present on the surface of the packing material. Bonding densities are defined by the number of ligands or antibodies linked to the surface of the packing material and are typically expressed in milligrams of ligand or antibody per gram of packing material. The total number of immobilized ligands or antibodies that actively participate in the partitioning process is given by the capacity of the extraction material.

The generally high recovery of the analyte is mainly affected by two parameters, the capacity of the extraction material and the affinity of the ligands or antibodies for the analyte. Regeneration of the extraction material can be very important if reusability is an issue. The eluting solvent must be chosen such that the probe molecules remain attached to the column and the activity is not irreversibly altered during elution. More information about (immuno)affinity extraction can be found elsewhere [73–75].

10.3.2.2 Practical Considerations for SPE on Chip

The retention behavior of analytes on the SPE material is best described by standard liquid chromatography theory, with a low number of theoretical plates because the sorbent bed is short [71, 76]. A detailed description of this theory is beyond the scope of this chapter. It is, however, important to mention some parameters that affect the performance of an SPE column. These are:

- *Flow rate.* The flow rate must not be so high that it will reduce partitioning of analyte between sample and extraction material, leading to low analyte recovery.

- *Particle size and porosity.* The size of the particles and diameter of the pores determine the surface area of the extraction material. In general, it can be stated that the higher the surface area of a material, the higher the number of theoretical plates and the better the efficiency of the extraction material is. However, pore size should not be smaller than the dimensions of the analytes.
- *Column length.* By increasing the length of the column in which the extraction takes place, the number of theoretical plates and capacity is increased as well. Other parameters that have an influence on the retention behavior of analytes are temperature, analyte properties, conditioning solvent, rinse solvent, and eluting solvent.
- *Surface functional groups.* The functional groups required for SPE need to be present on the packing material surface. Alternatively, it should be possible to derivatize surface groups such that the surface becomes suitable for SPE.
- *Temperature.* Partition constants are temperature dependent, with less extraction onto the solid phase at elevated temperatures. Hence, large temperature fluctuations should be avoided to prevent irreproducible extraction.
- *Extraction material.* The extraction material must be chosen such that the analyte of interest is compatible with the extraction material. Other matrix components present in the sample should be as incompatible with the extraction material as possible, to avoid a preconcentration of compounds that might interfere during the separation and/or detection. Flow charts such as that in Figure 10.12 help when making a first selection of extraction material.

Several approaches have been reported in the literature to introduce and pack beads in microfluidic channels. Most rely on the integration of weirs or wall structures in the microfluidic channels to act as physical barriers to retain particles but at the same time to allow solutions to pass through [77–79]. In studies by Oleschuk et al., particles were introduced via a separate bead introduction channel using electrokinetic pumping [79, 80]. In another approach, the channel was flushed with a suspension of 75-nm latex particles that absorb to the surface of silicon or pyrex microfluidic channels [81].

Particles may also be packed into microchannels without the necessity of a physical barrier, if tapered channel geometries are used. The tendency of small particles in a slurry to aggregate as they are forced into a narrowing channel geometry, and to clog this section of the channel as a result, is known as the *keystone effect*. The formation of the stationary particle plug in the taper allows the channel behind it to be packed with particles to lengths of at least several centimeters. In a recent example, 3-μm particles were trapped at the tapered end of a microchannel whose width was reduced from 70 to 16 μm over a short distance [19]. The application in this instance was

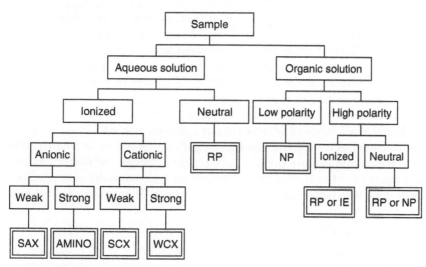

FIGURE 10.12 Method selection guide for the isolation of organic compounds from solution. SAX, strong anion exchanger; SCX, strong cation exchanger; WCX, weak cation exchanger; RP, reversed-phase sampling conditions; NP, normal-phase sampling conditions; IE, ion-exchange sampling conditions. (Adapted from C.F. Poole, A.D. Gunatilleka, R. Sethuraman, *J. Chromatogr. A* **2000**, 885, 17–39. Reprinted with permission. Copyright 2000 Elsevier Science.)

microchip electrochromatography, but as similar particles are often used for SPE, this approach should also be appropriate for realization of on-chip SPE.

The performance of an SPE column for a given analyte can be tested by measuring the breakthrough volume [70, 76], which can be measured by continually loading the extraction material with the sample and monitoring the analyte concentration at the end of the column. Breakthrough has occurred when the analyte concentration at the end of the SPE column begins to increase. The volume of sample required to achieve breakthrough is thus known as the breakthrough volume. A low breakthrough volume indicates a low level of analyte interaction with the extraction material. Analytes that exhibit high levels of interaction will be more actively retained, resulting in a larger volume of sample being required before sample is observed at the end of the SPE bed. Breakthrough of analytes can also be caused by overloading of the column with analytes or matrix components or by a low number of theoretical plates.

10.3.2.3 Examples of SPE on a Chip

10.3.2.3.1 Reversed-Phase SPE

Oleschuk et al. packed 1.5 to 4.0-µm octadecylsilane-(C-18)-coated silica beads as a reversed-phase material into a 10-µm-deep microfluidic channel by introducing a 9-μm-high physical barrier, or weir, in the channel. This approach is depicted in Figure 10.13. For the packing procedure, they used a separate bead introduction channel connected with the weir region, through which beads were electrokinetically pumped. A 1-nM BODIPY solution in buffer was loaded on the beads for different lengths of time at 1.2 nl/sec. Acetonitrile was used to elute the analyte from the column with a concentration enhancement of up to 500 times [79, 80].

Kutter et al. followed an open tubular approach with a C-18 coated glass microchannel. A coumarin C460 dye (8.7 nM) was dissolved in a buffer with 15% acetonitrile and electrokinetically pumped through the coated channel for extraction. After 160 sec the dye was eluted from the walls of the channel by flushing with buffer containing 60% acetonitrile. They achieved an 80-fold enrichment of a dye [82].

10.3.2.3.2 IP SPE
Soper and coworkers made a PMMA chip with IP SPE and integrated contact conductivity detection for the detection of oligonucleotides having lengths from 100 to 2000 base pairs [83]. The walls of the PMMA chips were coated with a C-18 stationary phase. The IP marker was triethylammonium acetate. This approach enabled the separation of a mixture of six DNA fragments with 100, 200, 400, 800, 1200, and 2000 base pairs.

10.3.2.3.3 Ion Exchange
Only a few examples exist where ion exchangers are immobilized in microfluidic channels. However, to the best knowledge of the authors, ion exchange on-chip has not yet been used for the preconcentration of analyte ions. To date, ion-exchange processes have been exploited in a chromatographic mode to separate different ions based on differences in their ion-exchange equilibrium. This is discussed below.

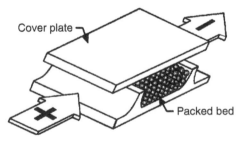

FIGURE 10.13 Cross-sectional drawing of a packed chamber, showing weir heights in relation to channel depth and particle size. Packing material is introduced via a third channel into the plane of the paper (not visible) by electrokinetic pumping. (Adapted from R.D. Oleschuk, A.B. Jemere, L.L. Schultz-Lockyear, F. Fajuyigbe, D.J. Harrison, *Proceedings of the μTAS 2000 Symposium*, Enschede, The Netherlands, May 14–18, 2000, van den Berg, A., Olthuis, W., Bergveld, P., Eds., 2000, Kluwer Academic Publishers: Dordrecht, pp. 11–14. Reprinted with permission. Copyright 2000 ACS.)

Glennon and co-workers performed open tubular ion-exchange chromatography, using microchannels as columns on chip. These channels were 23 cm long and were coated with 75-nm, quaternary ammonium-modified latex particles. Open tubular approaches on chip for liquid-phase separations are normally not desirable, due to the slow diffusion of analytes to the walls of the column. However, the width × height of the channels used in this work was $200 \times 3.6 \, \mu m$, which reduces the diffusion path length to the walls to a maximum of $1.8 \, \mu m$. Problems during the coating procedure of the walls of the channels were not reported. The adhesion of the particles to the walls of the chip (silicon covered with Pyrex) is such that a coating of the walls prior to introduction of the particles was not necessary. They demonstrated the separation of thiourea, NO_2^-, NO_3^-, and I^- [81].

10.3.2.3.4 Demonstrations of Chip-Based Immunoaffinity Extraction: Heterogeneous Immunoassay

At the time of writing, immunoaffinity extraction as a chip-based sample pretreatment method has not been reported on microfluidic devices, to the best of the authors' knowledge. However, an increasing interest in the integration of heterogeneous assays into microchannels has produced a number of related examples of immunoassay on chip. Probe molecules have either been immobilized on beads or on microchannel walls. In one example, polystyrene beads with a diameter of $45 \, \mu m$ were packed into 100-μm-deep, 200-μm-wide microfluidic channels for the determination of human secretory immunoglobulin A [77] and carcinoembryonic antigen [78]. The device comprised a single channel with a wall structure located midway down the channel perpendicular to fluid flow. The height of the wall in the channel was $90 \, \mu m$ so that a channel of $10 \, \mu m$ could be used to elute analyte over the packed bed of beads.

The first example of a heterogeneous immunoassay was performed on an electrokinetically operated glass chip using immobilized protein A (PA), which has a high affinity for fluorescently labeled (Cy5) rabbit immunoglobulin G (rIgG), the analyte in this case [84]. Figure 10.14 shows the layout of the glass chip, with the arrows indicating the direction of fluid movement. PA is physisorbed on the silanized walls of the double-T junction, which, with a volume of 165 pL, serves as the reaction chamber. Figure 10.15 is a trace of the photomultiplier signal during the different steps of the experiment. Of particular note is the elution of the fluorescent sample which appears as a sharp peak. In this case, a sample containing 1.5-μM fluorescently labeled (Cy5) rabbit immunoglobulin G(rIgG) in reservoir 5 was flushed through the chamber for 200 sec to the waste. This was long enough to saturate the immobilized PA with Cy5-labeled rIgG. Excess sample was then rinsed away with a Tris–HCl buffer (pH 7.5) in reservoir 3a. The sample was eluted by flushing the intersection with a glycine–HCl buffer of pH 2.0. Detection by fluorescence with a laser diode took place just below the double-T junction.

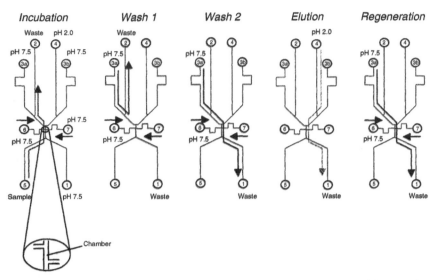

FIGURE 10.14 Layout of a microfluidic chip for heterogeneous assay, with its mode of operation. Overall outer dimensions of the microfluidic network are 6.7×2.8 cm^2 and channel dimensions are 50×20 μm^2. (Reprinted from A. Dodge, K. Fluri, E. Verpoorte, N.F. de Rooij, *Anal. Chem.* **2001**, *73*, 3400. With permission. Copyright 2001 ACS.)

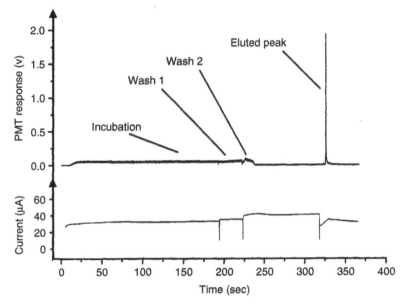

FIGURE 10.15 Photomultiplier response measured simultaneously with the current generated by electroosmotic pumping for a complete analysis cycle. The Cy5–rIgG concentration was 1.5 μ*M*. (Reprinted from A. Dodge, K. Fluri, E. Verpoorte, N.F. de Rooij, *Anal. Chem.* **2001**, *73*, 3400. With permission. Copyright 2001 ACS.)

The experiment was reproducible (~13%). The capacity of the column was estimated to be 5.1×10^9 IgG molecules (8×10^{-15} mol). Only 118 nl of sample was used in this experiment, with a total analysis time of less than 5 min. In fact, this device exploited an immunochromatography approach, and provides a nice illustration of how this may be implemented on a microfluidic device.

A disposable heterogeneous assay with a simple cross channel configuration in poly(dimethylsiloxane) (PDMS)/glass was used by Linder et al. for the determination of human IgG [85]. They used a special surface treatment, a three-layer sandwich coating, to avoid nonspecific adsorption of analytes on the PDMS surface in areas not containing immobilized antihuman IgG antibody. A similar surface treatment in a PDMS and glass structure was developed by Eteshola and Leckband [86].

10.3.3 ELECTROKINETIC-MOBILITY-BASED SAMPLE PRECONCENTRATION

Sample preconcentration comprises techniques where analyte molecules from a comparatively large sample volume are concentrated (or "stacked") into a much smaller volume, which is subsequently analyzed using techniques such as CE. The analyte concentration that the actual analysis process has to deal with is therefore much higher (5-fold to 10^6-fold) than the original concentration in the sample. This can alleviate the high performance requirements imposed on the detector instrumentation and widen the choice of suitable detector types for a particular application.

The techniques described in the following sections allow preconcentration of analyte molecules based on their electrokinetic mobility. This is generally achieved by creating a gradient at some point in the microfluidic system that changes the migration velocity of the analyte ions. As soon as they slow down with respect to overall flow velocity, they stack and preconcentrate.

After a general introduction into the molecular transport at such a gradient, different types of gradients are discussed. These basics lead to techniques such as microchip-based field-amplified injection (FAI), field-amplified sample stacking (FASS), and isotachophoresis (ITP).

10.3.4 CREATING ELECTROKINETIC VELOCITY GRADIENTS

Electrokinetic field gradients are the key element for all electrokinetic preconcentration techniques [87–93]. Molecules that travel through a gradient will experience a change in velocity that leads to preconcentration (slowing down) or dilution (speeding up). To understand the nature of these gradients, we will review the underlying principles of electroosmotic and electrophoretic flow components.

For simplicity, the general discussion will be focused on a straight, buffer-filled glass microchannel. Consider an analyte or buffer ion i migrating under the influence of an electric field E. The apparent migration velocity of the analyte is proportional to the sum of its electrophoretic mobility, μ_{ep}, and the electroosmotic mobility of the buffer, μ_{eo}, as outlined in Chapter 2:

$$v_i = E \underbrace{(\mu_{ep} + \mu_{eo})}_{\mu_i} \tag{10.7}$$

The observed migration direction of the ion depends on the sign of the sum of μ_{ep} and μ_{eo}, which is therefore also called observed mobility, μ_i.

If the ion can be approximated as a spherical particle, the electrophoretic mobility of an ion can be expressed by the Stokes–Einstein relation:

$$\mu_{ep} = \frac{q}{6\pi r \eta} \tag{10.8}$$

where q is the charge of the ionic cloud around the particle, r is the particle radius, and η is the buffer viscosity.

The electroosmotic mobility, μ_{eo}, on the other hand, is a function of buffer and capillary surface parameters:

$$\mu_{eo} = \frac{\varepsilon \zeta}{\eta} \tag{10.9}$$

where ε is the dielectric constant of the buffer, and ζ is the potential at the plane of shear between the stagnant and mobile layers of the double layer.

If one considers the adjustable parameters in Equations (10.7) to (10.9), a number of possibilities arise for the creation of electrokinetic velocity gradients. In order to increase v_i, one could locally:

(a) Increase the electrical field strength, E
(b) Reduce the buffer viscosity, η
(c) Increase the dielectric constant of the buffer, ε
(d) Increase the zeta potential, ζ (by modification of the channel wall or by changing the solution pH)

Of these options, no reports for (c) can be found in the literature, but all other techniques are used to create velocity gradients. The dominating technique today relies on altering the electrical field strength in a small portion of the channel by adjusting the concentration of buffer ions. The electrical field strength, E, in a portion of a microchannel depends on the electrical conductivity of the buffer, σ:

$$E = \frac{I}{A\sigma} \tag{10.10}$$

where I is the current in the capillary, A is the cross section of the capillary, and σ is the buffer conductivity.

While I is constant along the entire length of a uniform capillary having a constant A, the local electric field strength depends on σ, which in turn is a function of the buffer concentration:

$$\sigma = F \sum_{n} z_n \cdot \mu_{ep,n} \cdot c_n \tag{10.11}$$

where F is Faraday's constant (96,495 C/mol), z_n is the valence of species n, $\mu_{ep,n}$ is μ_{ep} of species n, and c_n is the concentration of species n.

Therefore, the channel section filled with a high concentration of buffer exhibits a lower electric field strength and thus lower ionic migration velocities than the low-conductivity portion, in which the higher electric field strength will lead to higher migration velocities.

A different approach to establishing a velocity gradient takes advantage of the temperature dependence of the electrical conductivity of an electrolyte solution [87]. Generally, solution conductivity, σ, increases with temperature due to reduced solution viscosity and can therefore be written as

$$\sigma(T) = \sigma_0 \cdot f_\eta(T) \cdot f(T) \tag{10.12}$$

where σ_0 is the conductivity at reference temperature (e.g., 20°C), f_η is the conductivity dependence on η normalized to 1 at the reference temperature, f is the conductivity dependence on all other factors normalized to 1 at the reference temperature, and T is the temperature.

The electric field strength therefore also depends on the temperature:

$$E(T) = \frac{E_0}{f_\eta(T) \cdot f(T)} \tag{10.13}$$

where E_0 is the field strength at reference temperature.

Unfortunately, the electrophoretic mobility, μ_{ep}, also depends strongly on the viscosity and thus on the temperature:

$$\mu_{ep}(T) = \mu_{ep,0} \cdot f_\eta(T) \cdot f_{ep}(T) \tag{10.14}$$

where $\mu_{ep,0}$ is the electrophoretic mobility at reference temperature, f_η is the mobility dependence on η normalized to 1 at the reference temperature, and f_{ep} is the mobility dependence on all other factors normalized to 1 at the reference temperature.

Combining Equations (10.13) and (10.14), we obtain the electrophoretic component of the analyte migration velocity, v_{ep}:

$$v_{ep}(T) = E(T) \cdot \mu_{ep}(T) = E_0 \cdot \mu_{ep,0} \cdot (f_{ep}(T)/f(T)) \tag{10.15}$$

This equation shows clearly that the temperature dependence of the electric field strength and of the electrophoretic mobility tends to cancel each other out, except for buffer solutions where $f_{ep}(T)/f(T)$ is not constant. As $f_{ep}(T)$ is constant for most buffers and analytes [88], buffers with a temperature dependent $f(T)$ have to be chosen to reliably generate velocity gradients by temperature changes. Ross and Locascio for instance, identified buffers made from Tris and boric acid to show a strong change by a factor of up to 1.8 when heating the solution from 20 to 80°C [87].

When comparing both examples to generate velocity gradients, it should be noted that the temperature method has the advantage of creating a locally stationary gradient at the point of heating or cooling. Inversely, gradients due to buffer solutions of different concentration generally migrate through the channel [89]. As the location of a stationary velocity gradient, and thereby the location of analyte preconcentration, remains constant, stationary gradients can be advantageous for long-term preconcentration processes to avoid uncontrolled migration of the preconcentrated zone through the device [87].

10.3.4.1 Molecular Transport at an Electrokinetic Velocity Gradient

To understand the preconcentration effect caused by a velocity gradient, let us consider now a fixed electric field gradient in the channel at zero EOF as depicted in Figure 10.16. The high and low field strength zones are assumed to be very long so that depletion effects are avoided and the width of the gradient between zones is small. For the following discussion, it is convenient to define γ as the ratio between the high and low field strengths (E_H and E_L) in the system:

$$\gamma = \frac{E_H}{E_L} = \frac{\sigma_{EL}}{\sigma_{EH}} \tag{10.16}$$

The field strengths are proportional to the buffer resistivities or inversely proportional to the buffer conductivities, σ_{EL} and σ_{EH}, where σ_{EL} is the conductivity of the low-field-strength segment and σ_{EH} is the conductivity of the high-field-strength segment. γ is therefore known as the *relative conductivity* [89].

Under an applied electric field, ions migrate from one side of the gradient to the other, the direction being dependent on the apparent velocity, v_i. For each species i, the ionic flow, F_i, through the cross section of the channel, A (in molecules per second) can be described as

$$F_i = c_i \cdot v_i \cdot A \tag{10.17}$$

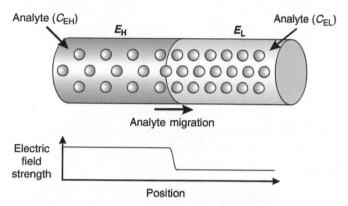

FIGURE 10.16 Analyte stacking at an electric field strength gradient. The analyte migrates from the high-field-strength region (E_H) into the low-field-strength region, E_L, thereby slowing down and increasing in concentration. Destacking (loss of analyte concentration) occurs if the migration direction is reversed.

In order to fulfill the mass-balance requirement, F_i has to be constant on both sides of the gradient. We can therefore write for the flows in the high- and low-field-strength regions, F_{iH} and F_{iL}, respectively, after eliminating the constant capillary cross section, A, and the constant electrophoretic mobility, μ_{ep}:

$$F_{iH} = F_{iL}$$
$$c_{iH} \cdot v_{iH} \cdot A = c_{iL} \cdot v_{iL} \cdot A$$
$$c_{iH} \cdot E_H \cdot \mu_{ep} = c_{iL} \cdot E_L \cdot \mu_{ep} \tag{10.18}$$
$$\frac{c_{iL}}{c_{iH}} = \frac{E_H}{E_L} = \gamma$$

The concentration of each ionic species therefore adjusts instantly when ions pass from one side of the gradient to the other. Consequently, the concentration is increased or decreased (depending on the migration direction) by the factor γ, which is therefore also called the *preconcentration factor*. It should be noted here that this relationship holds true for all ionic species in the channel, including buffer ions. In a practical implementation of the configuration in Figure 10.16, the sample is dissolved in a low-conductivity buffer (thereby creating a high field strength in this region) while the separation buffer has a high conductivity leading to a low field strength.

Let us now consider the FASS situation depicted in Figure 10.17. Contrary to the example in Figure 10.16, the high- and low-field-strength zones do not extend infinitely. Instead, a small sample plug of length L_{iEH}, containing analyte cations dissolved in a low-conductivity buffer, is injected into a microchannel filled with a high-conductivity separation buffer. We assume that the EOF is in its normal direction toward the cathode.

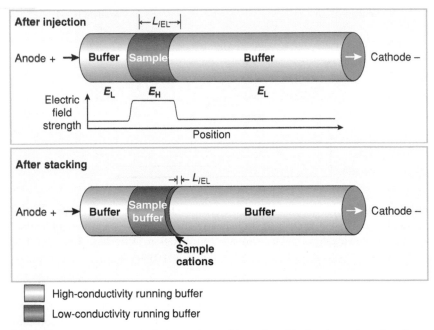

FIGURE 10.17 Field-amplified sample stacking of positively charged analyte ions. The electrical field strength is concentrated in the low-conductivity buffer plug and drives ions to one or the other end of the plug. Once they reach the high conductivity running buffer, they slow down and stack, due to the abrupt reduction in field strength.

As soon as the electric field is applied to the channel, the sample cations will be concentrated and simultaneously focused into a narrow zone of length $L_{i\text{EL}}$ once they cross the velocity gradient. $L_{i\text{EL}}$ can be drawn from the mass balance of the system as the original sample plug and its preconcentrated counterpart contain the same number of analyte molecules [89]:

$$L_{i\text{EH}} \cdot c_{i\text{H}} = L_{i\text{EL}} \cdot c_{i\text{L}} \tag{10.19}$$

where $c_{i\text{H}}$ is the analyte concentration prior to preconcentration (in the high-field-strength region) and $c_{i\text{L}}$ is the analyte concentration after preconcentration (in the low-field-strength region).

Therefore, by expressing the analyte concentration ratio in terms of γ, we obtain

$$L_{i\text{EL}} = L_{i\text{EH}} \cdot \frac{c_{i\text{EH}}}{c_{i\text{EL}}}$$

$$L_{i\text{EL}} = \frac{L_{i\text{EH}}}{\gamma} \tag{10.20}$$

We can conclude that FASS is beneficial for electrokinetic separation techniques in two ways: the analyte ions are preconcentrated (lower limits of detection) and simultaneously spatially focused (high separation resolution) [90].

10.3.4.2   Effects of Induced, Pressure-Driven Flow

Generally, if gradients of electrokinetic velocity are formed, the electroosmotic bulk flow is also changed. In the case of a field-strength gradient, the electroosmotic velocity will change proportionally to E. If the gradient is created by two different buffer concentrations, the electroosmotic mobility is further altered as a function of the ζ-potential, which depends on the ionic strength of the buffer solution (see Chapter 2). To maintain flow continuity in such a system, the difference in EOF velocity is compensated by a pressure-induced (Poiseuille) flow component [88]. The superposition of the flat EOF velocity profile and of the parabolic Poiseuille velocity profile give rise to increased dispersion of a well-defined sample plug, ultimately leading to sample dilution and reduced separation efficiency [91, 92]. These flow profiles have been observed both in capillaries [88] and on microfluidic devices [93].

10.3.5   Techniques for Electrokinetic Sample Preconcentration for Microchip Devices

The implementation of the various sample preconcentration techniques developed for capillary-based separations in microchip devices is straightforward in most cases. In addition, the benefits of microfluidic devices such as zero-dead-volume connections allow the integration of more precise and sometimes also more robust preconcentration schemes. However, one has to keep in mind that side effects such as pressure flows induced by local EOF mismatch can rapidly deteriorate flow patterns in complex, multichannel structures.

10.3.5.1   Field-Amplified Sample Stacking

As described for the situation depicted in Figure 10.17, preconcentration in FASS is achieved by generating a high local electric field within an injected sample plug, which results in sample ions being rapidly driven and stacked at the ends of that plug. This field amplification is created when the sample is dissolved in a buffer that has a much lower conductivity than the surrounding running buffer used for separation [89]. The electrophoretic velocity of each ionic species in the sample plug is proportional to the field strength, which leads to a rapid migration of anions to the back of the sample plug and cations to the front, assuming EOF toward the cathode. Once they reach the boundary to the high conductivity running buffer, the ions experience a sudden electric field drop, slow down, and form a zone of concentrated sample ions at the end of the sample plug. Generally, the preconcentration efficiency increases with increasing γ, but levels off for large γ-values (>250, depending also on the separation parameters). Differences in conductivity and ionic strength between sample and running buffer lead to EOF being much higher in the sample than in the running buffer. This induces hydrodynamic flow effects at large γ as the system adjusts to ensure continuity of flow, and broadening of the sample plug as a result [91, 92].

FASS was implemented on microdevices for stacking of 400-μm, volumetrically defined sample plugs. This yielded preconcentration efficiencies of up to 20-fold using fluorescently labeled amino acids as model analytes [93]. Injections were performed using a modified injection element, which allowed spatial confinement of the sample by two buffer streams from both sides without dilution of the sample plug. The advantages of volumetric definition are high injection reproducibility (1.1% RSD peak height) and the absence of an electrophoretic sample bias. The presence and effect of hydrodynamic pressure on the stacking efficiency could be verified experimentally by CE (Figure 10.18) and visually using fluorescence microscopy [93]. Beard et al. recently presented an injector geometry using narrow sample channels to reduce hydrodynamic effects during sample plug formation yielding a signal gain of up to 80-fold [94].

Higher preconcentration efficiencies require a larger amount of sample to be introduced into the system. Full-column stacking techniques using polarity

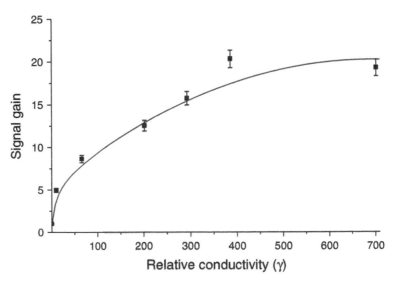

FIGURE 10.18 Signal gain as a function of relative conductivity for FASS on a microchip using 10-μM FITC-labeled Ser and Gly as sample. (Reprinted from J. Lichtenberg, E. Verpoorte, N.F. de Rooij, *Electrophoresis* **2001**, 22, 258. With permission. Copyright 2001 Wiley-VCH.)

reversal offer this possibility at the price of increased analysis time (1 to 2 min typically). Li et al. [95] used such a device to preconcentrate and focus trace-level protein digests as preparation for quadrupole time-of-flight MS (Figure 10.19a). After the entire 70-nl channel between sample reservoir and on-chip nanoelectrospray emitter was filled with sample solution, the polarity was reversed to remove the sample matrix while the analyte stacked at the boundary to the running buffer. The progress of matrix removal could be followed by current monitoring, until only a small amount of buffer was left in the channel. Finally, the preconcentrated sample zone was pumped by EOF into the electrospray emitter for MS analysis. The concentration detection limit could be enhanced from 3- to 50-fold, depending on the peptide.

A different configuration was used by Lichtenberg et al. for full-column stacking of fluorescently labeled amino acids combined with on-chip CE for separation [93]. The design, depicted in Figure 10.19(b), consists of a 55- to 170-mm-long, folded stacking channel connected to a 40-mm-long separation channel. In this way, high-efficiency stacking of a large sample volume could be combined with rapid separation in a short CE column. Filling of the stacking

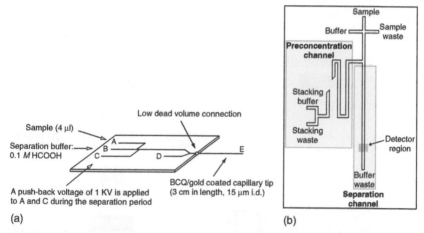

FIGURE 10.19 (a) Long-column stacking as preconcentration step before mass spectrometry. The channel between A and E is first filled completely with sample, then the sample is stacked by removing most of the sample buffer into reservoir C. (Reprinted from J. Li, C. Wang, J.F. Kelly, D.J. Harrison, P. Thibault, *Electrophoresis* **2000**, *21*, 198. With permission. Copyright 2000 Wiley-VCH.) (b) Schematic chip layout of a column-coupled system with a long preconcentration channel (left) and a shorter separation channel (right). Both sections are connected via a 9-mm-long common channel region, which serves as the injection "loop". Stacked sample is collected in this segment, to be injected into the separation channel for analysis. (Reprinted from J. Lichtenberg, E. Verpoorte, N.F. de Rooij, *Electrophoresis* **2001**, *22*, 258. With permission. Copyright 2001 Wiley-VCH.)

channel and the stacking itself were performed in 90 sec, while the separation took only 35 sec. The signal gain was found to be nearly linear with stacking channel lengths from 55 to 170 mm, yielding 60- to 95-fold increases.

10.3.5.2 Field-Amplified Injection

Prior to the chip-based FASS examples described above, FAI techniques on microchips were investigated by Jacobson and Ramsey [96] and Kutter et al. [97]. Figure 10.20 presents the differences between FASS and FAI in a microchip format. While FASS works for both negatively and positively charged ions, FAI has an electrophoretic bias depending on the direction of EOF. In this case, positive ions are stacked at the sample–buffer interface during the injection process. FAI can be easily implemented using time-based, gated injection methods proposed by Jacobson et al. [98]. FAI yielded a concentration enhancement for dansylated amino acids of up to 13.8-fold at a γ of 970. The efficiency was also investigated. Separations using stacked injection had efficiencies of 22,000 to 29,000 theoretical plates, which made them only 60 to 70% as efficient as CE on the same devices using a volume-defined, pinched injection scheme.

In combination with postcolumn complexation for detection, FAI was also used for the preconcentration of inorganic cations prior to CE [97], employing the same procedure as described above. The stacked injection mode improved

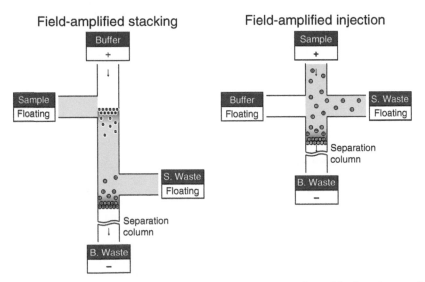

FIGURE 10.20 While on-chip FASS allows preconcentration of both anions and cations in volumetrically defined plugs, field-amplified injection enriches only one ionic species at the front boundary of the injection plug. (Reprinted from J. Lichtenberg, N.F. de Rooij, E. Verpoorte, *Talanta* **2002**, *56*, 233. With permission. Copyright 2002 Elsevier Science.)

the signal by a factor of 16, which corresponded to extrapolated detection limits of 18 ppb for calcium and 0.5 ppb for magnesium.

The difficulty of controlling the precise location of the stacked sample zone in conventional chip designs has been overcome by a device recently presented in Ref. [99]. The EOF in these devices was eliminated by coating the channel walls using a 0.1% solution of poly(allylglycidyl ether-co-N, N-dimethylacrylamide) copolymer. Initially, all the channels were filled with a high-conductivity buffer by a pressure-driven pump system, except for the sample loading side-channel, which was filled with a low-conductivity buffer. Upon application of an electric field, the sample ions travel from the sample reservoir until they reach the stationary buffer concentration boundary and stack at this point. Once the desired amount of sample is collected, separation is carried out in the main channel of the chip. One of the major advantages of the stationary concentration boundary is that analyte collection can be carried out without displacing the stacked plug, which led to a high preconcentration efficiency of more than 100-fold.

10.3.5.3 Thermal Gradient Focusing

As mentioned in Section 10.3.4, thermal gradients can cause electrokinetic velocity gradients, if temperature-induced changes in electrical conductivity and fluid viscosity do not cancel each other out. Ross and Locascio [87] deployed two techniques to generate the necessary, stationary temperature gradient, namely a narrow off-chip heating resistor in contact with the chip or, more elegant, internal Joule heating caused by a DC current flowing along the preconcentration microchannel. In the latter case, the temperature gradient was generated at a sudden widening of the fluidic channel with the narrow section being warmer than the wide one due to the higher dissipated electrical energy. Preconcentration efficiencies of up to 10,000-fold were reported for a fluorescent label. The method was shown to work well with a variety of samples, including amino acids, DNA, and polymer particles.

10.3.5.4 Stacking of Neutral Analytes

Though relatively easy to implement, methods like FASS or FAI are applicable to charged analytes only. Methods for stacking neutral analytes for separation by micellar electrokinetic chromatography in fused silica capillaries have also been the subject of a number of recent research efforts. One approach, termed *sweeping*, involves the injection of an analyte zone into a column of separation buffer containing a pseudostationary phase (micelles). Upon application of the electric field, the micelles are swept through the analyte zone, collecting analyte as they go and resulting in a unique focusing effect. Charged species could also be preconcentrated in this way. When first described [100], preconcentration factors of several thousand-fold were achieved. In a recent variation of sweeping, sensitivity increases approaching

1,000,000-fold have been reported [101]. Though not yet applied to micro-chips, it is clear that the use of this technique could be enormously advanta-geous, especially in conjunction with the integration of less-sensitive absorbance or electrochemical detection methods into microfluidic devices.

Another method of neutral-species stacking in micellar CE involves add-ition of salt (e.g., NaCl) to the sample matrix to increase its conductivity to levels two to three times higher than the separation buffer [102]. The latter typically contains negatively charged micelles of SDS or sodium chlorate. The result is a substantial drop in electric field strength at the sample plug–separ-ation buffer interfaces, and a field amplification effect analogous to the one described above. In this case, it is the negatively charged micelles, with elec-trophoretic mobilities counter to the EOF, that stack at the cathodic interface of the sample plug and the buffer. Analyte is collected and preconcentrated by the micelles as it is electroosmotically driven through this interfacial zone. The micelle zone is also effectively "locked" in place up against the higher con-centration chloride, until this latter zone has diffused sufficiently to let cholate anions enter. Hence, a sharp interface is maintained for analyte stacking, which makes this type of stacking very attractive for electrokinetic injection of the sample [103]. This technique lends itself very well to chip-based separation, with 20-fold peak height improvement being observed for an 80-sec injection of a solution containing 67 nM BODIPY [103].

10.3.6 ISOTACHOPHORESIS FOR SAMPLE PRECONCENTRATION AND CLEAN-UP

Extension of the FASS stacking concept to ternary buffer systems yields ITP, a separation technique that can be used for both sample preconcentration and sample clean-up. Furthermore, it is a quantitative analytical separation tech-nique in its own right, with a wide range of applications. However, the focus of this section will be on the sample preparation applications of ITP, which generally require coupled-column arrangements. These consist of a preparative microchannel network combined with a channel reserved for the actual analysis.

As depicted in Figure 10.21, two different background buffers are used in ITP. The sample plug is formed between a leading and a terminating buffer electrolyte, LE and TE, respectively, with the condition that the mobility of the LE, μ_{LE}, is larger than the μ_i of the fastest analyte in the sample, and μ_{TE} is smaller than the μ_i of the slowest. Upon application of the separation voltage, the sample constituents separate over a given period of time into distinct zones located between the LE and TE in order of descending mobi-lities. Once this steady state is reached, the boundaries between the analyte zones are very sharp, due to a self-focusing mechanism governed by the Kohlrausch regulating function [104, 105]. The latter ensures that ions dif-fusing out of the analyte zone are always recaptured due to the different field-strengths in the neighboring zones. To illustrate this, consider an ion that diffuses out of the frontal end of the analyte zone. It experiences a sudden

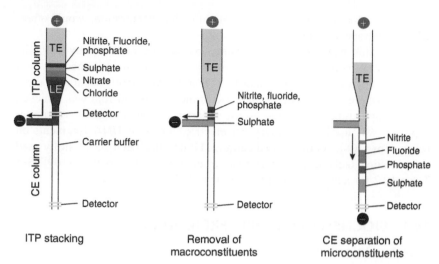

FIGURE 10.21 ITP–CE combination for ion analysis. During a first step the ITP steady state is established and the analyte zone concentration adjusts to that of the LE. Second, the high-concentration constituents chloride, nitrate, and sulfate are quantified and simultaneously removed from the column. Finally, the low-concentration constituents are separated by electrophoresis.

drop of the electric field strength as the higher mobility ions in the neighboring zone provide less resistance to current flow. The escaped ion therefore slows down and is quickly absorbed by its particular analyte zone. The same reasoning is also valid for the back end of the analyte zone after sign reversal.

Once the steady state is reached in ITP, the analyte concentration in the sample zones is equal to the concentration of the leading buffer. If ITP is used as an analytical separation technique, the original analyte concentrations in the sample can be deduced from the length of the respective zones in steady state. For sample preparation techniques, preconcentration or dilution can be tailored by adapting the LE composition accordingly.

Kaniansky et al. realized various ITP configurations on a poly(methyl methacrylate) (PMMA) chip with two coupled separation columns and on-chip conductivity detectors (Figure 10.21) [106]. One application involving a combination of sample pretreatment by ITP and separation by CE involved the analysis of low-concentration constituents (10 μM each of nitrite, phosphate, and fluoride) in a model mixture containing high background concentrations of ions like chloride (600 μM) and sulfate (800 μM). A sample fraction containing the low-concentration constituents could be extracted from the steady-state ITP zone system and electrophoretically transferred into the second column for CE separation. This technique, previously demonstrated by the same group in a capillary-based system, allows the analysis of complex mixtures, where constituents in high concentrations would other-wise mask trace analytes.

While this example focuses mainly on sample clean-up by ITP, the preconcentration aspect of the technique has also been investigated. By first stacking analytes into narrow bands by ITP and subsequently separating these by electrophoresis, a sensitivity increase of more than 400-fold was achieved for a sample mixture of fluorescent reporter molecules [107]. A similar approach was used for capillary gel electrophoresis of DNA fragments on microchips [108]. As mentioned earlier, ITP has also been successfully used as separation technique for quantitative analysis on microchips [109]. See Table 10.2 for a summary of the various techniques discussed.

10.4 BIOCHEMICAL SAMPLE PRETREATMENT

10.4.1 DNA Purification

After extraction from the living cell (see Section 10.2.3), a purification step is often necessary to isolate the nucleic acids from other interferents, such as proteins or metal complexes. In the early 1990s, the time necessary for DNA purification was greatly reduced by switching from phenol extraction proced-ures to chromatographic resins and SPE on silica. In particular, the latter approach, based on the method proposed by Boom et al. in 1990 [110], received particular interest because of its simplicity. The method is based on the lysing and nuclease-inactivation properties of a chaotropic agent, together with the nucleic acid-binding properties of silica particles in the presence of this agent. Chaotropic agents, such as urea, guanidine hydrochlor-ide, and sodium iodide, can reduce the activity of water by forming hydrated ions, leaving dehydrated biological molecules behind. In concentrated solu-tions, these agents can denature proteins and double-stranded DNA by redu-cing the hydrophobic interactions in their three-dimensional structure. Moreover, they can cause the lysis of mammalian cells, viruses, and gram-negative bacterial species, with consequent nucleic acid release in the solution and their binding on the silica surfaces [110]. The process of DNA adsorption on a silica surface is not completely understood. However, it seems that this adsorption of DNA is based on hydrogen bonds, which explains why the highest adsorption efficiencies are achieved in solutions having high ionic strength, low pH, and high chaotropic agent concentration [111, 112]. The first two conditions, in fact, cooperate to reduce the electrostatic repulsion between DNA and silica by decreasing the charge on the silica surface. The high concentration of the chaotropic agent rids the DNA and silica of their hydration shells and causes biomolecular denaturation, as mentioned above. In these conditions, the binding between silica and DNA becomes energetic-ally favorable. Unbound components can be removed from the silica by washing with solvents. DNA purification is finally accomplished by eluting

TABLE 10.2
Comparison of Different Sample Preconcentration Techniques Used on Microchips

Technique	Analyte	Preconcentration Achieved	Separation Technique	References
FASS of long sample plugs	FITC-labeled amino acids	Up to 20-fold	CE	[93]
FAI by gated injection	Dansylated amino acids	Up to 13.8-fold	CE	[96]
FAI by gated injection	Metal cations (Ca, Mg)	Up to 16-fold	CE with postcolumn complexation	[97]
FASS by sweeping	Protein digests	Up to 50-fold	Mass spectrometry	[95]
FASS by sweeping	FITC-labeled amino acids	Up to 95-fold	CE	[93]
FAI with suppressed EOF	Fluorescein and a fluorescein-labeled protein	Up to 100-fold	CE	[99]
Thermal gradient focusing	Fluorescent labels, amino acids, DNA, polymer particles	Up to 10,000-fold	None	[87]
Sweeping of neutral analytes	BODIPY (fluorescent label)	Up to 20-fold	CE	[103]
ITP	Fluorescent reporter molecules (eTags)	Up to 400-fold	CE	[107]

the nucleic acids from the surface with solutions of low ionic strength and quite basic pH (pH 8), or even water. This method is therefore simple and rapid, and it allows the recovery of undamaged and highly pure nucleic acids that can then be used as a reagent in molecular biological reactions, such as amplification by PCR or NASBA, digestion by restriction nucleases, and Southern blot. The data acquired are useful in diagnostics of bacterial and viral infections, as well as in gene polymorphism studies and prenatal diagnostics.

A larger number of the nucleic acid extraction kits available on the market are based on the Boom method and use beads as binding. This is because beads supply a large available surface area for nucleic acid binding. Usually centrifugation is chosen to separate the solid phase (i.e., particles) from the supernatant solution during the various sample loading, washing, and elution steps. Although centrifugation has been successfully integrated in microchips for multiple-assay microfluidic systems [113–118], the same goal (separation of stationary and mobile phase) can be achieved by driving different solutions through an extraction matrix, which is physically retained in a macro- or microfluidic structure. Alternatively, magnetic beads can be utilized as solid phase, and a magnetic field in this case is used to immobilize particles and separate them from the solution.

Tian et al. published the first miniaturized bead-based system for micro-SPE (μ-SPE) for adsorption of human DNA from different matrices [112]. The μ-SPE device contained 0.2 to 0.3 mg of silica particles, which were retained in a polyethylene sleeve between two glass fiber frits to define a total volume of around 500 nl. In spite of the small dimensions of the extractor, it turned out to be efficient for the adsorption and desorption of DNA in the pictogram-to-nanogram range. The protocol used for the extraction is based on the Boom method [110, 119]: guanidine HCl-based buffer, alcohols, and Tris–EDTA buffer were used for sample loading, washing, and elution, respectively. A syringe pump was used to move the solutions with a flow rate usually on the order of a few microliters per minute. Fluorescence spectroscopy was employed to analyze the DNA recovered from the solid phase, as well as the protein eluting during the loading and the washing phases from the column. PCR and CE were then used to evaluate the amplificability, and thus the purity, of the eluted DNA. A DNA recovery of roughly 70% from white cells was demonstrated, while more than 80% of the proteins was removed during the loading and washing steps, which took less than 10 min. Although this work was not done on a chip, the dimensions of the capillary used suggest that extraction can be transferred successfully to the microchip format.

On the basis of Tian's work, Landers' group have gone on to trap silica beads within a microchip for NA extraction [120]. In order to retain the particles in a microchamber, they worked with etched glass structures in which the beads are retained by a weir placed at the outlet of the 30-μm-deep cavity [79]. A gap of 12 μm remains between the weir and the cavity coverplate to allow the solution to flow through the bead bed. It was found that 15-μm silica beads alone can extract DNA with high efficiency, but with low reproducibility for repeated extractions on the same chip, and from chip to chip. The packed devices also had a short lifetime, arising from increased back pressure and decreased flow with time due to gradual compression of the bed with use. To overcome these problems, beads were stabilized with a sol solution, which was then converted into a gel matrix directly in the microchannel. With these silica bead or sol–gel microdevices, high extraction efficiency in stable and reproducible systems was achieved. The channel design was also modified by replacing the weir with a silica bead or sol–gel frit polymerized *in situ*. Nucleic acids were eluted in less than 10 μl, the extraction process took less than 30 min, and the DNA extracted was suitable for PCR amplification. The samples extracted were rather academic in nature, consisting of digested λ-phage DNA. Crude lysate was not tested.

A similar SPE approach, but in a PDMS rather than glass chip, was presented by Ceriotti et al. [121]. A 160-μm-deep, 320-nl channel was packed with silica particles which were retained at the end by a barrier or weir with a 10-μm gap. A two-layer master made in SU-8 epoxy was used to replicate channel devices, which were then irreversibly sealed to a PDMS slab after plasma oxidation. In a departure from the usual slurry-based methods, the channel was packed simply by drawing in dry, 15- to 30-μm silica particles using vacuum. For extraction, sample in $8M$ guanidine isothiocyanate was loaded into the device, followed by an ethanol wash. After drying with air, 10-μl plugs of elution buffer containing 1-μM YOYO-1, a dye for nucleic acid detection, were pumped at 2 μl/min through the packing and collected. Purified λ DNA samples and human cell lysate were analyzed. For a sample containing enough λ DNA (40 ng in 20 μl) to saturate the packed bed, 9 ng was recovered. This corresponded to a binding capacity of 24 ng of DNA per milligram of beads, in agreement with the value reported in Ref. [112]. Unfortunately, device lifetime was limited to three or four extractions, but chip-to-chip reproducibility for the first extraction was quite good, at about 12.5%. When cell lysate was extracted, 5 ng of nucleic acids were recovered, due perhaps to competition between the nucleic acids and other matrix components for available binding surface [112]. This extraction process represents somewhat of an enrichment step, since the nucleic acids in the initial lysate volume of 50 μl could be extracted and eluted in only 30 μl of buffer. More efficient preconcentration is possible through use of integrated electrodes or frits, as is described below.

The use of magnetic beads has in the last few years received great interest for applications in microfluidics, since the use of magnetic fields to trap magnetic beads in a precise place is an attractive solution. Paramagnetic beads were used to capture by affinity synthetic poly(A)-tailed DNA samples within a microfluidic device [122]. Jiang and Harrison published promising work in which paramagnetic oligo-(dT)25 beads (diameter of 2.8 μm) were used for mRNA isolation from total RNA in a microfluidic glass device [123]. The results suggest that rare RNA can be isolated on-chip in concentrations and with a quality good enough for constructing cDNA libraries.

Besides beads, cellulose packed into a microchamber has also been employed for nucleic acid extraction by the Boom method [124]. The microchamber was part of a multifunctional PC device developed for automated multistep genetic assays. DNA/RNA purification from serum lysate, RT-PCR, nested PCR, nucleic acid hybridization, Dnase fragmentation, and dephosphorylation, labeling, and washing steps could all be performed sequentially in a hands-off manner in this credit-card sized system.

Since in the presence of chaotropic agents nucleic acids bind on silica surfaces, simple open channels could work as extractor. However, the low surface area available for the binding limits this approach. Two different approaches were developed to increase surface area-to-volume ratios and improve extraction efficiency in open microstructures. The first approach used DRIE to fabricate pillar-like structures (200 μm high and 18 μm in diameter, for a total area of 0.36 cm^2) to use as binding substrate for DNA adsorption [125]. On-chip PCR and fluorescence detection were employed for eluted DNA evaluation. DNA quantities approaching the binding capacity of the system (40 ng/cm^2 as predicted for glass by Vogelstein and Gillespie [126]) were recovered. Although this approach shows some potential, it relies on expensive and complex technology for chip fabrication. The second approach was reported by Kim et al. They used photosensitive glass, which was easily structured to again realize pillar-like structures (200 μm high and 25 μm diameter for a total area of 2 cm^2). In this case, a binding capacity of 15 ng/cm^2 was reported for prepurified plasmid DNA. In both cases a syringe pump was used to move liquids through the extractor.

Recently, a new way of DNA sample preconcentration was introduced in the form of integrated porous membrane structures [127, 128]. These planar membranes allowed DNA fragments to be concentrated up to 100-fold before separation. The device structure, depicted in Figure 10.22(a), consists of a typical microchip CE layout with an additional U-shaped side channel, which is close to the injection element but not in direct contact (a 3- to 12-μm wide gap was maintained between the channels). Bonding of the structured glass wafer to the cover plate is achieved at low temperatures by spin-coating a thin layer of diluted potassium or sodium silicate solution onto the coverplate as adhesive. Due to its porosity, this layer also acts as a semipermeable membrane between the U-shaped side channel and the main channel in the region of the injection junction. Ionic current can flow over the junction, while larger molecules are retained. Preconcentration can be achieved by applying a voltage of 1 kV between the sample and side-channel reservoirs for a given time, which causes the analyte to migrate to the membrane region. Figure 10.22(b) shows three successive electropherograms obtained for different preconcentration times of a DNA–PCR marker, indicating a nonlinear

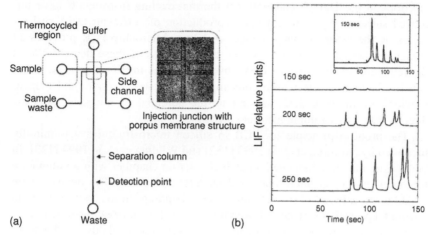

FIGURE 10.22 (a) Device layout for the CE chip with DNA preconcentration at a porous membrane at the injection junction. For preconcentration, an electric field of 1 kV is applied between sample and side channel reservoirs for 1 to 2 min to cause DNA to migrate and collect on the junction. (Reprinted from J. Khandurina, T.E. McKnight, S.C. Jacobson, L.C. Waters, R.S. Foote, J.M. Ramsey, *Anal. Chem.* **2000**, *72*, 2995. With permission. Copyright 2000 ACS.) (b) Electropherograms of a DNA PCR marker after different preconcentration times on a similar device. (Reprinted from J. Khandurina, S.C. Jacobson, L.C. Waters, R.S. Foote, J.M. Ramsey, *Anal. Chem.* **1999**, *71*, 1815. With permission. Copyright 1999 ACS.)

relationship between signal intensity and time. In a follow-up study, the porous membrane preconcentrator was used after on-chip PCR to reduce the number of thermal cycles necessary for product detection down to 10, leading to a total analysis time of less than 20 min [128]. Along the same lines, Lin et al. [129] recently developed a PMMA-based microchip with an in-channel metal electrode for DNA preconcentration without the use of a porous membrane.

10.4.2 NUCLEIC ACID AMPLIFICATION

The introduction of PCR by Mullis et al. proved to be a key event in the development of the genomics field [130]. The polymerase enzyme is the active component in this reaction, constructing DNA oligomer copies from starting templates and thereby amplifying them. Controlled temperature cycling is crucial, with three temperatures required for each reaction cycle (denaturation at ~90 to 95°C, extension [or DNA synthesis] at ~70°C, and annealing at ~50°C). The amount of DNA increases exponentially, doubling after each cycle, leading to enough DNA for further analysis after 20 to 30 cycles. PCR is thus generally viewed as a sample preparation technique. Once samples are amplified, they are generally analyzed by slab gel or capillary gel electrophoresis. Though conventional thermal cycling instruments have improved substantially in recent years, production of sufficient DNA still requires on the order of many minutes to hours. Microchip PCR provides an attractive alternative because the low thermal mass of these devices allows the reduction of cycle times from several minutes to 30 sec or less. Other benefits include the decreased consumption of power, sample, and reagent provided by these devices, as well as the potential to integrate the PCR directly with other sample handling steps [131].

The proof-of-principle for PCR in silicon microstructures was initially presented by Northrup et al. in 1993 [132] and Wilding et al. in 1994 [133]. In both cases, the chip contained an etched reaction chamber with a volume on the order of several microliters, and an inlet and outlet. The devices were sealed using a Pyrex glass plate and anodic bonding. In the former case, an integrated polysilicon heater allowed thermal cycling at rates two to ten times faster than conventional instruments, with comparable yields. In the work described in Ref. [133], heating and cooling were achieved by an external thermoelectric device on a copper block, yielding comparatively short cycling times in the order of 3 min. A concern when using silicon as substrate for high surface-to-volume ratio reactors is the material compatibility with the reagents used. Different silicon-based surface passivation layers for PCR chambers were evaluated in Ref. [134]. It was found that native Si and Si_3N_4 caused a high failure rate, while silanization followed by a polymer coating showed good but inconsistent amplification, and SiO_2 performed best. Similarly, alternatives to silanization for passivation of surfaces in glass PCR microchambers have been explored by Giordano et al. [135]. These researchers studied the surface interaction properties of the additives polyethylene glycol, polyvinylpyrrolidone, and hydroyethylcellulose, all of which act as dynamic coatings, and epoxy (poly)dimethylacrylamide, a compound that adsorbs to glass. Analysis of PCR products generated in the microchips under different conditions revealed that these coating agents adequately passivated chamber surfaces. Optimization of PCR reagent mixes for silicon-based devices has been published by Taylor et al. [136].

A distinctively different approach to on-chip PCR was introduced by Kopp et al. [137]. In this case, the chip is in contact with three side-by-side copper blocks, which provide three distinct constant-temperature zones. Instead of repetitively heating and cooling a sample contained in a microreactor, the sample plug is pumped at a constant velocity through a meander-like glass channel from one temperature zone to the other. In this way, the effect of thermal inertia is completely eliminated. Moreover, a large number of sample plugs separated by a spacer liquid can be amplified in parallel. Heating and cooling times were less than 100 msec, and total amplification times between 1.5 and 18.7 min were obtained, simply by changing the flow rate. The amplification gain decreased with decreased cycle time, but even for short, 12-sec cycles, a good fluorescence signal could be obtained after 20 cycles.

Since these initial examples, a significant number of micromachined PCR devices have been described in the literature. Many of these have been relatively simple microfluidic devices that rely on direct contact with external components such as conventional thermocyclers [35, 138] or Peltier elements [128, 139, 140]. To circumvent the technical issues associated with contact methods, Oda et al. have reported the use of a noncontact thermal cycling method, using an inexpensive infrared radiation source for heating and a solenoid-gated compressed air source for cooling [141]. Cycle times as fast as 17 sec could be achieved in glass microchambers. In a more recent polymer embodiment of the chip and using infrared-mediated temperature control, this group has achieved adequate amounts of PCR product after only 15 cycles for a total amplification time of only 240 sec [142].

Other devices have been more complex, integrating heaters and temperature sensors. In fact, the very first example of PCR on chip used a polysilicon layer as integrated heating element. This layer was deposited onto the silicon nitride membrane that formed the bottom of the silicon microchamber [132]. Similar devices were later used in portable systems [143, 144], as well as in a hybrid device combining a planar microchip for DNA separation and a PCR reactor [145]. In all but the first device, the polysilicon was doped with boron to improve the resistive heating characteristics of the material. Either gold or aluminum contacts were structured on the polysilicon by a photolithographic lift-off procedure, thereby defining the heater geometry. Whereas the first device contained only one heater, the devices in Refs. [143–145] were fabricated by bonding two processed wafers containing chamber-heater chips to form enclosed chambers with a symmetric cross section and heaters on both sides. This guaranteed excellent thermal uniformity throughout the microchambers and high-efficiency reactions [144].

The prototype devices in Refs. [132, 143] were further refined and used in a miniature analytical thermal cycling instrument, or MATCI [144, 146], which incorporated all the necessary electronics and optical components into a briefcase and was battery-operated. Real-time fluorescence detection of the PCR products of *Borellia burgdorferi*, the parasite responsible for Lyme's disease, was demonstrated. Template RNA from a standard HIV isolate (tissue culture supernatant) was also analyzed in a chip volume of 20 µl, in samples containing as few as 20 organisms [144]. The same MATCI was also used to demonstrate the feasibility of rapid detection of single-nucleotide polymorphism for diagnosis of infectious diseases and genetic disorders [146]. Belgrader et al. reported a more sophisticated instrument consisting of ten silicon reaction chambers with integrated thin-film resistive heaters for rapid, parallel pathogen detection by real-time PCR [147, 148]. Tests could be completed in as little as 16 min, with detection limits on the order of 10^2 to 10^4 organisms per milliliter.

The MATCI-type dual-heater device was also used in the first demonstration of combined microchip PCR–microchip electrophoresis for fast DNA analysis by Woolley et al. [145]. The device consisted of the hybrid assembly of a planar CE chip for DNA separation together with the PCR reactor, lined this time with a polypropylene insert. The PCR reactor served simultaneously for DNA amplification and subsequently as the sample reservoir for CE analysis. Genomic *Salmonella* DNA could be assayed in less than 45 min, demonstrating that quasi-real-time monitoring using this combined PCR–CE system was also possible.

Other silicon-based thermocyclers have been described in the literature. A device similar in design to that of Northrup et al. [132], but with Pt heaters and temperature sensors, was described by Lao et al. [149]. The temperature-sensing principle used relied on the fact that resistance is a temperature-dependent parameter. Measurement of the resistance of the Pt strip making up the sensor therefore allowed the temperature to be deduced. The process flow of this device is shown in Figure 10.23. This type of device was tested using four traditional Chinese medicine genes as templates, and found to yield DNA concentrations of 27 ng/μl or more, sufficient for further analysis [150]. Very accurate real-time temperature control to within $\pm 0.025°$C was demonstrated. This was accomplished with heating and cooling rates of 5 and 3.5°C/sec, respectively. Poser et al. integrated both a NiCr heater and Pt temperature sensor by thin-film processes onto the backside of a chip containing a chamber etched 450 μm into the silicon substrate [151]. The periphery of the chamber was defined by deep-etched grooves, which provided effective thermal isolation from the rest of the silicon substrate. This approach optimized temperature homogeneity in the chamber and facilitated the design of arrays of chambers. Notable about these 5- to 10-μl devices were the very fast heating and cooling rates achievable (up to 80 and 40°C/sec, respectively). Pt resistors formed by lift-off on a thin silicon nitride membrane at the bottom of an etched silicon chamber also served the dual purpose of heating and thermometry in Ref. [152]. Again, the heating and cooling rates achieved in this device were impressive, at between 60 and 90°C/sec for the former and 74°C/sec for the latter. No more than 2 W was required to heat the chamber when filled with 2 μl water and 1 μl oil.

A microassembly approach was taken by Zou et al. to realize a miniaturized multichamber thermal cycler capable of thermal multiplexing for high throughput PCR [153]. The temperature control layer was microfabricated separately and then later assembled with the reaction chamber array, which had been fabricated in low-cost polymer by hot embossing. For fabrication of the cycler, Al heaters, and temperature sensors were patterned on a thermally oxidized wafer and coated with a passivation layer of silicon nitride. Contacts to the underlying Al structures were patterned through the nitride. This wafer was then bonded by flip-chip bonding to a printed circuit board (PCB) containing electrical connection traces for the heaters and sensors on both sides. Because the thermal cycler was not integrated with the reaction chambers in this case, the former could be reused, while the latter was disposable.

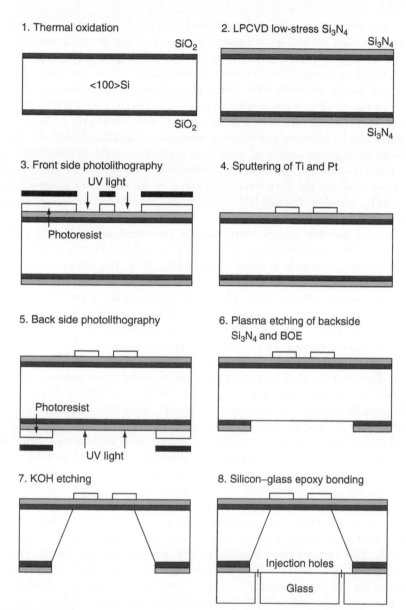

FIGURE 10.23 Process flow for fabrication of a microfluidic chamber for gas- and liquid-phase reactions requiring precise temperature control. 1 and 2: $\langle 1\ 0\ 0 \rangle$ wafers were first coated with 0.1 μm of thermal oxide and 1 μm of low-stress, LPCVD silicon nitride. 3 and 4: The temperature-regulating elements were then patterned in a Ti (10 nm)/Pt (100 nm) layer on the front side of the wafers using a lift-off process. 5: The backside silicon nitride was then patterned to define the reaction chamber area.

(continued on next page)

This very practical solution reaps simultaneously the dual benefits of ultrafast thermal cycling offered by integrated heaters, and the low cost associated with the use of a plastic microfluidic device. A 30-cycle PCR of plasmid DNA could be completed in less than 15 min. Another more integrated array approach was presented by Shinohara et al., and involved the fabrication of an 8×12 array of 1.3-μl wells in silicon [154, 155]. Twelve temperature sensor–heater pairs, one for each row of eight reactors, were formed on the backside of the device by alternately structuring layers of polyimide and titanium. This allowed temperature control of individual rows of reactors, as demonstrated in a DNA denaturing experiment. The formation of a temperature gradient of 60 to 90°C across the device was confirmed by the decreased fluorescence emitted by a double-stranded DNA-specific dye in wells having temperatures above which DNA is denatured.

Several research groups have recently pioneered sub-microliter PCR. Lagally et al. fabricated eight 280-nl PCR chambers directly coupled to electrophoresis channels in glass wafers [156, 157]. The device contained microfluidic valves and hydrophobic vents to immobilize the sample in the reaction chamber during PCR, which was driven using a resistive heater clamped onto the chip. Single-molecule DNA amplification and analysis was achieved [157]. This microdevice has been further refined to incorporate resistive heaters outside, and resistive temperature detectors (RTD) directly inside the chambers [158]. The RTDs were formed in a 200-nm Pt/10-nm Ti layer deposited on glass by patterning with photoresist and removal of the exposed metal using hot aqua regia (3:1 HCl:HNO$_3$, 90°C). They were subsequently covered with 200 nm of SiO$_2$. The heaters were etched into a similar Pt or Ti layer on the opposite side of the wafer by ion beam etching. Gold leads (5 μm thick) were electrodeposited on top of the heaters to provide a low-resistance electrical contact to these elements. The use of an integrated thermostatting approach yielded improved heating and cooling rates of 20°C/ sec. Successful sex determination from human genomic DNA was demonstrated in less than 15 min with this device. Liu et al. realized a nanoliter rotary device for PCR [131]. As the name suggests, the reaction was carried out in a circular channel that had an inlet and outlet directly opposite to one another. This channel was formed in PDMS, and had a 12-nl volume. Membrane pumps and valves were integrated into a second PDMS layer which covered the reaction channel, allowing the reaction loop to be sealed off from the inlet and outlet channels [159]. Three heaters were made of thin layers of

---

FIGURE 10.23 contd. 6: The exposed silicon nitride and oxide removed by plasma etching and wet etching in buffered HF, respectively. 7: The exposed silicon was then etched in 30% (w/v) KOH at 80°C straight down to the front-side membrane bearing the heater and temperature sensor. The resulting chamber had the pyramidal geometry typical of this anisotropic etching. 8: The chamber was sealed with a glass chip. (Reprinted from A.I.K. Lao, T.M.H. Lee, I.-M. Hsing, N.Y. Ip, *Sens. Actuators. A* **2000**, *84*, 11. With permission. Copyright 2000 Elsevier Science.)

tungsten on a glass microscope slide, laid out to form a circle with approximately the same diameter as the reaction loop. A lift-off procedure was used to manufacture these features, as well as the electrical leads in Al. The PDMS channel chip was then aligned and reversibly bonded to the heater chip. It was possible to generate three different temperature zones around the loop, so that a PCR reaction zone could be simply pumped around and around the channel from one zone to the next. This spatial thermal cycling principle is very similar to that proposed by Kopp et al. [137], and eliminates the need for precise temperature ramping. Alternatively, all three heaters could be simultaneously ramped to perform more conventional temporal cycling. Temperature calibration was carried out in part using thermochromic liquid crystals.

Although PCR is by far the most widely used nucleic acid amplification technique, others are advantageous for specific applications. Recently, a transcription-based amplification system specifically designed for various kinds of RNA was implemented on a silicon-glass-based microdevice [160]. Nucleic acid sequence-based amplification (NASBA), introduced in 1991 [161], relies on the simultaneous activity of three enzymes to produce over 10^9 copies during 90 min at isothermal conditions (41°C). Different from PCR systems, no thermal cycling is necessary, which facilitates the design of the chip and its associated control electronics. Gulliksen et al. have demonstrated NASBA amplification of single-stranded DNA and of a control oligonucleotide for the human papilloma virus in 10- to 50-nl reaction chambers with on-line optical monitoring using molecular beacons [160]. Similar to micromachined PCR chambers, a commercial organopolysiloxane surface coating was required to avoid adsorption of the NASBA reagents to the surface.

10.4.3 BIOMOLECULE FRAGMENTATION

For the analysis of biomacromolecules such as DNA or proteins, fragmentation of these species is often required. An example is the identification of proteins by MS, which is facilitated by digestion of the protein of interest by an enzyme (typically trypsin) prior to analysis. The resulting (poly)peptides are subsequently analyzed by a mass spectrometer and the original analyte structure is determined from the fragment distribution. Similar approaches are used for the analysis of recombinant DNA and for the preparation of DNA libraries. Both enzymatic and purely physical techniques have been implemented on microchips.

10.4.3.1 Enzymatic Digestion of DNA

Restriction enzymes are powerful tools for recombinant DNA analysis. The use of microfabricated devices to combine this sophisticated biochemical sample pretreatment process with CE separation has been demonstrated by Jacobson and Ramsey for DNA restriction fragment analysis [162]. On-chip, it took only 5 min at room temperature to digest the plasmid pBR322 by the enzyme *Hin*fI and to size it using a polymeric sieving matrix. Figure 10.24 shows the digestion procedure that takes place within a 0.7-nl reaction chamber on a glass-based chip device. EOF was minimized by covalent immobilization of linear polyacrylamide on the channel walls, and hydroxyethyl cellulose was used as sieving medium.

Initially, the reaction chamber was filled by electromigration toward the waste 1 reservoir of a mixture of the DNA sample (125 ng/µl) and the enzyme solution (4 units/µl). This reservoir was filled with a buffer mix containing cations necessary for the digestion, such as Mg^{2+}, which also migrate toward the reaction chamber. Digestion could be performed either in a transient, continuous-flow mode, yielding a reaction time of 9 sec using the voltages indicated in Figure 10.24, or in a static mode with all voltages switched off for a given digestion time. Peak height increased 10-fold by increasing the digestion period from 9 to 129 sec, but no further increase could be observed for longer reaction times. This effect could not be completely explained, though it was postulated that it might be due to completion of the digestion or loss of enzyme to the walls of the reaction chamber because of its high surface-to-volume ratio. However, the experiment showed that adsorption losses of DNA were not so significant.

10.4.3.2 Enzymatic Digestion of Proteins

Several devices reported recently embody a high degree of functionality, integrating protein digestion with subsequent analysis. One such device for combined digestion, separation, and postcolumn labeling of proteins and

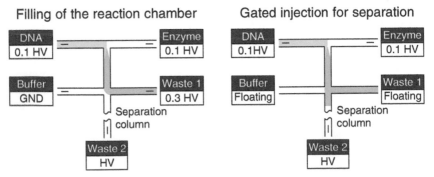

FIGURE 10.24 Operating procedure for DNA digestion and fragment sizing: (a) the reaction chamber is filled with enzyme and DNA, the excess being removed into the waste 1 reservoir; (b) after the digestion period, the DNA fragments are moved by a time-controlled, gated injection into the separation column; voltages are given relative to the voltage HV at the waste 2 reservoir, field strengths were varied between 190 and 690 V/cm. (Diagram adapted from S.C. Jacobson, J.M. Ramsey, *Anal. Chem.* **1996**, *68*, 720. Copyright 1996 American Chemical society.)

peptides using naphthalene-2,3-dicarboxaldehyde is described in Ref. [163]. A tryptic digest of oxidized insulin B-chain took 15 min to go almost to completion in a heated channel segment, followed by labeling and electrophoretic separation. The integration of proteolytic digestion on a microchip with mass spectrometric detection was demonstrated by Xue et al. [164] for the analysis of peptides generated by tryptic digestion of melittin. The digestion was performed in the sample reservoir to be able to monitor the process over time by repetitive MS analysis.

Harrison and co-workers [165] introduced a chip design featuring a 15-mm-long, 800-µm-wide, and 150-µm-deep bed of packed, 40- to 60-µm, trypsin-coated beads for on-chip protein digestion prior to introduction to an electrospray MS interface. The device also included a 45-mm-long separation channel. The rapid digestion, separation, and identification of proteins was demonstrated for melittin, cytochrome c, and bovine serum albumin. Generally, digestions took about 3 to 6 min. However, melittin was consumed within 5 sec, in sharp contrast to the 10 to 15 min required in a conventional cuvette. The use of an integrated digestion bed together with hydrodynamic flow effectively increased the ratio of trypsin to sample, to overcome the kinetic limitations often associated with digestion of small amounts (picomoles) of protein. Again, this is an example of the use of microfluidics to generate optimal conditions for a chemical reaction in a unique way not possible in macroscopic systems.

10.4.3.3 Physical Biomolecule Fragmentation

Recently, a new microfluidic device was reported that allows reproducible fragmentation of biomolecules such as DNA by subjecting them to rapid changes in flow velocity, leading to stretching (elongation) and eventually rupture [166, 167]. The chip translates a macroscopic technique widely used in biochemistry laboratories [168] into the microfluidic scale.

The macroscopic approach involves recirculation (up to 30 cycles) of the DNA solution through a tubing system with a 50-µm-diameter orifice using a syringe pump. Each time the solution passes the orifice and experiences elongational flow [169], the DNA strands uncoil and break in two if the stretching force is sufficient. On the microchip, time-consuming recirculation can be avoided by connecting up to 20 fragmentation units incorporating orifices, so that multiple fragmentation of the sample volume can be achieved in one single run through the device (Figure 10.25a).

The fragmentation device is fabricated in a single-lithography process using DRIE to form 50-µm-deep channels with rectangular cross sections and a number of orifices (5 to 50 µm wide and 5 to 800 µm long). Wafers were subsequently covered with 200 nm thermal oxide for surface passivation and anodically bonded to Pyrex 7740 glass wafers with drilled access holes. It was possible to break 48 kbp λ-phage DNA molecules down to fragments below 1500 bp in size (Figure 10.25b).

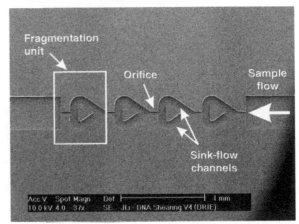

(a)

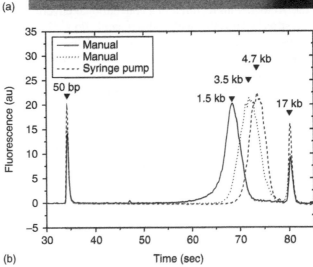

(b)

FIGURE 10.25 (a) An SEM photograph of a microfragmentation chip with five 25-μm-wide and 75-μm-long orifices. Access channels were 400 μm wide, the sink-flow channels of each fragmentation unit were 100 μm wide. (b) Fragmentation results obtained using manual pumping with a 0.5-ml syringe (full line: five 6-μm-wide constrictions, dotted line: five 10-μm-wide constrictions). The average flow rate was between 1000 and 2000 μl/min. Data are compared to a syringe pump at 500 μl/min (five 10-μm-wide constrictions). The gel electropherograms were obtained at Norchip AS, Klokkarstua, using a Bioanalyzer 2100 from Agilent Technologies. The right-most and left-most peaks are internal upper and lower ladder markers of 17,000 and 50 bp, respectively. (Measurements are courtesy of Anpan Han and Dr Laura Ceriotti, IMT Neuchâtel.)

10.5   CONCLUSION

Even very simple chemical analysis procedures require a certain degree of sample pretreatment before a sensitive and meaningful result can be obtained. A true micrototal chemical analysis system therefore requires the integration of at least some of these steps in order to provide full sample-to-answer functionality in a single device. The benefits of integrating sample pretreatment include lower labor cost per analysis, reduced risk of sample contamination, and faster analysis times. In cases where only minute amounts of sample are available, integrated sample pretreatment might be necessary, as conventional techniques cannot handle small volumes or suffer from unacceptable material loss from one pretreatment step to the next.

While most devices presented in the literature only integrate single pretreatment techniques in a device, a few demonstrators of fully integrated biochemical analysis systems have recently been published [124, 170]. These contain an impressive chain of sample pretreatment techniques, including cell capture and lysis, DNA purification, and PCR-based amplification, and rely on comparatively low-cost, polymer fabrication methodologies. Although the device presented by Anderson et al. [124] still relies on off-chip cell lysis, their on-chip preconcentration module allows the extraction of nucleic acids from the milliliter sample volumes pumped through the chip into a 10-μL volume for on-chip PCR. The device is fabricated using several layers of conventionally machined PC and thin silicone layers to provide integrated valves. Liu et al. have also developed a multilayer, PC hybrid device combining a microfluidic chip and electrical elements such as electrodes and heaters on a printed circuit board [170]. The microchip is capable of capturing target cells from a whole-blood sample by labeling the cells with a biotinylated antibody, and binding the resulting complex onto streptavidin-coated, paramagnetic beads by an immunoaffinity process. The cells can be recovered and lysed after washing the sample matrix out. Finally, the isolated DNA is amplified and analyzed using an electrochemical DNA microarray integrated on the chip. In order to provide means for liquid handling, this complex device includes a set of valves as well as electrochemical and thermopneumatic pumps.

Although the full integration of all processing steps necessary for a chemical analysis is desirable for a variety of reasons, the added complexity of the system also calls for a higher fabrication cost per device and for technically more advanced control instrumentation for the chip. The μ-TAS developer has to objectively evaluate the technical requirements for the chemical analysis task at hand, as well as the economic boundary conditions, to find a good balance between integrated and off-chip sample pretreatment steps. This is especially important for disposable analysis devices, typically used in medical diagnostics, as the fabrication cost directly affects the cost per analysis. However, as fabrication techniques become more refined and cost-efficient, and on-chip valving more commonplace, we will certainly see an evolution of increasingly complex analytical devices with integrated sample pretreatment techniques in the future.

ACKNOWLEDGMENTS

The authors would like to thank the past and present staff at the SAMLAB at the Institute of Microtechnology of the University of Neuchâtel, Switzerland, for their important help and support during the research years we spent there. Also, we would like to thank the editors of this book for their valuable input and efforts.

REFERENCES

1. A. Manz, N. Graber, H. M. Widmer, *Sens. Actuators B* **1990**, *1*, 244–248.
2. G. J. M. Bruin, *Electrophoresis* **2000**, *21*, 3931–3951.
3. J. P. Kutter, *Trac — Trends Anal. Chem.* **2000**, *19*, 352–363.
4. J. Lichtenberg, N. F. de Rooij, E. Verpoorte, *Talanta* **2002**, *56*, 233–266.
5. P.-A. Auroux, D. Iossifidis, D. R. Reyes, A. Manz, *Anal. Chem.* **2002**, *74*, 2637–2652.
6. D. R. Reyes, D. Iossifidis, P.-A. Auroux, A. Manz, *Anal. Chem.* **2002**, *74*, 2623–2636.
7. Y. Huang, E. L. Mather, J. L. Bell, M. Madou, *Anal. Bioanal. Chem.* **2002**, *372*, 49–65.
8. J. V. Zoval, M. J. Madou, *Proc. IEEE* **2004**, *92*, 140–153.
9. H. Andersson, W. van der Wijngaart, P. Enoksson, G. Stemme, *Sens. Actuators B* **2000**, *67*, 203–208.
10. P. Wilding, L. J. Kricka, J. Cheng, G. Hvichia, M. A. Shoffner, P. Fortina, *Anal. Biochem.* **1998**, *257*, 95–100.
11. M. Madou, *Fundamentals of Microfabrication*, 2nd ed., CRC Press: Boca Raton, FL, 2002.
12. E. L'Hostis, P. E. Michel, G. C. Fiaccabrino, D. J. Strike, N. F. de Rooij, M. Koudelka-Hep, *Sens. Actuators B* **2000**, *64*, 156–162.
13. A. P. Russo, D. Apoga, N. Dowell, W. Shain, A. M. P. Turner, H. G. Craighead, H. C. Hoch, J. N. Turner, *Biomed. Microdevices* **2002**, *4*, 277–283.
14. G. Ocvirk, E. Verpoorte, A. Manz, M. Grasserbauer, H. M. Widmer, *Anal. Methods Instrum.* **1995**, *2*, 74–82.
15. O. Aktas, N. R. Aluru, U. Ravaioli, *J. Microelectromech. Syst.* **2001**, *10*, 538–549.
16. P. A. Clerc, L. Dellmann, F. Gretillat, M. A. Gretillat, P. F. Indermuhle, S. Jeanneret, P. Luginbuhl, C. Marxer, T. L. Pfeffer, G. A. Racine, S. Roth, U. Staufer, C. Stebler, P. Thiébaud, N. F. de Rooij, *J. Micromech. Microeng.* **1998**, *8*, 272–278.
17. H. Andersson, W. van der Wijngaart, G. Stemme, *Electrophoresis* **2001**, *22*, 249–257.
18. G. L. Lettieri, A. Dodge, G. Boer, N. F. de Rooij, E. Verpoorte, *Lab Chip* **2003**, *3*, 34–39.

19. L. Ceriotti, N. F. de Rooij, E. Verpoorte, *Anal. Chem.* **2002**, *74*, 639–647.
20. A. Daridon, V. Fascio, J. Lichtenberg, R. Wütrich, H. Langen, E. Verpoorte, N. F. de Rooij, *Fresenius J. Anal. Chem.* **2001**, *371*, 261–269.
21. J. R. Anderson, D. T. Chiu, R. J. Jackman, O. Cherniavskaya, J. C. McDonald, H. K. Wu, S. H. Whitesides, G. M. Whitesides, *Anal. Chem.* **2000**, *72*, 3158–3164.
22. C. J. M. van Rijn, M. C. Elwenspoek, in *Proceedings of the IEEE 1995 Micro Electro Mechanical Systems Workshop (MEMS '95)*, Amsterdam, Netherlands, 1995, IEEE, pp. 83–87.
23. C. J. M. van Rijn, M. v. d. Wekken, W. Nijdam, M. C. Elwenspoek, *J. Microelectromech. Syst.* **1997**, *6*, 48–54.
24. S. Kuiper, H. van Wolferen, G. van Rijn, W. Nijdam, G. Krijnen, M. Elwenspoek, *J. Micromech. Microeng.* **2001**, *11*, 33–37.
25. B. He, L. Tan, F. Regnier, *Anal. Chem.* **1999**, *71*, 1464–1468.
26. G. Kittilsland, G. Stemme, B. Norden, *Sens. Actuators A* **1990**, *23*, 904–907.
27. W. H. Chu, R. Chin, T. Huen, M. Ferrari, *J. Microelectromech. Syst.* **1999**, *8*, 34–42.
28. J. K. Tu, T. Huen, R. Szema, M. Ferrari, *Biomed. Microdevices* **1999**, *1*, 113–120.
29. X. Xing, J. M. Yang, Y. C. Tai, C. M. Ho, *Sens. Actuators A* **1999**, *73*, 184–191.
30. H. Seidel, L. Csepregi, A. Heuberger, H. Baumgartel, *J. Electrochem. Soc.* **1990**, *137*, 3626–3632.
31. J. P. Brody, P. Yager, *Sens. Actuators A* **1997**, *58*, 13–18.
32. P. C. H. Li, D. J. Harrison, *Anal. Chem.* **1997**, *69*, 1564–1568.
33. E. A. Schilling, A. E. Kamholz, P. Yager, *Anal. Chem.* **2002**, *74*, 1798–1804.
34. H. Yu, P. Sethu, T. Chan, N. Kroutchinina, J. Blackwell, C. H. Mastrangelo, P. Grodzinski, in *Proceedings of the μTAS 2000 Symposium*, Enschede, The Netherlands, May 14–18, 2000, van den Berg, A., Olthuis, W., Bergveld, P., Eds., 2000, Kluwer Academic Publishers: Dordrecht, pp. 545–548.
35. L. C. Waters, S. C. Jacobson, N. Kroutchinina, J. Khandurina, R. S. Foote, J. M. Ramsey, *Anal. Chem.* **1998**, *70*, 158–162.
36. C. E. Sims, G. D. Meredith, T. B. Krasieva, M. W. Berns, B. J. Tromberg, N. L. Allbritton, *Anal. Chem.* **1998**, *70*, 4570–4577.
37. P. Belgrader, D. Hansford, G. T. A. Kovacs, K. Venkateswaran, R. J. Mariella, F. Milanovich, S. Nasarabadi, M. Okuzumi, F. Pourahmadi, M. A. Northrup, *Anal. Chem.* **1999**, *71*, 4232–4236.
38. P. Belgrader, M. Okuzumi, F. Pourahmadi, D. A. Borkholder, M. A. Northrup, *Biosens. Bioelectron.* **2000**, *14*, 849–852.
39. M. T. Taylor, P. Belgrader, B. J. Furman, F. Pourahmadi, G. T. A. Kovacs, M. A. Northrup, *Anal. Chem.* **2001**, *73*, 492–496.
40. D. C. Chang, *Guide to Electroporation and Electrofusion*, Academic Press: San Diego, 1992.
41. F. T. Han, Y. Wang, C. E. Sims, M. Bachman, R. S. Chang, G. P. Li, N. L. Allbritton, *Anal. Chem.* **2003**, *75*, 3688–3696.
42. M. A. McClain, C. T. Culbertson, S. C. Jacobson, N. L. Allbritton, C. E. Sims, J. M. Ramsey, *Anal. Chem.* **2003**, *75*, 5646–5655.
43. S. W. Lee, Y. C. Tai, *Sens. Actuators A* **1999**, *73*, 74–79.
44. M. A. McClain, C. T. Culbertson, S. C. Jacobson, J. M. Ramsey, *Anal. Chem.* **2001**, *73*, 5334–5338.

45. A. Y. Fu, H. P. Chou, C. Spence, F. H. Arnold, S. R. Quake, *Anal. Chem.* **2002**, *74*, 2451–2457.
46. A. Y. Fu, C. Spence, A. Scherer, F. H. Arnold, S. R. Quake, *Nat. Biotechnol.* **1999**, *17*, 1109–1111.
47. A. R. Wheeler, W. R. Throndset, R. J. Whelan, A. M. Leach, R. N. Zare, Y. H. Liao, K. Farrell, I. D. Manger, A. Daridon, *Anal. Chem.* **2003**, *75*, 3581–3586.
48. J. J. Abel, L. G. Rountree, B. B. Turner, *J. Pharm. Exp. Ther.* **1914**, *5*, 275–317.
49. U. Ungerstedt, C. Pycock, *Bull. Schweiz. Akad. Med. Wiss.* **1974**, *1278*, 1–13.
50. N. Torto, T. Laurell, L. Gorton, G. Marko-Varga, *Anal. Chim. Acta* **1999**, *379*, 281–305.
51. M. I. Davies, J. D. Cooper, S. S. Desmond, C. E. Lunte, S. M. Lunte, *Adv. Drug Deliver. Rev.* **2000**, *45*, 169–188.
52. J. F. van Staden, *Fresenius J. Anal. Chem.* **1995**, *352*, 271–302.
53. C. E. Lunte, D. O. Scott, P. T. Kissinger, *Anal. Chem.* **1991**, *63*, 773A–780A.
54. S. Böhm, W. Olthuis, P. Bergveld, *Sens. Actuators B* **2000**, *63*, 201–208.
55. D. P. Lundgren, *Ann. NY Acad. Sci.* **1960**, *87*, 904.
56. L. T. Skeggs, *Am. J. Clin. Pathol.* **1957**, *28*, 311.
57. H. K. Lonsdale, *J. Membr. Sci.* **1982**, *10*, 81.
58. P. M. Martin, D. W. Matson, W. D. Bennett, Y. Lin, D. J. Hammerstrom, *J. Vac. Sci. Technol. A* **1999**, *17*, 2264–2269.
59. C. Liu, S. A. Hofstadler, J. A. Bresson, H. R. Udseth, T. Tsukuda, R. D. Smith, *Anal. Chem.* **1998**, *70*, 1797–1801.
60. F. Xiang, Y. Lin, J. Wen, D. W. Matson, R. D. Smith, *Anal. Chem.* **1999**, *71*, 1485–1490.
61. N. Xu, Y. Lin, S. A. Hofstadler, D. Matson, C. J. Call, R. D. Smith, *Anal. Chem.* **1998**, *70*, 3553–3556.
62. Y. Jiang, P.-C. Wang, L. E. Locascio, C. S. Lee, *Anal. Chem.* **2001**, *73*, 2048–2053.
63. R. Freaney, A. McShane, T. V. Keaveny, M. McKenna, K. Rabenstein, F. W. Scheller, D. Pfeiffer, G. Urban, I. Moser, G. Jobst, A. Manz, E. Verpoorte, H. M. Widmer, D. Diamond, E. Dempsey, F. J. Saez de Viteri, M. Smyth, *Ann. Clin. Biochem.* **1997**, *34*, 291–302.
64. E. Dempsey, D. Diamond, M. R. Smyth, G. Urban, G. Jobst, I. Moser, E. Verpoorte, A. Manz, H. M. Widmer, K. Rabenstein, R. Freaney, *Anal. Chim. Acta* **1997**, *346*, 341–349.
65. H. A. Laitinen, W. E. Harris, *Chemical Analysis; An Advanced Text and Reference*, McGraw-Hill: New York, 1975.
66. D. G. Peters, J. M. Hayes, G. M. Hieftje, *Chemical Separations and Measurements*, W. B. Saunders Company: Philadelphia, 1974.
67. W. E. Harris, B. Kratochvil, *An Introduction to Chemical Analysis*, 1st ed., CBS College Publishing: New York, 1981.
68. M. Tokeshi, T. Minagawa, T. Kitamori, *Anal. Chem.* **2000**, *72*, 1711–1714.
69. J. R. Burns, C. Ramshaw, *Lab Chip* **2001**, *1*, 10–15.
70. E. Baltussen, C. A. Cramers, P. J. F. Sandra, *Anal. Bioanal. Chem.* **2002**, *373*, 3–22.
71. E. M. Thurman, M. S. Mills, *Solid Phase Extraction: Principles and Practice*, 1st ed., John Wiley & Sons: New York, 1998.
72. M. C. Carson, *J. Chromatogr. A* **2000**, *885*, 343–350.

73. N. Delaunay, V. Pichon, M.-C. Hennion, *J. Chromatogr. B* **2000**, *745*, 15–37.
74. N. H. H. Heegaard, S. Nilsson, N. A. Guzman, *J. Chromatogr. B* **1998**, *715*, 29–54.
75. N. H. H. Heegaard, R. T. Kennedy, *J. Chromatogr. B* **2002**, *768*, 93–103.
76. C. F. Poole, A. D. Gunatilleka, R. Sethuraman, *J. Chromatogr. A* **2000**, *885*, 17–39.
77. K. Sato, M. Tokeshi, H. Kimura, T. Kitamori, *Anal. Chem.* **2001**, *73*, 1213–1218.
78. K. Sato, M. Tokeshi, T. Odake, H. Kimura, T. Ooi, M. Nakao, T. Kitamori, *Anal. Chem.* **2000**, *72*, 1144–1147.
79. R. D. Oleschuk, L. L. Shultz-Lockyear, Y. Ning, D. J. Harrison, *Anal. Chem.* **2000**, *72*, 585–590.
80. R. D. Oleschuk, A. B. Jemere, L. L. Schultz-Lockyear, F. Fajuyigbe, D. J. Harrison, in *Proceedings of the µTAS 2000 Symposium*, Enschede, The Netherlands, May 14–18, 2000, van den Berg, A., Olthuis, W., Bergveld, P., Eds., 2000, Kluwer Academic Publishers: Dordrecht, pp. 11–14.
81. J. P. Murrihy, M. C. Breadmore, A. Tan, M. McEnery, J. Alderman, C. O'Mathuna, A. P. O'Neill, P. O'Brien, N. Advoldvic, P. R. Haddad, J. D. Glennon, *J. Chromatogr. A* **2001**, *924*, 233–238.
82. J. P. Kutter, S. C. Jacobson, J. M. Ramsey, *J. Microcolumn Sep.* **2000**, *12*, 93–97.
83. M. Galloway, W. Stryjewski, A. Henry, S. M. Ford, S. Llopis, R. L. McCarley, S. A. Soper, *Anal. Chem.* **2002**, *74*, 2407–2415.
84. A. Dodge, K. Fluri, E. Verpoorte, N. F. de Rooij, *Anal. Chem.* **2001**, *73*, 3400–3409.
85. V. Linder, E. Verpoorte, N. F. de Rooij, H. Sigrist, W. Thormann, *Electrophoresis* **2002**, *23*, 740–749.
86. E. Eteshola, D. Leckband, *Sens. Actuators B* **2001**, *72*, 129–133.
87. D. Ross, L. E. Locascio, *Anal. Chem.* **2002**, *74*, 2556–2564.
88. A. E. Herr, J. I. Molho, J. G. Santiago, M. G. Mungal, T. W. Kenny, M. G. Garguilo, *Anal. Chem.* **2000**, *72*, 1053–1057.
89. R. L. Chien, D. S. Burgi, *Anal. Chem.* **1992**, *64*, A489–A496.
90. M. B. Kerby, R. L. Chien, *Electrophoresis* **2002**, *23*, 3545–3549.
91. R. L. Chien, J. C. Helmer, *Anal. Chem.* **1991**, *63*, 1354–1361.
92. D. S. Burgi, R. L. Chien, *Anal. Chem.* **1991**, *63*, 2042–2047.
93. J. Lichtenberg, E. Verpoorte, N. F. de Rooij, *Electrophoresis* **2001**, *22*, 258–271.
94. N. R. Beard, C. X. Zhang, A. J. deMello, *Electrophoresis* **2003**, *24*, 732–739.
95. J. Li, C. Wang, J. F. Kelly, D. J. Harrison, P. Thibault, *Electrophoresis* **2000**, *21*, 198–210.
96. S. C. Jacobson, J. M. Ramsey, *Electrophoresis* **1995**, *16*, 481–486.
97. J. P. Kutter, R. S. Ramsey, S. C. Jacobson, J. M. Ramsey, *J. Microcolumn Sep.* **1998**, *10*, 313–319.
98. S. C. Jacobson, R. Hergenröder, L. B. Koutny, R. J. Warmack, J. M. Ramsey, *Anal. Chem.* **1994**, *66*, 1107–1113.
99. H. Yang, R.-L. Chien, *J. Chromatogr. A* **2001**, *924*, 155–163.
100. J. P. Quirino, S. Terabe, *Science* **1998**, *282*, 465–468.
101. J. P. Quirino, S. Terabe, *Anal. Chem.* **2000**, *72*, 1023–1030.
102. J. Palmer, N. J. Munro, J. P. Landers, *Anal. Chem.* **1999**, *71*, 1679–1687.

103. J. Palmer, D. S. Burgi, N. J. Munro, J. P. Landers, *Anal. Chem.* **2001**, *73*, 725–731.

104. R. A. Mosher, D. A. Saville, W. Thormann, *The Dynamics of Electrophoresis*, VCH: Weinheim, 1992.

105. F. M. Everaerts, J. L. Beckers, T. P. E. M. Verheggen, *Isotachophoresis: Theory, Instrumentation, and Applications*, Elsevier Scientific: Amsterdam, 1976.

106. D. Kaniansky, M. Masar, J. Bielcikova, F. Ivanyi, F. Eisenbeiss, B. Stanislawski, B. Grass, A. Neyer, M. Jöhnck, *Anal. Chem.* **2000**, *72*, 3596–3604.

107. A. Wainright, S. J. Williams, G. Ciambrone, Q. Xue, J. Wei, D. Harris, *J. Chromatogr. A* **2002**, *979*, 69–80.

108. Z. Q. Xu, T. Hirokawa, T. Nishine, A. Arai, *J. Chromatogr. A* **2003**, *990*, 53–61.

109. J. E. Prest, S. J. Baldock, P. R. Fielden, B. J. T. Brown, *Analyst* **2001**, *126*, 433–437.

110. R. Boom, C. J. A. Sol, M. M. M. Salimans, C. L. Jansen, P. M. E. Wertheim-van Dillen, J. van der Noordaa, *J. Clin. Microbiol.* **1990**, *28*, 495–503.

111. K. A. Melzak, C. S. Sherwood, R. F. B. Turner, C. A. Haynes, *J. Colloid Interface Sci.* **1996**, *181*, 635–644.

112. H. Tian, A. F. R. Hühmer, J. P. Landers, *Anal. Biochem.* **2000**, *283*, 175–191.

113. A. Eckersten, A. E. Örlefors, C. Ellström, K. Erickson, E. Löfman, A. Eriksson, S. Eriksson, A. Jorsback, N. Tooke, H. Derand, G. Ekstrand, J. Engström, A.-K. Honerud, A. Aksberg, H. Hedsten, L. Rosengren, M. Stjernström, T. Hultman, P. Andersson, in *Proceedings of the μTAS 2000 Symposium*, Enschede, The Netherlands, May 14–18, 2000, van den Berg, A., Olthuis, W., Bergveld, P., Eds., 2000, Kluwer Academic Publishers: Dordrecht, pp. 521–524.

114. D. C. Duffy, H. L. Gillis, J. Lin, N. F. Sheppard, Jr., G. J. Kellogg, *Anal. Chem.* **1999**, *71*, 4669–4678.

115. N. Thomas, A. Ocklind, I. Blikstad, S. Griffiths, M. Kenrick, H. Derand, G. Ekstrand, C. Ellström, A. Larsson, P. Andersson, in *Proceedings of the μTAS 2000 Symposium*, Enschede, The Netherlands, May 14–18, 2000, van den Berg, A., Olthuis, W., Bergveld, P., Eds., 2000, Kluwer Academic Publishers: Dordrecht, pp. 249–252.

116. G. Ekstrand, C. Holmquist, A. E. Örlefors, B. Hellman, A. Larsson, P. Andersson, in *Proceedings of the μTAS 2000 Symposium*, Enschede, The Netherlands, May 14–18, 2000, van den Berg, A., Olthuis, W., Bergveld, P., Eds., 2000, Kluwer Academic Publishers: Dordrecht, pp. 311–314.

117. A. Palm, S. R. Wallenborg, M. Gustafsson, A. Hedström, E. Togan-Tekin, P. Andersson, in *Proceedings of the μTAS 2001 Symposium*, Monterey, CA, USA, October 21–25, 2001, Ramsey, J. M., van den Berg, A., Eds., 2001, Kluwer Academic Publishers: Dordrecht, pp. 216–218.

118. G. Jesson, P. Andersson, in *Proceedings of the μTAS 2001 Symposium*, Monterey, CA, USA, October 21–25, 2001, Ramsey, J. M., van den Berg, A., Eds., 2001, Kluwer Academic Publishers: Dordrecht, pp. 551–552.

119. W. R. Boom, H. M. A. Adriaanse, T. Kievits, P. F. Lens, European Patent EP 0.389 063 A2, 1990.

120. K. A. Wolfe, M. C. Breadmore, J. P. Ferrance, M. E. Power, J. F. Conroy, P. M. Norris, J. P. Landers, *Electrophoresis* **2002**, *23*, 727–733.

121. L. Ceriotti, J. Lichtenberg, A. Dodge, N. de Rooij, E. Verpoorte, in *Proceedings of the μTAS 2002 Symposium*, Nara, Japan, November 3–7, 2002, Baba, Y.,

Shoji, S., van den Berg, A., Eds., 2002, Kluwer Academic Publishers: Dordrecht, pp. 175–177.

122. Z. H. Fan, S. Mangru, R. Granzow, P. Heaney, W. Ho, Q. Dong, R. Kumar, *Anal. Chem.* **1999**, *71*, 4851–4859.

123. G. Jiang, D. J. Harrison, *Analyst* **2000**, *125*, 2176–2179.

124. R. C. Anderson, X. Su, G. J. Bogdan, J. Fenton, *Nucleic Acids Res.* **2000**, *28*, e60.

125. L. A. Christel, K. Petersen, W. McMillan, M. A. Northrup, *J. Biomech. Eng.* **1999**, *121*, 22–27.

126. B. Vogelstein, D. Gillespie, *Biochemistry* **1978**, *76*, 615–619.

127. J. Khandurina, S. C. Jacobson, L. C. Waters, R. S. Foote, J. M. Ramsey, *Anal. Chem.* **1999**, *71*, 1815–1819.

128. J. Khandurina, T. E. McKnight, S. C. Jacobson, L. C. Waters, R. S. Foote, J. M. Ramsey, *Anal. Chem.* **2000**, *72*, 2995–3000.

129. Y.-C. Lin, H.-C. Ho, C.-K. Tseng, S.-Q. Hou, *J. Micromech. Microeng.* **2001**, *11*, 189–194.

130. K. Mullis, F. Faloona, S. Scharf, R. Saiki, G. Horn, H. Erlich, *Cold Spring Harbor Symp. Quant. Biol.* **1986**, *51*, 263–273.

131. J. Liu, M. Enzelberger, S. Quake, *Electrophoresis* **2002**, *23*, 1531–1536.

132. M. A. Northrup, M. T. Ching, R. M. White, R. T. Watson, in *Digest of Technical Papers: Transducers 1993*, Yokohama, Japan, June 7–10, 1993, **1993**, pp. 924–927.

133. P. Wilding, M. A. Shoffner, L. J. Kricka, *Clin. Chem.* **1994**, *40*, 1815–1818.

134. M. A. Shoffner, J. Cheng, G. E. Hvichia, L. J. Kricka, P. Wilding, *Nucleic Acids Res.* **1996**, *24*, 375–379.

135. B. C. Giordano, E. R. Copeland, J. P. Landers, *Electrophoresis* **2001**, *22*, 334–340.

136. T. B. Taylor, E. S. Winn-Deen, E. Picozza, T. M. Woudenberg, M. Albin, *Nucleic Acids Res.* **1997**, *25*, 3164–3168.

137. M. U. Kopp, A. J. de Mello, A. Manz, *Science* **1998**, *280*, 1046–1048.

138. L. C. Waters, S. C. Jacobson, N. Kroutchinina, J. Khandurina, R. S. Foote, J. M. Ramsey, *Anal. Chem.* **1998**, *70*, 5172–5176.

139. J. W. Hong, T. Fujii, M. Seki, T. Yamamoto, I. Endo, *Electrophoresis* **2001**, *22*, 328–333.

140. A. M. Chaudhari, T. M. Woudenberg, M. Albin, K. E. Goodson, *J. Microelectromech. Syst.* **1998**, *7*, 345–355.

141. R. P. Oda, M. A. Strausbauch, A. F. R. Huhmer, N. Borson, S. R. Jurrens, J. Craighead, P. J. Wettstein, B. Eckloff, B. Kline, J. P. Landers, *Anal. Chem.* **1998**, 4361–4368.

142. B. C. Giordano, J. Ferrance, S. Swedberg, A. F. Huhmer, J. P. Landers, *Anal. Biochem.* **2001**, *291*, 124–132.

143. M. A. Northrup, C. Gonzalez, D. Hadley, R. F. Hills, P. Landre, S. Lehew, R. Saiki, J. J. Sninsky, R. Watson, R. J. Watson, in *Digest of Technical Papers: Transducers 1995 and Eurosensors IX*, Stockholm, Sweden, June 25–29, 1995, Middelhoek, S., Cammann, K., Eds., 1995, pp. 764–767.

144. M. A. Northrup, B. Benett, D. Hadley, P. Landre, S. Lehew, J. Richards, P. Stratton, *Anal. Chem.* **1998**, *70*, 918–922.

145. A. T. Woolley, D. Hadley, P. Landre, A. J. de Mello, R. A. Mathies, M. A. Northrup, *Anal. Chem.* **1996**, *68*, 4081–4086.

146. M. S. Ibrahim, R. S. Lofts, P. B. Jahrling, E. A. Henchal, V. W. Weedn, M. A. Northrup, P. Belgrader, *Anal. Chem.* **1998**, *70*, 2013–2017.

147. P. Belgrader, W. Benett, D. Hadley, G. Long, R. J. Mariella, F. Milanovich, S. Nasarabadi, W. Nelson, J. Richards, P. Stratton, *Clin. Chem.* **1998**, *44*, 2191–2194.

148. P. Belgrader, W. Benett, D. Hadley, J. Richards, P. Stratton, R. J. Mariella, F. Milanovich, *Science* **1999**, *284*, 449–450.

149. A. I. K. Lao, T. M. H. Lee, I.-M. Hsing, N. Y. Ip, *Sens. Actuators. A* **2000**, *84*, 11–17.

150. T. M. Lee, I. M. Hsing, A. I. Lao, M. C. Carles, *Anal. Chem.* **2000**, *72*, 4242–4247.

151. S. Poser, T. Schulz, U. Dillner, V. Baier, J. M. Köhler, D. Schimkat, G. Mayer, A. Siebert, *Sens. Actuators A* **1997**, *62*, 672–675.

152. J. H. Daniel, S. Iqbal, R. B. Millington, D. F. Moore, C. R. Lowe, D. L. Leslie, M. A. Lee, M. J. Pearce, *Sens. Actuators A* **1998**, 81–88.

153. Q. Zou, Y. Miao, Y. Chen, U. Sridhara, C. S. Chong, T. Chai, Y. Tieb, C. H. L. Teh, T. M. Lim, C. Heng, *Sens. Actuators A* **2002**, *102*, 114–121.

154. E. Shinohara, S. Kondo, K. Akahori, K. Tashiro, S. Shoji, *IEICE Trans. Electron.* **2001**, *E84C*, 1807–1812.

155. K. Akahori, S. Kondo, E. Shinohara, K. Tashiro, S. Shoji, in *Proceedings of the μTAS 2000 Symposium*, Enschede, The Netherlands, May 14–18, 2000, van den Berg, A., Olthuis, W., Bergveld, P., Eds., 2000, Kluwer Academic Publishers: Dordrecht, pp. 493–496.

156. E. T. Lagally, P. C. Simpson, R. A. Mathies, *Sens. Actuators B* **2000**, *63*, 138–146.

157. E. T. Lagally, I. Medintz, R. A. Mathies, *Anal. Chem.* **2001**, *73*, 565–570.

158. E. T. Lagally, C. A. Emrich, R. A. Mathies, *Lab Chip* **2001**, *1*, 102–107.

159. M. A. Unger, H. P. Chou, T. Thorsen, A. Scherer, S. R. Quake, *Science* **2000**, *288*, 113–116.

160. A. Gulliksen, L. Solli, F. Karlsen, H. Rogne, E. Hovig, T. Nordstrom, R. Sirevag, *Anal. Chem.* **2004**, *76*, 9–14.

161. J. Compton, *Nature* **1991**, *350*, 91–92.

162. S. C. Jacobson, J. M. Ramsey, *Anal. Chem.* **1996**, *68*, 720–723.

163. N. Gottschlich, C. T. Culbertson, T. E. McKnight, S. C. Jacobson, J. M. Ramsey, *J. Chromatogr. B* **2000**, *745*, 243–249.

164. Q. Xue, Y. M. Dunayevskiy, F. Foret, B. L. Karger, *Rapid Commun. Mass Spectrom.* **1997**, *11*, 1253–1256.

165. C. Wang, R. Oleschuk, F. Ouchen, J. Li, P. Thibault, D. J. Harrison, *Rapid Commun. Mass Spectrom.* **2000**, *14*, 1377–1383.

166. J. Lichtenberg, L. Ceriotti, N. A. Lacher, S. M. Lunte, N. F. de Rooij, E. Verpoorte, in *Proceedings of the μTAS 2002 Symposium*, Nara, Japan, November 3–7, 2002, Baba, Y., Shoji, S., van den Berg, A., Eds., 2002, Kluwer Academic Publishers: Dordrecht, pp. 172–174.

167. A. Han, L. Ceriotti, J. Lichtenberg, N. F. d. Rooij, E. Verpoorte, in *Proceedings of the μTAS 2003 Symposium*, Squaw Valley, CA, October 5–9, 2003, Northrup, M. A., Jensen, K. F., Harrison, D. J., Eds., 2003, Transducers Research Foundation: Cleveland, OH, pp. 575–578.

168. P. J. Oefner, S. P. Hunicke-Smith, L. Chiang, F. Dietrich, J. Mulligan, R. W. Davis, *Nucleic Acids Res.* **1996**, *24*, 3879–3886.

169. T. T. Perkins, D. E. Smith, S. Chu, *Science* **1997**, *276*, 2016–2021.
170. R. H. Liu, J. N. Yang, R. Lenigk, J. Bonanno, P. Grodzinski, *Anal. Chem.* **2004**, *76*, 1824–1831.

11 Detection on Microchips: Principles, Challenges, Hyphenation, and Integration

Stephanie Pasas, Barbara Fogarty,
Bryan Huynh, Nathan Lacher, Brian Carlson,
Scott Martin, Walter Vandaveer IV, and Susan
Lunte

CONTENTS

11.1 INTRODUCTION TO DETECTORS FOR MICROCHIP-BASED SEPARATION TECHNIQUES

Detecting discrete separation bands that have been produced by microchip-based separation techniques is a challenging task. Most of the initial studies involving microchip CE focused on the fabrication, injection, and separation aspects, with detection being accomplished by laser-induced fluorescence (LIF).[1-4] More recently, researchers have realized the importance of other methods of detection for microchip CE and the challenges associated with them.[5,6] In addition, it has become evident that to increase the applicability and versatility of these systems, detectors other than LIF need to be explored. This chapter details the various detection modes that have been used for microchip CE and their unique considerations.

Although detection modes differ in terms of theory and operation, common analytical challenges exist for microchip applications. These challenges are mainly derived from the small volumes of sample that are injected and the small separation bands that are subsequently produced. To fully understand the various factors that affect the performance of a detector, the amount of band broadening that is produced during a CE-based separation must be considered. Since separation bands are Gaussian in shape, the efficiency of a column can be defined in terms of variance per unit length by the equation:

$$H = \frac{\sigma^2}{L} \tag{11.1}$$

where H is the total plate height and L is the length of the separation channel.[7] For a given microchip CE-based separation device, the total amount of variance (σ_{tot}^2) can be expressed by the individual components of the system using the equation:

$$\sigma_{tot}^2 = \sigma_{inj}^2 + \sigma_{det}^2 + \sigma_{diff}^2 + \sigma_{ads}^2 \tag{11.2}$$

where σ_{inj}^2 is the injection plug length, σ_{det}^2 is the detector observation length, σ_{diff}^2 is the axial diffusion, and σ_{ads}^2 is the analyte adsorption. Experimentally, band broadening is evaluated in terms of either plate height (H) or the number of theoretical plates (N), which are related by the following relationship:

$$H = \frac{L}{N} \tag{11.3}$$

11.1.1 Sample Volume Considerations

In microchip CE, channel dimensions range in size depending upon the substrate and the fabrication method, but are generally on the order of 10 to 100 μm in width and 1 to 50 μm in depth.[6] While a few studies have utilized separation channel lengths that are similar to those commonly encountered in conventional CE (25 to 30 cm),[8,9] most separation channels have been on the order of 5 to 10 cm in length. To minimize the contribution of the injector to

σ_{tot}^2, the injection volume must be very small. The amount of variance brought about by the injection methodology can be calculated using the equation:

$$\sigma_{inj}^2 = \frac{(\Delta t)^2}{12} \tag{11.4}$$

where Δt is the width of the band exiting the injector.

Although many methods exist for injecting discrete bands in microchip CE devices (see also Chapter 2),[10] channel cross sections and voltage control fix injection volumes from 10 to 400 pl and lead to minimal contributions to σ_{tot}^2. However, these small injection volumes also lead to very small amounts of analyte being injected into the separation channel. For example, if 100 pl of a 10 nM sample is injected into such a device, the subsequent separation band contains only 1 amol (amol = attomole) of analyte. Therefore, the detector that is employed must offer extreme sensitivity, especially considering the band broadening that can occur from other contributors in the system (Equation (11.2). Given the small volumes that are injected, caution must be exercised when comparing detectors for microchip-based separation methods to those for more traditional separation methods. Even though the absolute amounts (moles) that can be detected in microchips may be lower than in traditional separation approaches, the limits of detection with respect to concentration are often higher due to the small volumes utilized.

11.1.2 DETECTOR AND TIME CONSTANT

Another factor that needs to be considered is the contribution to band broadening from the detector itself (σ_{tot}^2), which can be calculated using a relationship similar to that given in Equation (11.4) (see also Section 2.7.3 of this book). The σ_{det}^2 term can be minimized by keeping the detection sensing area as small as possible. For example, with LIF detection this is accomplished by focusing the laser excitation spot size to 5 to 20 μm, or smaller than the length of the band that was injected into the system. For detection modes such as mass spectrometry and some forms of electrochemistry, the detector is not integrated inside the separation channel and is external to the chip. In these cases, σ_{det}^2 is largely due to off-channel band broadening, which is a function of the physical distance between the chip and the detector.

Depending upon the substrate, CE conditions, separation channel length, and field strength, peaks in microchip CE can elute very quickly. Typically, separation bands passing by the detector exhibit peak widths in the range of 0.1 to 2 sec. In extreme cases using very high field strengths (53 kV/cm) and short separation channels (200 μm), peaks as narrow as 0.3 msec have been reported.[11]

To adequately sample a peak as it passes by the detector and ensure satisfactory peak resolution, it has been shown in chromatography that a time constant approximately 10% of the minimum peak width at half-height is necessary.[12] The time constant of the detector is usually described in terms of how frequently the sensing region of the detector can be sampled. Another good guideline is to acquire roughly 10 data points across one peak width at half-height.[12] Therefore, depending upon the experimental conditions, the time constant of a detector may become a limiting factor in adequately sampling a separation band as it passes the detector. The time constant of the detector needs to be optimized for a given CE separation. If the time constant is too fast it will introduce excessive noise in the electropherogram, whereas a time constant that is too slow will distort peak shape and reduce peak resolution.[12]

11.1.3 SUBSTRATE MATERIAL

The last factor that should be considered involves the substrate that is used to make the chip. In terms of detection, the substrate should not interfere with the analytical signal. Therefore, with optical detection, a chip must be transparent to the wavelength of light used. For example, glass devices are transparent to the visible wavelengths typically used for LIF detection, but many polymers autofluoresce at these wavelengths or interact with the excitation radiation to produce additional fluorescence.[13] Electrochemical and mass spectrometric detection are not affected by the transparency of the substrate, although it has been shown that some polymer substrates lead to an increased background in mass spectrometry, mainly due to leaching of substrate components from the chip substrate.[14] Lastly, the σ_{ads}^2 term is greatly affected by the choice of substrate material used to fabricate the microfluidic device. It has been demonstrated that some polymer substrates, such as polydimethylsiloxane (PDMS), are hydrophobic, leading to adsorption of peptides and other hydrophobic compounds.[15]

11.1.4 DETECTION MODES

All detectors for microchip CE have some unique considerations due to the small dimensions involved. First, to have a truly integrated micro total analysis system (μTAS), the detector itself (or its main components) must also be miniaturized. The small dimensions of the separation channels are a limiting factor for optical methods that rely on Beer's law, as the effective pathlength becomes smaller and smaller. Although UV detection has been reported for microchip CE, modification of the chip is generally necessary to lengthen the effective optical pathlength.

Fluorescence continues to be the most popular detection method for microchip CE. Lasers are employed as high intensity excitation sources to facilitate improved sensitivity. Additionally, laser beams can be easily focused on the small channels used in microchip CE. A major consideration for both UV and fluorescence detection is the optical transparency of the substrate. Glass or quartz are generally the best substrates; however, many plastics can also be employed.

Mass spectrometric detection does not rely on transparent substrates, although device alignment and assembly is easier if the substrate is transparent. Mass spectrometric detection is becoming more popular for microchip separations. Although miniaturized mass spectrometers are being developed, the ability to perform fast parallel analyses is the driving force.

The detection of small currents and the decoupling of the separation voltage from the electrochemical detector are the major issues in electrochemical detection. Electrodes are relatively easily miniaturized and, in theory, should provide the same sensitivity as their larger counterparts. Although a variety of substrate materials can be used for microchip CE with electrochemical detection, there are limitations on the run buffers that can be employed. Conductivity detection is becoming a very popular method for use with microchip CE. It has replaced UV and refractive index as a "universal detection" mode for microchip electrophoresis. Contactless conductivity detection can also be performed by placing electrodes outside of the channel.

In the following pages, the various detection modes for microchip CE are described and the analytical performance of each is evaluated. These modes can be separated into three basic groups: optical, mass spectrometric, and electrochemical. This chapter is designed to give the reader experimental knowledge concerning the implementation of the various detection modes with microchip CE.

11.2   FLUORESCENCE DETECTION

11.2.1   Introduction

Fluorescence is one of the most sensitive and selective modes of detection for both liquid chromatography and capillary electrophoresis. For fluorescence to occur, a molecule must absorb a photon or photons to achieve an electronically excited singlet state. The linear relationship between absorbance and analyte concentration is predicted by the Beer–Lambert Law and can be used for quantitative analyses.

$$A = \varepsilon bc \tag{11.5}$$

For monochromatic radiation, the amount of radiation absorbed, A, is directly proportional to both the pathlength through the medium, b, and the concentration of the absorbing species, c.[16] When concentration units of molarity are used, ε defines the wavelength-dependent molar absorptivity coefficient with units of $M^{-1}\,cm^{-1}$. When the molecule returns to the ground state, it releases a photon of energy lower than that of the initial excited state (fluorescence). There are other mechanisms by which a molecule can return to the electronic ground state; these include triplet-state conversion and nonradiative transition. Fluorescence detection sensitivities are generally improved for molecules with a high quantum yield, defined as the ratio of the number of photons emitted to the number absorbed.

Many compounds are not naturally fluorescent and must be derivatized to be detected. Therefore, the appropriate selection of a fluorescent agent is critical. The ideal fluorophore displays high molar absorptivity and quantum yield, and should exhibit good photostability. A more detailed description of the basic principles of fluorescence spectroscopy can be found elsewhere.[17–19]

Fluorescence is the most commonly employed detection method for the microchip format, primarily due to its high sensitivity. Most fluorescence detection systems for microchip CE are much larger than the chip itself and are, therefore, not conducive to the development of a "true" miniaturized total analysis system. However, recent advances in the integration of microfabricated optics onto microchip devices are making the possibility of small, portable instruments more feasible.[20–22]

In this section, practical considerations for fluorescence detection will be presented. First, instrumental design considerations will be discussed. These include the type of excitation sources available, the detector design, and the configuration of the optics. Second, strategies for the detection of compounds that do not exhibit native fluorescence will be described. These include precolumn derivatization chemistries, on-chip strategies for extrinsic fluorophore tagging of nonfluorescent analytes, and indirect fluorescence detection. Lastly, present and future applications of optical detection modes for microchip CE are discussed.

11.2.2 INSTRUMENTAL DESIGN

For the inexperienced operator, one of the more challenging aspects of microchip CE with fluorescence detection is the setup and design of the detection system. In particular, the appropriate excitation sources and optical arrangement must be identified.

11.2.2.1 Excitation Sources

The two types of excitation sources that are commonly employed for fluorescence detection are lamps and lasers. Deuterium or xenon arc lamps offer continuous output at a wide range of spectral wavelengths (190 to 400 and 190 to 2000 nm, respectively). These lamps are often incorporated into fluorescence microscopes, which are frequently modified to act as fluorescence detectors for capillary electrophoresis. However, lasers are more commonly employed as high intensity excitation sources in CE because they generate narrow output wavelengths and can be very easily focused into small capillaries or channels. The use of lasers has become commonplace in spectroscopy, and many manufacturers offer durable lasers at a reasonable cost. In addition, the excitation wavelengths of many of the commercially available derivatization agents correspond very well with the spectral lines for Ar-ion and He–Cd lasers. A list of some of the most common agents employed for derivatization can be seen in Table 11.1.

The Ar-ion laser is available with either a fixed or adjustable wavelength. The most commonly employed wavelength is the 488 nm line, which is used for the excitation of fluorescein isothiocyanate (FITC) or Alexa Fluor 488. It can also be used for 3-(4-carboxybenzoyl)quinoline-2-carboxaldehyde (CBQA) and naphthalene-2,3-dicarboxaldehyde (NDA) derivatives. He–Ne lasers are popular excitation sources due to their low cost, compact size, and long operating lifetimes. There are several lines available, including 633 nm (red) and 594 nm (green), which correspond to the excitation wavelengths for the commonly employed derivatization reagents BODIPY (boron dipyrro-methane) and Texas Red (sulforhodamine 101 sulfonyl chloride), respectively. The He–Cd laser is available as a dual-line laser with 442 and 325 nm emission lines; it is a useful excitation source for analytes derivatized with NDA and o-phthalaldehyde (OPA), respectively.

Alternatives to gas lasers are the solid-state diode-pumped lasers, which come in a wide range of wavelengths. These include the IR (1053 to 1064 nm), green (527 to 532 nm), blue (400 to 440, 473 nm), UV (351 to 355 nm), and deep UV (263 to 266 nm) regions. Prices for diode-pumped lasers are comparable to the cost of the commonly used gas lasers, but the advantages of increased efficiency and smaller size should lead to a rise in demand.[23,24] Also noteworthy are the extremely small diode lasers made popular by the advent of compact disk players. These diodes are more amenable to incorporation into miniaturized devices and are commercially available at several wavelengths including 640 nm, 625 nm, 605 nm, 525 nm and 470 nm. There is a push to lower the lasing wavelength into the visible spectrum, as well as to increase the power, efficiency, and lifetime of these diode lasers.[23,25]

11.2.2.2 Detection

The most commonly used device for fluorescence detection with microchips is the photomultiplier tube (PMT). PMTs consist of a photocathode made of a thin film of photoemissive metal and a region of amplification dynodes. During operation, the photocathode and dynodes are held at negative potentials. When the photocathode is subjected to incident photons, electrons are ejected from the metal surface. These electrons accelerate toward the series of dynodes and are amplified upon collision with each sequential dynode until they reach the anode.

TABLE 11.1
Summary of Common Derivatization Reagents Used in Microchip CE-LIF

Analytes	Derivatization Reagent	Ex/Em (nm)	Excitation Source	Reactive Group
Proteins, peptides, amino acids	Alexa Fluor® 488 [a]	494/519	Ar+ laser, 488 nm line	1°, 2° amines
	BODIPY® TR [a]	589/617	He–Ne laser, 594 nm line	1°, 2° amines
	Dansyl chloride	340/515	He–Cd laser, 350 nm line	1°, 2° amines/phenols
	FITC	492/520	Ar+ laser, 488 nm line	1°, 2° amines
	Fluorescamine	390/475	Xenon, deuterium lamp	1°, 2° amines
	Nano Orange[c]	470/570	Ar+ laser, 472 nm line	—
	NBD	470/530	Ar+ laser, 472 nm line	Low MW 1° amines
	NDA	420/495	He–Cd laser, 442 nm line	Low MW 1° amines
	OPA	340/455	He–Cd laser, 350 nm line	Low MW 1° amines
	Rhodamine Red [b]	570/580	He–Ne laser, 594 nm line	1° amines
	TRITC	550/580	HeNe laser, 543 nm line	1°, 2° amines
	TAMRA [b]	555/580	HeNe laser, 543 nm line	1°, 2° amines
	SYPRO® Orange [c]	300, 470/570	Ar+ laser, 472 nm line	—
	SYPRO® Red [c]	300, 550/630	HeNe laser, 543 nm line	—
DNA, nucleic acids	Ethidium bromide [d]	514/605	Ar+ laser, 514 nm line	—
	POPO® [e]	434/456	Hg arc lamp, 436 nm line	—
	Propidium iodide [d]	535/617	He–Ne laser, 543 nm line	—
	TOTO® [e]	514/533	Ar+ laser, 514 nm line	—
	YOYO® [e]	497/533	Ar+ laser, 488 nm line	—

NBD, 7-halo-4-nitrobenzo-2-oxa-1,3-diazole; NDA, naphthalene-2,3-dicarboxaldehyde; FITC, fluorescein isothiocyanate; TAMRA, carboxytetramethylrhodamine; TRITC, tetramethylrhodamine isothiocyanate.

[a] Proprietary dyes can be available to be thio-, aldehyde-, or ketone-reactive, at various wavelengths.
[b] Dye can also be used for DNA and nucleic acid derivatization.
[c] Noncovalent protein-binding dye.
[d] Nucleic acid intercalator.
[e] Cell-impermeant nucleic acid dye.

At the anode, a current pulse results from these collisions. The magnitude of this pulse depends on the intensity and wavelength of the fluorescence as well as the overall voltage applied to the PMT. The higher the voltage used to operate the PMT, the greater the amplification, as the amount of electrons ejected from each dynode increases with voltage. However, if the operating voltage is too high, excessive photocurrents can damage the photocathode. This type of damage results in decreased amplification and the generation of excessive dark currents (residual current output in absence of incident light) that can increase background. Recommended voltages for PMTs are usually between 800 and 1200 V, depending on the type of photoemissive metal film in the cathode.[19]

In most cases, photomultiplier detectors measure radiant intensity in the analog mode, an approach that involves averaging the current generated from photoelectrons at the anode. Typically, current from the PMT is sent to a signal preamplifier that converts the current to voltage. This voltage is passed to an analog to digital (A/D) converter that is interfaced to a computer for data processing.

For applications that require higher sensitivity, such as single-molecule detection, avalanche photodiodes can be employed. These are solid-state detectors that operate in a manner similar to PMTs. In this case, the electrical current magnification is due to an impact ionization process that occurs within the photodiode itself resulting in high sensitivity.[26] Unfortunately, these photodetectors also have very small active areas, making optical alignment critical when they are used as detectors for CE.

Cameras using charge-coupled devices (CCD) are often used in conjunction with optical microscopes as fluorescence detectors for microchip CE due to their superior sensitivity to light. Most microscope manufacturers include ports for cameras and other optical components. CCD cameras can be interfaced to the PC using a video card and a microprocessor with enough memory for handling the large amount of information generated by the device. To monitor flow profiles, an expanded Ar-ion beam is focused onto a channel. The fluorescence is collected using an microscope objective, spectrally filtered and recorded.[27–32] Such imaging setups have been used for evaluation of injector configurations,[33–35] electroosmotic pumping in devices,[30,36,37] microfluidic focusing and fluid control,[38–40] microfluidic mixing,[29,41,42] and device performance.[4,11,27,43] Fluorescein itself is frequently employed for the visualization of reagent flow within the chip using a fluorescence microscope. Because the aqueous solubility of fluorescein is poor, the disodium salt is usually employed in these studies. Another substance that has been employed for flow visualization is rhodamine, a neutral dye whose fluorescence is not pH sensitive, an important consideration when measuring electroosmotic flow (EOF).[38,44] These types of studies have greatly contributed to the characterization of microchip CE in the past decade.

11.2.2.3 Optical Configurations

For home-built optical detection systems, considerations of microscope objective and filter characteristics are necessary. The most important requirements for the microscope objective are working distance and good collection efficiency. The working distance of a microscope objective is defined as the distance that the objective can image; the longer the working distance, the further the specimen to be imaged can be from the objective. Additionally, the working distance depends on the substrate thickness of the chip and can range from ~0.5 to 1.5 mm. Numerical aperture (N.A.) is a value assigned to a microscope objective to gauge its efficiency of collection. Typically, collection efficiencies increase as N.A. increases. For common microscope objectives, higher N.A. objectives generally have shorter working distances. Objectives with high N.A. and long working ranges are available, but can be quite expensive.

Once the fluorescence has been collected by the objective, spatial filtering prior to detection becomes important. By allowing focused light to pass through a precisely positioned pinhole aperture, stray light that can contribute to background is reduced.[45] Pinholes for spatial filtering can range in size from 1 to 1000 μm. Spectral filtering can also be used either to further select or eliminate collected emission as a function of wavelength using interference filters.

Two optical configurations are generally used when constructing a fluorescence detection system for CE microchips. The first configuration (Figure 11.1) involves directing a laser light orthogonal to the detector and focusing the beam onto the separation channel.[28,43] The ideal spot size is no larger than the width of the channel (20 to 30 μm). When the fluorophore passes through the detection region, it is excited and the resulting fluorescence is collected by the microscope objective. Spatial and spectral filters are used after the objective to reduce interference from scattered light before the emitted photons arrive at the detector.

The second approach utilizes confocal epifluorescence, which is illustrated in Figure 11.2. The excitation beam is focused and resulting fluorescence is collected using a single microscope objective. In this configuration, a laser beam is directed toward a dichroic mirror that reflects the beam 90° onto the microscope objective. The objective then focuses the beam into a 12 to 25 μm spot on the separation channel. The size of the spot is dependent on the type of objective employed. When a fluorophore is excited, the emission light is collected by the same objective and passes back through the dichroic mirror, which now acts as a filter. Pinhole and spectral filtering follows, and the light finally reaches the detector.[45,46]

It should also be noted here that although home-built systems are still employed, microscopes are also used as housed detection systems. Most modern microscopes include various ports and inserts that permit filters, detectors, and light sources to be changed. Lamps can also be utilized as

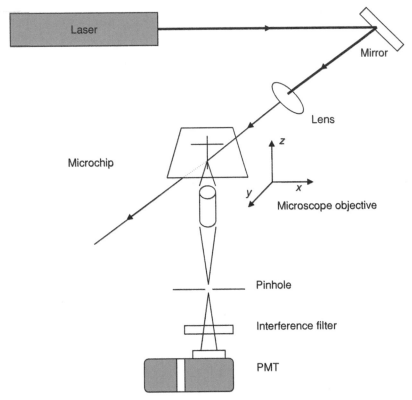

FIGURE 11.1 Scheme I. LIF detection for microchip CE.

excitation sources in fluorescence microscopes. Although cheaper, this system is less sensitive. Using these systems, light can easily be directed toward the chip using a fiber optic cable and focused onto the separation channel placed on a manipulative stage.

For applications involving parallel analysis (monitoring multiple channels) additional configurations should be considered at the design stage. A number of laser-scanning techniques exist for parallel analysis, of which acousto-optical deflection[47] and galvanometric optical scanning[48] have been demonstrated for microchip CE. Movement of the translational stage can be controlled for parallel analysis using a voice coil actuator.[49] In one study, a multichannel microfluidic device was fixed upon the stage, and sample injection was achieved by optical gating.[50,51] Confocal fluorescence detection of labeled amino acids was achieved by rotating the microfluidic device with the voice coil actuator. Figure 11.3A shows the setup and Figure 11.3B shows electropherograms of simultaneous analyses of three sample mixtures (1) Arg, Phe, Gly; (2) Arg, Phe, Gly, Glu; and (3) Arg, Phe, Glu in three separate channels using one detector.

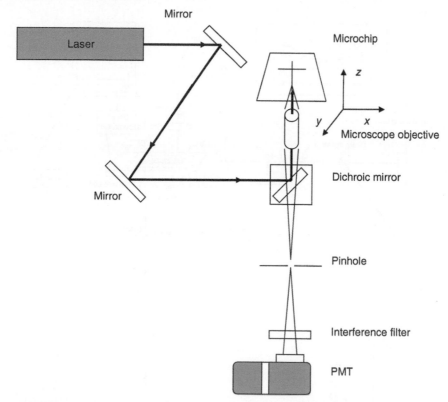

FIGURE 11.2 Scheme II: A confocal fluorescence configuration for microchip CE with LIF detection.

11.2.2.4 On-Chip Integration of Fluorescence Detection

Currently, analyses using microchip CE with fluorescence detection have been confined to the laboratory benchtop because the optical setups are large relative to the microfluidic device used for separation. However, to realize portability of a microchip CE-LIF device, the engineering of on-chip optics and detector components must be achieved. A fluorescence detector integrated onto a silicon CE device (Figure 11.4A) has been reported.[22] Integrated on the chip were the following: channels and electrodes for separation, silicon photodiodes for detection, a thin-film optical filter for blocking excitation radiation, and a thin transparent grounding layer of aluminum zinc oxide (AZO) to isolate the photodiode field from the separation field (Figure 11.4B). To complete the setup, a blue light-emitting diode (LED) was used for excitation with output amplification and a lock-in amplifier for photodiode current measurement. Separation of a plasmid digest prelabeled with SYBR Green I was carried out using this device.

(A)

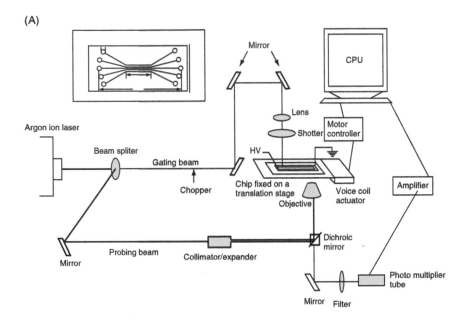

(B)

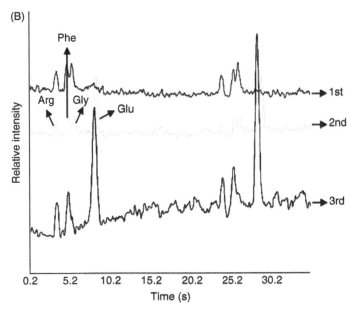

FIGURE 11.3 Setup and results of a parallel analysis experiment using microchip CE-LIF. In (A), a schematic of the gated injection scheme is shown. A laser beam is split into a gating and probing beam. Parallel detection is carried out confocally by rotating the stage that the multichanneled chip is on. The inset is a schematic of the multi-channel chip for analysis. In (B), the results of a simultaneous analysis of three sample mixtures in three different channels are shown. (Reprinted from Xu, H.; Roddy, T. P.; Lapos, J. A.; Ewing, A. G. *Anal. Chem.* 2002, *74*, 5517–5522. With permission.)

(A)

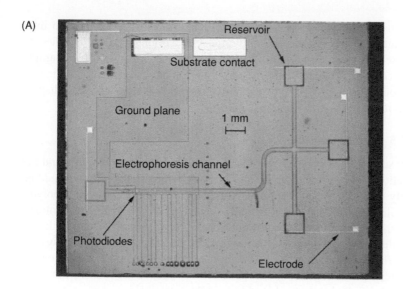

(B)

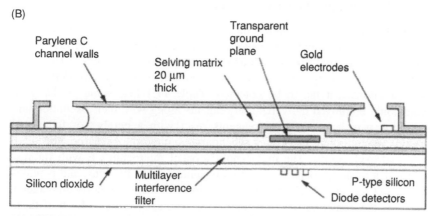

FIGURE 11.4 An optogram of the actual layout of the integrated device is shown in (A). In (B), a schematic of the component layers of the integrated chip is presented. (Reprinted from Roulet, J. C.; Volkel, R.; Herzig, H. P.; Verpoorte, E. M. J.; de Rooij, N. F.; Dandliker, R. *Anal. Chem.* 2002, *74*, 3400–3407. With permission.)

Additionally, a microoptical microfluidic system has been reported in which an array of microfabricated optical lenses was integrated onto a microfluidic platform.[21] A detection limit of 3.3 n*M* for a Cy5 solution in pH 7.4 phosphate buffer was obtained.[20] Chabinyc et al. developed a prototype device consisting of a disposable PDMS microfluidic channel layer and a reusable microavalanche photodiode detector.[52] An optical fiber embedded in the disposable layer delivered excitation light from a blue LED. Also integrated into the device was an additional polycarbonate layer designed to reduce scatter prior to detection. Using this device, separation of a sample mixture containing fluorescein-labeled proteins and 5-carboxyfluorescein was carried out. This approach illustrates the possibility of a portable and disposable device.

11.2.2.5 Other Considerations

Factors such as the optical properties of microchip substrate material and the elimination of background should also be considered. The substrate should be optically transparent: for example, glass, quartz, or the polymers polydimethylsiloxane (PDMS) and polymethylmethacrylate (PMMA). Furthermore, the material must not exhibit autofluorescence, which would result in dilution of the fluorophore's signal.[37] Other sources of background in the system include fluorescent impurities and light scattering. The addition of interference filters, polarizing films, and pinholes can reduce unwanted scatter, as can the use of shields or machined light-tight boxes.

11.2.3 Derivatization Chemistries

A number of substances exhibit native fluorescence. These include proteins and peptides containing tryptophan or tyrosine, some anticancer drugs, and many environmental toxins such as polyaromatic hydrocarbons.[19] However, there are many compounds that are not fluorescent and must be derivatized with a fluorescent tag to be detected. Table 11.1 lists some of the most commonly employed derivatization reagents along with their excitation and emission wavelengths. These reagents react with a number of functional groups, including primary and secondary amines, thiols, carboxylic acid, and hydroxyl groups. The main application of microchip CE with fluorescence detection has been the analysis of amino acids, proteins, and peptides (see also Chapter 12). Therefore, this section will focus primarily on reagents used for the derivatization of primary and secondary amines. Each of these reagents has its own unique characteristics, including reaction rate, stability of products, quantum yield, and background fluorescence of unbound reagent.

11.2.3.1 Derivatization Reagents

FITC, an amine-reactive isothiocyanate derivative of fluorescein, reacts with both primary and secondary amines. FITC derivatives exhibit large molar absorptivities and high quantum yields and are stable. This reagent, however, does have some disadvantages. The derivatization reaction is very slow (usually overnight reaction is needed) and, therefore, FITC must be used as a precolumn labeling reagent. Other potential drawbacks include rapid photobleaching and pH-sensitive fluorescence. Photobleaching results in a permanent loss of fluorescence due to the destruction of the fluorophore by incident irradiation. In addition, the reagent itself is also fluorescent, necessitating separation of excess FITC from the analytes of interest. Commercially, FITC is available in either the pure form or as a mixture of isomers. Isomer I is most commonly utilized for derivatization reactions. The excitation and emission maxima are 492 and 520 nm, respectively. This makes it compatible with the 488 nm line of the Ar-ion laser.

OPA and NDA are widely used for online precolumn (before separation) or postcolumn (after separation) derivatization of primary amines. They are particularly useful for microchip applications due to their fast reaction kinetics. Another advantage of these reagents is that they are fluorogenic and exhibit little to no emission at the excitation wavelength of the fluorescent product. OPA and NDA react with primary amines in the presence of excess thiol or cyanide (CN) to form fluorescent isoindole derivatives. The excitation wavelengths coincide well with the spectral lines of commercially available He–Cd lasers. The cyanobenzisoindole products created by the reaction of primary amines with NDA or CN are more stable than the thiolbenzisoindole derivatives produced by the reaction of OPA and mercaptoethanol. The products of both reagents exhibit good aqueous solubility and detection characteristics, with NDA possessing the better fluorescence quantum yield of the two.[53]

Other derivatization agents employed for the detection of amine-containing compounds include 7-halo-4-nitrobenzo-2-oxa-1,3-diazole (NBD), tetramethylrhodamine isothiocyanate (TRITC), and fluorescamine. NBD, which is used primarily for the detection of secondary amines, displays relatively faster reaction kinetics. TRITC (pH-insensitive derivative of rhodamine) and fluorescamine are used for the detection of both primary and secondary amines and exhibit fast reaction kinetics. Fluorescence detection of proteins can also be accomplished through the use of noncovalent dyes. These dyes bind to the hydrophobic regions of proteins, yielding a fluorescent product. Postcolumn derivatization of proteins using noncovalent dyes has been reported.[54,55]

11.2.3.2 Pre- and Postcolumn Derivatization Strategies

Derivatization reactions can be carried out before or after microchip separation. The advantages and disadvantages of these two approaches are outlined in Table 11.2. Precolumn derivatization can be carried out in either offline or online mode. In offline derivatization, the reaction is performed in a separate reaction vial prior to introduction into the microfluidic device. The present discussion focuses on online derivatization as this is most likely to lead to a totally integrated system.

In a typical microchip CE device, an injector precedes the separation channel. The most commonly used injector configurations involve either a cross or "double-T" structure. In the cross configuration, analytes can be injected into the separation channels by performing a simple injection[4,27,56] or

TABLE 11.2
Precolumn vs. Postcolumn Derivatization: Advantages and Disadvantages

	Precolumn (offline)	Precolumn (online)	Postcolumn
Advantages	No restriction on reaction conditions (buffer, reagents, etc.)	Possibility of offline contamination minimized	Side product formation minimized
	No need for additional reagent channels or reservoirs	Separation of short-lived fluorescent products	Analytes separated in underivatized form
	Lengthy, multistep derivatization can be performed		Possibility of offline contamination minimized
	Ease of derivatization		Separation of short-lived fluorescent products
Disadvantages	Contamination from derivatization possible	Change of electrophoretic properties of analyte	Restriction of reaction conditions
	Change of electrophoretic properties of analyte	Relatively fast labeling agent needed with low fluorescence in unbound state	Very fast labeling agent needed with low fluorescence in unbound state
	Side product formation	Some side product formation	

by a gated injection scheme.[55,57] For the double-T configuration, well-defined plugs can be introduced by using a pinched injection scheme (for more on injections see also Chapters 2 and 8).[34,58] For precolumn on-chip derivatization reactions, channels that couple the sample and derivatization reagent streams are fabricated before the injection cross. An example can be seen in Figure 11.5A where a gated scheme is used to inject the sample after derivatization. Because the reaction occurs before the injection, there is more time for the derivatization to occur, as compared to a postcolumn scheme. The time available for the derivatization can be optimized by changing the dimensions of the reaction column.[41]

For postcolumn reactions, an additional reagent reservoir is connected to the separation channel (see Figure 11.5B). The time required for derivatization is dependent on the length of the reaction channel and the field strength that is applied. The channel must be long enough to ensure complete reaction but short enough to minimize band broadening.[57] Therefore, extremely fast reaction kinetics are desirable for postcolumn configurations. The intersection geometries between the separation channel and the reagent introduction channel can affect separation efficiencies. A postcolumn mixing geometry has been used for the separation and derivatization of OPA-labeled amino acids.[42]

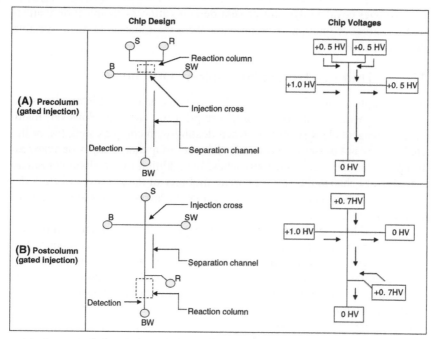

FIGURE 11.5 Schematic of pre- and postcolumn chip geometries and voltages needed for a gated injection scheme. Figure (A) illustrates sample and reagent mixing prior to injection and separation whereas (B) shows the incorporation of a reagent reservoir for derivatization after separation. The voltages were experimentally determined by Ramsey and coworkers. (Figure adapted and reprinted from Jacobson, S. C.; Hergenroder, R.; Moore, A. W.; Ramsey, J. M. *Anal. Chem.* 1994, *66*, 4127–4132 and Jacobson, S. C.; Koutny, L. B.; Hergenroder, R.; Moore, A. W.; Ramsey, J. M. *Anal. Chem.* 1994, *66*, 3472–3476.)

In addition to pre- and postcolumn derivatization, analyte tagging can be carried out during separation. One technique involves including a noncovalent fluorophore in the separation buffer. For protein-sizing applications, the fluorophore dynamically binds to a protein–sodium dodecyl sulfate (SDS) complex, allowing detection. Dynamic labeling has also been used for the detection of double-stranded (ds) DNA using CE-LIF.[59] Fluorescent dyes such as ethidium bromide, thiazole orange homodimer (TOTO), and thiazole orange thiazole indolenine heterodimer (butyl TOTIN) can noncovalently intercalate with or bind to ds DNA.

Protein-labeling has been demonstrated using SYPRO Orange during an SDS-PAGE separation in a polydimethylacrylamide-based sieving matrix. Dilution was used to reduce background produced by dye–SDS micelle interactions.[60] However, the use of the noncovalent dye NanoOrange removes the need for the dilution step.[59] Noncovalent binding dyes are generally favored over covalent dyes for protein derivatization as chemical derivatization can produce multiple photoproducts from the same protein and variable protein labeling.[60,61]

11.2.4 DIRECT VS. INDIRECT FLUORESCENCE DETECTION

In most applications of fluorescence detection, the analytes of interest must be derivatized with an extrinsic fluorophore prior to detection. However, many of these reagents lack specificity when dealing with complex samples, or the analyte may not possess a reactive functional group that is easily derivatized. In precolumn derivatization approaches, the addition of the fluorophore can significantly change the electrophoretic mobility of the analyte, affecting the separation. When such instances arise, it may be advantageous to use indirect detection.

Indirect detection is accomplished through the incorporation of an easily detectable molecule into the separation medium. Ideally the fluorophore will have an electrophoretic mobility similar to that of the analyte, resulting in symmetrical peaks[62] and a stable fluorescence background. When an analyte passes by the detector, the absence of a fluorescence signal causes a negative peak that is proportional to the analyte concentration. The detection limits for indirect detection are usually higher than those for direct detection due to the less sensitive nature of monitoring fluorescence displacement. When fluctuations in fluorescence occur, the detection sensitivity decreases further as it becomes more difficult to differentiate small displacements from an unstable background.[63]

One of the factors that can affect the stability of the background fluorescence is the fidelity of the excitation source. Instability in the voltages used for injection and separation can also produce background noise. The use of high counter or "pushback" voltages can cause backflow into the side channels, resulting in unstable background fluorescence.[58]

Optimization of the amount of fluorophore in the run buffer is necessary to obtain the appropriate limits of detection (LODs) and dynamic range required for the analysis. Employing high concentrations of fluorophore to generate greater signal levels does not necessarily deliver the best limits of detection. In fact, for cases where the molecule of interest is of low concentration, the best detection limits are obtained when the concentration of the signal generator is kept as low as possible. Microchip CE with indirect fluorescence detection has been used for the analysis of sugars, amino acids, and explosives.[58,64,65]

11.2.5 APPLICATIONS

The vast majority of microchip CE applications using fluorescence detection have involved the analysis of biologically important compounds that can be easily derivatized. These include DNA, proteins, peptides, and amino acids (see also Chapter 12).

11.2.5.1 DNA

The use of microchip CE with fluorescence detection for DNA analysis has been important for the Human Genome Project in both restriction fragment mapping and sequencing applications. A multicapillary array for the sequencing and genotyping of DNA was successfully transferred from conventional CE[66] to a microchip format, allowing faster analysis.[67] A 96-channel radial microplate was filled with a hydroxyethyl cellulose separation matrix and polymerase chain reaction (PCR)-amplified DNA samples derivatized with TOTO and TOTIN were separated (Figure 11.6A). Two-color genotyping was achieved using a radial confocal fluorescence scanner. Radiation from an Ar-ion laser at 488 nm was directed into a revolving microscope objective scanned radially across the channels on the microplate to excite the prelabeled DNA. Emission collected by the detection configuration shown in Figure 11.6B achieved a detection sensitivity of 10 pM (signal-to-noise ratio S/N = 2) at a laser power of 70 mW.

Rapid digest DNA restriction fragment analysis has also been achieved in a microfluidic device.[28] Two separate reservoirs for DNA and enzyme were connected by etched channels. When a potential was applied, separate streams of the DNA and enzyme entered a reaction chamber where digestion of the plasmid occurred. Digested DNA was then separated following a gated injection scheme. Derivatization was carried out by placing TOTO intercalating dye into the buffer waste reservoir. The dye migrated countercurrent to the DNA, allowing dye–DNA complexes to form prior to detection. The integrated device described here is very similar to the precolumn derivatization scheme seen in Figure 11.5A, which was developed by the same group and has been used to conduct other enzymatic assays.[68]

Efforts toward the development of a μTAS for integrated DNA purification, amplification, and separation with LIF detection have been described. DNA was purified from a biological sample using a microchip solid-phase extraction device that utilized silica beads immobilized in a sol-gel matrix.[69] The sample was then transferred to a second microchip device on which IR-mediated PCR amplification was performed.[70] Future work will involve the integration of all processes onto one device.

11.2.5.2 Protein, Peptides, and Amino Acids

Protein and peptide separations have been accomplished using microchip devices with fluorescence detection. Liu and Ramsey reported the detection of α-lactalbumin, β-lactoglobulin, and β-lactoglobulin B using postcolumn noncovalent protein labeling with NanoOrange.[55] Detection limits were approximately 30 amol of injected sample.

Another example of a peptide and amino acid separation is shown in Figure 11.7. Using a home-built system similar to that described in Figure 11.1, the separation and detection of the fluorescently labeled peptides

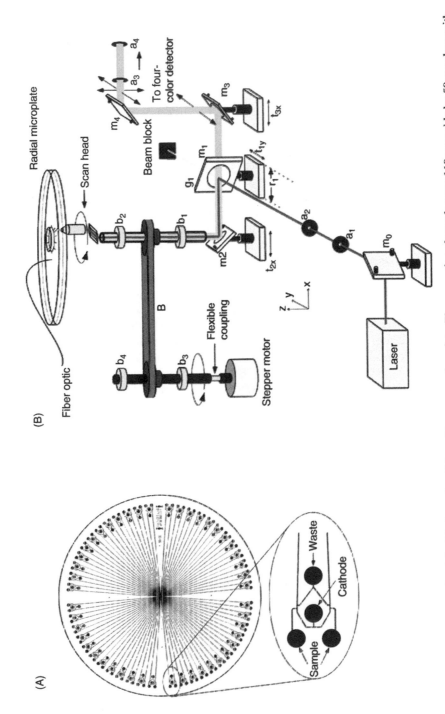

FIGURE 11.6 (A) Schematic of the 96-channel radial capillary array microplate. The separation channels were 110 μm wide by 50 μm deep with an effective length of 33 mm. (B) Confocal fluorescence scanner for DNA analysis is illustrated. (Reprinted from Shi, Y. N.; Simpson, P. C.; Scherer, J. R.; Wexler, D.; Skibola, C.; Smith, M. T.; Mathies, R. A. *Anal. Chem.* 1999, *71*, 5354–5361. With permission.)

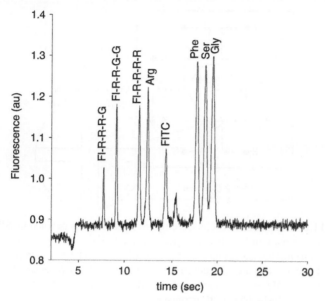

FIGURE 11.7 Separation and detection of FITC-labeled peptides and amino acids. The sample was injected using a twin-T design with an injection voltage of 2000 V for 30 sec. The separation was obtained by applying 3200 V with an antileak voltage of 1600 V. Lacher, N., et al. PhD thesis, University of Kansas, 2004.

and amino acids were achieved.[15] Other examples of protein and peptide separations have been described elsewhere.[54,71–73]

Amino acid separations on the microchip capillary electrophoresis format have been successfully carried out in many studies.[4,9,25,30,74] An example of a cyclic planar microstructure that incorporated a volume-defined injection scheme is shown in Figure 11.8.[31] This device was used to study the behavior of samples on cyclic devices and to conduct micellar electrokinetic chromatography separations of FITC-amino acids. In addition, the feasibility of direct injection of human urine and serum samples was demonstrated.

Reversed-phase electrochromatography has been used for the analysis of offline NDA-labeled peptides and amino acids using microchip CE-LIF.[75] Photopatterning was used to create the acrylic-based porous polymer monoliths in the channels of the device. The separations were faster and more efficient than conventional capillary-based separations.

11.2.6 OTHER OPTICAL DETECTION METHODS

11.2.6.1 Absorbance Detection

Absorbance detection in the ultraviolet (UV) and visible (vis) range is a well-established mode that has been applied to the determination of a wide variety

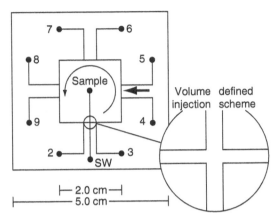

FIGURE 11.8 Schematics of cyclic CE chip. Sample waste is denoted by SW and the arrow marks the point of LIF detection. Reservoirs 3, 5, 6, and 8 and SW were where sample was electrokinetically pumped; reservoirs 2, 4, 7, and 9 and "sample" are those from which electrolytes are introduced into the rectangular separation channel. (Reprinted from von Heeren, F.; Verpoorte, E. M. J.; Manz, A.; Thormann, W. *Anal. Chem.* 1996, *68*, 2044–2053. With permission.)

of compounds with suitable chromophores. While UV–vis absorbance is one of the most popular detection modes for macroscale CE separations, it has not been used as extensively for microchip CE. Limiting factors include the reduced optical pathlengths and sample volumes characteristic of microscale devices, which can impact detection sensitivity. Strategies employed to improve absorbance detection limits in microchip systems have focused on increasing the effective pathlength. Early work investigating U-shaped flow cells achieved a tenfold increase in absorbance over conventional transverse channel detection.[76]

Multireflection cells have been used to increase optical pathlengths in microfabricated devices. While internal reflection has been achieved in silicon-based flow cells,[77,78] it is a less than ideal fabrication substrate for microchip CE due to the high voltages applied during separation. An alternative approach to multireflection cells (using glass substrates) is the alignment of mirrors alongside the flow channel to increase the optical pathlength (Figure 11.9).[79] Five- to tenfold increases in detection sensitivity were achieved using microfabricated sputter-deposited aluminum mirrors. The main advantage of the multireflection approach is that there is no need for an increase in detection cell volume, which can result in sample diffusion and band broadening. Some loss of signal intensity occurs with each internal reflection, limiting the number of reflections that can be performed; channel depth and beam divergence can also contribute to interference effects.

Waveguides have been integrated into CE microchips, permitting detection across the width of the separation channel. As the width often exceeds the

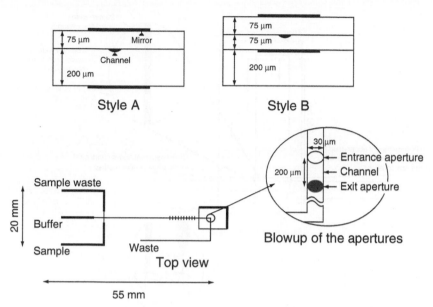

FIGURE 11.9 UV detector for microchip CE. (Bottom) Layout of the flow channels of the device with an inset showing arrangement of the apertures in the mirrors at the detector region. The detector was located 4.15 cm from the double-T style injector upstream of the corner and the effective buffer to waste reservoir distance was 6.78 cm. (Top) Devices fabricated in style A, in which the mirrors are on either side of the glass plates, and style B, in which a 200 or 5500 μm thick handle wafer was used to carry the thinner plates of glass during fabrication and one of the mirrors is sandwiched between the two plates. (Reprinted from Salimi-Moosavi, H.; Jiang, Y.; Lester, L.; McKinnon, G.; Harrison, D. J. *Electrophoresis* 2000, *21*, 1291–1299. With permission.)

channel depth, the effective pathlength is increased.[80] Integration of the waveguides allows alignment with separation channels during the fabrication step.[81,82] When used in combination with U-shaped detection cell designs, further gains in detection sensitivity can be achieved (Figure 11.10).[83]

The elimination of stray light affords improvements in sensitivity due to background reduction. To this end, optical slits designed to eliminate light scattered by the chip substrate have been fabricated into microchip devices. As an alternative approach to conventional "single-point detection," a linear imaging UV detector was investigated by Nakanishi et al.[84] A sputter-deposited Si layer that was used as an etching mask during chip fabrication which doubled as an optical slit and exposed only the area of the separation channel to the detector. The entire channel was then monitored, as opposed to a single spot, using a specially developed linear photodiode array (Figure 11.11). The resulting real-time output was described as analogous to that generated by slab-gel electrophoresis. Positioning of the slit was found to be optimal at the

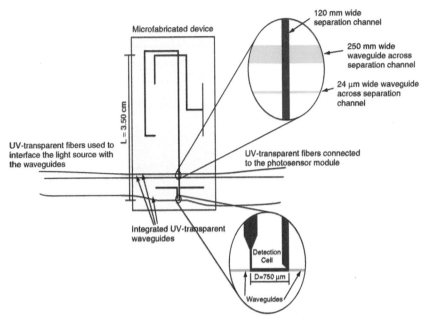

FIGURE 11.10 UV detector for microchip CE. Layout of microchip device with waveguides. (Reprinted from Petersen, N. J.; Mogensen, K. B.; Kutter, J. P. *Electrophoresis* 2002, *23*, 3528–3536. With permission.)

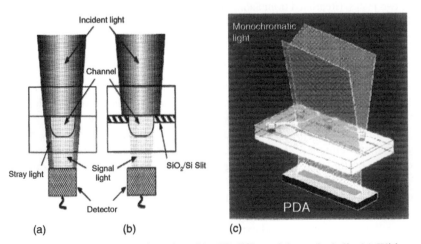

FIGURE 11.11 UV detector for microchip CE. Effect of the optical slit. (a) Without slit; (b) with slit; (c) location of microchip and linear photodiode array. (Reprinted from Nakanishi, H.; Nishimoto, T.; Arai, A.; Abe, H.; Kanai, M.; Fujiyama, Y.; Yoshida, T. *Electrophoresis* 2001, *22*, 230–234. With permission.)

bonding interface between chip layers where it was closest to the channel. Improvements in S/N ratios could be achieved through the use of repetitive scanning techniques and signal averaging. UV absorption imaging detection has also been investigated in conjunction with isoelectric focusing (IEF) using coated channels.[85]

Thermal lens spectroscopy may also offer advances in detection sensitivity for absorbance-based measurements. This technique, which monitors minute changes in temperature profiles following molecular absorbance, has been investigated previously for simple microscale flow-through devices.[86,87] With sub-zeptomole detection limits reported,[88] future applications to microchip CE may produce detection limits that rival those of LIF detection systems while allowing detection of nonfluorescent species. This less selective mode of detection will also have broader application but may suffer from background interferences.

In the move toward integration of detection components into the microchip platform, LEDs are proving to be compact and inexpensive light sources for absorbance detection in the visible range of the spectrum. With smaller components, the integration of detection units is more easily envisioned. High output LEDs have been used for spectrophotometric determinations on microchip CE devices, with connection to a detection unit via fiber optic cables. A detection limit of 0.1 μg/ml for orthophosphate was reported for a microflow injection analysis (μFIA) system with electrokinetic control of fluid.[89] The authors also used similar designs for the detection of nitrite[90] and nitrate,[91] achieving LODs of 0.2 and 0.51 μM, respectively. A microchip CE sensor was developed for the detection of uranium(IV) using a red LED as the light source.[92] Green LEDs have been employed for the detection of metals following complexation reactions with 4-(2-pyridylazo)resorcinol (PAR)[93] with reports of sub-ppm detection limits[92,93] and with 2-(5-bromo-2-pyridylazo)-5-(N-propyl-N-sulfopropylamino)phenol (5-Br-PAPS).[94]

11.2.6.2 Chemiluminescence

Chemiluminescence (CL) has emerged as a promising detection mode for use in conjunction with microchip CE applications.[95] The collection of photonic emissions generated as a result of chemical reaction can be used for quantitative determinations of analyte concentration. There are several advantages associated with CL detection for microchip CE, including wide dynamic range and selectivity of analysis. As no external light source is needed, background signal is minimal, permitting increased detection sensitivity. The cost of supporting instrumentation is reduced, and the need for fewer optical components facilitates a more realistic approach to the integration of detection onto a microchip platform. To exploit this mode of detection, target analytes must exhibit native luminescence following chemical reaction or play a role in the catalysis or suppression of the luminescence of other compounds.

Alternatively, analytes may be labeled with a luminescent tag that permits direct detection.

Solution-phase CL systems have been employed previously in conjunction with macroscale chromatographic and electrophoretic applications. The most popular systems involve redox reactions of luminol, peroxyoxalates, or tris(2,2′-bipyridine)ruthenium(II) to generate light emission.[96] Due to the speed of the reaction kinetics involved, postseparation mixing of analytes and reactants is usually performed to maximize collection of emission occurring within seconds of the reaction. While physically coupling the separation column to the mixing reactor is a challenge for macroscale systems, it is simplified by the move to planar microchip formats using multiple reservoirs and electrokinetic control of fluids.

CL is an established detection mode for immunoassays due to the sensitivity and short reaction times involved. Generally, direct chemiluminescent labels or enzyme-catalyzed reactions produce the CL signal. Detection of the products of an immunological reaction using the horseradish peroxidase (HRP)-catalyzed reaction of luminol and peroxide was the first demonstration of CL detection for an immunoassay performed on a microchip CE device.[97]

A peroxyoxalate CL reaction was employed for the microchip CE-CL detection of dansyl amino acids. Detection limits of $1 \times 10^{-5} M$ were reported, which were higher than those achieved using a macroscale CE system.[98,99] This was attributed to the smaller injection volumes of the microscale device. Metal ion analysis has also been performed on-chip using luminol[100,101] and peroxalate-based detection systems.[102] Detection of codeine was achieved by reduction of tris(2,2-bipyridyl)ruthenium(II), and the LOD was calculated to be $8.3 \times 10^{-7} \, mol/l$.[103]

While low background is characteristic of CL detection, optical slits have been employed to prevent stray light from reaching the detector. A silicone membrane positioned between the two layers of a glass chip decreased baseline noise by a factor of 10, allowing a detection limit of $1 \times 10^{-5} M$ for Co(II).[104] This LOD is comparable to that obtained with a macroscale CE instrument. While the majority of devices have been fabricated in quartz or glass, CE-CL has also been demonstrated using polymer-based microchip devices.[102]

Oxidation or reduction of appropriate CL reactants can also occur at an electrode; this has been termed electrogenerated chemiluminescence (ECL). The reaction can be controlled through varying the applied potential; this has the added benefit of rapid electrochemical recycling of reagents. As photonic emission is localized to the area around the electrode, efficient signal collection can also be achieved. Electrodes are also amenable to microfabrication and, therefore, complement the miniaturized dimensions of separation devices.

ECL has been demonstrated in a miniaturized flow cell format[105] following the reaction of tripropylamine and tris(2,2′-bipyridine)ruthenium(II) (TBR).

The resulting electron transfer causes excitation and emission from TBR molecules at 610 nm. A similar system has been applied to microchip CE for the direct detection of dichlorotris(2,2′-bipyridyl)ruthenium(II) hydrate and dichlorotris(1, 10-phenanthroline)ruthenium(II) hydrate.[106] A floating electrode was used to decouple the electrochemical response from the separation voltage. Detection limits in the low micromolar range were reported.

To broaden the application of ECL detection, a cascade approach was investigated that chemically decoupled the reporting and sensing elements of the system using separate channels. This allowed the detection of analytes not directly involved in the ECL reaction.[107] A subsequent dual channel design further enhanced the applicability of the technique, permitting the use of ECL-incompatible solvents by physical isolation of solutions while electrical coupling was maintained.[107,108] Although currently only demonstrated for a flow-through system, this approach could be a sensitive detection mode for microchip CE due to the reduced background and the minimal supporting instrumentation required.

11.2.6.3 Shah Convolution Fourier Transform

A multiple point detection system termed Shah convolution Fourier transform (SCOFT) has been demonstrated for use with microchip CE applications. A number of equally spaced detection slits that can be fabricated into the microchip design are located along the length of the separation channel. Analytes are continuously detected as they migrate along the separation channel. A single frequency analyte peak is then created from the sum of the peak areas collected during the separation process.

The transformation of the time-domain function of the electropherogram into a frequency signal can be used to interrogate sample mixtures. An advantage of this detection mode is the ability to isolate the analytical response from noise with an alternative frequency. To generate a single frequency peak, the analyte must maintain a constant migration rate throughout the separation process. SCOFT has been investigated for the detection of fluorescein and fluorescein isothiocyanate.[109] Comparison to a single point detection scheme illustrated effective isolation of noise, but a decrease in peak resolution was observed.

This detection mode has also been used to determine particle flow velocities for fluorescent microspheres in microchip CE.[110–113] An eightfold increase in S/N ratio over that of single point detection was reported.[113] Further improvements in S/N ratios were observed when a multiple sampling technique was employed.[112] Injection of three discrete sample plugs led to an increase in S/N because the sample amount loaded in the channel was increased without impacting resolution.

A variation on the original detection scheme termed Shah convolution differentiation Fourier transform (SCODFT) was applied to rear analysis on a microchip CE device.[111] A continuous sample injection filled the channel with sample with subsequent analysis performed on the "rear" of the analyte zone. The resulting signal output comprised discrete plateaus of analyte response. This approach eliminated the need for reproducible small plug injections and avoided electrokinetic sample bias. While Fourier transform (FT) has several advantages, it cannot provide information regarding the time domain of the separation. For this reason, wavelet (short piece of a waveform) transformation of signal has been investigated as it gives information on both frequency and time domains.[114] The substitution of a charged couple device for the fixed optical mask in combination with sine convolution of collected data was found to result in a 1000-fold improvement in S/N ratio for the detection of fluorescein.[115]

11.2.6.4 Refractive Index

Refractive index (RI) is a universal detection mode that is useful for the detection of compounds with poor absorbance and fluorescence characteristics. The ratio of the speed of light through a vacuum relative to that through the solution is calculated. A limitation of the technique is that it is extremely sensitive to changes in temperature, which restricts its application to temperature-controlled systems. In addition, sensitivity is pathlength-dependent; therefore, the use of RI detection for the analysis of small volumes is somewhat limited. Approaches to improve detection sensitivity have employed interferometric methods, which can report even slight changes in refractive index through the interrogation of generated interference patterns. A holographic forward-scatter RI detector was used for the microchip CE separation of sucrose, N-acetylglucosamine, and raffinose. Using a backscatter format and a multipass optical configuration allowed on-chip measurement of fluorescein with a reported mass detection limit of 3.0×10^{-13} mol.[116]

11.2.6.5 Raman Spectra

Detection in Raman spectroscopy is based on photon scattering by molecules. Resulting molecular vibrations cause a shift in the wavelength of incident monochromatic radiation and subsequent spectral analysis of scattered light can yield qualitative information on analyte structure. Raman spectra are useful for the identification of functional groups and fingerprint regions that facilitate compound identification. Raman spectroscopy is ideal for the analysis of aqueous samples, as water does not cause interference due to the weak spectrum it exhibits. This detection mode has been interfaced to microchip flow experiments.[117,118] The enhanced characterization ability of Raman along with the minimal interface requirements for coupling to microchip CE prompted an investigation of the technique for the analysis of pesticides.[119] Online sample concentration was performed using isotachophoresis with detection of pesticides at $2.3 \times 10^{-7} M$.

11.2.6.6   Near-IR Spectra

Detection in the near-IR region of the spectrum offers the advantage of reduced interference due to the limited number of compounds that exhibit native fluorescence in that region. Reduced scatter minimizes background effects and permits increased detection sensitivity. Near-IR detection on CE microchips has been demonstrated using a PMMA device for the separation of labeled oligonucleotides. Mass detection limits of 1.2 zmol (zmol = zeptomole) were reported.[120] DNA fragment analysis performed on a PMMA device using a near-IR dye yielded detection limits of 0.1 μg/ml.[121,122]

For further information on optical detection techniques employed in conjunction with microchip CE, the reader is referred to some relevant reviews of the area.[5,123–127]

11.2.7   Future Directions

Fluorescence has been the most popular form of optical detection for microchip CE due primarily to its high sensitivity. Currently, most supporting instrumentation is several times larger than the chip itself, which limits the portability of the system. For the fabrication of a truly μTAS system, miniaturization of the detection system and the ability to perform on-chip derivatization are needed. Microengineering, the development of higher throughput parallel analysis and multidetection setups may be a trend for the next decade of research. Future approaches to optical detection for microchip CE will also involve the investigation of other modes, permitting a broader range of application while maintaining sensitivity.

11.3   MASS SPECTROMETRIC DETECTION

11.3.1   Introduction

Mass spectrometry (MS) has become an increasingly popular detection mode for pharmaceutical, biomedical, and environmental analysis due to its high sensitivity and specificity. MS is a technique used to measure the mass-to-charge ratio (m/z) of individual compounds that have been converted to ions in the gas phase. A mass spectrum plots the mass-to-charge ratio (x-axis) versus relative or absolute intensity (y-axis). There are five stages of MS analysis: sample introduction (or source); ionization; mass analysis; ion detection; and data collection. The three main considerations for MS are the nature of sample introduction, ionization, and type of mass analyzer, each of which will be described in detail in the following sections.

MS detectors for liquid and gas chromatography are now commonplace in many analytical laboratories. Capillary electrophoresis has been interfaced to mass spectrometry (CE–MS) and has been used for a number of applications, for example, analysis of peptides and chiral compounds.[128–130] Despite the large disparity in size, microchip CE has also been interfaced to MS and has many distinct advantages as a detector in those cases where portability is not required.

The primary advantage of coupling a microchip CE device to MS is increased throughput. Due to the high selectivity of this detection mode, generally minimal separation of the analyte is needed. Therefore, if a sample preparation step is not necessary, microchip separation times are fast enough that throughput is limited only by the speed of sample introduction into the mass spectrometer. However, in those cases when sample preparation is necessary, the sample throughput will be lower. A major advantage of microchip CE coupled to MS is that sample preparation steps can be directly integrated onto the chip. This makes it possible to introduce sample directly into the mass spectrometer following sample processing and CE separation. In addition, the development of inexpensive and disposable microchip-based separation devices eliminates the possibility of sample cross-contamination prior to MS analysis. As a result of these advantages, there have been many reports of the development and use of microchip CE–MS devices in the literature.[131]

11.3.2 IONIZATION METHODS

For detection by MS, analytes must be ions in the gas phase; in addition, the solvent used for separation must be removed prior to introduction into the mass analyzer. The most common ionization techniques used to accomplish these goals include electron impact (EI), chemical ionization (CI), matrix-assisted laser desorption ionization (MALDI), fast atom bombardment (FAB), secondary ion mass spectrometry (SIMS), electrospray ionization (ESI), and atmospheric pressure chemical ionization (APCI).[132] Due to their own unique characteristics, ESI and MALDI are the only approaches that have been employed for microchip CE thus far.

11.3.2.1 Electrospray Ionization

ESI was first demonstrated in the early 1980s by John Fenn, and now is commonly used for the analysis of peptides, polypeptides, proteins, and oligonucleotides. As CE separates analytes based on differences in mass-to-charge ratios, it is a useful sample introduction technique for ESI-MS, which is most sensitive when analytes exist as ions in solution.

Following separation, the analyte solution is delivered through a channel interface to the mass analyzer. The outlet of the channel is under the influence of an electric field of either positive or negative polarity (about 6 kV). The field present at the tip of the channel attracts charged molecules to the edge of the sample solution, producing a spray of electrically charged droplets. Desolvation occurs as the sample passes through a heated gas (nebulizer gas), forming fine droplets with increased surface charge densities. Desolvated ions are discharged from the surface of the droplet when electrostatic forces become greater than the existing surface tension. The surface charge of these droplets is dependent on the polarity of the voltage placed across the interface channel. Therefore, after ionization, either positively or negatively charged ($[M + H]^+$ or $[M + H]^-$) analyte ions pass into the vacuum chamber of the MS where they are analyzed.

ESI is well suited for solution-phase sample introduction methods such as CE. As a soft ionization technique, ESI results in little to no fragmentation of the analytes. It is unlike other ionization techniques such as APCI, in that the analytes themselves do not have to be volatile. An additional benefit of ESI is the possible formation of multiply charged ions (e.g., $[M + H]^{2+}$), enabling the analysis of large peptides and proteins with a typical mass analyzer. The two main disadvantages of ESI are the inability to analyze nonpolar analytes and limitations on buffer selection. Due to possible ion-pairing effects with nonvolatile buffers, only volatile buffers such as ammonium formate and ammonium acetate can be used.

11.3.2.2 Matrix-Assisted Laser Desorption Ionization

MALDI is a soft ionization technique[133,134] that can be used for molecules having a molecular mass of approximately 100 to 1,000,000 Da (proteins and polymers). To perform MALDI, the sample is mixed with an organic matrix consisting of low molecular weight aromatic acids. Typical examples of matrices include nicotinic acid, benzoic acid derivatives, and cinnamic acid derivatives.[135] To perform the analyses, the matrix–analyte mixture is evaporated on a MALDI plate (metallic), and the solid mixture is then exposed to a pulsed laser beam. The organic matrix absorbs energy from the laser beam and enters the gas phase, bringing the analyte with it. The actual ionization is brought about in the solid phase by the transfer of a proton between the excited matrix and the analyte molecules. The ions are subsequently directed to a time-of-flight mass analyzer where they are identified by their mass-to-charge ratio.

MALDI exhibits very low sample consumption and, unlike other ionization approaches such as ESI, this method is tolerant of high salt concentrations, facilitating the analysis of complex samples such as plasma. As predominantly singly charged ions are produced ($[M+H]^{+/-}$), the types of mass analyzers that can be used are limited to those that detect a wide range of m/z ratios.

11.3.3 Mass Analyzers

Following ionization, analyte ions are introduced into a mass analyzer that must be capable of resolving small mass differences. Resolution, as it relates to mass analysis, is defined by the following equation:

$$R = m/\Delta m \qquad (11.6)$$

where resolution (R) is equal to the mass of the lower weight ion (m) divided by the mass difference (Δm) between the two ions that are being resolved. Generally, ion peaks are considered to be resolved if the height of the valley between the peaks is less than 10% of peak intensity. The required resolution will depend on the desired application. Several different analyzers have been described for MS. These include magnetic sector instruments (single-focusing), double-focusing spectrometers, quadrupole mass spectrometers, quadrupole ion trap mass spectrometers, time-of-flight (TOF) mass analyzers, and Fourier transform instruments (based on ion cyclotron resonance).[136] Of these, the quadrupole and TOF analyzers are most often used in conjunction with conventional and microchip CE.

11.3.3.1   Quadrupole Mass Spectrometer

A quadrupole mass spectrometer consists of four parallel cylindrical rods that serve as electrodes (Figure 11.12). Ions from the source are accelerated through the region between the rods with an applied potential. The rods directly opposite to one another are connected electrically to both a direct current (DC) and a radio-frequency (RF) generator. The trajectory of each ion is controlled by the ratio of DC/RF applied to the quadrupole. Ions that collide with the rods of the quadrupole are neutralized by an electron transfer, removed under vacuum, and cannot be detected. For a specific DC/RF ratio,

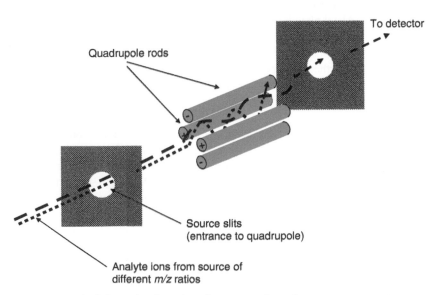

FIGURE 11.12 Schematic of quadrupole mass analyzer for mass spectrometry.

only those ions with the selected m/z do not collide with the rods and are able to travel through the entire length of the mass analyzer to the transducer (detector). In this way, the quadrupole acts as a filter, and only ions of the correct m/z ratio are able to reach the detector. The most commonly employed transducer is an electron multiplier that amplifies the electrical signal, which is then processed and recorded. In addition, by varying the AC and DC voltages, a mass range can be scanned.

11.3.3.2 Time-of-Flight Mass Analyzer

In TOF spectrometry, ions from the source are accelerated into a field-free drift tube using an electric field pulse. Ideally, all ions entering the tube will have the same kinetic energy and, therefore, their velocity in the drift tube will be inversely proportional to their mass. As a result, ions with lower mass arrive at the detector prior to heavier ions, with typical flight times of 30 μsec.[137]

There are three types of TOF — standard, reflectron, and multiturn — which differ in the configuration of the field-free drift tube. Standard TOF has a straight tube and compounds are separated based on their mass-to-charge ratio. In reflectron TOF, ions pass through the drift tube toward the reflectron, which is a series of electric fields. The ions are focused by the reflectron based on their velocity, compensating for differences in kinetic energies. The ions are then reflected back down through the drift tube toward the detector. This approach generally results in greater resolution than standard TOF. In multiturn TOF, the drift tube has a multiturn or serpentine configuration with multiple reflectrons employed to increase resolution.[138–141]

11.3.3.3 Tandem Mass Spectrometry

Tandem mass spectrometry (MS/MS) is accomplished by coupling multiple mass analyzers together. In MS/MS, an ion of interest (precursor ion) is isolated by the first mass analyzer before passing into a field-free collision cell. In the collision cell, ions collide with atoms of an inert gas (usually argon or helium) and undergo fragmentation in a process known as collision-induced dissociation (CID). A second mass analyzer separates the fragments (product ions) generated by the collisions. Three combinations of mass analyzers commonly used for MS/MS analysis are the tandem quadrupole, a quadrupole/time-of-flight (Q-TOF), and quadrupole/ion trap.

The tandem quadrupole is composed of two quadrupoles separated by a collision cell. The ions leave the source and travel through the first quadrupole (MS1), resulting in mass separation. Ions of selected m/z enter the collision cell, which contains gas molecules of either argon or nitrogen (collision gas). Ions collide with the neutral gas molecules, resulting in bond cleavage and fragmentation. The fragmented ions migrate to the second quadrupole (MS2), where they are further separated due to differences in their m/z. The tandem quadrupole is excellent for analyte quantitation due to its sensitivity, selectivity, and large dynamic range. For Q-TOF analysis, the second quadrupole of the tandem quadrupole is replaced by TOF. Q-TOF provides excellent mass accuracy and sensitivity while scanning the entire mass range. This makes it ideal for providing qualitative and structural information. A quadrupole ion-trap mass spectrometer consists of two hemispherical electrodes separated by a ring electrode. Ions can be trapped and accumulated by applying the appropriate potentials to the electrodes. Ions are ejected from the trap selectively to obtain a mass spectrum. In an ion trap mass spectrometer, the entire process occurs in one mass analyzer, and the events are separated in time, not space. This approach is the least expensive and is capable of MS^n.

MS/MS is extremely useful in deducing structural information and gives increased selectivity and sensitivity when used for quantitation. CE, GC, or LC are often used for sample introduction and may not be necessary for separation. In each of the cases described above, final detection of the ions occurs using an electron multiplier or photomultiplier tube.

11.3.4 MICROCHIP SUBSTRATES FOR CE–MS

Microchip CE–MS does not require the substrate to be optically transparent and, therefore, a broader range of materials can be used for fabrication. Microfluidic CE chips have generally been fabricated from two classes of materials, glass and polymers. Glass (Pyrex, soda lime, and quartz) has been the most popular substrate for chips coupled to MS detection because it has electroosmotic flow (EOF) properties similar to that of fused silica. The surface chemistry is stable and can be used with buffers required for CE–MS analyses. In addition, analyte adsorption to glass surfaces is less an issue than it is for polymer-based devices. However, covalent and noncovalent coatings have been employed to reduce adsorption of basic proteins. Coatings that have been employed include [(acryloylamino)propyl]-trimethylammonium chloride (BCQ),[142–144] 3-aminopropylsilane,[145,146] and polybrene.[147]

Several different polymers have been successfully employed for microchip CE–MS; these include parylene,[148] copolyester,[149] polycarbonate,[150–152] poly(dimethylsiloxane) (PDMS),[14,153] poly(methylmethacrylate) (PMMA),[149,154–156] and Zeonor.[157] An important consideration is the stability of the microchip substrate under the conditions necessary for separation and ionization. It is critical that the microchip exhibits EOF so that the sample will reach the end of the separation channel and be transported into the MS source. The pH of the run buffer usually affects the EOF and also has a role in defining the electrophoretic mobility of the analyte. If a peptide has a large electrophoretic mobility in the direction opposite to the interface, the peptide may never reach the mass analyzer for detection.[147]

Buffer pH determines the analyte charge, which will then affect the selection of the ionization mode (positive or negative). High pH applications generally require negative ion mode while lower pH applications normally require positive ion mode to optimize the signal. Therefore, a great advantage of using a microfabricated CE device for separation is that the voltages needed to carry out the CE separation can also be used for sample introduction to the mass spectrometer.

11.3.5 INTERFACING MICROCHIP CE TO MS

11.3.5.1 Microchip CE-Electrospray Ionization

Ionization of analytes for MS analysis can be performed both off-chip and on-chip. In general, off-chip ionization is easier to implement and has thus been more common. In the context of this discussion, off-chip ionization involves the connection of an external fitting while on-chip ionization is achieved directly from the microchip device.

11.3.5.1.1 Off-Chip Electrospray Ionization

The simplest way to accomplish off-chip ionization is to attach a capillary to the end of the separation channel of the microdevice. In one approach, a capillary puller was used to produce either a micro- or nano-ESI tip. This tip was then coated with metal (Ag-conductive glue) to ensure uniform conductivity for ionization. Dead volume was eliminated by attaching the capillary directly to the microchip device using UV glue.[154] Nano-tips can also be prepared by sputtering the capillary tip with a gold coating, which is then covered with a silver-conductive adhesive and finally electroplated with gold. This electroplating procedure greatly improves tip performance.[158] In a report by Lazar and coworkers, nanospray tips were inserted into a hole drilled at the end of the separation channel, producing subattomole sensitivity for peptide and protein samples.[159] With these techniques, external sources (pumps or vacuum) are not necessary because the fluid delivery occurs due to electrostatic forces and capillary action.

A more elegant method used to perform off-chip ionization has been accomplished using a liquid-junction microelectrospray interface.[160] A transfer capillary was used to deliver the fluid flow from the separation channel to the micro-ESI liquid junction. A sheath capillary was employed to aid in the alignment of the transfer capillary with the end of the separation channel. The use of Teflon tubing and finger-tight fittings surrounding the fluid transfer interface improved the durability of the device. The transfer capillary was then attached to a micro-ESI liquid junction, which directed the flow into the MS for analysis. Figure 11.13 shows the use of a nonpulled capillary and sheath flow interface to perform microchip CE–MS.[161]

11.3.5.1.2 On-Chip Electrospray Ionization

On-chip ESI eliminates many of the problems that can arise when trying to interface microchip devices with MS. Ionization occurs at the end of the separation channel without the use of a capillary interface, making fabrication easier.[152,162,163] Advantages of this approach include reduced dead volume

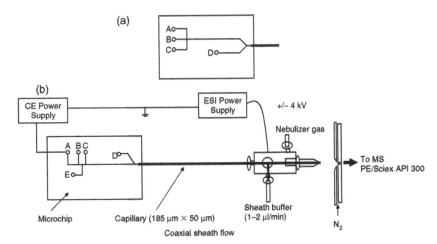

FIGURE 11.13 Schematic representation of the chip-CE configuration for MS detection using (a) a disposable nanoelectrospray emitter and (b) a sheath flow ESI–MS interface. (Reprinted from Li, J. J.; Thibault, P.; Bings, N. H.; Skinner, C. D.; Wang, C.; Colyer, C.; Harrison, D. J. *Anal. Chem.* 1999, *71*, 3036–3045. With permission.)

and band broadening, faster analysis times, and increased sample throughput. However, generating electrospray from the edge of the microchip can result in an increased threshold voltage due to fluid spreading over the surface, reducing sensitivity and efficiency.[162]

Many different approaches for performing electrospray with microchip devices have been proposed but not yet implemented with CE. For example, researchers have shown that it is possible to form electrospray tips mechanically at the edge of the microchip. A trimming method,[164,165] two-layer photoresist,[165,166] and a resin-casting method[165] have been used to fabricate an ESI emitter from PDMS. Electrospray arrays have been fabricated in PMMA in an octagonal pattern and arranged with one channel ending at each point of the octagon.[156] An electrospray nozzle with high aspect ratios has been constructed using deep-reactive ion etching (DRIE).[167] A sample reservoir was produced on the backside of the nozzle pictured in Figure 11.14. The dead volume was less than 25 pl, which should make this nozzle compatible with microchip CE devices. Arrays have been fabricated using polymer resin, with a total of 96 ESI channels or emitters produced for high-throughput analysis. This design, coupled to microtiter plates, could allow consecutive analyses with a high degree of throughput.

11.3.5.2 Microchip CE-MALDI

Although not as popular as ESI, MALDI has also been used as an ionization method for microchip CE. A new technique known as rapid open-access

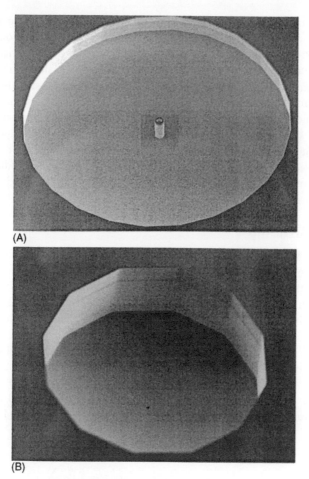

(A)

(B)

FIGURE 11.14 SEM images of a microfabricated silicon electrospray device. (A) shows an electrospray nozzle tilted at a 30° angle. The nozzle has dimensions of 10 μm, inner diameter, and 20 μm, outer diameter. Also shown is the recessed annular region surrounding the nozzle etched to a depth of 50 μm and diameter of 600 μm. (B) shows the reservoir etched 100 μm into the backside of the substrate. The 10 μm channel extending from the tip of the nozzle to the reservoir is visible as a dark circle in the middle of the reservoir. (Reprinted from Schultz, G. A.; Corso, T. N.; Prosser, S. J.; Zhang, S. *Anal. Chem.* 2000, *72*, 4058–4063. With permission.)

channel electrophoresis (ROACHE), which involves the direct coupling of microchip CE to MALDI-MS, has been described.[168] Analytes are separated in open microchannels and the organic matrix is added to the run buffer prior to separation. At the end of the channel a laser pulse initiates ionization. Ions are then directed to the mass spectrometer for analysis.

Many interesting advances have recently been made relating to microtechnology and MALDI-MS that could be employed in conjunction with CE separations. The most notable is the development of a microfabricated toolbox. This toolbox includes MALDI target nanovial plates, microenzyme reactors, and a piezoelectric microdispenser fabricated using lithography and anisotropic etching of silicon (Figure 11.15).[168–170] Silicon, as a substrate, cannot be used for CE unless an insulating layer is deposited on the microchip surface. Nanovial target plates have also been produced using polymer substrates.[171,172] Additionally, microreactors have been integrated onto a target plate for MALDI-MS analysis.[173] Reaction products were moved from the chip to the channel outlet where they were identified in real time using MALDI-TOF MS.

11.3.6 APPLICATIONS

Microchip MS continues to evolve as new approaches to interfacing chip-based CE to MS are investigated. Recently, it is the field of proteomics that has driven the development of microchip CE–MS interface applications.

Microchip CE–MS has also been explored for the analysis of small molecules in human plasma and urine.[143,157,174] Carnitines, which play a role in the metabolism of fatty acids,[175] were analyzed using microchip CE–MS.[174] In another study, a polymer hybrid microchip coupled to ESI-MS was used for the detection of phenobarbital and barbiturates. The same procedure was used for the study of aflatoxins.[149]

Sample preparation techniques such as preconcentration and tryptic digestion prior to separation have been investigated for making MS analysis

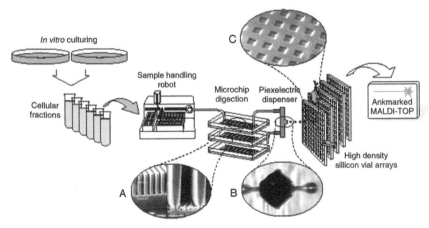

FIGURE 11.15 Platform for protein expression analysis, as realized in the microtechnology format. (Reprinted from Laurell, T.; Nilsson, J.; Marko-Varga, G. *Trends Anal. Chem.* 2001, *20*, 225–231. With permission.)

easier and more efficient. The on-chip digestion of proteins prior to MS analysis has also been reported. One approach involved on-chip proteolytic digestion in a sample reservoir.[176] A mass spectrum of digested human hemoglobin obtained in this manner is shown in Figure 11.16. A second approach involved fabrication of an enzyme reaction bed using trypsin beads.[144] In this case, the sample was loaded in the microfluidic chip along with digestion buffer. Digestion continued until the addition of formic acid terminated the reaction. An important attribute of MS is that sequencing can be performed, allowing an amino acid sequence of a peptide to be determined. Proteins can then be identified using a protein database. It is anticipated that many of the applications that currently use conventional CE–MS or LC–MS will be transferred to a microchip format, leading to higher throughput and reduced analysis costs.

11.3.7 FUTURE DIRECTIONS

MS continues to evolve as an important method of detection for microchip CE. Current research is focused primarily on the development of a useful and rugged interface between microchips and MS. Many different materials have been investigated for the fabrication of microchip CE–MS devices, with each of the materials having unique properties. Evaluation of new chip substrates is necessary to ensure that background noise from the microchip device does not interfere with the detection of an analyte at very low concentrations (amol/μl). The eventual development of a truly miniaturized CE–MS device

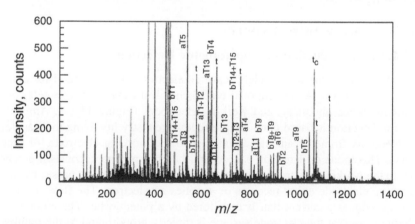

FIGURE 11.16 Microchip CE with nanospray ESI/TOF. Mass spectrum of a tryptic digest of 0.24 μM human hemoglobin extracted from blood. Abbreviations: t, trypsin autolysis products; t_c, trypsin contaminants. (Reprinted from Lazar, I. M.; Ramsey, R. S.; Ramsey, J. M. *Anal. Chem.* 2001, *73*, 1733–1739. With permission.)

that can perform sample isolation, digestion, and separation with sensitive detection for proteomic and bioanalytical applications is a goal of current research.

11.4 ELECTROCHEMICAL DETECTION

11.4.1 Introduction to Electrochemical Detection

Electrochemical (EC) detection is ideally suited for the microchip format for several reasons. Another advantage is that the response in electrochemical detection is independent of pathlength, therefore the electrodes used for detection can be miniaturized without loss of sensitivity. This is possible because, although microelectrodes generate extremely small currents, the background current is further reduced after miniaturization, resulting in an increase in signal-to-noise ratio. Additionally, the microelectrodes can be directly fabricated onto the chip using common photolithographic techniques, leading to a fully integrated system.

There are three general modes of EC detection: amperometry, conductimetry, and potentiometry. In amperometry, a potential is applied to a working electrode. Electroactive compounds are either oxidized or reduced at the working electrode, and the resulting current is monitored. Conductimetry measures differences in the conductivity of the bulk solution compared to that of the analyte zones. Potentiometry is a static (zero current) technique in which information about the sample composition is obtained from measurements of the potential between two electrodes established across an ion-selective membrane. Each of these modes will be discussed fully in the following sections.

11.4.2 Amperometric Detection

11.4.2.1 Introduction to Amperometric Detection

Amperometry is the most widely reported electrochemical method for chip-based separations. A conventional electrochemical three-electrode cell contains reference, auxiliary, and working electrodes (Figure 11.17). Amperometric detection is based upon simple redox chemistry involving the transfer of electrons from the analyte of interest to the working electrode, or vice versa. This mode of detection is accomplished by applying a constant potential to the working electrode. As the electroactive compound(s) passes by, the working electrode either receives or supplies an electron. The flow of electrons produces a current that is monitored by a potentiostat. The amount of charge passed at the electrode surface is directly proportional to the number of moles of analyte, according to Faraday's law. This is demonstrated in the equation below:

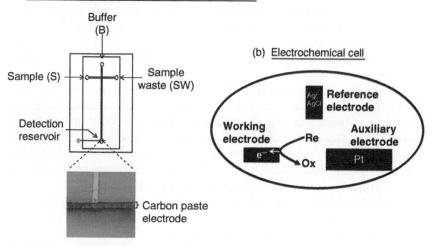

FIGURE 11.17 (See color insert following page 208) (a) A typical microchip CE configuration for amperometric detection with end-channel detection. (b) A three-electrode electrochemical cell which contains a reference, auxiliary, and working electrode.

$$i_t = \frac{dQ}{dt} = nF\frac{dN}{dt} \qquad (11.7)$$

where i_t is the current generated at the electrode at time t, Q is the charge at the electrode surface, t is time, n is the number of moles of electrons transferred per mole of analyte, N is the number of moles of analyte oxidized or reduced, and F is the Faraday constant.[177] The reference electrode (e.g., Ag/AgCl) acts as a zeroing point for the electrochemical cell. The circuit is completed by the auxiliary electrode (e.g., platinum wire); if oxidation is occurring at the working electrode, reduction occurs at the auxiliary electrode.

Amperometry is a popular EC detection mode due to its ease of operation and minimal background current contributions. In addition, the selectivity is tunable by adjusting the potential placed across the working electrode. A variety of working electrode materials are available depending on the nature of the redox reaction being monitored. For those compounds that are not natively electroactive, derivatization is possible. For further information on amperometry used with microchip CE, the reader is referred to review articles by Schwarz and Hauser,[5] Vandaveer et al.,[178] and Lacher et al.[179]

11.4.2.2 Instrumental Design

11.4.2.2.1 Electrode Materials and Fabrication

The electrode material most commonly employed for amperometric detection with CE and other separation methods is carbon; this is due to the broad range of compounds that can be analyzed. Some analytes that have been detected with carbon-based electrodes and microchip CE include catechols,[180–182] phenols,[183] neurotransmitters,[181,182] peptides,[184,185] and nitroaromatic compounds.[186] Several forms of carbon electrodes have been reported, including carbon fibers, carbon paste, and screen-printed carbon electrodes.

The majority of microchip CE studies employing carbon electrodes have used carbon paste or carbon fibers. There are some disadvantages to these carbon-based working electrodes, as opposed to other materials, for microchip applications. Carbon fiber and paste electrodes can be integrated into the microchip device, but are currently not amenable to microfabrication and mass production. Additionally, carbon fibers are very brittle and can break easily. Pyrolyzed photoresist films (PPFs) have recently been shown to offer a more robust approach for electrode fabrication and to exhibit the same electrochemical properties as glassy carbon electrodes.[187] These electrodes are produced by the deposition of photoresist directly onto the microchip substrate, which is then exposed to high temperatures (up to 1000°C). PPE electrodes have been employed for sinusoidal voltammetric detection of dopamine on microchips.[188] More recently these electrodes have also been used for the amperometric detection of ascorbic acid and catecholamines by microchip CEEC (Fisher et al., in press).[225] As an alternative to integration, Wang and coworkers have developed screen-printed carbon electrodes that are more rugged and easily produced, but are external to the microchip CEEC device (Figure 11.18B).[181]

External metal wires and integrated microfabricated metal electrodes are alternatives that have been employed in microfluidic CE devices. Some metals that have been used as working electrodes in amperometric detection for microchip CE are gold, platinum, and copper. Gold electrodes have been used to detect a variety of compounds including catechols,[189] phenols,[190] amino acids,[191] neurotransmitters,[192,193] and nitroaromatics.[194,195] Platinum electrodes have been used to detect catechols,[196] neurotransmitters,[196] uric acid,[197] ascorbic acid,[192] and PCR products,[198] while copper has been employed predominantly for the detection of amino acids[192] and carbohydrates.[199] Gold–mercury amalgam electrodes have been utilized to detect thiols.[200]

Chemically modified working electrodes have been used for CE. These electrodes lower the potential required for detection and are usually carbon based. At the lower potential, fewer compounds are able to undergo an electron transfer reaction, leading to fewer interferences and a decrease in background noise. These two factors result in an increase in both sensitivity and selectivity. While modified electrodes have been employed for conventional CEEC, only one method has been transferred to the microchip CE format thus far. Martin et al. have used carbon electrodes modified with cobalt phthalocyanine (carbon CoPC) with microchip CE for the detection of thiols.[186]

11.4.2.2.2 Decouplers and Electrode Alignment

Isolation of the electrochemical detector from the separation voltage is an important consideration for microchip CE applications. Effective isolation prevents the grounding of the separation voltage through the detector, which can damage the detector electronics. Three approaches have been developed for this purpose: end-channel, in-channel, and off-channel detection (Figure 11.18). These techniques are described in more detail below.

End-Channel Detection: End-channel is the most common configuration of amperometric detection (Figure 11.18A and B).[74,181,192,196,199,200] The working electrode is placed 5 to 20 μm from the end of the separation channel. While this distance effectively isolates the separation voltage from the working electrode, alignment on the micron scale can be difficult and requires a degree of skill.

Because the CE system is also grounded within the detection reservoir, the working electrode still experiences some of the electric field. Therefore, end-channel detection generally exhibits a higher background than off- or in-channel detection, resulting in a decrease in sensitivity. The redox potential can also be influenced by the separation voltage; therefore a hydrodynamic voltamogram (HDV) must be run for each analyte. If the electrode is placed too close to the end of the channel, small fluctuations in the separation voltage can cause noise at the detector. In the worst case, the separation voltage can ground through potentiostat, destroying the electronics. Additionally, dispersion of the analyte occurs in the space between the separation channel and the working electrode, leading to a loss in separation efficiency.

In-Channel Detection: In-channel detection involves placing the working electrode directly within the separation channel (Figure 11.18C). This has been made possible by the development of an electrically isolated potentiostat.[180] A 9 V battery powers this "floating" potentiostat, and the collected data are optically transmitted to an A/D converter. The in-channel configuration helps to eliminate some of the disadvantages associated with an end-channel configuration, such as difficulty in aligning working electrodes and analyte diffusion. It was shown that both EC and LIF detection exhibited similar separation performances in terms of plate height and peak symmetry with this configuration.

Off-Channel Detection: Off-channel amperometric detection has also been employed to overcome the disadvantages of end-channel detection. Reports in the literature have focused on the use of a decoupler to isolate the separation voltage from the detector (working electrode). A decoupler allows the separation voltage to be grounded prior to the working electrode. This minimizes the interference from the separation field, leading to an increase in sensitivity. Pressure-induced flow then carries the analyte bands to the working electrode for detection.

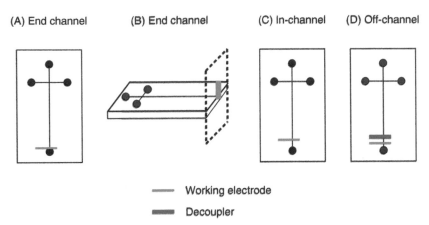

FIGURE 11.18 Electrode configurations for amperometric detection: (A) end-channel with an internal electrode, (B) end-channel with an external screen-printed carbon electrode, (C) in-channel, and (D) off-channel. (Reprinted from Vandaveer IV, W. R.; Pasas, S. A.; Martin, R. S.; Lunte, S. M. *Electrophoresis* 2002, *23*, 3667–3677. With permission.)

Two approaches to decoupling have been reported in the literature. The first method involves the creation of holes in the separation channel prior to the working electrode. This provides the separation voltage a pathway to ground, and has been achieved using both glass[201] and polymer[183] devices. The second method involves the fabrication of an additional integrated electrode that acts as a decoupler. Palladium[202,203] and platinum[204] have been used as decoupling electrodes. Both metals are able to absorb hydrogen generated due to the electrolysis of water in the background electrolyte.

11.4.2.2.3 External vs. Integrated Electrodes for Microchip CE

The fabrication and use of microelectrodes is well established for both scientific and engineering applications. Microelectrodes have been an essential component for the advancement of computer technology. Standard photolithography and microfabrication techniques have been adapted for the production of microelectrodes for electrochemical analysis.[205]

There have been several attempts to transfer amperometric detection from a conventional to a microchip CE format. The simplest and least labor-intensive of these is the use of an external electrode. Literature has documented the use of both screen-printed carbon electrodes[206] and metal wires[197,200] that are aligned off-chip for the detection of a variety of electroactive compounds. The external electrodes are mounted at the end of the separation channel in an end-channel configuration. The use of external electrodes in microchip CEEC detection requires less time in the cleanroom and electrodes can be replaced and realigned easily, which enables a researcher to investigate a variety of compounds utilizing one microfabricated device. However, if a truly portable device is to be developed, integration of the electrode directly onto the microchip itself is necessary.

In addition to external electrodes used for amperometric detection, electrodes have been fully integrated onto the microfluidic devices during the fabrication procedure(s). Metals can be deposited by thermal evaporation or sputtered onto the microfluidic substrate. Following deposition, metal microelectrodes can be produced utilizing a variety of photolithographic patterning techniques. These steps are generally performed within a clean room to reduce the possibility of contamination.[207] The working electrode and the high-voltage separation electrodes can be microfabricated simultaneously[208] with this approach and are amenable to mass production.

Carbon-based electrodes have been integrated onto PDMS microfluidic devices.[184,185] In this case, an electrode channel with defined dimensions is created in a layer of PDMS substrate and carbon paste or a carbon fiber is placed within this channel. Following the fabrication of the carbon electrode, the electrode is aligned with the end of the separation channel, and both layers of the device are sealed together (either reversibly or irreversibly).

11.4.2.2.4 Dual Electrode Amperometric Detection

Dual electrode amperometric detection permits the detection of electroactive compounds that undergo reversible redox reactions as shown below:

$$W_1 \quad Red + e^- \rightarrow Ox \tag{11.8}$$

$$W_2 \quad Ox \rightarrow e^- + Red \tag{11.9}$$

where W_1 is the first working electrode and W_2 is the second working electrode. Dual electrode detection greatly increases the selectivity of an assay due to the fact that the oxidized and reduced forms of a compound can be detected during the same separation. The fabrication of a dual electrode detection configuration on a conventional CE system can be labor-intensive and requires a high degree of skill. In addition, the electrodes are generally not robust. Many of these disadvantages arise from difficulties associated with the fabrication of the electrodes and their attachment and alignment with the capillary.

Fabrication and alignment on a planar microchip device is much easier and requires no more labor than for a single electrode detection design. Figure 11.19 illustrates the use of a dual gold electrode microchip CE system for the detection of tyrosine, HIAA, and catechol.[207] Because catechol is the only compound of the three that undergoes a reversible redox reaction, it is the only analyte detected at the second working electrode, leading to an increase in selectivity of the assay for this compound.

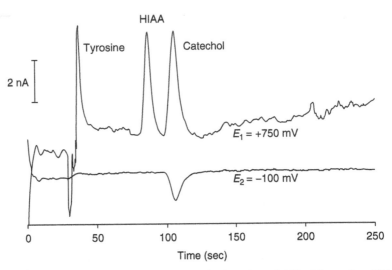

FIGURE 11.19 Dual electrode amperometric detection of $100\,\mu M$ tyrosine, HIAA, and catechol using Au microelectrodes. (Reprinted from Martin, R. S.; Gawron, A. J.; Lunte, S. M.; Henry, C. S. *Anal. Chem.* 2000, *72*, 3196–3202. With permission.)

In another study, dual electrode amperometric detection on a microchip CE device was used to detect the nonelectroactive peptide des-Tyr leu-enkephalin.[185] Prior to injection, the peptide was derivatized using the biuret reaction. This method derivatizes the amide backbone of peptides and proteins with Cu(II) without chemically modifying the compound. Using a carbon fiber electrode, the Cu(II)-peptide complex was oxidized at the first working electrode ($E_1 = +750\,\text{mV}$) and the resulting Cu(III)-peptide complex was reduced at the second working electrode ($E_2 = +100\,\text{mV}$).

11.4.2.2.5 Dual Detection Systems

The use of two different detection modes simultaneously can increase the sensitivity and selectivity of an assay in addition to facilitating the confirmation of peak identity. Figure 11.20A illustrates the chip layout of a dual detection scheme in which a combination of LIF and amperometric detection was used.[196] For LIF detection, an Ar-ion laser beam was focused at the end of the separation channel. Amperometric detection was carried out using an end-channel electrode alignment. The detection of a five-component sample that contained dopamine, catechol, and NBD-labeled amino acids was accomplished and is shown in Figure 11.20B.

A microchip CE system using both contactless conductivity and amperometric detection has also been developed for the detection of nitroaromatic and ionic explosives.[209] A gold-modified thick-film carbon electrode was used for the amperometric detection of the electroactive compounds. Two aluminum foil electrodes used as sensing electrodes (contactless conductivity

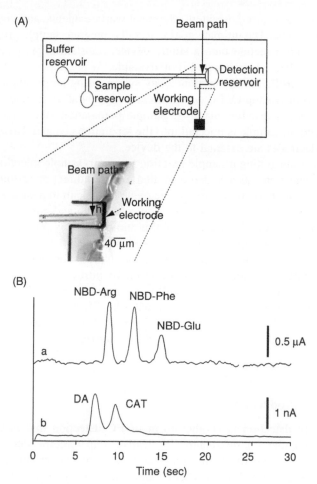

FIGURE 11.20 The chip layout and results for dual EC/LIF detection analysis for CE. (A) Schematic of the chip. The alignment of the working electrode is shown relative to the site of LIF detection. (B) Simultaneous separation of NBD-labeled amino acids and the EC active dopamine and catechol. (Reprinted from Lapos, J. A.; Manica, D. P.; Ewing, A. G. *Anal. Chem.* 2002, *74*, 3348–3353. With permission.)

detection) were attached to the top of a glass cover for the detection of the ionic species (explosives). Conductivity detection will be discussed in detail in the next section.

11.4.2.3 Applications

As stated previously, the main incentive for coupling microfluidic devices with amperometric detection is the possibility of developing fully integrated and portable devices for the detection of a variety of compounds. Possible analytes include neurotransmitters, chemical warfare agents, and compounds of clinical interest. The ultimate goal is to fully integrate sample preparation, separation, and detection onto a single device. Substantial progress toward this goal has been demonstrated. For example, Wang and coworkers have reported a microfluidic CE immunoassay system using amperometric detection.[210] This microchip CEEC design includes a separation channel, working electrodes, and a reaction chamber. Sample preparation takes place in this chamber and the sample is injected into the separation channel. However, the working electrodes are external to the device.

A device integrating a sample reaction reservoir, sample injection, separation, and detection was also demonstrated for the indirect detection of nitric oxide.[211] Due to its short half-life, nitric oxide is difficult to measure directly. Therefore, monitoring the breakdown products, nitrite and nitrate, is a useful method for the indirect detection of this biologically important molecule. Using an integrated carbon fiber electrode, the concentration of nitrite was determined by direct amperometric detection. Nitrate concentration was determined indirectly after on-chip reduction to nitrite using copper-coated cadmium granules. Due to the versatility of amperometry and the ability to construct the electrode simultaneously with the CE microfabrication of the separation device, this mode of electrochemical detection has advanced the field of microfluidic devices toward total miniaturization.

11.4.3　Conductivity Detection

11.4.3.1　Introduction

Conductivity detection is another form of EC detection that is becoming extremely popular for microchip CE. Unlike amperometric detection, any analyte with a conductivity different than that of the background electrolyte can be detected regardless of whether the analyte contains an electroactive functional group, making this a more universal mode.[212] Conventional conductivity detection is based on a two-electrode system in which electrodes are in direct contact with the electrolyte solution and are arranged in either an on-column or end-column configuration. An AC potential is applied across the two electrodes to avoid any Faradaic reactions (or electron-transfer reaction as in amperometry). The signal is due to the conductivity of the bulk solution. As the analyte zones pass by the electrodes and displace the buffer ions in the background electrolyte solution, a change in conductivity occurs, resulting in either a positive or negative deviation from baseline. The change of conductivity is monitored and can be directly related to the concentration of analyte as demonstrated by the equation below:

$$L = \frac{A}{l} \sum (\lambda_i c_i) \qquad (11.10)$$

where L is the conductivity of the solution, A is the electrode area, I is the distance between the two electrodes, c_i is the concentration of the ion, i, and λ_i is its molar conductivity.[213] Because the response is characteristic of the bulk solution itself and not that of a reaction occurring at the surface of the electrodes (such as electrolysis), this mode of detection is less likely to be affected by background interference.

An important consideration for conductivity detection is the electrolyte solution chosen for the separation of the analytes being detected. As stated previously, the detection of the analytes depends on a difference in conductivity between the bulk solution and the analyte zones. However, highly conductive electrolytes contain ions that can interfere with the electrical response and lead to a high background signal. Therefore, it is imperative to ensure that the background electrolyte solution has a conductivity that is different from that of the analytes being detected without compromising the sensitivity of the assay.[214]

Conductivity detection in the microchip format has been coupled to several different modes of capillary electrophoresis, including capillary zone electrophoresis (CZE), isotachophoresis (ITP), micellar electrokinetic chromatography (MEKC), and electrochromatography. Further information about the use of conductivity detection for microchip CE can be found in reviews by Schwarz and Hauser[5] and Tanyanyiwa et al.[213]

11.4.3.2 Instrumental Design and Considerations

11.4.3.2.1 System Design
The conventional instrumental design for conductivity detection involves the placement of two electrodes, usually platinum wires, directly in contact with the electrolyte solution. However, contactless conductivity detection has also been developed for CE.[213] Both of these system designs are described below.

11.4.3.2.2 Contact Conductivity Detection
Galvanic contact, or sensing, conductivity detection utilizes the simplest and most straightforward design and instrumental setup involving direct contact of the electrodes with the electrolyte solution. Originally, contact conductivity detection was performed in an on-column configuration, but end-column or wall-jet configurations have also been reported. For conventional CE, the on-column configuration requires drilling two holes through the capillary wall. Two platinum wire electrodes are then placed within the holes and sealed into place. An AC voltage is applied between the two electrodes and the conductivity differences between the bulk solution and the analyte zones are monitored.[215] On-column conductivity detection is much easier to fabricate in a microchip CE format. Because of the planar geometry of the device, the electrodes can easily be positioned directly in the separation channel. There is no need to drill holes as in the conventional capillary electrophoresis system. A typical on-column microchip CE contact conductivity detection system is shown in Figure 11.21.

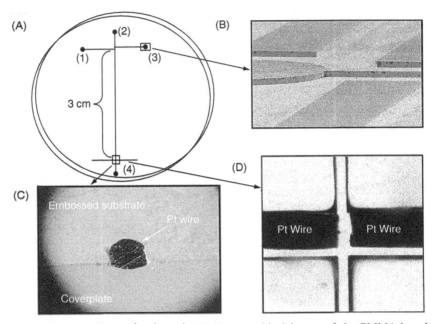

FIGURE 11.21 (See color insert) (A) Topographical layout of the PMMA-based microchip with an integrated conductivity detector. (1) sample reservoir, (2) buffer reservoir, (3) waste reservoir, and (4) receiving reservoir. (B) SEM of the Ni electroform embossing die used to make the chip. (C) Optical micrograph of PMMA microchip that was assembled with a coverplate and electrodes and then cut down the center of the fluidic channel. (D) Optical micrograph of the conductivity detector (T-cell, electrode gap ~20 μm) integrated to the PMMA microfluidic device. (Reprinted from Galloway, M.; Styjewski, W.; Henry, A.; Ford, S. M.; Llopis, S.; McCarley, R. L.; Soper, S. A. *Anal. Chem.* 2002, *74*, 2407–2415. With permission.)

An additional end-column electrode configuration for conductivity detection involves the placement of one of the electrodes directly outside the channel outlet. The other electrode is placed a slight distance away in the outlet reservoir. The system for this mode of conductivity detection is easier to fabricate compared to on-column detection with a conventional CE system.[215] However, both are easily fabricated with either external metal wires or thermal deposition of metals directly onto the microfluidic substrate.

11.4.3.2.3 Contactless Conductivity Detection

Contactless capacitively coupled or, more simply, contactless conductivity detection has become a very popular mode for microchip CE. Contactless conductivity detection was first developed on a conventional CE system to improve the sensitivity of conductivity detection. The system design on a conventional CE system requires two electrodes positioned millimeters apart that are attached to the capillary by either a conductive varnish or metal tubes surrounding the capillary.[215] Contactless conductivity detection has gained much popularity within the microchip CEEC community over the past few years due to its many advantages over the closely related contact detection.[213] The arrangement of the electrodes in an on-channel or on-column configuration leads to an increase in sensitivity for two reasons. First, because the electrodes are not in direct contact with the electrolyte solution within the channel, the background noise is decreased, leading to lower limits of detection. Second, the on-channel configuration produces higher separation efficiencies, resulting in a more sensitive assay (Figure 11.22).

11.4.3.2.4 Integration into Microchip CE Format

The integration of contact conductivity detection onto a microchip CE format has been a relatively easy transition.[216–219] Both glass- and polymer-based substrates have been utilized for the fabrication, making disposable devices

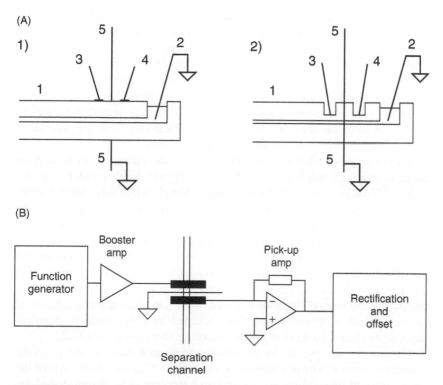

FIGURE 11.22 Microchip CE configuration for contactless conductivity detection. (A) Cross-sectional view of the two cell arrangements without (1) and with (2) troughs for the detector electrodes. (B) Block diagram of the electronic circuitry: (1) detector end of the separation device, (2) electrophoretic ground electrode, (3) actuator electrode for detection, (4) pick-up electrode for detection, and (5) Faraday shield. (Reprinted from Tanyanyiwa, J.; Hauser, P. C. *Anal. Chem.* 2002, *74,* 6378–6382. With permission.)

possible. The electrodes, usually platinum wires, are positioned on the microfluidic substrate anywhere from a few to a few hundred micrometers from the end of separation channel. Following alignment of the detection electrodes, a cover plate is bonded on top of the separation plate to seal the device and avoid contamination and evaporation. Soper and coworkers[216,217] have demonstrated a straightforward method of electrode integration for contact conductivity detection with a microchip CE device. The polymer poly(methylmethacylate) (PMMA) was used as the microchip device substrate. In addition to a separation channel and sample or buffer reservoirs, two guide channels for the detection electrodes located a few micrometers away from the waste reservoir (or receiving reservoir) were fabricated. Two platinum wires were manipulated into the guide channels and then a PMMA cover was thermally sealed on top of the microchip device.

Contactless conductivity detection is coupled to microchip CE in a fashion similar to that described above for contact conductivity detection as illustrated in Figure 11.22.[220-223] There have been numerous methods employed to couple contactless conductivity detection to microchip CE. One method, demonstrated by Wang and colleagues[222,223] involves placing aluminum foil strips on top of a 125 μm thick PMMA coverslip that is used to seal the separation channel. The electrodes are a few hundred micrometers away from the end of the separation channel. This fabrication method is easy, inexpensive, and takes only a small amount of time to assemble. However, due to the large distance that separates the electrodes from the separation channel, the sensitivity of the assay is compromised.

Hauser et al. developed a chip that enabled the electrodes to be aligned closer to the bulk solution but still not come into direct contact with the buffer.[220] Troughs were drilled into a glass cover plate that was sealed directly above the separation channel. In Figure 11.22A, the placement of the detection electrodes within the troughs is shown. Additionally, Figure 11.22B demonstrates the electronic circuitry of the contactless conductivity detector. Placing the sensing electrodes in troughs greatly increased the ruggedness of the microfluidic device. In addition to the troughs, a Faraday shield between the two detection electrodes was used to reduce the coupling of the first detecting electrode to the second electrode (e.g., crosstalk), increasing the sensitivity of the assay. Both of these modifications resulted in a higher electrochemical response and lower background interference, as shown in Figure 11.23.

Despite the success of the previously mentioned methods for coupling contactless conductivity detection to a microchip CE device, both involve the use of external metal wires as electrodes. Verpoorte et al. demonstrated the ability to produce a fully integrated device where both the electrodes and separation channels are fabricated by common microfabrication techniques and wet etching of the glass substrate.[221] Recesses for the electrodes and channels are created simultaneously in one substrate. Then, Pt for the detection electrodes is deposited in the recesses. The Pt detection electrodes and

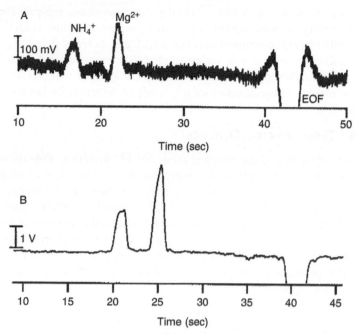

FIGURE 11.23 Electropherograms of ammonium and magnesium (200 μM) obtained for the cell configuration given in Figure 6.22: (A) electrodes placed on an unaltered chip at 1 mm distance from separation channel and (B) electrodes placed in electrode wells at 0.2 mm from the separation channel. (Reprinted from Tanyanyiwa, J.; Hauser, P. C. *Anal. Chem.* 2002, *74*, 6378–6382. With permission.)

the separation channel are separated by 15 to 20 μm of glass substrate. The resulting device was used to detect a variety of cations such as K^+, Na^+, and Li^+. The simultaneous fabrication of detection and separation electrodes on one substrate demonstrates the feasibility of mass production of these types of microfluidic devices.

11.4.3.3 Applications

Anions and cations, have been detected using microchip CZE with conductivity.[209,217,220,222,223] In addition to small cations and anions (such as K^+ and Cl^-), larger charged species such as amino acids and peptides can be detected. Isotachophoresis (ITP) has been used for sample concentration and improved detection limits with CE-CD.[218,219] However, cations and anions must be separated and preconcentrated separately when using ITP, which is a limiting factor.

For neutral and hydrophobic compounds, micellar electrokinetic chromatography (MEKC) can be used to aid separation. The detection of a protein mixture has been demonstrated.[217] Finally, reversed-phase ion-pair electro-chromatography on a microchip was used to separate nucleic acids[217] and PCR-amplified double-stranded (ds) DNA with conductivity detection.[216]

The ability to detect a wide variety of charged compounds and the relative ease of the microfabrication of the microchip CE systems will certainly lead to a great number of applications for this mode of detection for lab-on-a-chip devices.

11.4.4 POTENTIOMETRIC DETECTION

Potentiometry is one of the simplest modes of EC detection. Potentiometric detection utilizes a two-electrode configuration: an ion-selective electrode (ISE) and a reference electrode. The reference electrode can be either external (e.g., Ag/AgCl electrode) or internal. The ISE is usually membrane based and is semipermeable to selected ions. The sample containing the ions of interest passes through the ISE and a charge separation is created, resulting in a measurable Nernst potential. The electrical response is monitored by a volt-meter. The potential observed can be related to concentration of the ionic species by the equation below:

$$E = E^{\circ} + \frac{RT}{F} \ln \left(\sum_i K_i^{\text{pot}} c_i^{1/z_i} \right) \qquad (11.11)$$

where E is the electrode potential, E° is a constant, R is the gas constant, T is temperature, F is Faraday's constant, K_i^{pot} is the selectivity coefficient for species i, and c_i^{1/z_i} is the concentration of the ionic species i of charge z.[213] A unique aspect of potentiometric detection is the logarithmic electrical response, which differs from the linear response associated with both ampero-metric and conductivity detection. The logarithmic equation used in potentio-metric detection results in a large linear range of the assay, but can also lead to a higher standard deviation and a decrease in precision for each measurement.

Potentiometry has been used to detect a variety of inorganic and organic ions as well as compounds such as amino acids, fatty acids, and neurotransmit-ters. The selectivity of the electrode is dependent on the ionophore of the ISE utilized. There are typically two types of ionophores used in the fabrication of electrodes for potentiometric detection: charged ionophores and electrically neutral ionophores. Charged ionophores are positively or negatively charged, highly lipophilic compounds within the membrane of the ISE. The presence of charged ionophores results in the uptake of oppositely charged ions within the sample. The selectivity of this variety of electrode is limited to the charge and lipophilicity of the analytes (ions) of interest. Electrically neutral (zwitterionic) ionophores tend to be more selective due to the fact that they are usually specific for a certain functional group or compound(s).[213]

An important issue in potentiometric detection is selection of the run buffer used during separation by CE. To obtain a background signal, the electrolyte must exhibit a slight response to the electrode. However, if the background electrolyte exhibits too great a potentiometric response, this will cause excessive background noise and lead to decreased sensitivity of the assay.[213]

Currently, no one has attempted to couple potentiometric detection to microchip CEEC. Tantra and coworkers have created a flow-based chip sensor in which an ISE is placed on the microfluidic substrate.[224] Although, this mode of detection has not been used with a separation-based microchip device, these results demonstrate the feasibility of using potentiometry with a miniaturized analytical system.

11.4.5 FUTURE DIRECTION OF ELECTROCHEMICAL DETECTION

The future direction of EC detection for microchip CE is the production of a fully integrated detection and separation system contained within a miniaturized format. Researchers have demonstrated the ability to integrate sample preparation, separation, and detection electrodes directly onto the microchip CE–EC device. Future directions will involve the integration of all components onto a single platform towards the development of a truly μTAS system. Due to the demand for point-of-care and on-site analysis, the focus of many research groups will continue to be the development of such devices for complex environmental, biological, and pharmaceutical applications.

11.5 CONCLUSIONS

The principles and applications of detectors for microchip CE have been discussed in this chapter. Due to its high sensitivity, fluorescence continues to be the most popular form of optical detection. Other optical methods for microchip CE are under development and will provide a more universal mode of detection. Electrochemical detection for microchip CE is closest to realizing the fabrication of a fully integrated detection and separation system due to the ease of microfabricating both the separation and detection electrodes. Current research employing mass spectrometry is focused primarily on the development of a useful and rugged interface between microchips and the mass analyzer. Each of these detection modes is moving toward the development of a fully μTAS system for applications in the pharmaceutical, bionanalytical, and environmental sciences.

ACKNOWLEDGMENTS

This work was supported by the National Science Foundation, National Institutes of Health, National Cancer Institute Training Grant, American Heart Association Pre-doctoral Fellowship, Biotechnology Training Grant, and Merck Tuition Fellowship. The authors thank Joshua Cooper and Nancy Harmony for their assistance in the preparation of the manuscript.

REFERENCES

1. Jacobson, S. C.; Koutny, L. B.; Hergenroder, R.; Moore, A. W.; Ramsey, J. M. *Anal. Chem.* 1994, *66*, 1107–1133.
2. Jacobson, S. C.; Hergenroder, R.; Koutny, L. B.; Ramsey, J. M. *Anal. Chem.* 1994, *66*, 1114–1118.
3. Harrison, D. J.; Manz, A.; Fan, Z.; Luedi, H.; Widmer, H. M. *Anal. Chem.* 1992, *64*, 1926–1932.
4. Effenhauser, C. S.; Manz, A.; Widmer, H. M. *Anal. Chem.* 1993, *65*, 2637–2642.
5. Schwarz, M. A.; Hauser, P. C. *Lab on a Chip* 2001, *1*, 1–6.
6. Dolnik, V.; Liu, S. R.; Jovanovich, S. *Electrophoresis* 2000, *21*, 41–54.
7. Giddings, J. C. *Unified Separation Science*; John Wiley & Sons: New York, NY, 1991.
8. Paegel, B. M.; Emrich, C. A.; Wedemayer, G. J.; Scherer, J. R.; Mathies, R. A. *Proc. Natl. Acad. Sci. USA* 2002, *99*, 574–579.
9. Culbertson, C. T.; Jacobson, S. C.; Ramsey, J. M. *Anal. Chem.* 2000, *72*, 5814–5819.
10. Polson, N. A.; Hayes, M. A. *Anal. Chem.* 2001, *73*, 312A–319A.
11. Jacobson, S. C.; Culbertson, C. T.; Daler, J. E.; Ramsey, J. M. *Anal. Chem.* 1998, *70*, 3476–3480.
12. Hinshaw, J. V. *LC GC* 2001, *19*, 1136–1140.
13. Wabuyele, M. B.; Ford, S. M.; Stryjewski, W.; Barrow, J.; Soper, S. A. *Electrophoresis* 2001, *22*, 3939–3948.
14. Chan, J. H.; Timperman, A. T.; Qin, D.; Aebersold, R. *Anal. Chem.* 1999, *71*, 4437–4444.
15. Lacher, N. A.; de Rooij, N. F.; Verpoorte, E.; Lunte, S. M. *J. Chromatogr. A* 2003, *1004*, 225–235.
16. Skoog, D. A.; West, D. M.; Holler, F. J. *Fundamentals of Analytical Chemistry*, 7th ed.; Saunders College Publishing: Orlando, FL, 1996.
17. Sharma, A.; Schulman, S. G. *Introduction to Fluorescence Spectroscopy*; John Wiley & Sons: New York, NY, 1999.
18. Menzel, E. R. *Laser Spectroscopy: Techniques and Applications*; Marcel Dekker: New York, NY, 1995.
19. Lakowicz, J. R. *Principles of Fluorescence Spectroscopy*; Penum Press: New York, NY, 1983.
20. Roulet, J. C.; Volkel, R.; Herzig, H. P.; Verpoorte, E. M. J.; de Rooij, N. F. *J. Microelectromech. Systems* 2001, *10*, 482–491.
21. Roulet, J. C.; Volkel, R.; Herzig, H. P.; Verpoorte, E. M. J.; de Rooij, N. F.; Dandliker, R. *Anal. Chem.* 2002, *74*, 3400–3407.
22. Webster, J. R.; Burns, M. A.; Burke, D. T.; Mastrangelo, C. H. *Anal. Chem.* 2001, *73*, 1622–1626.
23. Melles Griot Laser Group. *An Introduction to Lasers*; Melles Griot: Irvine, CA, 1999.
24. *Product Sheets: SSDP Lasers*; Photonics Industries International, Inc., 2002.
25. Ceriotti, L.; Verpoorte, E. M. J.; De Rooij, N. F., Proceedings of Micro-TAS 2000, Enschede, Netherlands; Kluwer Academic, 2000; pp. 225–228.
26. *Avalanche Photodiodes: A User's Guide*; Perkin–Elmer Optoelectronics, 1998–2001.

27. Harrison, D. J.; Fluri, K.; Seiler, K.; Fan, Z. H.; Effenhauser, C. S.; Manz, A. *Science* 1993, *261*, 895–897.
28. Jacobson, S. C.; Ramsey, J. M. *Anal. Chem.* 1996, *68*, 720–723.
29. Jacobson, S. C.; McKnight, T. E.; Ramsey, J. M. *Anal. Chem.* 1999, *71*, 4455–4459.
30. Seiler, K.; Fan, Z. H. H.; Fluri, K.; Harrison, D. J. *Anal. Chem.* 1994, *66*, 3485–3491.
31. von Heeren, F.; Verpoorte, E. M. J.; Manz, A.; Thormann, W. *Anal. Chem.* 1996, *68*, 2044–2053.
32. Oleschuk, R. D.; Shultz-Lockyear, L. L.; Ning, Y.; Harrison, D. J. *Anal. Chem.* 2000, *72*, 585–590.
33. Zhang, C. X.; Manz, A. *Anal. Chem.* 2001, *73*, 2656–2662.
34. Shultz-Lockyear, L. L.; Colyer, C. L.; Fan, Z. H.; Roy, K. I.; Harrison, D. J. *Electrophoresis* 1999, *20*, 529–538.
35. Ermakov, S. V.; Jacobson, S. C.; Ramsey, J. M. *Anal. Chem.* 2000, *72*, 3512–3517.
36. Attiya, S.; Jemere, A. B.; Tang, T.; Fitzpatrick, G.; Seiler, K.; Chiem, N.; Harrison, D. J. *Electrophoresis* 2001, *22*, 318–327.
37. Fan, Z. H.; Harrison, D. J. *Anal. Chem.* 1994, *66*, 177–184.
38. Jacobson, S. C.; Ramsey, J. M. *Anal. Chem.* 1997, *69*, 3212–3217.
39. Culbertson, C. T.; Ramsey, R. S.; Ramsey, J. M. *Anal. Chem.* 2000, *72*, 2285–2291.
40. Polson, N. A.; Hayes, M. A. *Anal. Chem.* 2000, *72*, 1088–1092.
41. Jacobson, S. C.; Hergenroder, R.; Moore, A. W.; Ramsey, J. M. *Anal. Chem.* 1994, *66*, 4127–4132.
42. Fluri, K.; Fitzpatrick, G.; Chiem, N.; Harrison, D. J. *Anal. Chem.* 1996, *68*, 4285–4290.
43. Seiler, K.; Harrison, D. J.; Manz, A. *Anal. Chem.* 1993, *65*, 1481–1488.
44. Jacobson, S. C.; Ermakov, S. V.; Ramsey, J. M. *Anal. Chem.* 1999, *71*, 3273–3276.
45. Ocvirk, G.; Tang, T.; Harrison, D. J. *Analyst* 1998, *123*, 1429–1434.
46. Munro, N. J.; Snow, K.; Kant, E. A.; Landers, J. P. *Clin. Chem.* 1999, *45*, 1906–1917.
47. Huang, Z.; Munro, N.; Huhmer, A. F. R.; Landers, J. P. *Anal. Chem.* 1999, *71*, 5309–5314.
48. Cheng, S. B.; Skinner, C.; Taylor, J.; Attiya, S.; Lee, W. E.; Picelli, G.; Harrison, D. J. *Anal. Chem.* 2001, *73*, 1472–1479.
49. Xu, H.; Roddy, T. P.; Lapos, J. A.; Ewing, A. G. *Anal. Chem.* 2002, *74*, 5517–5522.
50. Monnig, C. A.; Jorgenson, J. W. *Anal. Chem.* 1991, *63*, 802–807.
51. Moore, A. W.; Jorgenson, J. W. *Anal. Chem.* 1993, *65*, 3550–3560.
52. Chabinyc, M. L.; Chiu, D. T.; McDonald, J. C.; Stroock, A. D.; Christian, J. F.; Karger, A. M.; Whitesides, G. M. *Anal. Chem.* 2001, *73*, 4491–4498.
53. *Molecular Probes Catalogue,* 9th ed.; Molecular Probes: Eugene, OR, 2002.
54. Colyer, C. L.; Mangru, S. D.; Harrison, D. J. *J. Chromatogr. A* 1997, *781*, 271–276.
55. Liu, Y. J.; Foote, R. S.; Jacobson, S. C.; Ramsey, R. S.; Ramsey, J. M. *Anal. Chem.* 2000, *72*, 4608–4613.
56. Effenhauser, C. S.; Paulus, A.; Manz, A.; Widmer, H. M. *Anal. Chem.* 1994, *66*, 2949–2953.

57. Jacobson, S. C.; Koutny, L. B.; Hergenroder, R.; Moore, A. W.; Ramsey, J. M. *Anal. Chem.* 1994, *66*, 3472–3476.
58. Wallenborg, S. R.; Bailey, C. G. *Anal. Chem.* 2000, *72*, 1872–1878.
59. Jin, L. J.; Giordano, B. C.; Landers, J. P. *Anal. Chem.* 2001, *73*, 4994–4999.
60. Bousse, L.; Mouradian, S.; Minalla, A.; Yee, H.; Williams, K.; Dubrow, R. *Anal. Chem.* 2001, *73*, 1207–1212.
61. Pinto, D.M.; Arriaga, E. A.; Craig, D.; Angelova, J.; Sharma, N.; Ahmadzadeh, H.; Dorchi, N. J.; Boulet, C. A. *Anal. Chem.* 1997, *69*, 3015–3021.
62. Baker, D. R. *Capillary Electrophoresis: Techniques in Analytical Chemistry*; John Wiley & Sons: New York, NY, 1995.
63. Sweedler, J. V. In *CRC Handbook of Capillary Electrophoresis*; Landers, J. P., Ed.; CRC Press: Boca Raton, FL, 1994.
64. Monahan, J.; Gewirth, A. A.; Nuzzo, R. G. *Electrophoresis* 2002, *23*, 2347–2354.
65. Munro, N. J.; Huang, Z.; Finegold, D. N.; Landers, J. P. *Anal. Chem.* 2000, *72*, 2765–2773.
66. Huang, X. C.; Quesada, M. A.; Mathies, R. A. *Anal. Chem.* 1992, *64*, 2149–2154.
67. Shi, Y. N.; Simpson, P. C.; Scherer, J. R.; Wexler, D.; Skibola, C.; Smith, M. T.; Mathies, R. A. *Anal. Chem.* 1999, *71*, 5354–5361.
68. Hadd, A. G.; Raymond, D. E.; Halliwell, J. W.; Jacobson, S. C.; Ramsey, J. M. *Anal. Chem.* 1997, *69*, 3407–3412.
69. Breadmore, M. C.; Wolfe, K. A.; Arcibal, I. G.; Leung, W. K.; Dickson, D.; Giordano, B. C.; Power, M. E.; Ferrance, J. P.; Feldman, S. H.; Norris, P. M.; Landers, J. P. *Anal. Chem.* 2003, *75*, 1880–1886.
70. Ferrance, J. P.; Wu, Q.; Giordano, B.; Hernandez, C.; Kwok, Y.; Snow, K.; Thibodeau, S.; Landers, J. P. *Anal. Chim. Acta* 2003, *500(1–2)*, 223–236.
71. Jiang, G. F.; Attiya, S.; Ocvirk, G.; Lee, W. E.; Harrison, D. J. *Biosens. Bioelectron.* 2000, *14*, 861–869.
72. Gottschlich, N.; Culbertson, C. T.; McKnight, T. E.; Jacobson, S. C.; Ramsey, J. M. *J. Chromatogr. B* 2000, *745*, 243–249.
73. Cohen, C. B.; Chin-Dixon, E.; Jeong, S.; Nikiforov, T. T. *Anal. Biochem.* 1999, *273*, 89–97.
74. Fang, Q.; Wang, F. R.; Wang, S. L.; Liu, S. S.; Xu, S. K.; Fang, Z. L. *Anal. Chim. Acta* 1999, *390*, 27–37.
75. Throckmorton, D. J.; Shepodd, T. J.; Singh, A. K. *Anal. Chem.* 2002, *74*, 784–789.
76. Liang, Z.; Chiem, N.; Ockvirk, G.; Tang, T.; Fluri, K.; Harrison, D. J. *Anal. Chem.* 1996, *68*, 1040–1046.
77. Chau, L. K.; Osbourn, T.; Wu, C. C.; Yager, P. *Anal. Sci.* 1999, *15*, 721–724.
78. Verpoorte, E.; Manz, A.; Ludi, H.; Bruno, A. E.; Maystre, F.; Krattiger, B.; Widmer, H. M.; van der Schoot, B. H.; de Rooij, N. F. *Sensors Actuators B Chem.* 1992, *6*, 66–70.
79. Salimi-Moosavi, H.; Jiang, Y.; Lester, L.; McKinnon, G.; Harrison, D. J. *Electrophoresis* 2000, *21*, 1291–1299.
80. Splawn, B. G.; Lytle, F. E. *Anal. Bioanal. Chem.* 2002, *373*, 519–525.
81. Mogensen, K.; Petersen, N.; Hubner, J.; Kutter, J. *Electrophoresis* 2002, *22*, 3930–3938.
82. Mogensen, K.; El-Ali, J.; Wolff, A.; Kutter, J. *Appl. Opt.* 2003, *42*, 4072–4079.

83. Petersen, N. J.; Mogensen, K. B.; Kutter, J. P. *Electrophoresis* 2002, *23*, 3528–3536.
84. Nakanishi, H.; Nishimoto, T.; Arai, A.; Abe, H.; Kanai, M.; Fujiyama, Y.; Yoshida, T. *Electrophoresis* 2001, *22*, 230–234.
85. Mao, Q.; Pawliszyn, J. *Analyst* 1999, *124*, 637–641.
86. Sorouraddin, H. M.; Hibara, A.; Proskurnin, M. A.; Kitamori, T. *Anal. Sci.* 2000, *16*, 1033–1037.
87. Tokeshi, M.; Uchida, M.; Hibara, A.; Sawada, T.; Kitamori, T. *Anal. Chem.* 2001, *73*, 2112–2116.
88. Sato, K.; Kawanishi, H.; Tokeshi, M.; Kitamori, T.; Sawada, T. *Anal. Sci.* 1999, *15*, 525–529.
89. Doku, G. N.; Haswell, S. J. *Anal. Chim. Acta* 1999, *382*, 1–13.
90. Greenway, G. M.; Haswell, S. J.; Petsul, P. H. *Anal. Chim. Acta* 1999.
91. Petsul, P. H.; Greenway, G. M.; Haswell, S. J. *Anal. Chim. Acta* 2001, *428*, 155–161.
92. Collins, G. E.; Lu, Q. *Sensors Actuators B Chem.* 2001, *76*, 244–249.
93. Lu, Q.; Collins, G. E. *Analyst* 2001, *126*, 429–432.
94. Deng, G.; Collins, G. *J. Chromatogr. A* 2003, *436*, 311–316.
95. Garcia-Campana, A. M.; Gamiz-Gracia, L.; Baeyens, W. R. G.; Barrero, F. A. *J. Chromatogr. B* 2003, *793*, 49–74.
96. Birks, J. W. *Chemiluminescence and Photochemical Reaction Detection in Chromatography*; VCH Publishers: New York, NY, 1989.
97. Mangru, S. D.; Harrison, D. J. *Electrophoresis* 1998, *19*, 2301–2307.
98. Hashimoto, M.; Tsukagoshi, K.; Nakajima, R.; Kondo, K.; Arai, A. *Chem. Lett.* 1999, 781–782.
99. Hashimoto, M.; Tsukagoshi, K.; Nakajima, R.; Kondo, K.; Arai, A. *J. Chromatogr. A* 2000, *867*, 271–279.
100. Huang, X. J.; Pu, Q. S.; Fang, Z. L. *Analyst* 2001, *126*, 281–284.
101. Tsukagoshi, K.; Hashimoto, M.; Nakajima, R.; Arai, A. *Anal. Sciences* 2000, *16*, 1111–1112.
102. Liu, B. F.; Ozaki, M.; Utsumi, Y.; Hattori, T.; Terabe, S. *Anal. Chem.* 2003, *75*, 36–41.
103. Greenway, G. M.; Nelstrop, L. J.; Port, S. N. *Anal. Chim. Acta* 2000, *405*, 43–50.
104. Tsukagoshi, K.; Hashimoto, M.; Suzuki, T.; Nakajima, R. *Anal. Sci.* 2001, *17*, 1129–1131.
105. Arora, A.; de Mello, A. J.; Manz, A. *Anal. Commun.* 1997, *34*, 393–395.
106. Arora, A.; Eijkel, J. C.; Morf, W. E.; Manz, A. *Anal. Chem.* 2001, *73*, 3282–3288.
107. Zhan, W.; Alvarez, J.; Crooks, R. M. *Anal. Chem.* 2002, *75*, 313–318.
108. Zhan, W.; Alvarez, J.; Crooks, R. M. *J. Am. Chem. Soc.* 2002, *124*, 13265–13270.
109. Crabtree, H. J.; Kopp, M. U.; Manz, A. *Anal. Chem.* 1999, *71*, 2130–2138.
110. Kwok, Y. C.; Jeffery, N. T.; Manz, A. *Anal. Chem.* 2001, *73*, 1748–1753.
111. Kwok, Y. C.; Manz, A. *J. Chromatogr. A* 2001, *924*, 177–186.
112. Kwok, Y. C.; Manz, A. *Electrophoresis* 2001, *22*, 222–229.
113. Kwok, Y. C.; Manz, A. *Analyst* 2001, *126*, 1640–1644.
114. Eijkel, J. C.; Kwok, Y. C.; Manz, A. *Lab on a Chip* 2001, *1*, 122–126.

115. McReynolds, J. A.; Edirisinghe, P.; Shippy, S. A. *Anal. Chem.* 2002, *74*, 5063–5070.
116. Burggraf, N.; Krattiger, B.; de Mello, A. J.; de Rooij, N. F.; Manz, A. *Analyst* 1998, *123*, 1443–1447.
117. Pan, D.; Mathies, R. A. *Biochemistry* 2001, *40*, 7929–7936.
118. Pan, D.; Ganim, Z.; Kim, J. E.; Verhoeven, M. A.; Lugtenburg, J.; Mathies, R. A. *J. Am. Chem. Soc.* 2002, *124*, 4857–4864.
119. Walker, P. A., III; Morris, M. D.; Burns, M. A.; Johnson, B. N. *Anal. Chem.* 1998, *70*, 3766–3769.
120. Soper, S. A.; Ford, S. M.; Xu, Y. C.; Qi, S. Z.; Mc Whorter, S.; Lassiter, S.; Patterson, D.; Bruch, R. C. *J. Chromatogr. A* 1999, *853*, 107–120.
121. Chen, Y. H.; Chen, S. H. *Electrophoresis* 2000, *21*, 165–170.
122. Lee, G. B.; Chen, S. H.; Huang, G. R.; Lin, Y. H.; Sung, W. C. *4177(Microfluidic Devices and Systems III)* 2000; 112–121.
123. Swinney, K.; Markov, D.; Bornhop, D. J. *Anal. Chem.* 2000, *72*, 2690–2695.
124. Guzman, N. A.; Majors, R. E. *LC GC* 2001, *June*, 1–9.
125. Auroux, P. A.; Reyes, D. R.; Iossifidis, D.; Manz, A. *Anal. Chem.* 2002, *74*, 2637–2652.
126. Bruin, G. J. M. *Electrophoresis* 2000, *21*, 3931–3951.
127. Willauer, H. D.; Collins, G. E. *Electrophoresis* 2003, *24*, 2193–2207.
128. Banks, J. F. *Electrophoresis* 1997, *18*, 2255.
129. Figeys, D.; Aebersold, R. *Electrophoresis* 1998, *19*, 885.
130. Tomlinson, A. J.; Guzman, N. A.; Naylor, S. *J. Capillary Electrophor.* 1995, *2*, 247.
131. Limbach, P. A.; Meng, Z. J. *Analyst* 2002, *127*, 693–700.
132. Venn, R. F. *Principles and Pratice of Bioanalysis*; Taylor and Francis: London, UK, 2000.
133. Karas, M.; Hillencamp, F. *Anal. Chem.* 1988, *60*, 2299.
134. Tanaka, H.; Waki, Y.; Ido, S.; Akita, Y.; Yoshidda, T. *Rapid Commun. Mass Spectrom.* 1988, *2*, 151.
135. Karas, M.; Bahr, U. *Trends Anal. Chem.* 1990, *9*, 321–325.
136. Skoog, D. A.; Holler, F. J.; Nieman, T. A. *Principles of Instrumental Analysis*, 5th ed.; Saunders College Publishing: Philadelphia, PA, 1998.
137. Delgass, W. N.; Cook, R. G. *Science* 1987, *235*, 545.
138. Poschenreider, W. P. *J. Mass Spectrom. Ion Phys.* 1972, *9*, 357.
139. Wollnik, H.; Przewloka, M. *J. Mass Spectrom. Ion Processes* 1990, *96*, 267.
140. Matsuo, T.; Sakurai, T.; Ishihara, M. *J. Mass Spectrom.* 1997, *32*, 1179–1185.
141. Toyoda, M.; Ishihara, M.; Yamaguchi, S.-I.; Ito, H.; Matsuo, T.; Roll, R.; Rosenbauer, H. *J. Mass Spectrom.* 2000, *35*, 163–167.
142. Li, J. J.; Wang, C.; Kelly, J. F.; Harrison, D. J.; Thibault, P. *Electrophoresis* 2000, *21*, 198–210.
143. Deng, Y. Z.; Henion, J.; Li, J. J.; Thibault, P.; Wang, C.; Harrison, D. J. *Anal. Chem.* 2001, *73*, 639–646.
144. Wang, C.; Oleschuk, R.; Ouchen, F.; Li, J. J.; Thibault, P.; Harrison, D. J. *Rapid Commun. Mass Spectrom.* 2000, *14*, 1377–1383.
145. Figeys, D.; Ducret, A.; Yates III, J. R.; Aebersold, R. *Nat. Biotechnol.* 1996, *14*, 1579–1583.
146. Figeys, D.; Ning, Y. B.; Aebersold, R. *Anal. Chem.* 1997, *69*, 3153–3160.

147. Pinto, D. M.; Ning, Y.; Figeys, D. *Electrophoresis* 2000, *21*, 181–190.
148. Licklider, L.; Wang, X. Q.; Desai, A.; Tai, Y. C.; Lee, T. D. *Anal. Chem.* 2000, *72*.
149. Jiang, Y.; Wang, P. C.; Locascio, L. E.; Lee, C. S. *Anal. Chem.* 2001, *73*, 2048–2053.
150. Xu, N. X.; Lin, Y. H.; Hofstadler, S. A.; Matson, D.; Call, C. J.; Smith, R. D. *Anal. Chem.* 1998, *70*, 3553–3556.
151. Xiang, F.; Lin, Y. H.; Wen, J.; Matson, D. W.; Smith, R. D. *Anal. Chem.* 1999, *71*, 1485–1490.
152. Wen, J.; Lin, Y. H.; Xiang, F.; Matson, D. W.; Udseth, H. R.; Smith, R. D. *Electrophoresis* 2000, *21*, 191–197.
153. Liu, H. H.; Felten, C.; Xue, Q. F.; B.L., Z.; Jedrzejewski, P.; Karger, B. L.; Foret, F. *Anal. Chem.* 2000, *72*, 3303–3310.
154. Chen, S. H.; Sung, W. C.; Lee, G. B.; Lin, Z. Y.; Chen, P. W.; Liao, P. C. *Electrophoresis* 2001, *22*, 3972–3977.
155. Meng, Z. J.; Qi, S. Z.; Soper, S. A.; Limbach, P. A. *Electrophoresis* 2001, *73*, 1286–1291.
156. Yuan, C. H.; Shiea, J. *Anal. Chem.* 2001, *73*, 1080–1083.
157. Kameoka, J.; Craighead, H. G.; Zhang, H. W.; Henion, J. *Anal. Chem.* 2001, *73*, 1935–1941.
158. Li, J. J.; Kelly, J. F.; Chernushevich, I.; Harrison, D. J.; Thibault, P. *Anal. Chem.* 2000, *72*, 599–609.
159. Lazar, I. M.; Ramsey, R. S.; Sundberg, S.; Ramsey, J. M. *Anal. Chem.* 1999, *71*, 3627–3631.
160. Figeys, D.; Gygi, S. P.; McKinnon, G.; Aebersold, R. *Anal. Chem.* 1998, *70*, 3728–3734.
161. Li, J. J.; Thibault, P.; Bings, N. H.; Skinner, C. D.; Wang, C.; Colyer, C.; Harrison, D. J. *Anal. Chem.* 1999, *71*, 3036–3045.
162. Ramsey, R. S.; Ramsey, J. M. *Anal. Chem.* 1997, *69*, 1174–1178.
163. Xue, Q.; Foret, F.; Dunayevskiy, Y. M.; Zavracky, P. M.; McGruer, N. E.; Karger, B. L. *Anal. Chem.* 1997, *69*, 426.
164. Kim, J.-S.; Knapp, D. R. *J. Am. Soc. Mass Spectrom.* 2001, *12*, 463.
165. Kim, J. S.; Knapp, D. R. *J. Chromatogr. A* 2001, *924*, 137–145.
166. Kim, J. S.; Knapp, D. R. *Electrophoresis* 2001, *22*, 3993–3999.
167. Schultz, G. A.; Corso, T. N.; Prosser, S. J.; Zhang, S. *Anal. Chem.* 2000, *72*, 4058–4063.
168. Liu, J.; Tseng, K.; Garcia, B.; Lebrilla, C. B.; Mukerjee, E.; Collins, S.; Smith, R. *Anal. Chem.* 2001, *73*, 2147–2151.
169. Laurell, T.; Nilsson, J.; Marko-Varga, G. *J. Chromatogr. B* 2001, *752*, 217–232.
170. Laurell, T.; Nilsson, J.; Marko-Varga, G. *Trends Anal. Chem.* 2001, *20*, 225–231.
171. Ekström, S.; Nilsson, J.; Marko-Varga, G. *Electrophoresis* 2001, *22*, 3984–3992.
172. Marko-Varga, G.; Ekström, S.; Helldin, G.; Nilsson, J.; Laurell, T. *Electrophoresis* 2001, *22*, 3978–3983.
173. Brivio, M.; Fokkens, R. H.; Verboom, W.; Reinhoudt, D. N.; Tas, N. R.; Goedbloed, M.; van den Berg, A. *Anal. Chem.* 2002, *74*, 3972–3976.
174. Deng, Y. Z.; Zhang, H. W.; Henion, J. *Anal. Chem.* 2001, *73*, 1432–1439.
175. Bremer, J. *Physiol. Rev.* 1983, *63*, 1420–1424.

176. Lazar, I. M.; Ramsey, R. S.; Ramsey, J. M. *Anal. Chem.* 2001, *73*, 1733–1739.
177. Manz, A.; Graber, N.; Widmer, H. M. *Sensors Actuators B Chem.* 1990, *1*, 244–248.
178. Vandaveer IV, W. R.; Pasas, S. A.; Martin, R. S.; Lunte, S. M. *Electrophoresis* 2002, *23*, 3667–3677.
179. Lacher, N. A.; Garrison, K. E.; Martin, R. S.; Lunte, S. M. *Electrophoresis* 2001, *22*, 2526–2536.
180. Martin, R. S.; Ratzlaff, K. L.; Huynh, B. H.; Lunte, S. M. *Anal. Chem.* 2002, *74*, 1136–1143.
181. Wang, J.; Pumera, M.; Chatrathi, M. P.; Escarpa, A.; Konrad, R.; Griebel, A.; Dorner, W.; Lowe, H. *Electrophoresis* 2002, *23*, 596–601.
182. Wang, J.; Pumera, M.; Chatrathi, M. P.; Rodriguez, A.; Spillman, S.; Martin, R. S.; Lunte, S. M. *Electroanalysis* 2002, *14*, 1251–1255.
183. Rossier, J. S.; Ferrigno, R.; Girault, H. H. *Electroanal. Chem.* 2000, *492*, 15–22.
184. Martin, R. S.; Gawron, A. J.; Fogarty, B. A.; Regan, F. B.; Dempsety, E.; Lunte, S. M. *Analyst* 2001, *126*, 277–280.
185. Gawron, A. J.; Martin, R. S.; Lunte, S. M. *Electrophoresis* 2001, *22*, 242–248.
186. Wang, J.; Tian, B.; Sahlin, E. *Anal. Chem.* 1999, *77*, 5436–5440.
187. Kim, J.; Song, X.; Kinoshita, K.; Madou, M.; White, R. *J. Electrochem. Soc.* 1998, *145*, 2314–2319.
188. Hebert, N. E.; Snyder, B.; McCreery, R. L.; Kuhr, W. G.; Brazill, S. A. *Anal. Chem.* 2003, *75*.
189. Manica, D. P.; Ewing, A. G. *Electrophoresis* 2002, *23*, 3735–3743.
190. Wang, J.; Tian, B.; Sahlin, E. *Anal. Chem.* 1999, *71*, 3901–3904.
191. Wang, J.; Chatrathi, M. P.; Ibanez, A.; Escarpa, A. *Electroanalysis* 2002, *14*, 400–404.
192. Schwarz, M. A.; Galliker, B.; Fluri, K.; Kappes, T.; Hauser, P. C. *Analyst* 2001, *126*, 147–151.
193. Schwarz, M. A.; Hauser, P. C. *J. Chromatogr. A* 2001, *928*, 225–232.
194. Hilmi, A.; Luong, J. H. T. *Environ. Sci. Technol.* 2000, *34*, 3046–3050.
195. Hilmi, A.; Luong, J. H. T. *Anal. Chem.* 2000, *72*, 4677–4682.
196. Lapos, J. A.; Manica, D. P.; Ewing, A. G. *Anal. Chem.* 2002, *74*, 3348–3353.
197. Fanguy, J. C.; Henry, C. S. *Electrophoresis* 2002, *23*, 767–773.
198. Woolley, A. T.; Lao, K.; Glazer, A. N.; Mathies, R. A. *Anal. Chem.* 1998, *70*, 684–688.
199. Hebert, N. E.; Kuhr, W. G.; Brazill, S. A. *Electrophoresis* 2002, *23*, 3750–3759.
200. Pasas, S. A.; Lacher, N. A.; Davies, M. I.; Lunte, S. M. *Electrophoresis* 2002, *23*, 759–766.
201. Osbourn, D. M.; Lunte, C. E. *Anal. Chem.* 2003, *75*, 2710–2714.
202. Chen, D. C.; Hsu, F. L.; Zhan, D. Z.; Chen, C. H. *Anal. Chem.* 2001, *73*, 758–762.
203. Lacher, N. A.; Lunte, C. E.; Martin, R. S. *Anal. Chem.* 2004, *76(9)*, 2482–2491.
204. Wu, C. C.; Wu, R. G.; Huang, J. G.; Lin, Y. C.; Chang, H. C. *Anal. Chem.* 2003, *75*, 947–952.
205. Madou, M. J. *Fundamentals of Microfabrication*; CRC Press: Boca Raton, FL, 2002.
206. Wang, J.; Chatrathi, M. P.; Tian, B. *Anal. Chim. Acta* 2000, *416*, 9–14.

207. Martin, R. S.; Gawron, A. J.; Lunte, S. M.; Henry, C. S. *Anal. Chem.* 2000, *72*, 3196–3202.
208. Baldwin, R. P.; Roussel, T. J. J.; Crain, M. M.; Bathlagunda, V.; Jackson, D. J.; Gullapalli, J.; Conklin, J. A.; Pai, R.; Naber, J. F.; Walsh, K. M.; Keynton, R. S. *Anal. Chem.* 2002, *74*, 3690–3697.
209. Wang, J.; Pumera, M. *Anal. Chem.* 2002, *74*, 5919–5923.
210. Wang, J.; Ibanez, A.; Chatrathi, M. P.; Escarpa, A. *Anal. Chem.* 2001, *73*, 5323–5327.
211. Kikura–Hanajiri, R.; Martin, R. S.; Lunte, S. M. *Anal. Chem.* 2002, *74*, 6370–6377.
212. Landers, J. P. *Handbook for Capillary Electrophoresis*, 2nd ed.; CRC Press: Boca Raton, FL, 1997.
213. Tanyanyiwa, J.; Leuthardt, S.; Hauser, P. C. *Electrophoresis* 2002, *23*, 2659–3666.
214. Zemann, A. J. *Trends Anal. Chem.* 2001, *20*, 346–354.
215. Zemann, A. J.; Schnell, E.; Volgger, D.; Bonn, G. K. *Anal. Chem.* 1998, *70*, 563–567.
216. Galloway, M.; Soper, S. A. *Electrophoresis* 2002, *23*, 3760–3768.
217. Galloway, M.; Styjewski, W.; Henry, A.; Ford, S. M.; Llopis, S.; McCarley, R. L.; Soper, S. A. *Anal. Chem.* 2002, *74*, 2407–2415.
218. Masar, M.; Zuborova, M.; Bielcikova, J.; Kaniansky, D.; Johnck, M.; Stanislawski, B. *J. Chromatogr. A* 2001, *916*, 101–111.
219. Prest, J. E.; Baldock, S. J.; Fielden, P. R.; Goddard, N. J.; Brown, B. J. T. *Analyst* 2002, *127*, 1413–1419.
220. Tanyanyiwa, J.; Hauser, P. C. *Anal. Chem.* 2002, *74*, 6378–6382.
221. Lichtenberg, J.; De Rooij, N. F.; Verpoorte, E. *Electrophoresis* 2002, *23*, 3769–3780.
222. Pumera, M.; Wang, J.; Opekar, F.; Jelinek, I.; Feldman, J.; Lowe, H.; Hardt, S. *Anal. Chem.* 2002, *74*, 1968–1971.
223. Wang, J.; Pumera, M.; Collins, G. E.; Mulchandani, A. *Anal. Chem.* 2002, *74*, 6121–6125.
224. Tantra, R.; Manz, A. *Anal. Chem.* 2000, *72*, 2875–2878.
225. Fischer, D. J.; Vandaveer, W. R. IV; Grigsley, R. J.; Lunte, S. M. *Electroanalysis*, in press.

12 Analytical Applications on Microchips

Shaorong Liu and Vladislav Dolnik

CONTENTS

12.1 INTRODUCTION

Researchers have been reporting progress in microchip technology development for more than a decade. Considerable progress in microchip applications has been made. Many companies (Caliper, Aclara, Gyros, Cepheid, Fluidigm, Micronics, and Agilent, to name a few) have been involved in commercialization of microchip technology, and several commercial products are available now. Microfluidic functional components, such as microfluidic pump, microvolume reactor, microseparation column, and on-chip detectors, have been developed. Various sample treatment techniques, including sample concentration and dilution, sample purification and filtering, sample preseparation, cell culture and cell handling, etc., have been demonstrated.

Analytical separations of various samples have been performed. Previous chapters have covered the fabrication and operation of microchip functional components and various techniques of sample treatment. In this chapter, we will focus on separations of various analytes, such as DNA and oligonucleotides, protein and peptides, and low molecular weight compounds. Representative examples will be discussed.

12.2 ANALYSIS OF DNA

One of the leading applications of microchip electrophoresis is the analysis of DNA. The analyses are extremely rapid, from less than 1 min for oligonucleotides [1] to less than 20 min for DNA sequencing [2]. Microchip separation has proved to be at least 18 times faster than conventional capillary electrophoresis (CE) [3]. A parallel design for DNA separation based on a 96-channel array has been developed and tested [4]. The analysis of 96 alleles in parallel has been achieved, which has demonstrated the feasibility of performing high-throughput genotyping separation with this device. Microchips have also been used to analyze oligonucleotides and RNA, and genotype and sequence DNA. DNA is typically detected with laser-induced fluorescence (LIF), but electrochemical detection has been applied as well [5]. Microchips are being developed for commercial applications such as genotyping medically important loci and sequencing human genomic DNA. Microchip electrophoresis has been used to screen common mutations in breast cancer-susceptible genes by a single-strand conformation polymorphism analysis [6]. There are several commercial products available for DNA and RNA separations by Caliper, Agilent, Shimadzu, and Hitachi. Several review articles [7–9] have been published on applications of microchips for DNA analysis.

12.2.1 DNA FRAGMENT SIZING

Separations of large double-stranded DNA longer than 40,000 base pairs (bp) are generally performed with pulsed-field gel electrophoresis. However, these processes are slow and recovering the separated DNA from gels is complex. Fast separations of fluorescently labeled oligonucleotides were first demonstrated on a poly(dimethylsiloxane) microchip using hydroxypropyl cellulose (HPC, average MW 100,000) as sieving matrix in 1994 [1]. The device consisted of an electrophoresis channel with a cross-channel injector. With the application of high electric fields of up to 2300 V/cm, the DNA fragments (single-stranded) were separated in <45 sec using an effective separation distance of 3.8 cm. Column efficiencies of up to 200,000 theoretical plates were obtained. In the same year, oligonucleotides of more than 1 kbp were separated on a glass microchip [10]. ϕX174 *Hae*III restriction fragments, ranging from 72 to 1353 bp, were separated with high resolution using hydroxyethyl cellulose (HEC, average MW 438,000) as sieving matrix. The fragments were fluorescently labeled with an intercalating dye contained in the sieving matrix. By use of a channel with an effective length of 3.5 cm, the separation of φX174 *Hae*III DNA fragments was completed in 120 sec. The effects of the intercalating dye, channel dimensions, and other separation conditions on separation were discussed. Separations of polymerase chain reaction (PCR)-amplified HLA-DQ α-alleles were also demonstrated.

Plastic microchips were also used to separate φX174 *Hae*III DNA fragments [11]. An injection-molding process was used to produce these microchips. Because the cost of mass production of plastic devices is generally low, these microchips can be made as single-use devices. Using a separation channel with ~5 cm effective length and HEC (average MW 90,000 to 105,000) as sieving matrix, high-resolution separations of double-stranded DNA fragments were demonstrated and separation times of less than 3 min were achieved on these plastic microchips.

One of the major advantages of microchip electrophoresis is the capability of housing parallel structures to perform parallel assays on a microchip. Figure 12.1 presents a channel arrangement of a 12-channel microchip and Figure 12.2 shows the electropherograms of the separations of 12 pBR322 *Msp*I samples in parallel. These separations were completed in less than 160 sec [12]. An LIF confocal scanner was used to detect the separated bands at a sampling frequency of 3 Hz. Parallel separations of pBR *Hae*III digest were also conducted on an eight-channel microchip [13] equipped with an innovative LIF scanning detection scheme. The laser beam scanning was driven by an acousto-optical defective (AOD, see Chapter 6 for details) with a scanning frequency of 10 Hz. When HEC was used as sieving matrix and an electric field of 165 V/cm was applied, the separation was complete within 200 sec in a microchannel having an effective channel distance of 4.2 cm. Separations of pBR322 *Msp*I DNA fragments were also performed on a 96-channel radial chip [4]. Using a separation channel of 3.3 cm and an electric field strength of 200 V/cm, the separations were carried out in ~2 min. HEC was used as the sieving matrix.

The concept of "lab-on-a-chip" is to integrate sample preparation, separation, and detection on a single microchip device to perform complex assays. The conventional biochemical DNA analysis, such as Sanger sequencing reaction, polymerase-assisted amplification, and restriction endonuclease digestion, typically requires several linked steps to proceed from an unknown sample to base pair information. Toward an integrated device, Figure 12.3 presents a photograph of a chip that performs a chemical reaction followed by an electrophoretic separation [14]. The device mixes a DNA sample with a restriction enzyme in a 0.7-nl reaction chamber and, after a digestion period, injects the fragments onto a 6.7-cm-long CE channel for sizing. This device can be a very useful tool for forensic analysis if on-chip PCR amplification of DNA can be carried out.

In the same year, a microfabricated silicon PCR reactor was integrated with microchip electrophoresis [15]. Figure 12.4 presents a schematic dia-

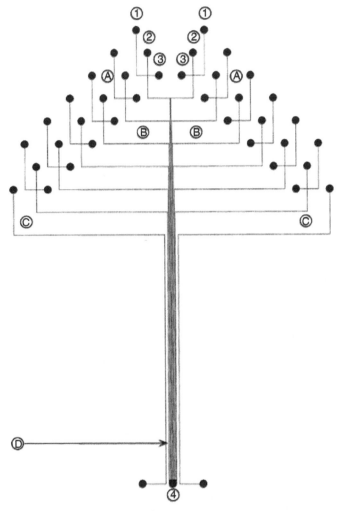

FIGURE 12.1 Photomask design used to photolithographically pattern a 12-channel CAE chip: (A) injection channels, 8-mm long; (B) separation channels, 6-cm long; (C) optical alignment channels; (D) detection region, ~4.5 cm from the junctions of the injection and separation channels. (1) Injection waste reservoirs. (2) Ground reservoirs. (3) Injection reservoirs. (4) High-voltage anode reservoir. (From Woolley, A.; Sensabaugh, G.F.; Mathies, R.A., *Anal. Chem.* 1997, 69, 2181. With permission.)

gram of the integrated PCR–CE microdevice. The device combined the rapid thermal cycling capabilities of microfabricated PCR devices (e.g., 10°C/sec heating, 2.5°C/sec cooling) with the high-speed DNA separation of microchip electrophoresis. The PCR chamber and the CE chip were directly linked through a photolithographically fabricated channel filled with HEC. Rapid

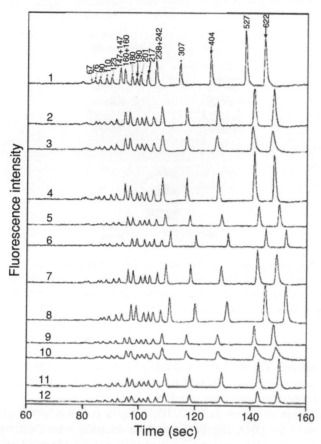

FIGURE 12.2 Electropherograms of pBR322 MspI separations. (From Woolley, A.; Sensabaugh, G.F.; Mathies, R.A., *Anal. Chem.* 1997, 69, 2181. With permission.)

PCRs of a β-globin target cloned in M13 and the genomic *Salmonella* DNA were performed in the PCR chamber and the reaction products were immediately injected into the microchip channel and analyzed.

Two years later, an integrated nanoliter DNA analysis device was developed [16]. After introduction of DNA and reagent solutions to this device, electronic signals corresponding to genetic information were directly produced. The device automatically performed all solution metering, chemistry, separation, and detection. Figure 12.5 presents a schematic diagram of the device that includes a nanoliter liquid injector, a sample mixing and positioning system, a temperature-controlled reaction chamber, an electrophoretic separation system, and a fluorescence detector. All the components of the system, with the exception of the excitation source, pressure source, and control circuitry, were microfabricated and contained on a single glass–

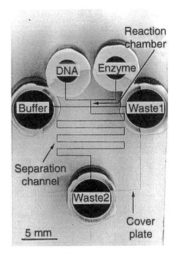

FIGURE 12.3 Photograph of a chip with an integrated precolumn reactor (0.7-nl volume) and 6.7-cm serpentine separation column. The channels and reservoirs are filled with black ink to provide contrast for the picture. The reservoirs are labeled by the various solutions normally contained. (From Jacobson, S.C.; Ramsey, J.M., *Anal. Chem.* 1996, 68, 720. With permission.)

silicon substrate. The device is capable of measuring aqueous reagent and DNA-containing solutions, mixing the solutions, amplifying or digesting the DNA, and separating and detecting DNA fragments.

Sieving matrices, such as HPC, HEC, linear polyacrylamide (LPA), are normally used for DNA fragment sizing in microfabricated electrophoresis channels because high-resolution separations can be obtained using these matrices. Meanwhile, these viscous sieving matrices offer an additional benefit by suppressing the siphoning effect on microchips. When low-viscous solutions are used to carry out microchip electrophoresis, solutions in all reservoirs are required to be at the same level to prevent the siphoning effect. Fine adjustments of these liquid levels can be tedious and time consuming.

Dilute and low-viscous hydroxypropylmethyl cellulose-50 (HPMC-50, average MW 11,500) solutions containing polyhydroxy additives (e.g., mannitol) have been used as separation media for DNA separations [17], because filling the microchannels with these low-viscous solutions is convenient. High-performance separations of DNA restriction fragments ranging in size from 72 to 1353 bp were achieved in a 3-cm-long channel of a polymethyl-methacrylate microchip at an electric field strength of 300 V/cm in 170 sec. Polyhydroxy additives improved the separation presumably due to formation of hydrogen-bonding interactions of polyhydroxy additives with the HPMC-50 matrix and DNA so as to increase the coupling interactions between matrix and DNA molecules during electrophoresis. Experimental variables, such as applied

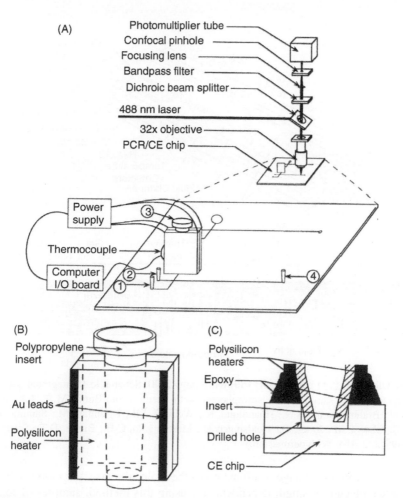

(A)

Photomultiplier tube
Confocal pinhole
Focusing lens
Bandpass filter
Dichroic beam splitter
488 nm laser
32x objective
PCR/CE chip

Power supply
Thermocouple
Computer I/O board

(B)
Polypropylene insert
Au leads
Polysilicon heater

(C)
Polysilicon heaters
Epoxy
Insert
Drilled hole
CE chip

FIGURE 12.4 Schematic of the integrated PCR–CE microdevice: (A) laser-excited confocal fluorescence detection apparatus and an integrated PCR–CE microdevice; (B) expanded view of the microfabricated PCR chamber; (C) expanded cross-sectional view of the junction between the PCR and CE devices. (From Woolley, A.T.; Hadley, D.; Landre, P.; deMello, A.J.; Mathies, R.A.; Northrup, M.A., *Anal. Chem.* 1996, 68, 4081. With permission.)

electric field strength, fluorescent intercalator (YOPro-1) concentration, polymer concentration, and additive concentration, were thoroughly investigated.

Nongel, microchip-based methods were developed to separate DNA mixtures [18]. In one example [18], the separation was performed on a multiple channel nanofluidic device with many entropic traps. When driven by an electric field, DNA molecules traveled through alternating thick and thin regions and their conformations changed alternatively. Since the extent of

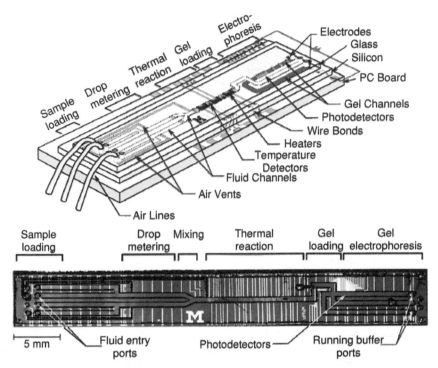

FIGURE 12.5 (See color insert following page 208) Schematic of integrated device with two liquid samples and electrophoresis gel present. (From Burns, M.A.; Johnson, B.N.; Brahmasandra, S.N.; Hhandique, K.; Webster, J.R.; Krishnan, M.; Sammarco, T.S.; Man, P.M.; Jones, D.; Heldsinger, D.; Mastrangelo, C.H.; Burke, D.T., *Science* 1998, 282, 484. With permission.)

the conformation change is a function of the molecular size, the mobility of a DNA molecule is length dependent. By using this method, samples of long DNA molecules (500 to ~160,000 bp) were efficiently separated into resolved bands in 1.5-cm-long channels. The T2 (164 kbp) and T7 (37.9 kbp) mixture can be separated within 30 min at 21 V/cm. In another example [18], long DNA molecules were separated using a chip containing an array of 2-μm pillars with 2-μm spacings arranged in a hexagonal lattice. Taking advantage of the difference of Brownian mobility, the T4 (168.9 kbp) and λ (48.5 kbp) DNAs could be resolved in ~10 sec.

12.2.2 DNA GENOTYPING

High throughput and low cost are desired for DNA genotyping analysis. Very often, the cost goes down as the throughput increases. To improve the throughput, capillary array electrophoresis (CAE) microchips have been used in the analysis of restriction fragment markers from the HLA-H gene, a candidate gene for the diagnosis of hereditary hemochromatosis since 1997 [12]. In this work, 12 samples were analyzed in parallel with an LIF scanner and an intercalating dye, thiazole orange. To further improve the throughput, a 96-sample analysis chip was developed. Figure 12.6 presents the channel geometry of this chip that contains 96 sample reservoirs and 48 separation channels [19]. The channel layout of this chip inherited, more or less, the chip design of the previous work [12]. Using this design, 48 separation channels were close to the maximum number of channels that could be arranged on a 4-in.-diameter chip device. Two samples shared one separation channel for serial separation and detection. This design risked sample cross-contamination. For genotyping using this device, 96 hemochromatosis samples were analyzed in less than 8 min. Apparently, the throughput increased with the number of the separation channels.

An elegant radial chip design illustrated in Figure 12.7 was developed to eliminate the risk of sample cross-contamination and boost the sample

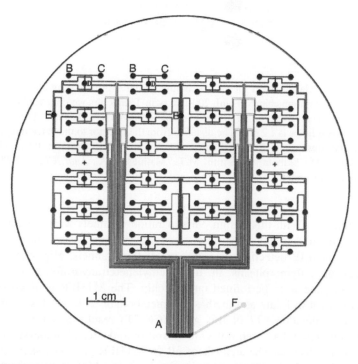

FIGURE 12.6 Photomask pattern for a 96-sample capillary array electrophoresis chip. A is the detection region, B and C are sample reservoirs, D are waste reservoirs, E are cathode reservoirs, and F is the anode reservoir. (From Shi, Y.N.; Simpson, P.C.; Scherer, J.B.; Wexler, D.; Skibola, C.; Smith, M.T.; Mathies, B.A., *Anal. Chem.* 1999, 71, 5354. With permission.)

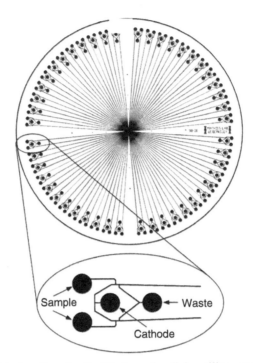

FIGURE 12.7 Mask pattern for the 96-channel radial capillary array electrophoresis microplate. Separation channels with 200-μm twin-T injectors were masked to 10-μm width and then etched to form 110-μm wide by ~50-μm deep channels. The diameter of the reservoir holes is 1.2 mm. The distance from the injector to the detection point is 33 mm. The substrate is 10 cm in diameter. (From Shi, Y.N.; Simpson, P.C.; Scherer, J.B.; Wexler, D.; Skibola, C.; Smith, M.T.; Mathies, B.A., *Anal. Chem.* 1999, 71, 5354. With permission.)

throughput [4]. With this design, 96 separation channels were conveniently arranged on a 4-in.-diameter chip. A rotary scanner was constructed to facilitate the LIF detection of DNA in all 96 channels. High-throughput analyses of polymorphisms in the methylenetetrahydrofolate reductase (MTHFR) gene were performed on this chip. The MTHFR gene codes for protein critical in folate and methionine metabolism. A C $\longrightarrow$ T substitution mutation at position 677 of this gene (C677T) results in an Ala $\longrightarrow$ Val conversion, which has been linked to neural tube defects, homocysteinemia, and vascular disease. The analysis of all 96 MTHFR genotype samples was complete in ~1.5 min. The genotyping analysis of MTHFR polymorphisms was described earlier [20] using multiple-injection CE and one-color LIF detection. With a single capillary, as many as ten samples were serially injected and separated in ~16 min. In comparison, the results obtained from the 96-channel radial chip presented a more than 100-fold improvement in throughput. Applications of the same radial chip have also been performed for single-nucleotide polymorphism typing of a hereditary hemochromatosis mutation [21]. The 201-bp variant was successfully differentiated from the 211-bp wild type in 2 min.

Fast allelic profiling assays were performed on microchips in 1997 [22]. Electrophoretic separations of single-locus short tandem repeat (STR) samples were baseline resolved and complete in 30 sec. The separations of PCR samples containing the four loci *CSF1PO, TPOX, THO1*, and *vWA* (abbreviated as *CTTv*) were performed in less than 2 min, which constitutes a 10- to 100-fold improvement in speed relative to capillary or slab gel systems. The separation device consisted of a 45-μm-deep, 100-μm-wide, and 2.6-cm-long electrophoresis channel with a cross-injector. The channels were filled with a replaceable polyacrylamide matrix under denaturing conditions at 50°C. A fluorescently labeled STR ladder was used as an internal standard for allele identification. Samples were prepared by standard procedures and only 4 μl was required for each analysis. The device is capable of repetitive operation and is suitable for automated high-speed and high-throughput applications.

Microchip electrophoresis was also used for sizing microsatellites suitable to genetic, clinical, and forensic applications [23]. The samples used for this evaluation included the D19S394 tetranucleotide (AAAG) repeat characterized by a wide variation in the repeat number (1 to 17) and a short recombination distance from the low-density lipoprotein (LDL)-receptor gene that makes it suitable to cosegregation analysis of familial hypercholesterolemia (FH). Seventy carriers of two LDL-receptor mutations and 100 healthy controls were used in this research. The PCR amplification products were directly analyzed on a microchip with LIF detection. The results demonstrated that the microchip can distinguish 17 microsatellite alleles varying from 0 to 17 repeats. Many of these alleles were quite rare, but the seven more abundant ones accounted for over 70% of allele distribution in the control population. The standard deviation in the sizing of the most abundant alleles ranged from ± 0.60 to ± 0.75 bp. This indicated that the size attribution to a conventional allele using the ± 1 bp range allowed a confidence limit above 80%.

Locus-specific, multiplex PCR products have been separated on a silicon–glass microchip [24]. In this work, random amplification of the human genome using the degenerate oligonucleotide primed–polymerase chain reaction (DOP–PCR) was performed using a silicon–glass chip. An aliquot of the DOP–PCR amplified genomic DNA was then introduced into a different silicon–glass chip for a locus-specific, multiplex PCR of the dystrophin gene exons in order to detect deletions that cause Duchenne/Becker muscular dystrophy. Whole genome amplification products obtained by DOP–PCR were proved to be suitable templates for multiplex PCR as long as the amplicon size was <250 bp.

12.2.3   DETECTION OF DNA MUTATION

A primary focus of functional genomic studies in the postgenomic era is the analysis of gene variants in human populations for the purpose of disease diagnosis, prognosis, and management. Many specialized electrophoretic methodologies, such as single-strand conformation polymorphism (SSCP), allele-specific PCR (AS-PCR), and heteroduplex analysis (HDA), have been developed to detect DNA mutations. One of the first reports using these techniques on microchip describes a method for SSCP analysis of mutations in the tumor susceptibility genes BRCA1 and BRCA2 [25], in which a small number of specific mutations have been found at relative high frequency in certain ethnic populations. For example, mutations 185delAG/5382insC in BRCA1 and 6147delT in BRCA2 were identified as common mutations in the Ashkenazi Jewish population. SSCP electrophoresis technique takes advantage of the electrophoretic mobility differences due to mutation-induced conformational changes in single-stranded DNA. DNA fragments housing these mutations were PCR amplified using labeled primers, denatured, and separated. The separations were complete in less than 120 sec on a microchip with an effective separation length of 3 cm. Accurate temperature control is a must with slab-gel SSCP, but it was not an issue with microchip separation, suggesting that dramatic temperature fluctuations did not occur in the separation channels.

The HDA alone [26] or in combination with AS-PCR [27] has been used to detect mutations in the BRCA1 and BRCA2 tumor susceptibility genes. In HDA, duplex DNA is formed by denaturing the double-stranded PCR products and allowing the strands to reanneal. When mutations are present, homoduplexes are formed between like strands, but heteroduplex DNA is formed between wild-type and mutant PCR fragments. These heteroduplexes can be easily identified by their change in electrophoretic mobility, resulting in multiple peaks in the subsequent microchip separation and allows for discrimination of wild type from mutated DNA. In one example, HDA analysis was performed on a microchip using six heterozygous mutations, 185delAG, E1250X(3867GT), R1443G(4446CG), 5382insC, 5677insA in BRCA1, and 6174delT in BRCA2. Figure 12.8 presents six electrophoregrams of fast mutation detection using microchip HAD. All separations were complete in less than 130 sec. The sensitivity of this method was adequate to detect mutations present in concentrations as low as 1% of the total DNA concentration with an analysis time of less than 130 sec. HDA has also been used in combination with AS-PCR to detect the mutations in BRCA1, BRCA2, and PTEN [27]. This technique utilized three primers, one of which was constructed to amplify a fragment only if the mutation was present. The resulting amplified products were then separated, and detection of three heterozygous mutations (an insertion, a deletion, and a substitution mutation) was accomplished in 180 sec or less.

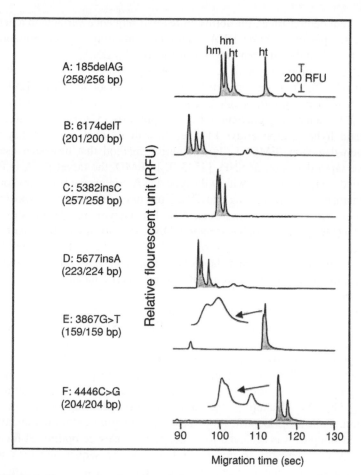

FIGURE 12.8 Fast mutation detection via heteroduplex analysis on a microfabricated electrophoretic chip. The heterozygous mutants are specified in the figure. The separation buffer was 2.5% HEC containing 10% glycerol in $1 \times$ TBE (pH 8.6) for (*A–D*) and 4.5% HEC containing 10% glycerol and 15% urea in $1 \times$ TBE (pH 8.6) for (*E–F*). The microchannel on the chip was coated with polyvinylpyrrolidone (M_r 1,000,000) and detection mediated by laser-induced fluorescence (em/ex 520 nm/488 nm). The PCR products were injected into the channel for 100 sec at 333 V/cm (effective microchannel length of 5.5 cm), and the separation voltage was 573 V/cm. (From Tian, H.; Brody, L.C.; Landers, J.P., *Genome Res.* 2000, 10, 1403. With permission.)

A polymer microfluidic system has been used for the detection of DNA point mutations via temperature gradient gel electrophoresis [28]. The method used the principle of SSCP for DNA mutation detection. In this work, a temporal thermal gradient was induced in the microfluidic network by controlling a heating block placed against the surface of the microfluidic device.

The system demonstrated the ability to resolve sequence variants in a hetero-duplex sample of a 100-bp DNA fragment containing a point mutation at position 72 (numbered from one end of the 100-bp DNA).

The most common type of genetic diversity is single nucleotide polymor-phorism (SNP). There are many different methods to identify SNPs [29], such as DNA sequencing [30], single-stranded conformational polymorphism an-alysis [31], denaturing gradient gel electrophoresis [32], mass spectrometry [33], and hybridization arrays [34], each with its own merits. Enzymatic mutation detection (EMD) is an alternative method that has been used to achieve rapid detection of SNPs [35–37]. In EMD, the target DNA is PCR-amplified and then mixed with wild-type DNA. Heteroduplexes are formed by denaturing and renaturing the DNA mixture, followed by exposing to enzymatic cleavage by endonuclease VII. This enzyme recognizes structural changes ("loops") in double-stranded DNA due to single base pair mis-matches and cleaves within 6 bp on the 3′-side of the mutation. The digest is then analyzed by denaturing slab gel electrophoresis, revealing not only the presence or absence of a mutation but also its specific location within a given target sequence.

The SNP sites in the p53 suppressor gene have been determined using microchip electrophoresis combined with EMD [38]. Using clinical samples, robust assays were developed with quality factors as good as conventional electrophoresis in ~100 sec, representing a 10 and 50 times speed increase, compared to capillary and slab gel electrophoresis, respectively. The method was highly accurate with an average error of mutation site measurement of only ± 5 bp. No cleanup of the digestion mixtures was needed prior to injection. Following identification, absolute mutation determination of the screened samples was achieved in a second microdevice optimized for four-color DNA sequencing. In this second device, the sequencing took about 25 min and the sequencing data were in full agreement with ABI Prism® 377 sequencing runs that required 3.5 h.

In about the same time, an SNP typing assay was developed and evaluated on a CAE microchip system [21]. Using fluorescently labeled allele-specific primers, the S65C (193A $\longrightarrow$ T) substitution associated with hereditary hemo-chromatosis in the HFE gene was genotyped. The covalently labeled PCR products were separated on a microfabricated radial CAE chip using non-denaturing gel media in under 2 min. Detection was accomplished with a laser-excited rotary confocal scanner. The Rox-labeled A-allele specific amplicon (211 bp) was differentiated from the R110-labeled T-allele specific amplicon (201 bp) by both size and color. High-throughput SNP genotyping was performed using a 15-cm-diameter 96-channel CAE chip [39]. The assay targeted the three common variants at the HFE locus associated with the genetic disease hereditary hemochromatosis (HHC) and employed AS-PCR for the C282Y (845G $\longrightarrow$ A), H63D (187C $\longrightarrow$ G), and S65C (193A $\longrightarrow$ T) variants using fluorescently labeled energy-transfer (ET) allele-specific primers. Using a 96-channel radial CAE microplate, the labeled AS-PCR products generated from 96 samples in a reference Caucasian population were simultaneously separated with a single base pair resolution and genotyped in less than 10 min.

A microfluidic approach has been reported for allele-specific extension of fluorescently labeled nucleotides for scoring of SNP [40]. The method takes advantage of the fact that the reaction kinetics differs between matched and mismatched configurations of allele-specific primers hybridized to DNA template. By combining reaction kinetics of the extension reaction and flow kinetics of a microfluidic device, mismatched configurations can be discriminated from matched configurations. In this example [40], all three possible variants of an SNP site at codon 72 of the p53 gene were scored using the above approach. The result demonstrated the possibility of scoring SNP by allele-specific extension of fluorescently labeled nucleotides in a microfluidic flow-through device.

Detection of single nucleotide incorporation using the DNA pyrosequencing technique in a microfluidic flow-through device has been developed as well [41]. Pyrosequencing is based on the detection of released pyrophosphate during DNA synthesis. The microfluidic platform allows detection of single base incorporation, enabling base-by-base DNA sequencing. The miniaturization has lead to a 1000-fold reduction in reagent consumption compared to the present standard volume in pyrosequencing and will hopefully enable reading of longer DNA sequences. This work [41] demonstrated the possibility of performing DNA pyrosequencing in a microfluidic flow-through device. It is possible to detect single base incorporation in the chip, enabling base-by-base DNA sequencing.

The separation of the homoduplex and heteroduplex PCR products has been explored by using the chip-based temperature gradient electrophoresis [42]. A two-dimensional electrophoresis chip incorporated with the temperature gradient feature was designed to increase the throughput and further enhance the separation efficiency. Using this chip-based temperature gradient electrophoresis method, lots of premixed DNA samples were analyzed for their SNPs in a single run, which simplified the analysis process and shortened the total time needed for analysis.

12.2.4 DNA SEQUENCING

Microchip DNA sequencing was first demonstrated in 1995 [43]. Soda-lime microscope slides were used to produce the electrophoresis chip devices. The fabrication process used photoresist as an etching mask and channels up to 8-μm-deep were made. DNA sequencing fragment ladders fluorescently labeled with energy transfer dye primers were separated using a denaturing polyacrylamide sieving matrix and a confocal LIF detection system. Single base resolution reached 150 to 200 bases in 10 to 15 min with an effective separation channel distance of 3.5 cm [43]. The sequencing quality was significantly improved after changes and optimization of some experimental parameters [2]. Readlengths of over 500 bases were achieved in about 20 min at 99.4% accuracy [2]. The major changes included the extension of the separation channel length (to 7 cm) to promote the separation resolution, application of optimized LPA to improve the separation quality, increase of the electrophoresis channel depth (to 50 μm) to enhance the fluorescence signal, and use of low fluorescence-background borofloat glass wafers to reduce the detection noise. This work demonstrated the feasibility of high-speed, high-throughput, four-color DNA sequencing using CE on microchips [2]. To improve the sample throughput, a 16-channel microchip was fabricated on a 10-cm-diameter wafer with an effective channel length of 7.6 cm, and parallel sequencing was performed [44]. Using a four-color confocal fluorescent detection, the system routinely yields more than 450 bases in 15 min in all 16 channels. In the best cases, up to 543 bases have been called with an automated base-calling program at an accuracy of greater than 99% and in less than 18 min. Effects of temperature, template concentration, chip design, and injection time on separations were studied. In a separate group, optimization of DNA sequencing was performed on a chip device containing 150-μm twin-T injector and an 11.5-cm-long separation channel [45]. Using a separation matrix composed of 3.0% (w/w) 10 MDa plus 1.0% (w/w) 50 kDa LPA, elevated microdevice temperature (50°C), and 200 V/cm, high-speed DNA sequencing of 580 bases was obtained in 18 min with a base-calling accuracy of 98.5%. Readlengths of 640 bases at 98.5% accuracy were achieved in around 30 min by reducing the electric field strength to 125 V/cm.

A detailed investigation of the dependence of resolution on various separation parameters was carried out for DNA sequencing on microfabricated electrophoretic devices [46]. The various parameters included the selectivity, diffusion, injector size, channel length, and channel folding. One of the conclusions of this investigation was that "DNA sequencing to more than 400 bases on microfabricated devices much shorter than 10 cm in length is unlikely, assuming linear PAA (polyacrylamide) as sieving matrix and a minimum resolution criterion of $R = 0.5$ for all bases." It indicated the importance of channel distance to DNA sequencing resolution, although the exact numbers stated in conclusion may vary under different experimental conditions, such as increased temperature, decreased or even pulsed field strength, varied buffer composition, etc.

Large plates were employed [47, 48] to stretch the channel length so as to increase the sequencing readlengths. These plates had dimensions of 53 to 58 cm by 7.6 to 13 cm, and special equipment was used for their fabrication. Parallel straight channels were densely packed on these plates. Access holes were drilled at the ends of each channel to serve as cathode and anode reservoirs. The following describes a brief operation procedure for sequencing using these large plates: filling channels with sieving matrix, cleaning cathode reservoirs, loading sample to cathode reservoirs and running buffer to anode reservoirs, executing sample injection, cleaning cathode reservoirs again, loading running buffer to cathode reservoirs, and finally performing the electrophoretic separation. Readlengths of 400 to 640 bases were reported at an accuracy of 98% [47, 48]. While direct injection from sample reservoirs to separation channels enabled simple channel layout and device compactness, it could not take advantage of a chip injector, such as tolerance of high concentrations of templates in samples, uniform signal intensity, etc., as described in the literature [44].

The uniqueness of a chip injector offers many advantages for DNA sequencing. First, a uniform signal intensity profile is usually obtained over a wide range of fragment sizes. A typical signal trace from a conventional capillary gel electrophoresis (CGE) has an exponential profile. Base-calling usually fails due to a low signal-to-noise ratio in the long fragment region. Using a chip (cross or twin-T) injector, the "differential concentration effect" [44] enhances the signal intensity in the long fragment region and consequently more bases may be called. Second, a chip injector introduces narrower sample plugs into separation channels than conventional direct injection methods. Using a pinched injection scheme [49], ≤ 100-μm sample plugs are routinely obtained. This is about an order of magnitude lower compared to conventional CGE in which the sample band usually stretches to ≥ 1 mm. As a result, higher resolution can be achieved on chips, especially when the separation channel is short. Third, DNA template can be removed on-chip using a chip injector [44]. Removal of these large molecules has been reported essential to achieve high-quality separations [50, 51]. In CGE, they are removed offline with cumbersome methods that employ membrane filters. Additionally, a chip injector can potentially reduce the sample volume to submicroliter levels. Combined with small-volume sample preparation techniques, the cost of sequencing may be significantly reduced.

Recently, 32 identical and isolated channels, each with a 15-μm twin-T injector, were fabricated on a 50-cm-long and 25-cm-wide glass chip device [52]. Separation distances of these channels were about 40 cm. When DNA sequencing was performed using one of these channels, an average readlength of up to 800 bases at an accuracy of 98% was obtained for either M13 standards or DNA sequencing samples from the Whitehead Institute Center for Genomic Research production line. However, this was achieved with dramatically longer separation times (80 min). A 384-channel sequencing device with 40-cm-long separation channels was fabricated and reported as well [53, 54].

Folding the channel is another method to increase the separation channel length on a small microchip. Simply folding the channel generally degrades the separation due to both the race-track effect and electric field distortion at the turns [55]. A hyper-turn, which uses a very narrow connection channel for the turn, was created to minimize these effects and promising results were obtained [56–58]. Using a 96-channel CAE chip, fabricated on a 15-cm-diameter glass plate, an average readlength of 430 bases was obtained [59]. In this chip, every channel folded twice and contained four super turns, and the effective separation length increased to 15.9 cm.

An alternative means to extend channel length is to couple a microchip to a capillary since the capillary provides virtually unlimited separation distance. Several methods can be used to couple capillaries to microfluidic devices [60, 61]. These methods, however, usually bring large dead volumes. Recently, another method was developed to connect a capillary to a microfluidic device [62]. A hole was drilled into the edge of the chip device to meet a microchannel inside and then a capillary was inserted into the hole and glued in position. Using this device, good separations were obtained for amino acids. However, the fabrication process was delicate and required very accurate drill alignments, which lead to practical challenges for mass production. Additionally, a microchip channel usually has a trapezoidal profile while the connection capillary has a round profile. This profile disruption at the joint makes it uncertain for resolution-demanding DNA sequencing.

An innovative method was developed to eliminate the delicate drilling process and minimize profile disruption [63]. This method consists of two main steps. In the fist step, round channels of two different diameters are fabricated on a chip. Figure 12.9 presents a schematic fabrication process to make a round channel in a microchip. Using a photomask of very narrow linewidth, isotropic photolithographic etching will create a semicircular groove on a wafer. Round channels are formed when two etched wafers are face-to-face aligned and bonded. To make channels of two different diameters, a two-photomask process is developed. Referring to Figure 12.10 as an example [63], Mask 1 was used to make semicircular twin-T injector grooves

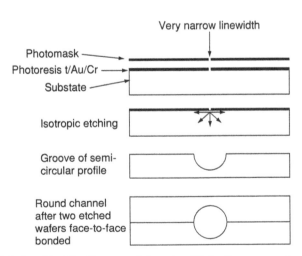

FIGURE 12.9 A schematic diagram of the microfabrication process to make round channels on a chip.

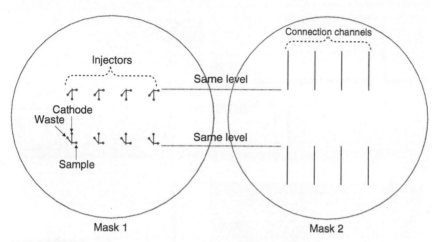

FIGURE 12.10 The two masks used in this experiment to make round channels of two different diameters.

having a diameter that matches the bore of a separation capillary. After twin-T grooves were made, Mask 2 was used to align the connection-channel grooves to the injector grooves and etch the connection-channel grooves to have a diameter that matches the outer diameter of the separation capillary. When two of such etched wafers were aligned face-to-face and bonded, round channels of two different diameters were formed. In the second step, the finished chip was diced across the connection channel and a separation capillary was inserted into the connection channel and secured in position with adhesives. Shown in Figure 12.11 are pictures of the real devices before and after the assembling. The resulting hybrid device had very small connection dead-volumes since the bore of the capillary was flush with the injector channels on the chip.

DNA sequencing has been performed on this hybrid device using M13 standard samples [63] or real-sequencing samples [64] from Joint Genome Institute (JGI). Using standard M13 samples, a readlength of 812 bases was obtained at an accuracy of 98.5% in 56 min when a 200-μm twin-T injector was coupled with capillaries of 20-cm effective separation distance. With an increased separation field strength and a diluted sieving matrix (2.5% LPA), the separation time was reduced to 20 min with a readlength of 700 bases at 98.5% accuracy. Using real-world samples from JGI, an average Phred20 readlength of 675 bases was achieved in about 70 min with a success rate of 91% [64].

It is worth pointing out that DNA sequencing can be anticipated in channels of less than or equal to 1.5 cm when a low electric field is applied for separation [65]. Although DNA sequencing was not performed in these channels, this work demonstrated impressive separations of single-stranded

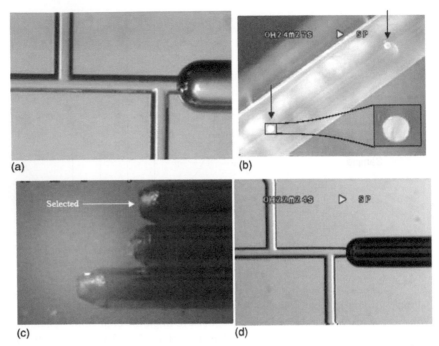

FIGURE 12.11 (See color insert) Pictures of real devices: (a) a 400-μm twin-T injector; (b) cross section profile of a round channel; (c) ground capillary tips; and (d) a twin-T injector incorporated with a capillary.

DNA (under denaturing conditions) in a 1.5-cm channel at a field strength of 20 V/cm or less. While the analysis time was long by microchip standards (85 min out to 400 bases) and the resolution unacceptable (10 base resolution) for sequencing applications, the peak shape was impressive, showing much less peak broadening than would be expected with such low fields.

In a recent review article [66], it was anticipated that a highly integrated microfluidic and electrophoresis device can take a single transformed cell directly to the called bases using less than 1 μl of reagent in total. In the last few years, microfabricated DNA sequencing, sample processing, and analysis devices have advanced rapidly toward the goal of a "sequencing lab-on-a-chip." Cell sorting and lysing, DNA extraction and purification, PCR reaction, as well as sequencing separation have all been performed on microchips. In a "sequencing lab-on-a-chip," the sequencing processes and the sample and reagent transportation between processes will be realized. These processes will include "nanoliter-scale enzymatic reactions that are completed in minutes and linked directly to product purification, followed by high-speed microchip electrophoretic analysis" [66]. Analysis times can be expected to drop an order of magnitude and reaction volumes can be expected to drop two or more orders of magnitude.

12.3 ANALYSIS OF PROTEINS AND PEPTIDES

12.3.1 CAPILLARY SDS ELECTROPHORESIS IN SIEVING MATRIX

Capillary SDS gel electrophoresis as a method for separation proteins according to their size has a successful history. First, the separation was performed in capillaries filled with cross-linked gel [67, 68]. Later, when replaceable matrices were introduced to CE [69, 70], it was only a question of time when they will be applied to SDS electrophoresis of proteins. Various polymers have been applied as a sieving matrix including LPA [71, 72], dextran [71], poly(ethylene oxide) [73, 74], and guaran [75]. Three sieving matrices have been developed into commercial products and are currently sold by Coulter-Beckman, Bio-Rad, and Sigma. Capillary SDS electrophoresis has been performed in regular instrumentation for CE with UV detection.

SDS microchip electrophoresis demands eliminating electroosmotic flow (EOF) and increasing detection sensitivity. Complete elimination of EOF has been a challenging task, while improvement of the detection sensitivity is achieved primarily by using LIF detection.

SDS microchip electrophoresis was first demonstrated by the group of Peter Schultz [76].

A commercial system for SDS electrophoresis of proteins on a microchip has been developed by Caliper Technologies [77]. The microchip contains 16 wells for both samples and electrolytes (Figure 12.12). Well D4 contains SDS

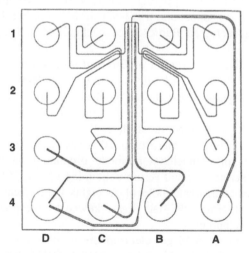

FIGURE 12.12 Chip design for SDS electrophoresis. Well D4 is the SDS dilution well connected to both sides of the dilution intersection. Wells A4 and C4 are the separation buffer and waste wells; B4 and D3 are load wells. All other wells contain samples. (From Bousse, L.; Mouradian, S.; Minalla, A.; Yee, H.; Williams, K.; Dubrow, R., *Anal. Chem.* 2001, 73, 1207. With permission.)

dilution solution to increase detection sensitivity (see below). A4 and C4 serve as electrode chambers for the separation. B4 and D3 are loading wells. All other 11 wells (A1 to A3, B1 to B3, C1 to C3, D1, and D2) are sampling wells. The chips are made from soda-lime glass. Channels (12.5-mm long, 36-μm wide, and 13-μm deep) are made by etching. The wells have diameter of 2 mm and after dicing the size of each chip is about 17.5 mm^2. No static wall coating is prepared in the channel since the presence of polydimethyla-crylamide (PDMA) serving as a sieving matrix reduces the EOF to 0.5 $\times 10^{-9}$ m^2 V^{-1} sec^{-1}.

The background electrolyte contains 42 mM Tris and 120 mM Tricine at pH 7.6 as a buffer, 0.25% (8.7 mM) SDS as a protein complexing detergent, 3.25% PDMA [77, 78] as a sieving polymer, and 4 μM intercalating dye with excitation–emission maximum at 650/680 nm (without disclosed structure, called Agilent dye). Samples are prepared by mixing in a 2:1 ratio with denaturing buffer containing 4% SDS, 3% β-mercaptoethanol, and 290 mg/l myosin as an internal standard. Heating the mixture at 100°C for 5 min denatures the proteins. Before the analysis the samples may be diluted approximately 1:10.

Separation and detection were performed in an Agilent 2100 Bioanalyzer. The separation was carried out at the electric field strength of 200 to 300 V/cm. To reduce fluorescence background caused by the interaction of the intercal-ating dye with SDS micelles and increase the detection sensitivity, two chan-nels containing only buffer without SDS are connected to the separation channel (Figure 12.13). By applying voltage to those two channels, Tricine enters the separation channel and this leads to SDS dilution. In the detection cell it results in decreased fluorescence background and higher detection sensitivity.

The method allows separation of Bio-Rad standard proteins (lysozyme, soybean trypsin inhibitor, carbonic anhydrase, ovalbumin, bovine serum albumin, phosphorylase b, β-galactosidase, and myosin) for SDS CGE in 40 sec (Figure 12.14). The separation efficiency for the larger peaks ranges from 7 $\times$ 10^4 to 1.2 $\times$ 10^5 theoretical plates [77].

The above-described system (Protein 200 LabChip® Assay Kit) was used for determination of monoclonal antibody production in hybridoma cell cultures in both serum-free and serum-containing media [79]. In serum-free media, it provided comparable molecular weights and concentrations whereas in the presence of serum, the data were skewed by the presence of larger amount of serum albumin.

Another system for SDS gel electrophoresis on a microchip used 12% crosslinked polyacrylamide gel and separated 6-TRITC-labeled molecular-weight markers (20.2 to 25 kDa) in less than 30 sec in an effective length of ca. 2 mm [80].

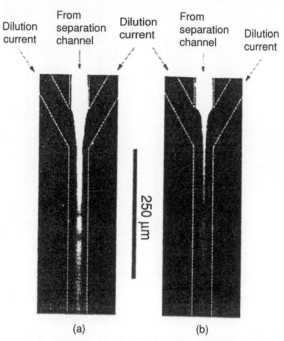

FIGURE 12.13 DS dilution process at dilution ratio 4 (a) and 12 (b). (From Bousse, L.; Mouradian, S.; Minalla, A.; Yee, H.; Williams, K.; Dubrow, R., *Anal. Chem.* 2001, 73, 1207. With permission.)

12.3.2 CAPILLARY ISOELECTRIC FOCUSING

Capillary isoelectric focusing (CIEF) of proteins is a successful method for separation of proteins. Typically, proteins migrate in a pH gradient generated by carrier ampholytes and focus at pH corresponding to their isoelectric point (pI).

A microchip device for IEF of proteins has been constructed and the three most common methods of mobilization of separated proteins were compared [81]. It was found that electroosmotically driven mobilization, which occurs simultaneously with the focusing, is the most suitable for miniaturization because of high speed, compatibility of EOF, and minimum instrumentation requirements. Upon optimization, Cy5-labeled peptides can be focused in less than 30 sec in a 7-cm-long channel (width 200 μm, depth 10 μm) with a total analysis time of less than 5 min. The maximum peak capacity has been estimated to be 30 to 40 peaks.

CIEF has been adapted to microchips using a separation channel (50-μm deep × 120 μm wide) [82]. In a device with focusing path of 47 mm,

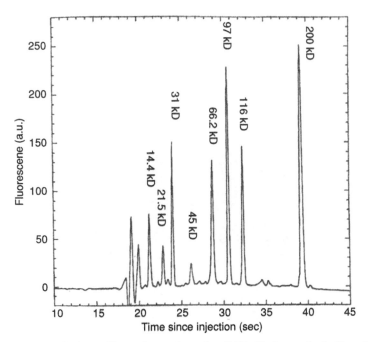

FIGURE 12.14 SDS capillary electrophoresis of Bio-Rad protein ladder. (From Bousse, L.; Mouradian, S.; Minalla, A.; Yee, H.; Williams, K.; Dubrow, R., *Anal. Chem.* 2001, 73, 1207. With permission.)

separation efficiency for lysozyme was 4.7×10^5 theoretical plates. In a channel 12-mm long, proteins with pI 5.4 to 11 were focused, mobilized, and detected in 150 sec. The system was also used to study protein–protein interactions between immunoglobulin G and protein G, and between anti-6x-histidine and 6xHis-tagged green fluorescent protein (Figure 12.15). The detection limit in this case was 50 fmol of proteins [82].

Several papers were also devoted to CIEF where the pH gradient is generated by a compound, the pK of which strongly depends on the temperature in a temperature gradient. Tris has been used as the compound with temperature-sensitive pK. In this work, the temperature gradient was created when one end of the channel was heated to 80°C and the other one cooled to 10°C, which allowed separation of green fluorescent protein into three fractions in Tris–borate buffer [83, 84].

Another approach is CIEF in a tapered channel. In a tapered channel the electric field strength changes progressively and so produces Joule heat and temperature in the channel. With Tris or any other electrolyte with temperature-sensitive pK, a pH gradient is formed in the channel and can be used for CIEF. The method has been used to focus hemoglobins of various origins [85, 86].

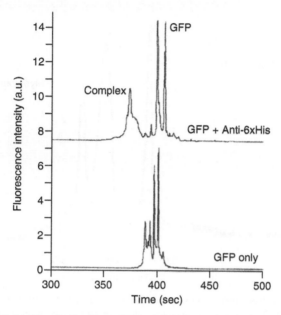

FIGURE 12.15 Isoelectric focusing of green fluorescent protein (GFP) tagged with 6xHis and anti-6xHis (upper electropherogram) and 6xHis-tagged GFP only (lower electropherogram). (From Tan, W.; Fan, Z.H.; Qui, C.X.; Ricco, A.J.; Gibbons, I., *Electrophoresis* 2002, 23, 3636. With permission.)

In slab gel IEF-immobilized pH gradients dominate the field. A method of CIEF on microchip in an immobilized pH gradient has been developed [87]. The system consists of 96 minichambers (75 μl each) allowing simultaneous separation of 8 samples into 12 fractions. The minichambers in each separation dimension are separated by immobilized pH gels of a specific pH to form a pH gradient. The collected fractions can be then analyzed by mass spectrometry. The method provided much better separation of peptides as compared to ion-exchange chromatography.

A microchip for CIEF with direct electrospray ionization-mass spectrometry consisting of three pieces has been constructed on polycarbonate plates [88]. In this experiment, the separation channel (50 μm × 30 μm × 16 cm) was fabricated on carbonate plate with excimer laser. Separation by CIEF was performed using Pharmalyte (p*I* 3 to 10) as carrier ampholytes with 10% acetic acid and 0.3% ammonia as anolyte and catholyte. Focused zones were mobilized by a run with 0.3% ammonia added to the anode reservoir. The electrospray emitter was positioned 2 to 3 mm from the heated capillary inlet of the mass spectrometer. The electrospray ionization was stable and the performance was further improved with sheath gas and sheath liquid, as demonstrated by CIEF of carbonic anhydrase and myoglobin (Figure 12.16).

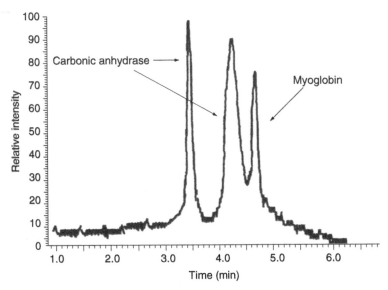

FIGURE 12.16 Isoelectric focusing of carbonic anhydrase (p*I* 5.9 and 6.8) and myoglobin (p*I* 7.2). Focused proteins were immobilized with 0.5 psi nitrogen (0.05 μl/min) and 0.2 μl/min sheath liquid assistance. (From Wen, J.; Lin, Y.H.; Xiang, F.; Matson, D.W.; Udseth, H.R.; Smith, R.D., *Electrophoresis* 2000, 21, 191. With permission.)

A rather different approach was applied when peptides were separated by CIEF in a free-flow electrophoresis device with a flow perpendicular to the CIEF [89]. The system allowed isoelectric focusing of model peptides in less than 1 sec.

CIEF has been performed on a microchip with a separation channel of only 4 mm using the system with CCD camera without protein mobilization of focused proteins [80].

A system for high-throughput analysis of proteins has been developed to separate proteins by isoelectric focusing, digest them on immobilized trypsin, concentrate them by solid-phase extraction on C_{18}, and eventually inject the obtained peptide fragments into a mass spectrometer [90].

12.3.3 MULTIDIMENSIONAL ELECTROPHORESIS OF PROTEINS

Two-dimensional electrophoresis of proteins is a laborious method for high-resolution separation of proteins. The method is very difficult to automate and its transfer to a microchip format would probably solve many problems. There is a strong wish to achieve this task and several groups work hard on it. Becker et al. designed and produced a chip with two rectangular separation dimensions and 500 sub-micrometer channels in the second dimension [91].

A prototype for a two-dimensional CE system has been produced [92, 93]. Mike Ramsey's group has developed a two-dimensional electrophoresis system combining micellar electrokinetic chromatography (MEKC) and capillary zone electrophoresis (CZE) and demonstrated its function by separating fluorescently labeled tryptic digest of cytochrome c [94]. The chip was made from soda-lime glass and its design combined two separation channels and reservoirs for sample, two buffers, and three for waste (Figure 12.17). The channels were 35-μm wide and 10-μm deep. The channel for MEKC in the first dimension was 69-mm long; the channel for CZE was 10-mm long. The microchip was equipped with two gate valves. The background electrolyte for MEKC contained 50 mM triethylamine, 25 mM acetic acid, and 10 mM SDS; the background electrolyte for CZE in the second dimension contained 50 mM triethylamine and 25 mM acetic acid. Proteins were first reductively alkylated with dithiothreitol and iodoacetic acid and then digested with trypsin. Tryptic digest peptides from cytochrome c were fluorescently labeled with mixed isomer 5(6)-carboxytetramethylrhodamine succinimidyl ester (5(6)-TAMRA) and the undigested protein was removed by size exclusion chromatography using HiTrap desalting cartridges. Tryptic peptides were separated when the effluent from the first dimension was sampled to the second dimension every 3 sec. The overall time of analysis was less than 10 min and two-dimensional electrophoresis provided more spots than the theoretically expected number

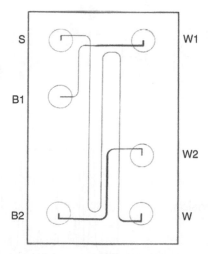

FIGURE 12.17 Schematic diagram of the microchip used for two-dimensional separations. The first dimension extends 69 mm from the first cross-intersection to the second. The second separation channel extends 10 mm from the second cross to the point of detection. S = sample, B1 = buffer 1, B2 = buffer 2, W1 = waste 1, W2 = waste 2, W = waste. (From Rocklin, R.D.; Ramsey, R.S.; Ramsey, J.M., *Anal. Chem.* 2000, 72, 5244. With permission.)

of tryptic peptides (17 for cytochrome c) and provided spots for unreacted derivatizing reagent and labeled free lysine (Figure 12.18). The total peak capacity has been estimated to be 500 to 1,000 peaks (20 to 40 for MEKC and 25 for CZE). Although the separation was performed using peptides only, the work represents the best two-dimensional separation on chip combining electrophoretic methods published so far. The authors converted the collected data into a two-dimensional electropherogram (Figure 12.18) [94]. In a modified system, a 20-cm serpentine channel was used in the first dimension and a 13-mm one in the second dimension [95]. The separation efficiencies were 200,000 and 25,000 theoretical plates for MEKC and CZE, respectively. The system achieved peak capacity of up to 4,500 peaks with analysis time of 15 min.

Another attempt to separate proteins by two-dimensional separation combined ion-exchange chromatography in the first dimension and isoelectric focusing in the second dimension [96]. Using this method TRITC-labeled BSA was focused.

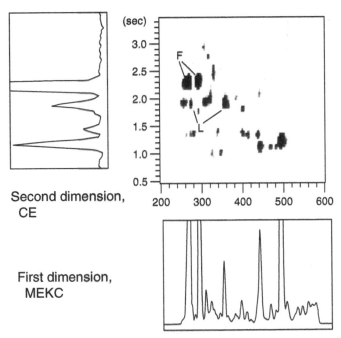

FIGURE 12.18 One- and two-dimensional electropherograms of peptides from a tryptic digest of cytochrome c fluorescently labeled with 5,6-TAMRA. F, unreacted derivatizing reagent; L, labeled lysine. Injections into the second dimensions were made every 3 sec. (From Rocklin, R.D.; Ramsey, R.S.; Ramsey, J.M., *Anal. Chem.* 2000, 72, 5244. With permission.)

12.3.4 ZONE ELECTROPHORESIS

Model proteins albumin, α_1-antitrypsin, transferrin, and IgG, simulating the electrophoretic pattern of human serum proteins, have been separated by CE on microchip after labeling with 2-toluidinonaphtalene-6-sulfonate [97, 98]. Real-world samples of human serum proteins did not provide, however, five traditional peaks due to the poor sensitivity of the label for several of the serum proteins.

CZE was used to analyze low-density lipoproteins: 40 mM N-methylglucamine, 40 mM Tricine, pH 9, in a glass microchip without a wall coating [99]. N-Methylglucamine interacted with the channel wall improving separation efficiency that achieved the value of 2.2×10^7 theoretical plates per meter.

Human γ-interferon has been determined by immunoassay on a microchip with branching multichannels. The immunoassays are typically based on analysis of an antigen and its complex with an antibody. Whereas the analysis time for one sample was 35 min, application of branching multichannels resulted in an analysis time of 50 min for four samples [100].

Antihuman IgG labeled with fluorescein isothiocyanate (FITC) was analyzed on a microchip in a 6-cm-long channel with the effective length of 2.8 cm [101]. The electrophoretic analysis was achieved in less than 16 sec. Monoclonal mouse anti-BSA IgG was determined in mouse ascites fluid by a direct microchip-based CE immunoassay [102]. The calibration curve is linear up to at least 135 mg/l. The method can be used to measure the antigen–antibody interactions between BSA and anti-BSA, and to calculate stoichiometry and equilibration constants.

A free-flow electrophoretic system has been constructed for continuous separation and micropreparation of proteins on chip [103, 104]. BSA, bradykinin, and ribonuclease A have been used as the model mixture to test the instrument. The fraction can be collected as shown on rat plasma.

12.3.5 ENZYMATIC ASSAYS

Enzymatic assays may be used both for determination of enzymatic activity and for quantitation of compounds that can be a substrate or an inhibitor of enzymatic reaction [105]. The enzymatic reaction may be performed prior, during, or after the capillary electrophoretic separation. The quantitated substrate can be even a nonionogenic compound (e.g., glucose) provided the product of enzymatic reaction is an ionic compound.

Microchips offer an opportunity to downscale enzymatic assays to nanoliter levels and consumption of enzymes, cofactors, and substrates by approximately four orders of magnitude [106]. Microchips were used primarily for reactions of soluble enzymes. There are three basic reaction schemes relating enzymatic reaction and electrophoretic separation and we consider precolumn, on-column, and postcolumn reactions. In precolumn reactions, an enzyme catalyzes the reaction of a substrate that may be a neutral molecule not migrating in the electric field. Here, oxidases and dehydrogenases can be used as the converting enzyme. Glucose or ethanol can be determined by reaction of glucose oxidase and alcohol dehydrogenase [107–109].

A layout of a chip for precolumn reactions is shown in Figure 12.19. An enzyme assay, where electromigration is used to make zones migrate and enter the detection cell without separating analytes electrophoretically, is performed on a microchip measuring the enzymatic activity of β-galactosidase [110]. The system allows detecting phenylethyl β-D-thiogalactoside, a competitive inhibitor of galactosidase. Inhibitors of acetylcholinesterase can be monitored from the reduced fluorescence signal of a thiocholine product (Figure 12.20) [111]. The negative peaks were proportional to competitive and irreversible inhibition.

An on-column reaction can be performed without pre- or postcolumn mixers and reactors. Substrates and enzymes are mixed in the channel simply by differences in electrophoretic mobilities. However, increased electric-field strength reduces the efficiency of the enzymatic reaction. Therefore, a microchip has been designed to measure electrochemically glucose in the presence of urate and ascorbate, based on an enzymatic reaction with glucosoxidase

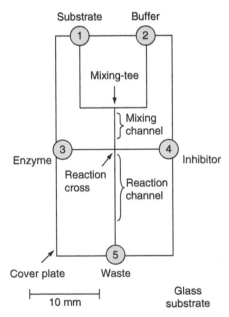

FIGURE 12.19 Schematic of the enzyme-based chip for measuring inhibitors. The channels terminate at reservoirs containing the indicated solutions. (From Hadd, A.G.; Raymond, D.E.; Halliwell, J.W.; Jacobson, S.C.; Ramsey, J.M., *Anal. Chem.* 1997, 69, 3407. With permission.)

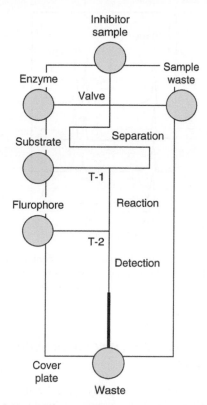

FIGURE 12.20 Schematic of the microchip used for analysis of acetylcholinesterase inhibitors. The fluid reservoirs were filled with AChE for enzyme, acetylthiocholine chloride for substrate, and coumarinylphenylmaleinimide for fluorophore. The sample waste and waste reservoirs contained background electrolyte. (From Hadd, A.G.; Jacobson, S.C.; Ramsey, J.M., *Anal. Chem.* 1999, 71, 5206. With permission.)

and electrophoretic separation of neutral peroxide from anionic urate and ascorbate (Figure 12.21) [107].

Clinically important enzymes leucine aminopeptidase [112] and protein kinase A [113] can be quantitated by analysis on a chip with an on-column reaction.

Postcolumn reactions are suitable for "class" enzymes such as tyrosinase or amino acid oxidase reacting with substrates separated by electrophoresis. Band broadening and incomplete enzymatic reaction are typical drawbacks of this type of reaction. The postcolumn reaction must be fast with an efficient mixing of substrate and enzyme to minimize these negative factors. Postcolumn reactions were used to monitor immunological reactions in connection with the use of an alkaline phosphatase tag [114].

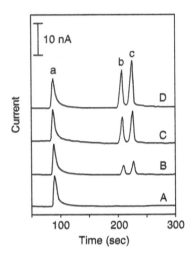

FIGURE 12.21 Electropherogram of 1 mM glucose containing increasing levels of ascorbic acid (AA) and uric acid (UA): (A) 0 mM, (B) 0.2 mM, (C) 0.5 mM, and (D) 0.7 mM. (From Wang, J.; Chatrathi, M.P.; Tian, B.M.; Polsky, R., *Anal. Chem.* 2000, 72, 2514. With permission.)

12.4 ANALYSIS OF LOW MOLECULAR WEIGHT COMPOUNDS

A number of low molecular compounds have been separated by CE and related methods on microchip. In most cases, a model mixture was separated to demonstrate the separation power of the given system. Here we will focus on practical applications where separation on a chip can serve as a real practical tool in everyday practice.

12.4.1 Explosive Residues and Warfare Agents

Explosives were analyzed by MEKC on a chip with indirect laser-induced fluorescence detection [115]. Five micromoles of Cy7 was added to the background electrolyte to achieve indirect fluorescence detection when excitation was provided by a near-IR diode laser operating at 750 nm. An EPA 8330 mixture containing 14 explosives could be analyzed at the parts per million level in less than 1 min in 65-mm-long capillary. Ten peaks corresponding to trinitrobenzene, dinitrobenzene, trinitrotoluene, tetryl, 2,4-DNT, 2,6-DNT, 2-amino-4,6-DNT, 4-amino-2,6-DNT were obtained when 2-, 3-, and 4-nitrotoluene were not resolved and nitramines HMX and RDX showed rather high detection limits.

Organophosphonate nerve agents, such as Sarin, Soman, and VX and their degradation products, were detected by CE on a microchip with contactless conductivity detection [116]. The method demonstrated detection limits ranged from 48 to 86 ng/l, reproducibility from 3.8 to 5.0%, and linearity over the 0.3 to 100 mg/l range.

Portable μChemlab™ system was tested for separations of isoforms of protein biotoxins, from *Ricinus communis* and *Staphylococcus enterotoxins*. The toxins were separated after labeling with fluorescamine by CZE using 10 mmol/l phytic acid, 2 mmol/l laurylsulfobetaine, pH 9.5, as the background electrolyte, or by CGE using 14–200 Beckman gel [117].

12.4.2 PHARMACEUTICALS AND DRUGS OF ABUSE

A method of determination of theophylline, a drug for treatment of asthma, in serum samples via immunoassay by CE on-chip was developed [118]. Labeled theophylline is mixed with a sample containing unlabeled theophylline and with theophylline antibody. The two compete for the limited amount of the antibody. CE separation provides a peak for free theophylline and a peak for theophylline–antibody complex. As the content of theophylline in the sample increases the signal for free-labeled theophylline increases as well. The limit of detection is 1.25 μg/l in diluted serum, with a separation time of about 40 sec.

Serum thyroxine (T4), which is a marker in clinical diagnostics of thyroid function, can be analyzed by immunoassay using CE and microchip [119]. Specific amounts of fluorescein-labeled thyroxin and antithyroxine polyclonal antibody are added to the serum samples. Over 90% of serum thyroxine is bound to serum proteins and therefore a thyroxine-displacing reagent has to be added to measure the total thyroxine in serum. The electrophoretic separation is performed in a channel with an effective length of 22 mm using 20 mM TAPS, 2-amino-2-methyl-1,3-propanediol, pH 8.8, as a background electrolyte. Applying a voltage of 1200 V, the separation can be completed in about 15 sec for serum samples. The fluorescent signal of the free-labeled antigen is used for quantitation (Figure 12.22).

Serum cortisol has been quantitated by electrophoretic immunoassay on microchip [120] in the concentration range of clinical interest (10 to 600 μg/l) without any need for extraction when fluorescein was used as in internal standard. Serum cortisol labeled with fluorescein and detected by LIF could be analyzed using background electrolyte comprising 20 mmol/l TAPS, 2-amino-2-methyl-1,3-propanediol, pH 8.8, in less than 30 sec (Figure 12.23). The separation channel was 28-μm wide and 66-μm deep, its effective length was 22 mm, and the applied voltage was 2 kV. Since many proteins bind cortisol, to determine total serum cortisol the releasing agent 8-anilino-1-naphtalenesulfonic acid was added to the serum during incubation. Reproducibility of the quantitation was typically 1 to 2% for aqueous samples and 3 to 6% for serum samples.

A microchip-based electrophoretic immunoassay has been developed for determination of serum cortisol [120]. The assay can be performed over the range of clinical interest (10 to 600 mg/l) without any preconcentration.

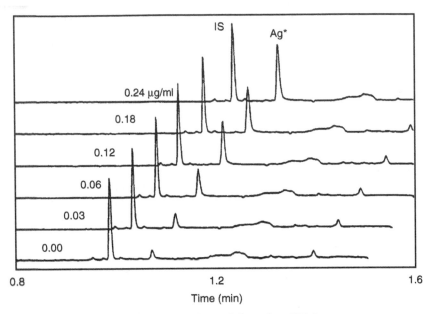

FIGURE 12.22 Immunoassay separations of thyroxine (T4) in serum at concentrations of 0 to 0.24 µg/ml. (From Schmalzing, D.; Koutny, L.B.; Taylor, T.A.; Nashabeh, W.; Fuchs, M., *J. Chromatogr. B* 1997, 697, 175. With permission.)

Amphetamine and its analogs methamphetamine, 3,4-methylenedioxy-methamphetamine, and β-phenylethylamine were analyzed in human urine after FITC derivatization by CZE and MEKC in synchronized cyclic mode [121]. These compounds stimulate the central nervous system and are useful in medical care but are also abused. Using solid-phase extraction to concentrate urine, the limit of identification was found to be about 10 mg/l, a value that is too high for practical purposes. The synchronized cyclic mode of MEKC was applied to the separation of mode mixture of FITC-labeled amino acids [122].

A racemic mixture of gemifloxacin, an antibacterial agent, has been separated by CE with 50 mmol/l Bis-Tris, citric acid, pH 4.0. Crown ether, (+)-(18-crown-6)-tetracarboxylic acid, was added as a chiral selector at concentration of 0.2 to 0.5 mmol/l. EDTA (10 to 15 mmol/l) was added to BGE to improve peak shape and separation efficiency. The resolution of both enantiomers was 1.9. Gemifloxacin was detected by LIF with an excitation at 270 nm and an emission at 405 nm [123].

12.4.3 SMALL MOLECULES IN BODY FLUIDS

Homocysteine, a sulfur-containing amino acid, which is involved in the metabolism of the essential amino acid methionine, can be analyzed by

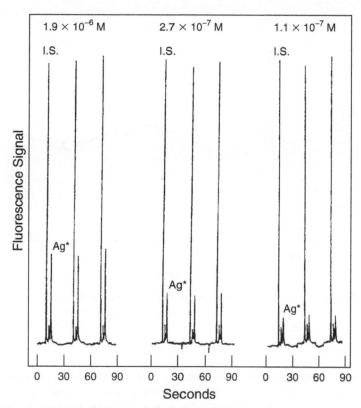

FIGURE 12.23 Immunoassay separations of cortisol in serum samples. Three injections for each of three samples with cortisol level of $1.9 \times 10^{-6} M$, $2.7 \times 10^{-7} M$, and $1.1 \times 10^{-7} M$. (From Koutny, L.B.; Schmalzing, D.; Taylor, T.A.; Fuchs, M., *Anal. Chem.* 1996, 68, 18. With permission.)

CZE in capillary and on microchip with electrochemical detection (Figure 12.24) [124]. In capillary with off-column amperometric detection, the detection limit was 500 nmol/l and the detector response was linear in the range of 1 to 100 μmol/l. A model mixture of 200 μmol/l homocysteine and reduced glutathione was separated by CZE on microchip using 20 mmol/l TES, pH 8.0, in 80 sec. Protein-bound homocysteine in human serum was determined after isolation of serum proteins by ultrafiltration and release of all-bound thiols by an addition of reducing tris-(2-carboxyethyl)phosphine. The protein-bound homocysteine was found in the range of 2.8 ± 0.3 μmol/l. The total homocysteine in human plasma was determined by the method of standard addition after reduction of all thiols with tris-(2-carboxyethyl)phosphine. The level of total homocysteine in human serum was found to be 3.4 ± 0.2 μmol/l.

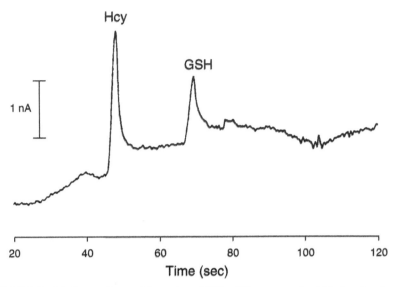

FIGURE 12.24 Separation and detection of 200 μ*M* homocysteine (Hcy) and reduced glutathione (GSH) at pH 8.0 with electrochemical detection. (From Pasas, S.A.; Lacher, N.A.; Davies, M.I.; Lunte, S.M., *Electrophoresis* 2002, 23, 759. With permission.)

Urine oxalate has been determined by CZE on microchip with conductivity detection [125]. Using 15-m*M* propionic acid, ε-aminocaproic acid, pH 4.0, as background electrolyte with 0.05% methylhydroxyethyl cellulose the detection limit was 8×10^{-8} mol/l for a 500-nl sample volume in the presence of 3.5×10^{-3} mol/l chloride. In real samples of urine, oxalate can be determined in the concentration range of 1 to 5×10^{-4} mol/l (Figure 12.25). The analysis takes about 280 sec, the reproducibility of migration times is 0.5 to 1.0%, and the relative standard deviation of the peak area varies from 2 to 5%.

Uric acid in human urine was also determined by CE on a microchip with amperometric detection using an off-chip working platinum electrode [126]. CE performed in 25 mmol/l MES, pH 5.5, separated uric acid from dopamine in 30 sec. The detection limit of the method was 1 μmol/l, and the calibration curve was linear in the range of 15 to 110 μmol/l. The recovery of uric acid spiked in diluted urine was 103.3%. The method was compared with a standard enzymatic assay of uric acid with uricase and a provided value of 96.9% for the uricase method.

12.4.4 MISCELLANEOUS

Here we include separations that we consider significant but were not appropriate to be discussed in the previous subsections.

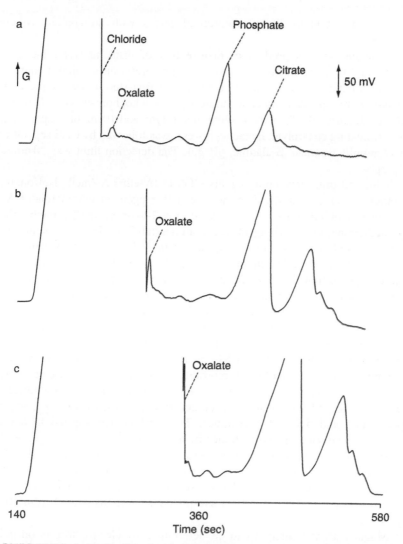

FIGURE 12.25 Analysis of oxalate in human urine — impact of destacking process on resolution of oxalate and chloride at dilution (a) 100-fold, (b) 50-fold, (c) 25-fold. (From Žúborová, M.; Masár, M.; Kaniansky, D.; Jöhnck, M.; Stanislawski, B., *Electrophoresis* 2002, 23, 774. With permission.)

Amino-acid enantiomers labeled with FITC were separated by MEKC in the presence of cyclodextrin [127]. The background electrolyte consisted of 100 mM borate, pH 9.4, 30 mM SDS, and 10 mM γ-cyclodextrin. The analysis time ranged from 75 sec for the most basic amino acids to 160 sec for the most acidic ones. In a separation channel 70-mm long (effective length 4.8 cm), 60-μm wide, and 15-μm deep, the separation efficiencies range from 7,000 to 28,000 theoretical plates with an applied voltage of 1.5 kV.

Bromate, as a by-product of disinfection of drinking water has been determined by zone-electrophoresis–isotachophoresis on a column-coupling chip with conductivity detection [128]. In the first step, ionic sample components were separated and concentrated by isotachophoresis with 10 mmol/l HCl, β-alanine, pH 3.2, as the leading electrolyte, and 10 mmol/l aspartate as the terminating electrolyte. Isotachophoresis was followed by CZE separation in 15 mmol/l aspartate, β-alanine, pH 3.4. The detection limit was 20 nmol/l (2.5 ppb).

Film and print developing agents CD3 (4-(N-ethyl-N-(methylsulfonami-doethyl)-2-methylphenylenediamine, the active agent in commercial RA-4 developer solutions)) and CD4 (4-(N-ethyl-N-2-hydroxyethyl)-2-methylphe-nylenediamine, the active agent in a commercial C-41 developer solutions) were analyzed by CE on a microchip in a channel 10-μm deep, 40-μm wide, and 12-mm effective length in less than 30 sec [129, 130]. Using 5 mM fluorescein, 100 mM Na_2SO_3, 60 mM phosphate buffer, pH 11.9, as a background electrolyte, CD-3 and CD-4 were detected with indirect UV detection with the detection limits of 0.17 mM and 0.39 mM, respectively.

Biogenic amines (putrescine, cadaverine, spermidine, spermine, histamine, tyramine, tryptamine, and phenylethylamine) can by analyzed after derivatization with FITC by MEKC on microchip [131]. The analysis time is about 80 sec, the run-to-run reproducibility of migration times varies between 0.15 and 0.54%, day-to-day reproducibility of migration times is between 0.40 and 0.80%, and the detection limit varies between 2.94 μM (putrescine) and 6.57 μM (spermine). The method was applied to a real sample by analyzing biogenic amines in soy sauce.

Herbicides paraquat and diquat were analyzed by isotachophoresis on a microchip with Raman spectroscopy detection [132]. The Raman band of the counter ion, sulfate, at 980 cm^{-1} was used as an internal standard to correct for instrument variations. Working sample concentrations ranged from 10^{-5} to $10^{-7} M$.

Magnesium and calcium can be determined by electrophoresis on chip after sample stacking and on-chip complexation [133]. After a gated injection, the sample is concentrated in the stacking channel and is derivatized on-chip with 8-hydroxyquinolin-5-sulfonic acid. The limit of detection is 18 ppb (0.45 μM) for calcium and 0.5 ppb (21 nM) for magnesium.

12.5 CONCLUSIONS

Microfluidic research has expanded tremendously over the last few years, although the field is still in its developmental stages. Microchip CE was the first of the microchip methods to be demonstrated and has been the most universally applicable method for all the methods described to date in the literatures. There have been many noteworthy advances made recently, especially with respect to surface engineering in microchannels and large-scale integration. Microfluidic devices are becoming increasingly complex in their layouts, as researchers gradually gain experience in how to better control fluids in these valveless systems. This will lead to more devices, which are not based solely on CE, and a more varied palette of analytical microchips as a result.

Industry acceptance to microchips plays a critical role in the development of microchip or "lab-on-a-chip" technology. Despite considerable progress in microchip applications, few commercial products based on this technology are available. The main reason is that no "killer application," i.e., an application that provides a solution to an unsolved problem and a solution that brings benefits overwhelmingly superior to alternative solutions, has been developed. The authors believe that some killer applications will be developed first in the genomic and proteomic research area, high-throughput drug screening, and clinical diagnosis.

REFERENCES

1. Effenhauser, C. S.; Paulus, A.; Manz, A.; Widmer, H. M., *Anal. Chem.* 1994, 66, 2949.
2. Liu, S. R., Shi, Y. N., Ja, W. W., Mathies, R. A., *Anal. Chem.* 1999, 71, 566–573.
3. Ueda, M.; Kiba Y.; Abe, H.; Arai, A.; Nakanishi, H.; Baba, Y., *Electrophoresis* 2000, 21, 176–180.
4. Shi, Y. N.; Simpson, P. C.; Scherer, J. B.; Wexler, D.; Skibola, C.; Smith, M. T.; Mathies, B. A., *Anal. Chem.* 1999, 71, 5354–5361.
5. Woolley, A. T.; Lao, K. Q.; Glazer, A. N., Mathies, R. A., *Anal. Chem.* 1998, 70, 684–688.
6. Tian, H. J.; Jaquins-Gerstl, A.; Munro, N.; Trucco, M.; Brody, L. C.; Landers, J. P., *Genomics* 2000, 63, 25–34.
7. Pierre-Alain, A.; Dimitri, I.; Darwin, R. R.; Andreas, M., *Anal. Chem.* 2002, 74, 2637–2652.
8. Verpoorte, E., *Electrophoresis* 2002, 23, 677–712.
9. Landers, J. P., *Anal. Chem.* 2003, 75, 2919–2927.
10. Woolley, A. T.; Mathies, R. A., *Proc. Natl. Acad. Sci. U. S. A.* 1994, 91, 11348–11352.
11. McCormick, R. M.; Nelson, R. J.; Alonso-Amigo, M. G.; Benvegnu, J.; Hooper, H. H., *Anal. Chem.* 1997, 69, 2626–2630.
12. Woolley, A.; Sensabaugh, G. F.; Mathies, R. A., *Anal. Chem.* 1997, 69, 2181–2186.
13. Huang, Z.; Munro, N.; Huhmer, A. F. R.; Landers, J. P., *Anal. Chem.* 1999, 71, 5309–5314.
14. Jacobson, S. C.; Ramsey, J. M., *Anal. Chem.* 1996, 68, 720–723.
15. Woolley, A. T.; Hadley, D.; Landre, P.; deMello, A. J.; Mathies, R. A.; Northrup, M. A., *Anal. Chem.* 1996, 68, 4081–4086.

16. Burns, M. A.; Johnson, B. N.; Brahmasandra, S. N.; Hhandique, K.; Webster, J. R.; Krishnan, M.; Sammarco, T. S.; Man, P. M.; Jones, D.; Heldsinger, D.; Mastrangelo, C. H.; Burke, D. T., *Science* 1998, 282, 484–487.
17. Xu, F.; Jabasini, M.; Baba, Y., *Electrophoresis* 2002, 23, 3608–3614.
18. Duffy, D. C.; McDonald, J. C.; Schueller, O. J. A.; Whitesides, G. M., *Anal. Chem.* 1998, 70, 4974–4984.
19. Simpson, P. C.; Roach, D.; Woolley, A. T.; Thorsen, T.; Johnston, R.; Sensabaugh, G. F.; Mathies, R. A., *Proc. Natl. Acad. Sci. U. S. A.* 1998, 95, 2256–2261.
20. Ulvik, A.; Refsum, H.; Kluijtmans, L. A. J.; Ueland, P. M., *Clin. Chem.* 1997, 43, 267–272.
21. Medintz, I.; Wong, W. W.; Sensabaugh, G.; Mathies, R. A., *Electrophoresis* 2000, 21, 2352–2358.
22. Schmalzing, D.; Koutny, L.; Adourian, A.; Belgrader, P.; Matsudaira, P.; Ehrlich, D., *Proc. Natl. Acad. Sci. U. S. A.* 1997, 94, 10273–10278.
23. Cantafora, A.; Blotta, I.; Bruzzese, N.; Calandra, S.; Bertolini, S., *Electrophoresis* 2001, 22, 4012–4015.
24. Cheng, J.; Waters, L. C.; Fortina, P.; Hvichia, G.; Jacobson, S. C.; Ramsey, J. M.; Kricka, L. J.; Wilding, P., *Anal. Biochem.* 1998, 257, 101–106.
25. Tian, H.; Jaquins-Gerstl, A.; Munro, N.; Trucco, M.; Brody, L. C.; Landers, J. P., *Genomics* 2000, 63, 25–34.
26. Tian, H.; Brody, L. C.; Landers, J. P., *Genome Res.* 2000, 10, 1403–1413.
27. Tian, H.; Brody, L. C.; Fan, S.; Huang, Z.; Landers, J. P., *Clin. Chem.* 2001, 47, 173–185.
28. Buch, J. S.; Rosenberger, F.; DeVoe, D.; Lee, C. S., *Proceedings of the µTAS 2002 Symposium*, Eds. Baba, Y.; Shoji, S.; van den Berg, A., Kluwer Academic Publishers, Dordrecht, The Netherlands, 2002, pp. 233–235.
29. Prosser, J., *Trends Biotechnol.* 1993, 11, 238–246.
30. Rao, V. B., *Anal. Biochem.* 1994, 216, 1–14.
31. Orita, M.; Iwahana, H.; Kanazawa, H.; Hayashi, K.; Sekiya, T., *Proc. Natl. Acad. Sci. U. S. A.* 1989, 86, 2766–2770.
32. Fischer, S. G.; Lerman, L. S., *Proc. Natl. Acad. Sci. U. S. A.* 1983, 80, 1579–1583.
33. Haff, L.; Smirnov, I. P., *Genome Res.* 1997, 7, 378–388.
34. Wang, D. G.; Fan, J. B.; Siao, C. J.; Berno, A.; Young, P.; Sapolsky, R.; Ghandour, G.; Perkins, N.; Winchester, E.; Spencer, J. et al. *Science* 1998, 280, 1077–1082.
35. Del Tito, B. J., Jr.; Poff, H. E.; Novotny, M. A.; Cartledge, D. M.; Walker, R. I.; Earl, C. D.; Bailey, A. L., *Clin. Chem.* 1998, 44, 731–739.
36. Abon, J. J.; McKenzie, M.; Cotton, R. G. H., *Electrophoresis* 1999, 20, 1162–1170.
37. Youil, R.; Kemper, B. W.; Cotton, R. G. H., *Proc. Natl. Acad. Sci. U. S. A.* 1995, 92, 87–91.
38. Schmalzing, D.; Belenky, A.; Novotny, M. A.; Koutny, L.; Salas-Solano, O.; El-Difrawy, S.; Adourian, A.; Matsudaira, P.; Ehrlich, D., *Nucleic Acids Res.* 2000, 28, e43.
39. Medintz, I.; Wong, W. W.; Berti, L.; Shiow, L.; Tom, J.; Scherer, J.; Sensabaugh, G.; Mathies, R. A., *Genome Res.* 2001, 11, 413–421.

40. Russom, A.; Andersson, H.; Nilsson, P.; Ahmadian, A.; Stemme, G., *Proceedings of the μTAS 2002 Symposium*, Eds. Baba, Y.; Shoji, S.; van den Berg, A., Kluwer Academic Publishers, Dordrecht, The Netherlands, 2002, pp. 218–220.

41. Russom, A.; van der Wijngaart, W.; Ehring, H.; Tooke, N.; Andersson, N.; Stemme, G., *Proceedings of the μTAS 2002 Symposium*, Eds. Baba, Y.; Shoji, S.; van den Berg, A., Kluwer Academic Publishers, Dordrecht, The Netherlands, 2002, pp. 308–310.

42. Liu, P.; Xing, W. L.; Liang, D.; Huang, G. L.; Cheng, J., *Proceedings of the μTAS 2002 Symposium*, Eds. Baba, Y.; Shoji, S.; van den Berg, A., Kluwer Academic Publishers, Dordrecht, The Netherlands, 2002, pp. 311–313.

43. Woolley, A. T.; Mathies, R. A., *Anal. Chem.* 1995, 67, 3676–3680.

44. Liu, S.; Ren, H.; Gao, Q.; Roach, D. J.; Loder, R. T., Jr.; Armstrong, T. M.; Mao, Q.; Blaga, I.; Barker, D. L.; Jovanovich, S. B., *Proc. Natl. Acad. Sci. U. S. A.* 2000, 97, 5369–5374

45. Salas-Solano, O.; Schmalzing, D.; Koutny, L.; Buonocore, S.; Adourian, A.; Matsudaira, P.; Ehrlich, D., *Anal. Chem.* 2000, 72, 3129–3137.

46. Schmalzing, D.; Adourian, A.; Koutny, L.; Ziaugra, L.; Matsudaira, P.; Ehrlich, D., *Anal. Chem.* 1998, 70, 2303–2310.

47. Backhouse, C.; Caamano, M.; Oaks, F.; Nordman, E.; Carrillo, A.; Johnson, B.; Bay, S., *Electrophoresis* 2000, 21, 150–156

48. Davidson, C.; Balch, J.; Brewer, L.; Kimbrough, J.; Swierkowski, S.; Nelson, D.; Madabhushi, R.; Pastrone, R.; Lee, A.; McCready, P.; Adamson, A.; Bruce, R.; Mariella, R.; Carrano, A., *DOE Human Genome Program Contractor-Grantee Workshop VI*, November 9–13, 1997, Santa Fe, NM.

49. Jacobson, S. C.; Hergenroder, R.; Koutny, L. B.; Warmack, R. J.; Ramsey, J. M., *Anal. Chem.* 1994, 66, 1107–1113

50. Swerdlow, H.; Gesteland, R., *Nucleic Acids Res.* 1990, 18, 1415–1419.

51. Ruiz-Martinez, M. C.; Carrilho, E.; Berka, J.; Kieleczawa, J.; Miller, A. W.; Foret, F.; Carson, S.; Karger, B. L., *Biotechniques* 1996, 20, 1058–1069.

52. Koutny, L.; Schmalzing, D.; Salas-Solano, O.; El-Difrawy, S.; Adourian, A.; Buonocore, S.; Abbey, K.; McEwan, P.; Matsudaira, P.; Ehrlich, D., *Anal. Chem.* 2000, 72, 3388–3391.

53. Schmalzing, D.; Tsao, N.; Koutny, L.; Chrisholm, D.; Srivastava, A.; Adourian, A.; Linton, L.; McEwan, P.; Matsudaira, P.; Ehrlich, D., Toward real-world sequencing by microdevice electrophoreis. In: *Sixth International Automation in Mapping and DNA Sequencing Conference*, Hinxton, Cambridge, UK, September 1–4, 1999.

54. Ehrlich, D.; Adourian, A.; Barr, C.; Breslau, D.; Buonocore, S.; Burger, R.; Carey, L.; Carson, S.; Chiou, J.; Dee, R.; Desmarais, S.; El-Difrawy, S.; King, R.; Koutny, L.; Lam, R.; Matsudaira, P.; Mitnik-Gankin, L.; O'Neil, T.; Novotny, M.; Saber, G.; Salas-Solano, O.; Schmalzing, D.; Srivastava, A.; Vazquez, M., Biomems–768 DNA sequencer. *Proceedings of the μTAS 2001 Symposium*, Eds. Ramsey, J. M.; van den Berg, A., Kluwer Academic Publishers, Dordrecht, The Netherlands, October 21–25, 2001, pp. 16–18.

55. Culbertson, C. T.; Jacobson, S. C.; Ramsey, J. M., *Anal. Chem.* 1998, 70, 3781–3789.

56. Paegel, B. M.; Hutt, L. D.; Simpson, P. C.; Mathies, R. A., *Anal. Chem.* 2000, 72, 3030–3037.
57. Paegel, B. M.; Emrich, C. A.; Blazej, R. G.; Elkin, C. J.; Scherer, J. R.; Mathies, R. A., *Proceedings of the μTAS 2001 Symposium*, Eds. Ramsey, J. M.; van den Berg, A., Kluwer Academic Publishers, Dordrecht, The Netherlands, October 21–25, 2001, pp. 462–464.
58. Molho, J. I.; Herr, A. E.; Mosier, B. P.; Santiago, J. G.; Kenny, T. W.; Brennen, R. A.; Gordan, G. B.; Mohammadi, B., *Anal. Chem.* 2001, 73, 1350–1360.
59. Paegal, B. M.; Emrich, C. A.; Weyemayer, G. J.; Scher, J. R.; Mathies, R. A., *Proc. Natl. Acad. Sci. U. S. A.* 2002, 99, 574–579.
60. Figeys, D., *Proceedings of the μTAS 1998 Workshop*, Eds. Harrison, D. J.; van den Berg, A., Kluwer Publishing, Dordrecht, Netherlands, October 13–16, 1998, pp. 457–462.
61. Figeys, D.; Aebersold, R., *Anal. Chem.* 1998, 70, 3721–3724.
62. Bings, N. H.; Wang, C.; Skinner, C. D.; Colyer; C. L.; Thibaut, P.; Harrison, D. J., *Anal. Chem.* 1999, 71, 3292–3296.
63. Liu, S., *Electrophoresis*, submitted.
64. Liu, S.; Elkin, C.; Kapur, H., *Electrophoresis*, submitted.
65. Ugaz, V. M.; Brahmasandra, S. N.; Burke, D. T.; Burns, M. A., *Electrophoresis* 2002, 23, 1450–1459.
66. Paegel, B. M.; Blazej, R. G.; Mathies, R. A., *Curr. Opin. Biotechnol.* 2003, 14, 42–50.
67. Cohen, A. S.; Karger, B. L., *J. Chromatogr.* 1987, 397, 409–417.
68. Dolnik, V.; Cobb, K. A.; Novotny, M., *J. Microcolumn Sep.* 1991, 3, 155–159.
69. Zhu, M.; Hansen, D. L.; Burd, S.; Gannon, F., *J. Chromatogr.* 1989, 480, 311–319.
70. Hjerten, S.; Valtcheva, L.; Elenbring, K.; Eaker, D., *J. Liq. Chromatogr.* 1989, 12, 2471.
71. Ganzler, K.; Greve, K. S.; Cohen, A. S.; Karger, B. L.; Guttman, A.; Cooke, N. C., *Anal. Chem.* 1992, 64, 2665–2674.
72. Best, N.; Arriaga, E.; Chen, D. Y.; Dovichi, N. J., *Anal. Chem.* 1994, 66, 4063–4067.
73. Guttman, A.; Nolan, J. A.; Cooke, N., *J. Chromatogr.* 1993, 632, 171–175.
74. Benedek, K.; Guttman, A., *J. Chromatogr. A* 1994, 680, 375–381.
75. Dolnik, V.; Gurske, W. A.; Padua, A., U.S. Patent Application 20020049184.
76. Yao, S.; Anex, D. S.; Caldwell, W. B.; Arnold, D. W.; Smith, K. B.; Schultz, P. G., *Proc. Natl. Acad. Sci. U. S. A.* 1999, 96, 5372–5377.
77. Bousse, L.; Mouradian, S.; Minalla, A.; Yee, H.; Williams, K.; Dubrow, R., *Anal. Chem.* 2001, 73, 1207–1212.
78. Dubrow, R. S., U.S. Patent 5,948,227.
79. Ohashi, R.; Otero, J. M.; Chwistek, A.; Hamel, J.-F., *Electrophoresis* 2002, 23, 3623–3629.
80. Han, J.; Singh, A. K., *Micro Total Analysis Systems 2002*, Eds. Baba, Y.; Shoji, S.; van den Berg, A., Kluwer Academic Publishers, Dordrecht, The Netherlands, 2002, p. 596–598.
81. Hofmann, O.; Che, D. P.; Cruickshank, K. A.; Muller, U. R., *Anal. Chem.* 1999, 71, 678–686.
82. Tan, W.; Fan, Z. H.; Qui, C. X.; Ricco, A. J.; Gibbons, I., *Electrophoresis* 2002, 23, 3636–3645.

83. Ross, D.; Locascio, L. E., *Anal. Chem.* 2002, 74, 2556–2564.
84. Ross, D.; Locascio, L. E., *Micro Total Analysis Systems 2002*, Eds. Baba, Y.; Shoji, S.; van den Berg, A., Kluwer Academic Publishers, Dordrecht, The Netherlands, 2002.
85. Huang, T. M.; Pawliszyn, J., *Analyst* 2000, 125, 1231–1233.
86. Huang, T. M.; Wu, X. Z.; Pawliszyn, J., *Anal. Chem.* 2000, 72, 4758–4761.
87. Tan, A.; Pashkova, A.; Zang, L.; Foret, F.; Karger, B. L., *Electrophoresis* 2002, 23, 3599–3607.
88. Wen, J.; Lin, Y. H.; Xiang, F.; Matson, D. W.; Udseth, H. R.; Smith, R. D., *Electrophoresis* 2000, 21, 191–197.
89. Xu, Y.; Zhang, C. X.; Manz A., *Micro Total Analysis Systems 2002*, Eds. Baba, Y.; Shoji, S.; van den Berg, A., Kluwer Academic Publishers, Dordrecht, The Netherlands, 2002, pp. 539–541.
90. Taylor, J.; Wang, C.; Harrison, D. J., *Micro Total Analysis Systems 2002*, Eds. Baba, Y.; Shoji, S.; van den Berg, A., Kluwer Academic Publishers, Dordrecht, The Netherlands, 2002, p. 344–346.
91. Becker, H.; Lowack, K.; Manz, A., *J. Micromach. Microeng.* 1998, 8, 24.
92. McDonald, J. C.; Chabinyc, M. L.; Metallo, S. J.; Anderson, J. R.; Stroock, A. D.; Whitesides, G. M., *Anal. Chem.* 2002, 74, 1537–1545.
93. Chen, X. X.; Wu, H. K.; Mao, C. D.; Whitesides, G. M., *Anal. Chem.* 2002, 74, 1772–1778.
94. Rocklin, R. D.; Ramsey, R. S.; Ramsey, J. M., *Anal. Chem.* 2000, 72, 5244–5249.
95. Jacobson, S. C.; Ramsey, J. D.; Culbertson, C. T.; Ramsey, J. M., *Micro Total Analysis Systems 2002*, Eds. Baba Y.; Shoji, S.; van den Berg, A., Kluwer Academic Publishers, Dordrecht, The Netherlands, 2002, pp. 608–610.
96. Yamada, M., Scki, M., *Micro Total Analysis Systems 2002*, Eds. Baba, Y.; Shoji, S.; van den Berg, A., Kluwer Academic Publishers, Dordrecht, The Netherlands, 2002, p. 611–613.
97. Colyer, C. L.; Mangru, S. D.; Harrison, D. J., *J. Chromatogr. A* 1997, 781, 271–276.
98. Colyer, C. L.; Tang, T.; Chiem, N.; Harrison, D. J., *Electrophoresis* 1997, 18, 1733–1741.
99. Ceriotti, L.; Shibata, T.; Folmer, B.; Weiller, B. H.; Roberts, M. A.; de Rooij, N. F.; Verpoorte, E., *Electrophoresis* 2002, 23, 3615–3622.
100. Sato, K.; Yamanaka, M.; Takahashi, H.; Tokeshi, M.; Kimura, H.; Kitamori, T., *Electrophoresis* 2002, 23, 734–739.
101. Rodriguez, I.; Zhang, Y.; Lee, H. K.; Li, S. F. Y., *J. Chromatogr. A* 1997, 781, 287–293.
102. Chiem, N. H.; Harrison, D. J., *Electrophoresis* 1998, 19, 3040–3044.
103. Raymond, D. E.; Manz, A.; Widmer, H. M., *Anal. Chem.* 1994, 66, 2858–2865.
104. Raymond, D. E.; Manz, A.; Widmer, H. M., *Anal. Chem.* 1996, 68, 2515–2522.
105. Bao, J. M.; Regnier, F. E., *J. Chromatogr.* 1992, 608, 217–224.
106. Wang, J., *Electrophoresis* 2002, 23, 713–718.
107. Wang, J.; Chatrathi, M. P.; Tian, B. M.; Polsky, R., *Anal. Chem.* 2000, 72, 2514–2518.
108. Wang, J.; Chatrathi, M. P.; Tian, B. M., *Anal. Chem.* 2001, 73, 1296–1300.
109. Wang, J.; Pumera, M.; Chatrathi, M. P.; Escarpa, A.; Musameh, M., *Anal. Chem.* 2002, 74, 1187–1191.

110. Hadd, A. G.; Raymond, D. E.; Halliwell, J. W.; Jacobson, S. C.; Ramsey, J. M., *Anal. Chem.* 1997, 69, 3407–3412.
111. Hadd, A. G.; Jacobson, S. C.; Ramsey, J. M., *Anal. Chem.* 1999, 71, 5206–5212.
112. Zugel, S. A.; Burke, B. J.; Regnier, F. E.; Lytle, F. E., *Anal. Chem.* 2000, 72, 5731–5735.
113. Cohen, C. B.; ChinDixon, E.; Jeong, S.; Nikiforov, T. T., *Anal. Biochem.* 1999, 273, 89–97.
114. Wang, J.; Ibanez, A.; Chatrathi, M. P.; Escarpa, A., *Anal. Chem.* 2001, 73, 5323–5327.
115. Wallenborg, S. R.; Bailey, C. G., *Anal. Chem.* 2000, 72, 1872–1878.
116. Wang, J.; Pumera, M.; Colllins, G. E.; Mulchandani, A., *Anal. Chem.* 2002, 74, 6121–6125.
117. Fruetel, J. A.; Horn, B. A.; West, J. A. A.; Stamps, J. F.; Vandernoot, V. A.; Stoddard, M. C.; Renzi, R. F.; Padgen, D., *Micro Total Analysis Systems 2002*, Eds. Baba, Y.; Shoji, S.; van den Berg, A., Kluwer Academic Publishers, Dordrecht, The Netherlands, 2002, pp. 524–526.
118. Chiem, N. H.; Harrison, D. J., *Clin. Chem.* 1998, 44, 591–598.
119. Schmalzing, D.; Koutny, L. B.; Taylor, T. A.; Nashabeh, W.; Fuchs, M., *J. Chromatogr. B* 1997, 697, 175–180.
120. Koutny, L. B.; Schmalzing, D.; Taylor, T. A.; Fuchs, M., *Anal. Chem.* 1996, 68, 18-22.
121. Ramseier, A.; vonHeeren, F.; Thormann, W., *Electrophoresis* 1998, 19, 2967–2975.
122. vonHeeren, F.; Verpoorte, E.; Manz, A.; Thormann, W., *Anal. Chem.* 1996, 68, 2044–2053.
123. Cho, S. I.; Lee, K.-N.; Kim, Y.-K.; Jang, J.; Chung, D. S., *Electrophoresis* 2002, 23, 972–977.
124. Pasas, S. A.; Lacher, N. A.; Davies, M. I.; Lunte, S. M., *Electrophoresis* 2002, 23, 759–766.
125. Žúborová, M.; Masár, M.; Kaniansky, D.; Jöhnck, M.; Stanislawski, B., *Electrophoresis* 2002, 23, 774–781.
126. Fanguy, J. C.; Henry, C. S., *Electrophoresis* 2002, 23, 767–773.
127. Rodriguez, I.; Jin, L. J.; Li, S. F. Y., *Electrophoresis* 2000, 21, 211–219.
128. Bodor, R.; Kaniansky, D.; Masar, M.; Silleova K.; Stanislawski, B., *Electrophoresis* 2002, 23, 3630–3637.
129. Sirichai, S.; de Mello, A. J., *Electrophoresis* 2001, 22, 348–354.
130. Sirichai, S.; deMello, A. J., *Analyst* 1999, 125, 133–137.
131. Rodriguez, I.; Lee, H. K.; Li, S. F. Y., *Electrophoresis* 1999, 20, 118–126.
132. Walker, P. A.; Morris, M. D.; Burns, M. A.; Johnson, B. N., *Anal. Chem.* 1998, 70, 3766–3769.
133. Kutter, J. P.; Ramsey, R. S.; Jacobson, S. C.; Ramsey, J. M., *J. Microcolumn Sep.* 1998, 10, 313–319.

Index

Printed and bound by CPI Group (UK) Ltd, Croydon, CR0 4YY

24/10/2024

01778302-0018